WAVEGUIDE JUNCTION CIRCULATORS

WAVEGUIDE JUNCTION CIRCULATORS
THEORY AND PRACTICE

J. Helszajn
Heriot-Watt University, Edinburgh, UK

JOHN WILEY & SONS
Chichester • New York • Weinheim • Brisbane • Singapore • Toronto

Copyright © 1998 by John Wiley & Sons Ltd
Baffins Lane, Chichester,
West Sussex, PO19 1UD, England

National   01243 779777
International (+44) 1243 779777
e-mail (for orders and customer service enquiries): cs-books@wiley.co.uk
Visit our Home Page on http://www.wiley.co.uk
or http://www.wiley.com

All Rights Reserved. No part of this publication may be reproduced, stored in a retrieval system, or transmitted, in any form or by any means, electronic, mechanical, photocopying, recording, scanning or otherwise, except under the terms of the Copyright, Designs and Patents Act 1988 or under the terms of a licence issued by the Copyright Licensing Agency, 90 Tottenham Court Road, London, W1P 9HE, UK, without the permission in writing of the Publisher.

Other Wiley Editorial Offices

John Wiley & Sons, Inc., 605 Third Avenue,
New York, NY 10158-0012, USA

WILEY-VCH Verlag GmbH, Pappelallee 3,
D-69469 Weinheim, Germany

Jacaranda Wiley Ltd, 33 Park Road, Milton,
Queensland 4064, Australia

John Wiley & Sons (Asia) Pte Ltd, 2 Clementi Loop #02-01,
Jin Xing Distripark, Singapore 129809

John Wiley & Sons (Canada) Ltd, 22 Worcester Road,
Rexdale, Ontario, M9W IL1, Canada

Library of Congress Cataloging-in-Publication Data
Helszajn, J. (Joseph)
  Waveguide junction circulators : theory and practice
/ J. Helszajn.
     p. cm.
  Includes bibliographical references and index.
  ISBN 0-471-98252-0 (case : alk. paper)
  1. Circulators, Wave-guide. 2. Junctions, Wave-guide.
I. Title.
TK7871.65.H48   1998
621.381′331—dc21   98–15975
                                                   CIP

British Library Cataloguing in Publication Data
A catalogue record for this book is available from the British Library

ISBN 0 471 98252 0

Typeset in 10/12pt Times from the author's disks by The Florence Group, Stoodleigh.
Printed and bound in Great Britain by Antony Rowe Ltd, Chippenham.
This book is printed on acid-free paper responsibly manufactured from
sustainable forestry in which at least two trees are planted for each
one used for paper production.

To Nick Vouloumanos
in appreciation of his support throughout

CONTENTS

Preface ix

1. Architecture of Symmetrical Waveguide Junction Circulators — 1
2. Scattering Matrix of m-Port Junction — 23
3. Eigenvalue Adjustment of 3-Port Circulator — 39
4. Impedance Matrix of Junction Circulator — 57
5. The Post Gyromagnetic Resonator — 77
6. Okada Resonator — 89
7. Isotropic, Anisotropic and Gyromagnetic Circular Waveguides — 109
8. Isotropic and Anisotropic Open Circular Waveguides — 139
9. The Dielectric Cavity Resonator — 155
10. The Gyromagnetic Cavity Resonator — 179
11. Impedance in Rectangular, Ridge and Radial Waveguides — 199
12. Junction Circulator Using Post Resonators — 229
13. Complex Gyrator Circuit of a Waveguide Junction Circulator using an Okada Resonator — 255
14. Degree-1 and 2 Okada Circulators — 279
15. An Evanescent Mode Okada Junction Circulator — 297
16. Complex Gyrator Circuit of an H-Plane Junction Circulator using E-Plane Turnstile Resonators — 311
17. Complex Gyrator Circuit of an Evanescent-Mode E-Plane Junction Circulator using H-Plane Turnstile Resonators — 339
18. Waveguide Circulators using Triangular and Prism Resonators — 359
19. Synthesis of Quarter-Wave Coupled Junction Circulators with Degrees 1 and 2 Complex Gyrator Circuits — 379
20. The 4-Port Single Junction Waveguide Circulator — 399
21. Microwave Switching using Junction Circulators — 415
22. Insertion Loss of Waveguide Circulators — 431
23. Synthesis of Stepped Impedance Transducers — 445
24. Experimental Evaluation of Junction Circulators — 471
25. Circulator Specifications — 489
26. Gyromagnetic Effect in Magnetic Insulator — 511

Index 537

PREFACE

The ideal circulator discussed in this book is defined as a non-reciprocal m-port lossless passive network, in which a signal introduced at one port is transferred wholly to an adjacent one while decoupling it from the other ports. Circulators with three or more ports have nonreciprocal properties that are invaluable in many microwave systems. It consists, in its simplest form, of a suitably magnetized gyromagnetic resonator at the junction of three waveguides or transmission lines. The adjustment of this class of circuit requires an appreciation of the gyromagnetic effect in a magnetic insulator, an understanding of the eigenvalue problem met in connection with symmetric networks and junctions, the nature of nonreciprocity in circuit theory, microwave engineering and the concept of gain-bandwidth met in filter theory. The main purpose of the text is to bridge the important interface between theory and practice of this important topic in the case of the waveguide arrangement. As such it will be of interest to both the academic worker and the industrial engineer. The former will find a firm foundation of the theoretical aspects of this class of devices and the latter experimental platforms for its commercial design.

The text is divided into one block of chapters which deals with the properties and adjustment of the 3-port junction circulator, another which concentrates on the gyromagnetic resonator, still another which describes the design of some practical classic circulator arrangements. Individual chapters are devoted to the 4-port single junction circulator and switches. Practical aspects such as insertion loss and commercial specifications are given particular consideration. The origin of the gyromagnetic effect upon which the class of devices dealt with in this book rests is also given special attention. The text concludes with a block of chapters on some aspects of filter theory met in the design of the classic 3-port junction circulator.

The waveguide junction device discussed here is, of course, only one possible type of circulator. It may also be constructed using either the principle of Faraday rotation in a longitudinally magnetized ferrite medium or nonreciprocal phase shifters in rectangular waveguides in conjunction with 3 dB hybrids and magic tees.

One typical application of the junction circulator is as a duplexer in which a single antenna is connected to both a transmitter and a receiver, another as an isolator, still another as a switch between two transmitters. It is also met in the design of 1-port negative resistance amplifiers and other microwave equipments.

1

ARCHITECTURE OF SYMMETRICAL WAVEGUIDE JUNCTION CIRCULATORS

1.1 Introduction

The 3-port circulator is a unique non-reciprocal symmetrical junction having one typical input port, one output port and one decoupled port. The fundamental definition of the junction circulator has its origin in energy conservation. It states that the only matched symmetrical 3-port junction corresponds to the definition of the circulator. In such a junction a wave incident at port-1 is emergent at port-2, one at port-2 at port-3 and so on in a cyclic manner. One possible model of a circulator is a magnetized ferrite or garnet gyromagnetic resonator having 3-fold symmetry connected or coupled to three transmission lines or waveguides. The purpose of this introductory chapter is to provide one phenomenological description of the operation of this sort of device, summarize some of the more common resonator geometries encountered in this class of device and indicate some of its uses. The introduction of any such resonator at the junction of three E or H-plane waveguides or fine line circuits readily produces a degree-1 circulation solution. In practice the gyromagnetic resonator is embedded in a filter circuit in order to produce a degree-2 frequency response. This problem is dealt with in some detail elsewhere in this text. Since a matched 3-port junction is a circulator by definition matching such a magnetized resonator is both necessary and sufficient for design purpose. The phenomenological adjustment of this class of device involves, under some simplifying conditions, the removal of the degeneracy of a pair of counter-rotating modes under the influence of a direct magnetic field and the rotation of a figure-of-eight standing wave pattern in such a way as to locate a null in its pattern at one of the three ports of the circulator. The rotation of the standing wave pattern under the application of a direct magnetic field may be understood by decomposing the linearly polarized radio frequency magnetic field on the axis of the resonator into counter-rotating ones which are then split by its gyrotropy. Since the direction in which the standing wave pattern

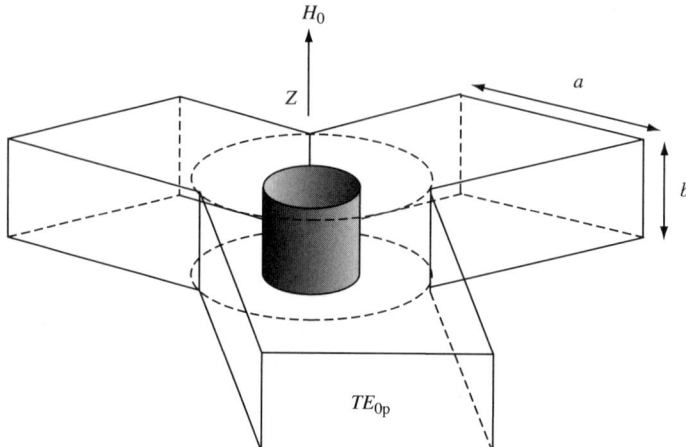

Figure 1.1 Schematic diagram of waveguide circulator using simple ferrite post resonator

in a gyromagnetic resonator is fixed by that of the direct magnetic field the junction may be utilized to realize an electrically actuated waveguide switch.

1.2 Operation of Waveguide-circulator using Post-resonator

The most general form of the 3-port junction circulator consists of a magnetized ferrite resonator with three-fold symmetry at the junction of three waveguides. The most simple geometry is that of a ferrite post between the top and bottom walls of the waveguide with no variation of the radio frequency fields along the axis of the post. A schematic diagram of this arrangement is illustrated in Figure 1.1. The ferrite resonator is magnetized perpendicularly to the plane of the post by a static magnetic field. A property of such a magnetized gyromagnetic medium is that it displays different scalar permeabilities under the influence of counter-rotating alternating magnetic fields. In this device, power entering port 1 emerges from port 2 and so on, in a cyclic manner. An important property of the device is that a perfect circulator is obtained when it is matched. For a 3-port junction this requires two independent variables. Strictly speaking, the construction of any 3-port junction circulator necessitates the adjustment of the phase angles of one inphase and two counterrotating field patterns of a ferrite resonator with three-fold symmetry. However, under certain simplifying conditions the electric field of the inphase mode is zero or negligible at the terminals of the resonator. If this condition is satisfied then the adjustment of any 3-port circulator may be described in terms of the first pair of split degenerate counterrotating field patterns of the gyromagnetic resonator. When the junction is unmagnetized, the resonant frequencies of the two field patterns are the same. When it is magnetized, the degeneracy is removed, by the gyrotropy, and the standing wave pattern within the resonator is rotated. One circulation condition is established by operating

1.3 GYROTROPY IN MAGNETIC INSULATORS

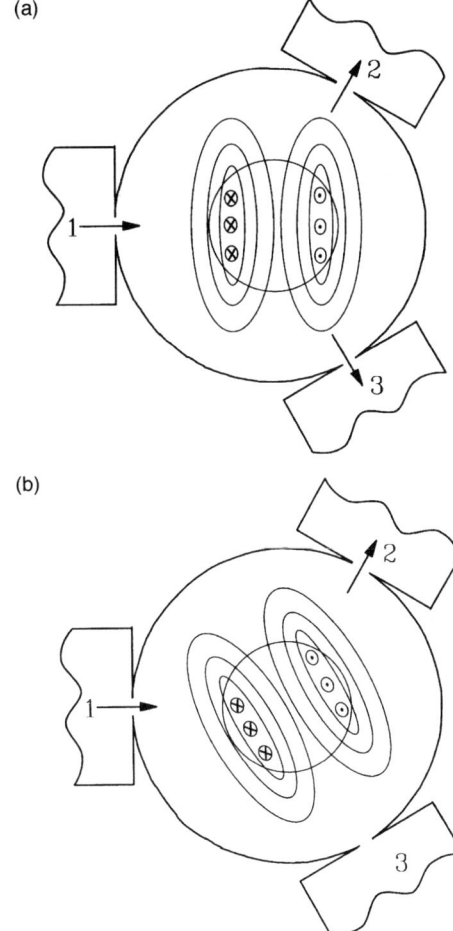

Figure 1.2 Phenomenological operation of waveguide circulator using post resonator

between the two split frequencies. This requirement essentially fixes the dimensions of the ferrite resonator. The second circulation condition is met by adjusting the splitting between the degenerate modes, until the standing wave pattern is rotated through 30°. From symmetry, port 3 is then situated at a null of the standing wave pattern and is therefore isolated. The 3-port junction then behaves as a transmission line resonant cavity between ports 1 and 2. This condition is met by the magnitude of the direct magnetic field: Figure 1.2(a) gives the field pattern in a demagnetized 3-port junction using a post resonator while Figure 1.2(b) depicts the same field pattern rotated through 30° to form an ideal circulator. Figure 1.3 shows one practical arrangement.

1.3 Gyrotropy in Magnetic Insulators

The origins of the magnetic effects or magnetization in magnet insulators are due to the effective current loops of electrons in atomic orbits and the effects of

4   1   ARCHITECTURE OF SYMMETRICAL WAVEGUIDE JUNCTION CIRCULATORS

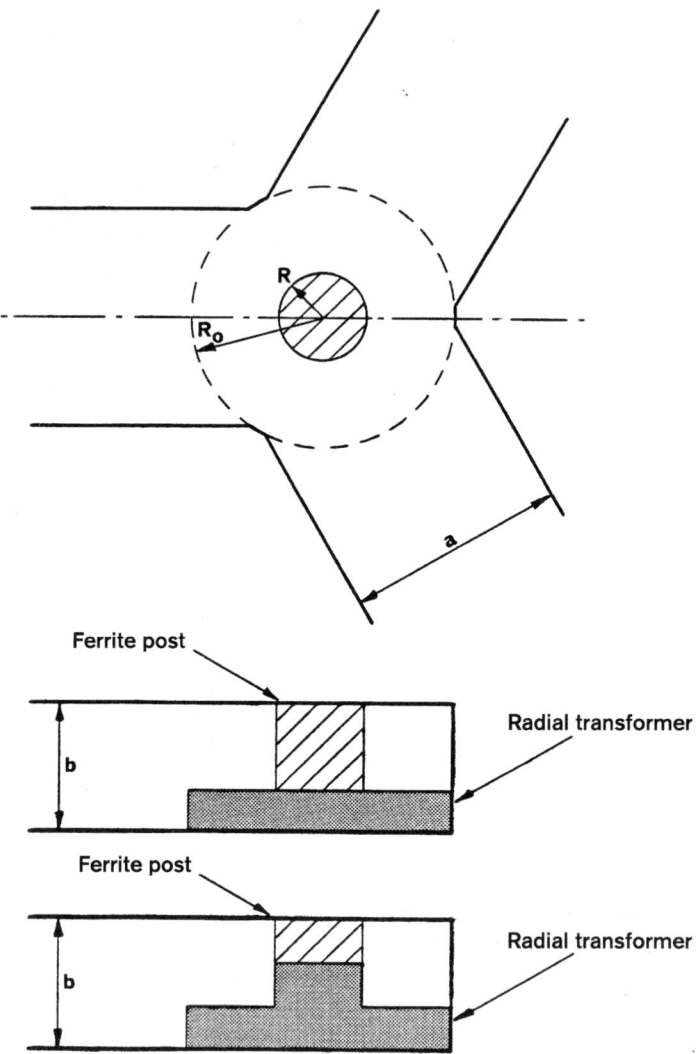

Figure 1.3 Schematic diagram of a waveguide circulator using radial transformer

electron spin and atomic nuclei (Figure 1.4). Each of these features produces a magnetic field that is equivalent to that arising from a magnetic dipole; the total magnetic moment is the vector sum of the individual moments. In ferromagnetic insulators the predominant effect is due to the electron spin. If the direction of the polarization of the alternating radio frequency wave is in the same direction as that of an electron spin under the influence of a direct magnetic field then one sort of interaction occurs and one value of scalar permeability is realized; if it is in the opposite direction, then a second interaction takes place and a different value of permeability is manifested. Figure 1.5 summarizes the two gyromagnetic states met in a gyromagnetic waveguide biased along its axis. The scalar permeabilities ($\mu \pm \kappa$) displayed by such waveguides under those circumstances are

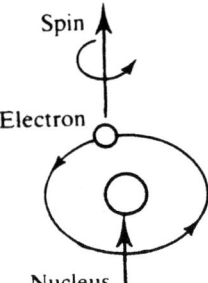

Figure 1.4 Spin motion in magnetic insulator

simple linear combinations of the diagonal (μ) and off-diagonal (κ) entries of the tensor permeability of the magnetized gyromagnetic medium. The absolute values of these quantities are essentially fixed by the frequency of the alternating radio magnetic field and the direct magnetization of the magnetic insulator and its direct magnetic field. At microwave frequencies μ equals unity and κ is usually bracketed between $0.30 \leq \kappa \leq 0.70$. The relative dielectric constant of a typical ferrite or garnet material is bracketed between 8 and 15.

1.4 The Turnstile Junction Circulator

The original waveguide junction circulator, known as the turnstile one, was first described by Tor Schaug-Pattersen. It consists of a circular guide containing a longitudinally magnetized ferrite section at the junction of three rectangular waveguides. This arrangement relies for its operation on the well-known Faraday rotation principle along a magnetized ferrite loaded circular waveguide. Figure 1.6 depicts one topology.

The operation of the turnstile junction may be understood by having recourse to superposition. It starts by decomposing a single input wave at port 1 (say) into a linear combination of voltage settings at each port

$$\begin{bmatrix} 1 \\ 0 \\ 0 \end{bmatrix} = \frac{1}{3}\begin{bmatrix} 1 \\ 1 \\ 1 \end{bmatrix} + \frac{1}{3}\begin{bmatrix} 1 \\ \alpha \\ \alpha^2 \end{bmatrix} + \frac{1}{3}\begin{bmatrix} 1 \\ \alpha^2 \\ \alpha \end{bmatrix}$$

where $\alpha = \exp(j120)$ and $\alpha^2 = \exp(j240)$. A scrutiny of the first, so called in-phase generator settings, indicates that it produces an electric field along the axis of the circular waveguide which does not couple into it. The reflected waves at the three ports of the junction are therefore in this instance unaffected by the details of the gyromagnetic waveguide. A scrutiny of the second and third, so-called counter-rotating generator settings, indicates, however, that these establish counter-rotating circularly polarized alternating magnetic fields at the open face of the circular gyromagnetic waveguide which readily propagate. The fields produced at the axis of the junction by each of these three possible generator

1 ARCHITECTURE OF SYMMETRICAL WAVEGUIDE JUNCTION CIRCULATORS

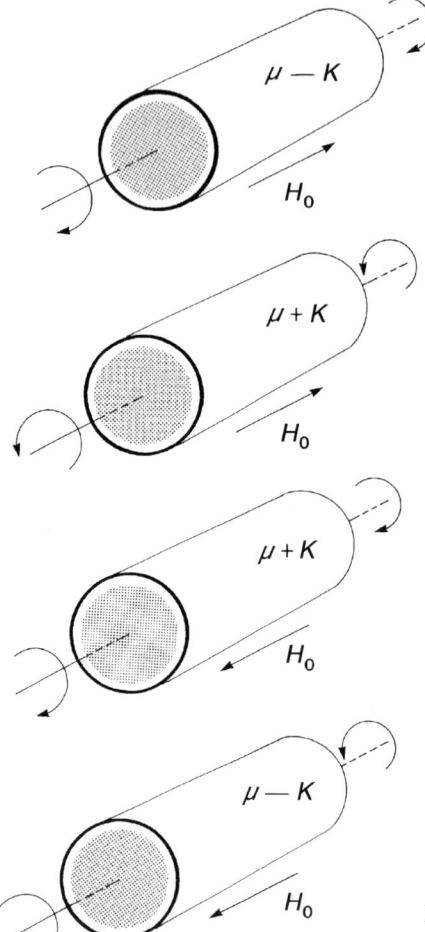

Figure 1.5 Normal modes on longitudinally magnetized circular gyromagnetic waveguide

settings are illustrated in Figure 1.7. Since a characteristic of a gyromagnetic waveguide is that it has different scalar permeabilities under the two arrangements it provides one practical means of removing the degeneracy between the reflected waves associated with these two generator settings.

A typical reflected wave at any port is constructed by adding the individual ones due to each possible generator setting. A typical term is realized by taking the product of a typical incident wave and a typical reflection coefficient.

$$\begin{bmatrix} b_1 \\ b_2 \\ b_3 \end{bmatrix} = \frac{\rho_0}{3} \begin{bmatrix} 1 \\ 1 \\ 1 \end{bmatrix} + \frac{\rho_-}{3} \begin{bmatrix} 1 \\ \alpha \\ \alpha^2 \end{bmatrix} + \frac{\rho_+}{3} \begin{bmatrix} 1 \\ \alpha^2 \\ \alpha \end{bmatrix}$$

An ideal circulator is now defined as

1.4 THE TURNSTILE JUNCTION CIRCULATOR

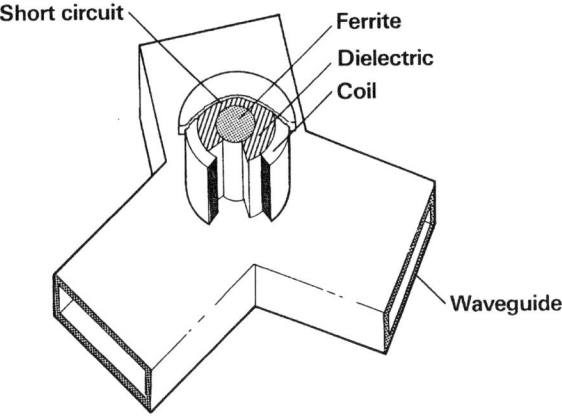

Figure 1.6 Schematic diagram of H-plane turnstile calculator

$$\frac{\rho_0 + \rho_- + \rho_+}{3} = 0$$

$$\frac{\rho_0 + \alpha\rho_- + \alpha^2\rho_+}{3} = -1$$

$$\frac{\rho_0 + \alpha^2\rho_- + \alpha\rho_+}{3} = 0$$

To adjust this, and other circulators, requires a 120° phase difference between the reflection coefficients of the three different ways it is possible to excite the three rectangular waveguides. One solution is

$$\rho_+ = \exp[-j2(\theta_1 + \theta_+ + \pi/2)]$$
$$\rho_{-1} = \exp[-j2(\theta_1 + \theta_- + \pi/2)]$$
$$\rho_0 = \exp[-j(2\theta_0)]$$

provided that $\theta_1 = \theta_0 = \pi/2$ and $\theta_+ = -\theta_- = \pi/6$. The required phase angles of the three reflection coefficients are established by adjusting the length of the demagnetized ferrite section so that the angle between the in-phase and counter rotating reflection coefficients is initially 180°. The degenerate phase angles of the counter rotating reflection coefficient are then separated by 120° by magnetizing the ferrite region, thereby producing the ideal phase angles of the circulator. These two steps represent the necessary and sufficient conditions for the adjustment of this class of circulator.

Since the relationship between the incident and reflected waves at the terminals of a network or junction is often described in terms of a scattering matrix it is appropriate to reduce the result established here in that notation. Figure 1.8 indicates the nomenclature entering into the definition of this matrix in the case

8 1 ARCHITECTURE OF SYMMETRICAL WAVEGUIDE JUNCTION CIRCULATORS

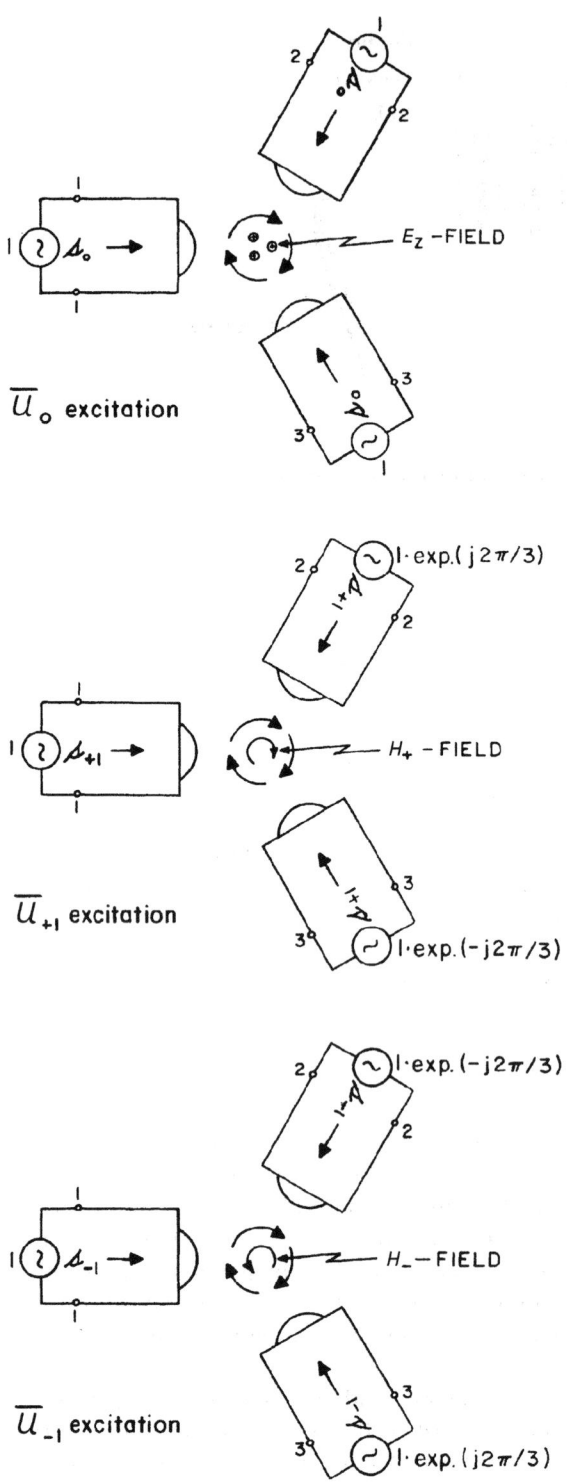

Figure 1.7 In-phase and counter-rotating field patterns of junction circulator

1.4 THE TURNSTILE JUNCTION CIRCULATOR

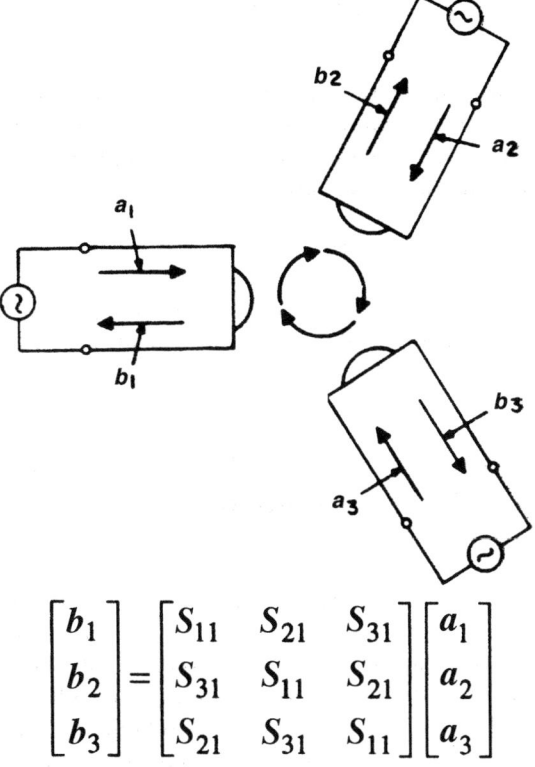

$$\begin{bmatrix} b_1 \\ b_2 \\ b_3 \end{bmatrix} = \begin{bmatrix} S_{11} & S_{21} & S_{31} \\ S_{31} & S_{11} & S_{21} \\ S_{21} & S_{31} & S_{11} \end{bmatrix} \begin{bmatrix} a_1 \\ a_2 \\ a_3 \end{bmatrix}$$

Figure 1.8 Definition of a scattering matrix

of a 3-port network with three-fold symmetry. Its entries relates incident and reflected waves at suitable terminal planes of the circuit

$$S_{11} = (b_1/a_1), \quad a_2 = a_3 = 0$$

$$S_{21} = (b_2/a_1), \quad a_2 = a_3 = 0$$

$$S_{31} = (b_3/a_1), \quad a_2 = a_3 = 0$$

A scrutiny of these definitions indicates that the entries of the scattering matrix may be readily evaluated once the reflected waves at all the ports due to an incident wave at a typical port are established. Taking a_1 as unity and making use of the results for b_1, b_2 and b_3 gives the required parameters without ado.

$$S_{11} = \frac{\rho_0 + \rho_+ + \rho_-}{3}$$

$$S_{21} = \frac{\rho_0 + \alpha \rho_+ + \alpha^2 \rho_-}{3}$$

$$S_{31} = \frac{\rho_0 + \alpha^2 \rho_+ + \alpha \rho_-}{3}$$

The entries of the scattering matrix are therefore linear combinations of the reflection variables at any port associated with each possible family of generator settings. One definition of an ideal circulator which is on keeping with the description of the turnstile junction circulator is therefore

$$S_{11} = 0; \quad S_{21} = -1; \quad S_{31} = 0$$

This solution may be separately established by having recourse to the unitary condition and may therefore be taken as a universal definition of a 3-port lossless junction circulator.

1.5 H- and E-Plane Junctions using Re-entrant Resonators

While the operation of the 3-port circulator using either a post resonator or a turnstile resonator may be readily visualized, neither is the most important practical realization. Commercial waveguide circulators more often than not use quarterwave coupled quarterwave long open triangular or circular resonators opencircuited at one end and shortcircuited at the other or open resonators open-circuited at both ends. Figure 1.9 depicts schematic diagrams of the three arrangements met in practice. Introduction of an image plane in the configuration in Figure 1.9(a) indicates that it is dual in every respect to those in Figure 1.9(b) and (c) except that the susceptance slope parameter of the arrangement in Figure 1.9 is twice that of the others. A quarterwave long magnetized ferrite resonator shortcircuited at one end, and opencircuited or loaded by an image wall at the other is therefore a suitable prototype for the construction of this class of device. The operating frequency of the H-plane circuit corresponds to the odd solution of two coupled open dielectric resonators constructed from sections of a gyromagnetic waveguide propagating the hybrid HE_{11} mode.

The partial-height or re-entrant H-plane junction circulator can be visualized as a 7-port network consisting of a 3-port H-plane junction symmetrically coupled along its axis to cylindrical gyromagnetic waveguides each supporting 2-orthogonal ports. The network is reduced to a 3-port one by closing the gyromagnetic waveguides by shortcircuit pistons. This sort of circulator has its origin in the turnstile structure indicated in Figure 1.6. The duality between the two arrangements may be achieved by replacing the electric wall in the circular waveguides of the original turnstile junction by open or magnetic ones. The first circulation condition in this sort of circulator coincides with that for which the inphase and degenerate counterrotating reflection coefficients are in anti-phase. The second one is established by splitting the degeneracy between the counterrotating ones by a suitable direct magnetic field.

A useful assembly technique at millimeter bands is separately illustrated in Figure 1.10.

1.5 H- AND E-PLANE JUNCTIONS USING RE-ENTRANT RESONATORS

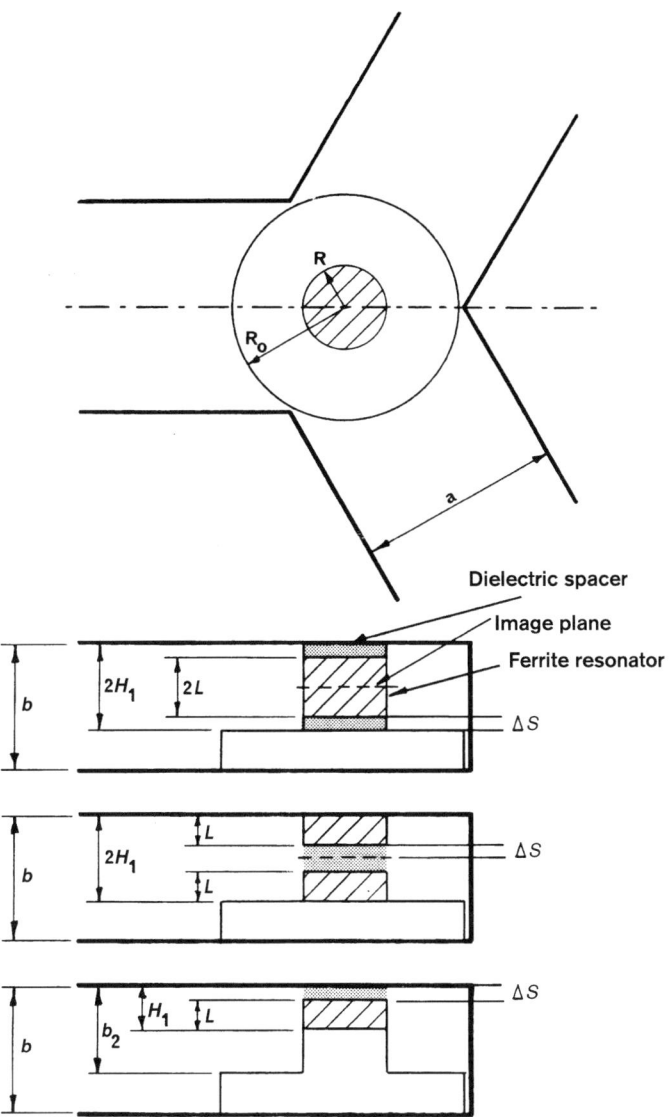

Figure 1.9 Schematic diagrams of H-plane waveguide circulators using re-entrant turnstile junctions

Another important waveguide circulator topology is the E-plane one. It is realized by locating quarterwave long open resonators at the junction of three E-plane instead of H-plane waveguides In the E-plane device, with the direction of propagation taken along the z axis, H_z rather than H_x is perpendicular to the symmetry axis of the device. The magnetic fields corresponding to the counter-rotating eigenvectors are therefore circularly polarized at the side instead of the top and bottom walls of the waveguide. Figure 1.11 illustrates a 5-port junction using discrete 3-port E-plane circulators.

12 1 ARCHITECTURE OF SYMMETRICAL WAVEGUIDE JUNCTION CIRCULATORS

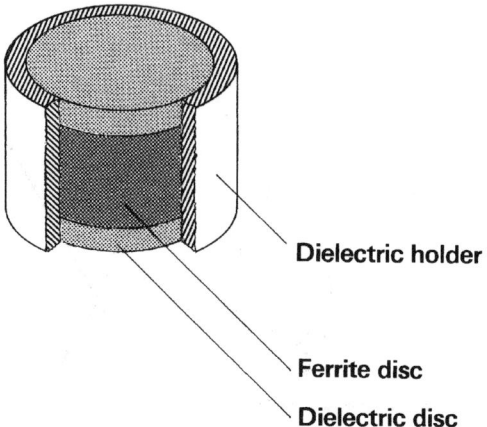

Figure 1.10 Assembly technique for half-wave long resonator

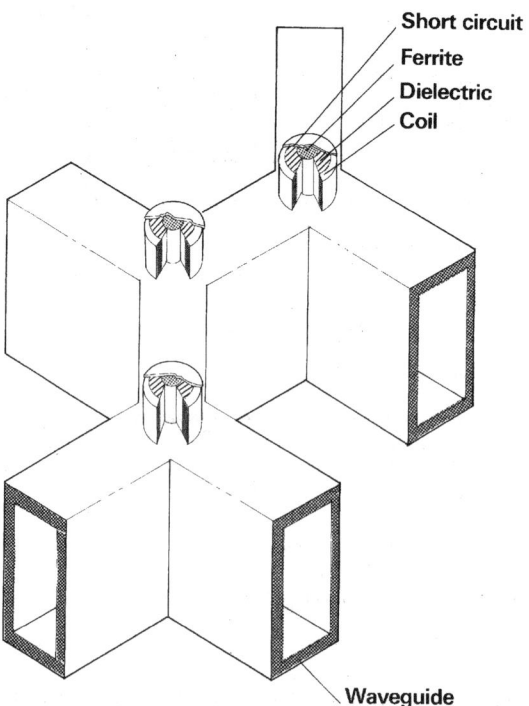

Figure 1.11 Tandem connection of 3-port E-plane turnstile circulators

1.6 Switched Resonators

Since the direction of circulation of a circulator is determined by that of the direct magnetic field it may be employed to switch an input signal at one port to either one of the other two. Switching is achieved by replacing the permanent magnet by an electromagnet or by latching the microwave ferrite resonator directly by embedding a current carrying wire loop within the resonator. Some possiblities are shown in Figure 1.12. This sort of switch is particularly useful in the construction of Butler type matrices in phased array systems. The standing wave solutions in triangular and wye resonators are separately illustrated in Figure 1.13.

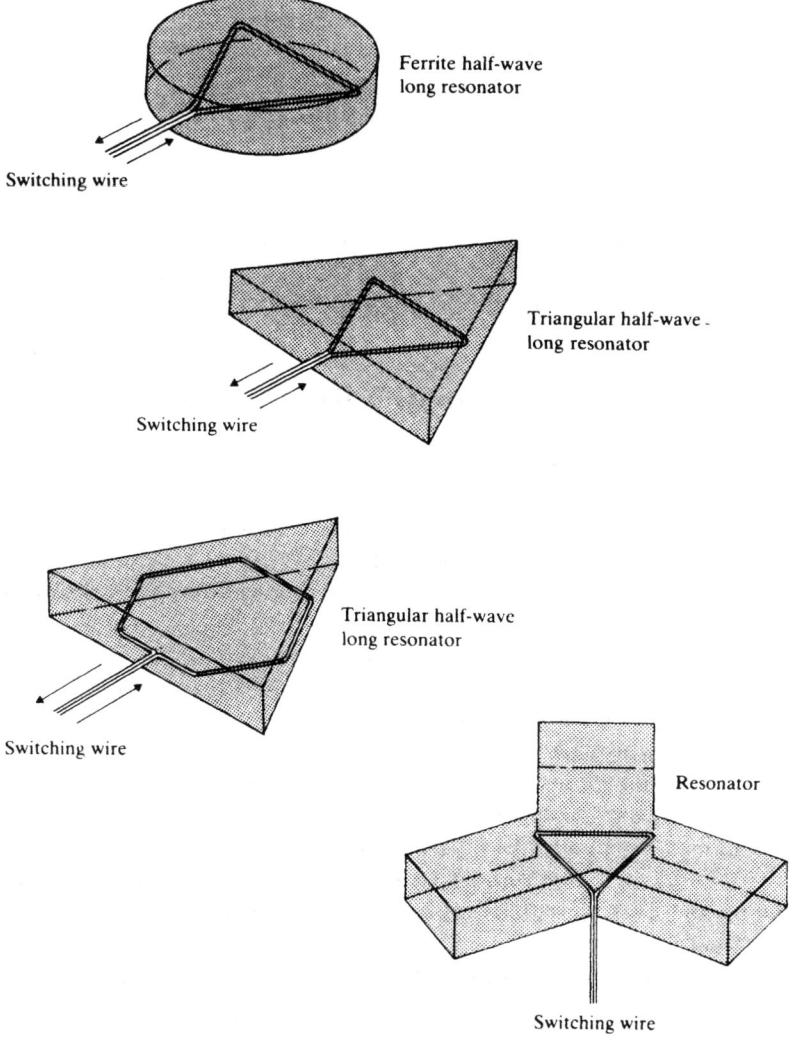

Figure 1.12 Details of switching wire embedded in half-wave circular, triangular and wye resonator

14   1   ARCHITECTURE OF SYMMETRICAL WAVEGUIDE JUNCTION CIRCULATORS

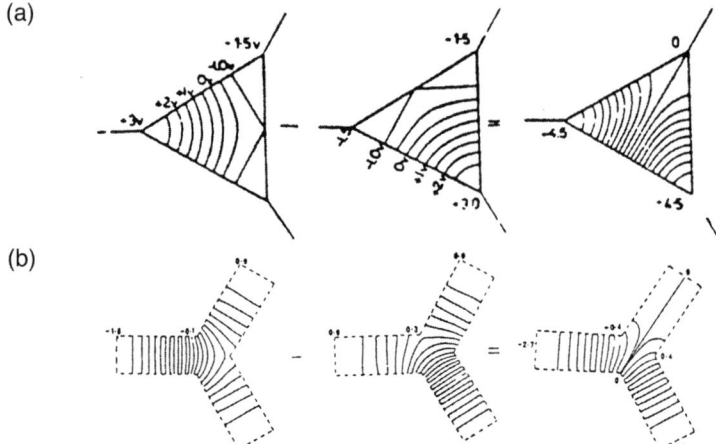

Figure 1.13   Standing wave solution of triangular (a) and wye (b) post resonators

The switching power necessary to actuate this sort of circuit is determined by the stored energy in the magnetic circuit and the switching time. Switching times between 100 ns and 1 ms are achievable depending upon whether the gyromagnetic circuit is internally and externally magnetized.

1.7   Composite Resonators

The conventional waveguide junction relies on an open dielectric resonance in a ferrite or garnet disc for one of its two circulation conditions. The maximum average power that such a circulator can handle is determined by the temperature drop across the thin dimension of the ferrite disc. Since the thermal conductivity of the ferrite material is relatively low the temperature drop across the disc can be reduced by replacing part of it by dielectric material that has a higher thermal conductivity, without altering the first circulation condition. For instance, the thermal conductivity of WESGO AL-995 ceramic is 29.31 W/m deg C (70×10^{-3} cal/c s deg C) and that of beryllia oxide is 219.81 W/m deg C (525×10^{-3} cal/cm s deg C). For ferrites it is 2.09 W/deg C (5×10^{-3} cal/cm s deg C). One way to overcome this difficulty is to employ a composite resonator in the design. These types of resonators have been utilized in the construction of devices capable of handling a number of hundreds of Watts of mean power. Figure 1.14 depicts one planar and one radial configuration. A further advantage of this class of resonator is that the temperature stability is improved because many dielectric materials are temperature stable.

The second circulation condition relates the gyrator conductance, the susceptance slope parameter and the split frequencies of the junction to the specifications of the device; a large separation between the split frequencies being essential for the realization of high quality circulators. Replacing part of the ferrite material by a dielectric reduces the difference between the split frequencies of

1.8 Okada Resonator

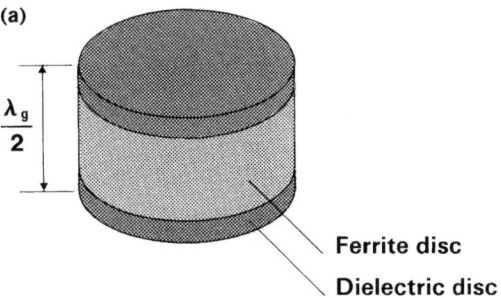

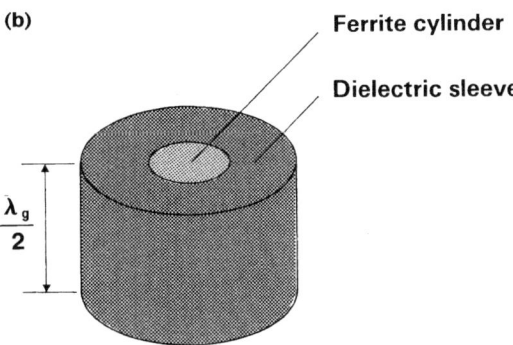

Figure 1.14 Schematic diagrams of (a) axial and (b) radial composite resonators

the junction, a compromise is therefore necessary between the microwave specification and the power rating of the device. This problem is discussed in terms of the filling factor and magnetization of the resonator.

1.8 Okada Resonator

One possible way to increase the power rating of a waveguide circulator is to have recourse to the Okada resonator. This resonator consists of stacked circular metal plates on which are mounted ferrite disks separated by a dielectric or free space region. The schematic diagram of a single chamber arrangement is indicated in Figure 1.15. It may be visualized as a composite post resonator. One property of this sort of arrangement is that it is characterized by an effective dielectric constant which is bracketed between that of the ferrite region and that of the gap. One effect of this feature is that the surface area of the resonator region is increased. This makes it particularly attractive in the design of high average power devices.

Some of the considerations entering into the description of this sort of resonator are average and peak power ratings, cooling provisions, magnetic and dielectric losses, gain bandwidth or quality factor. In this sort of device, the filling factor is

16 1 ARCHITECTURE OF SYMMETRICAL WAVEGUIDE JUNCTION CIRCULATORS

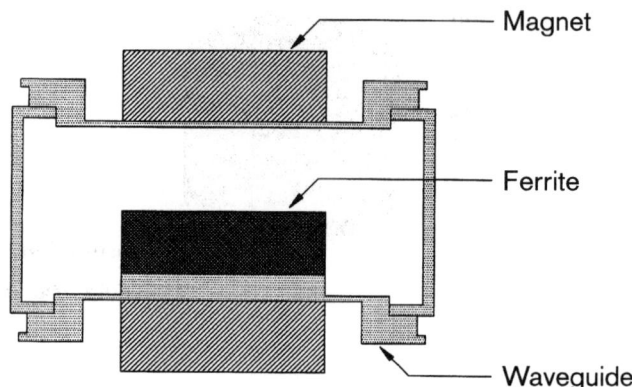

Figure 1.15 Schematic diagram of Okada resonator using a single chamber

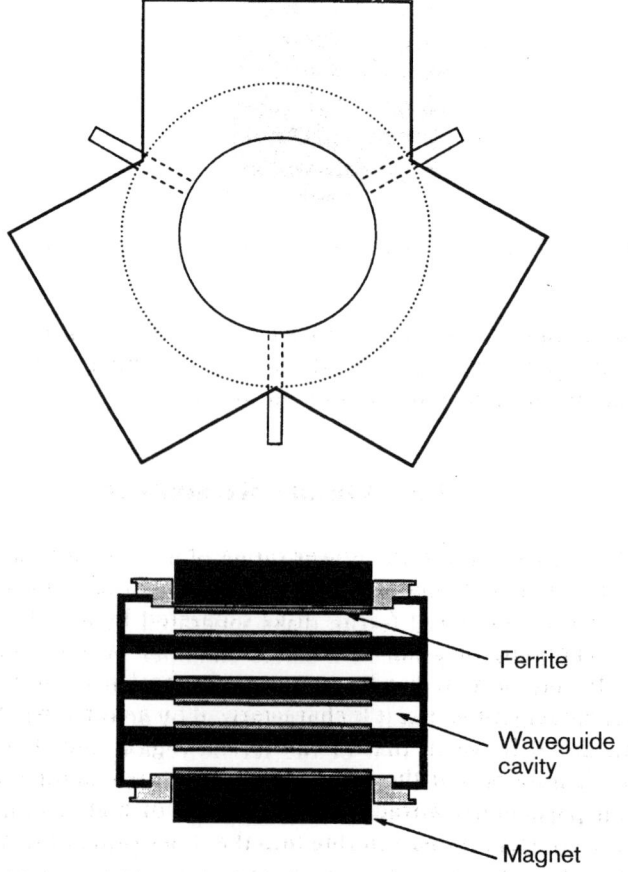

Figure 1.16 Schematic diagram of Okada resonator using four double chambers

fixed by the quality factor of the resonator, the magnetic parameters of the ferrite material by the magnetic losses, the gap between the ferrite layers by the peak power specification. The number of chambers by the average power rating. The radial filling factor is utilized as the independent variable. An arrangement using four chambers is shown in Figure 1.16.

1.9 Power Rating of Gyromagnetic Resonators

Important aspects of microwave components are their peak and average power ratings. The peak power rating is usually fixed by arcing and by nonlinear effects in magnetic insulators due to spin wave instabilities. The average power rating is restricted by the temperature rise of the device which may to some extent be mitigated by cooling it by forced air or water. The power dissipated in the device has its origin in dielectric losses, linear and nonlinear magnetic ones, and waveguide ones.

The choice of resonator in any particular situation is dictated by one or both of these difficulties. The family of planar resonators is in general suitable for the construction of devices with a large mean power rating, the Okada resonator is usually employed in the design of very large mean power devices, the re-entrant or turnstile one is appropriate in the design of devices with large peak power specifications. The family of composite resonators is used where a compromise between mean and peak power is necessary. Insertion losses of the order of 0.06 dB to 0.12 dB are not in practice unusual.

Since the power rating of a typical circulator is dependent upon the choice of resonator, the waveguide size, the frequency, the bandwidth, the available ferrite material, the cooling arrangement, the temperature range, pressurization and the outline drawing to mention but some considerations that enter into any design it is difficult to compile precise recommendations in any particular situation. Some representative ratings at say 14 GHz in WR 75 waveguide are 250 000 W peak and 2000 W average at a single frequency with appropriate pressurization and cooling; 50 000 W peak and 100 W average in a turnstile junction over some 20% bandwidth and a return loss of 23 dB; 1000 W average in a planar circulator and 2,000 W average in an Okada one over similar commercial bandwidths.

1.10 Quarter-Wave Coupled Circulator

Any gyromagnetic resonator with three-fold symmetry suitably placed at the junction of three waveguides may, as is now understood, be adjusted to form an ideal 3-port circulator at a single frequency. Practical circulators, however, have to operate over finite frequency intervals with a specified ripple level or retun loss and isolation. One way to realize a classical frequency response is to absorb each port of the gyromagnetic resonator into a 2-port filter or matching network. A knowledge of the i-port complex gyrator circuit of the gyromagnetic junction at a typical port is sufficient for this purpose. Figure 1.17 indicates one schematic arrangement. While a host of network solutions are in practice possible the

18 1 ARCHITECTURE OF SYMMETRICAL WAVEGUIDE JUNCTION CIRCULATORS

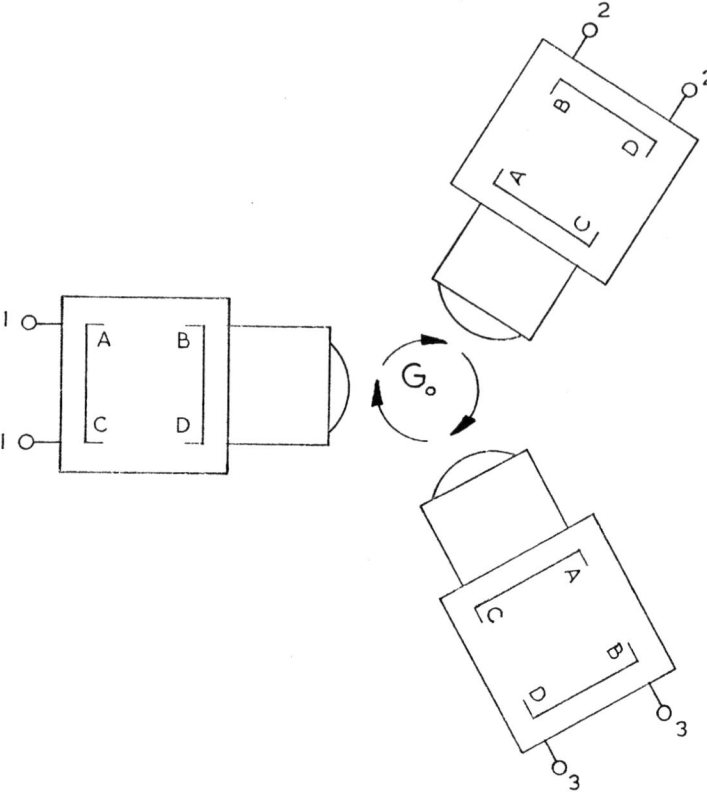

Figure 1.17 Topology of 3-port circulators using 2-port filter circuits

most common arrangement consists of one or more quarter-wave long impedance transformers. A typical frequency response of such a circulator using a single quarter-wave long transformer at each port of the gyromagnetic resonator is indicated in Figure 1.18. This sort of topology is a classic network problem in the literature and much of modern circulator practice rests on an understanding of this topic. The actual gain-bandwidth of any circulator is in practice fixed by the quality factor of the gyromagnetic resonator and the topology of the filter circuit. It fixes the relationship between the isolation on return loss in dBs of the circulator and its bandwidth. A return loss or isolation of 23 dB (say) and a bandwidth of typically 25% is in practice realizable in conjunction with a single quarter-wave long impedance transformer. A similar return loss or isolation specification is readily realizable over a bandwidth of between 40 and 66% with 2 or 3 impedance transformers. The geometries of some practical transformer structures are illustrated in Figure 11.1 in Chapter 11. The ridge one is perhaps the recommended arrangement at this time.

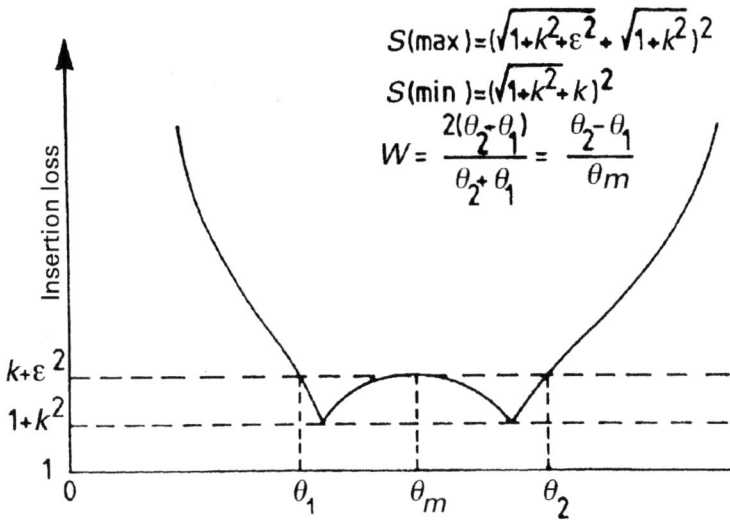

Figure 1.18 Frequency response of quarter-wave coupled junction circulation

1.11 Single Port Amplifiers Using Junction Circulators

A simplified schematic of a reflection tunnel diode amplifier (TDA) which utilizes a circulator to separate the input signal from the amplified one is shown in Figure 1.19. Since the gain of the amplifiers may be comparable to the isolation of the junction, such amplifiers are often used in conjunction with 5-port circulators, as depicted in Figure 1.20. The input and output junctions are in this arrangement connected as isolators in order to minimize gain variations due to source and load impedance variations. The magnetic field for the input circulator is here sometimes supplied by an electromagnet. The reversal of the magnetic field in this circuit allows the TDA to be protected from radio frequency leakage during the transmitting period by reversing the direction of circulation during this interval. Figure 1.21 outlines an m-port amplifier chain using junction circulators.

1.12 Duplexing using Junction Circulators

The ferrite circulator may also be used in duplexing systems for simultaneous transmissions and reception of microwave energy with a single antenna. Ferrite circulators here replace conventional types of duplexing and are suited for both high- and low-power systems. The arrangement used here is shown in Figure 1.22. Circulators are also employed in communication systems to eliminate mutual interference between closely separated transmitters. Figure 1.23 gives an example of a single antenna being shared by a number of transmitters with the help of

Figure 1.19 Single port amplifier using 3-port circulator

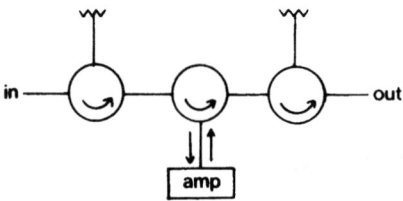

Figure 1.20 Single port amplifier using 5-port circulator

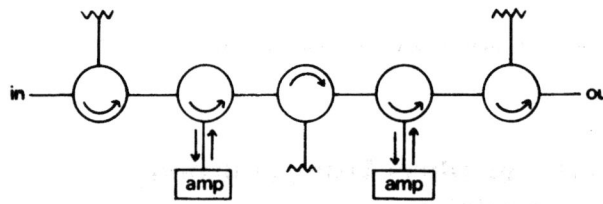

Figure 1.21 m-Port amplifier chain using junction circulators

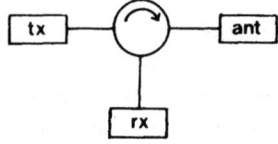

Figure 1.22 Low-power duplexer using 3-port circulator

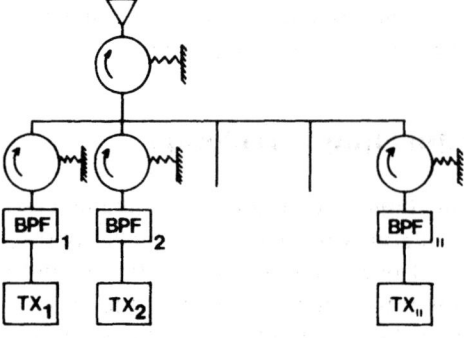

Figure 1.23 Duplexing between closely spaced transmitters

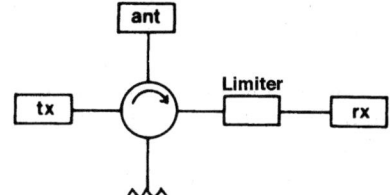

Figure 1.24 High-power duplexer using 4-port circulators

circulators and bandpass filters (BPF). Figure 1.24 illustrates a 4-port high-power duplexer which utilizes a reflection limiter to protect the receiver during the transmission interval.

References and Further Reading

Akaiwa, Y. (1974) Operation modes of a waveguide Y-circulator, *IEEE Trans.*, **MTT-22**, 954–959.

Auld, B. A. (1959) The synthesis of symmetrical waveguide circulators, *IRE Trans. Microwave Theory Tech.*, **MTT-7**, 238–246.

Butterweck, H. J. (1963) The Y-circulator, *Arch. Elek. Ubertragung*, **17**(4), 163–176.

Chait, H. N. and Curry, T. R. (1959) Y-circulator, *J. Appl. Phys.*, **30**, 152.

Collins, J. H. and Watters, A. R. (1963) A field map for the waveguide Y-junction circulator, *Electron. Engr. London*, **35**, 540–543.

Davies, J. B. (1962) An analysis of the *m*-port symmetrical *H*-plane waveguide junction with central ferrite post, *IRE Trans. Microwave Theory Tech.*, **MTT-10**, 596–604.

Denlinger, E. J. (1974) Design of partial-height ferrite waveguide circulators, *IEEE Trans.*, **MTT-22**, 810–813.

Fay, C. E. and Comstock, R. L. (1965) Operation on the ferrite junction circulator, *IEEE Trans. Microwave Theory Tech.*, **MTT-13**, 15–27.

Hauth, W. (1981) Analysis of circular waveguide cavities with partial-height ferrite insert, In *Proc. Eur. Microwave Conf.*, 383–388.

Hauth, W., Lenz S. and Pivit, E. (1986) Design and realization of very-high-power waveguide junction circulators, *Frequency*, **40**, 2–11.

Helszajn, J. (1967) An *H*-plane high-power TEM ferrite junction circulator, *Radio and Electron. Eng.*, **33**, 257–261.

Helszajn, J. (1969) Frequency and Bandwidth and H-plane TEM Junction Circulator, *Proc. IEE*, **117**(7), 1235–1238.

Helszajn, J. (1974) Common waveguide circulator configurations, *Electron. Eng.*, 66–68.

Helszajn, J. (1974) Composite-junction circulators using ferrite disks and dielectric rings, *IEEE Trans.*, **MTT-22**, 400–410.

Helszajn, J. and Tan, F. C. C. (1975) Mode charts for partial-height ferrite waveguide junction circulators; *Proc. IEE*, **122**, 34–36.

Helszajn, J., James, D. S. and Nisbet, W. T. (1979) Circulators using planar triangular resonators, *IEEE Trans. on Microwave Theory Tech.*, **MTT-27**, 188–193.

Helszajn, J., Guixa, R., Girones, J., Hoyos, E. and Garcia Taheno, J. (1995) Adjustment of Okada resonator using composite chambers, *IEEE Trans Microwave Theory, Tech.*, **MTT-43**.

Konishi, Y. (1969) A high power UHF circulator, *IEEE Trans. on Microwave, Theory, Tech.*, **MTT-15**, 700–708.

Okada, F. and Ohwit, K. (1975) The development of a high power microwave circulator for use in breaking of concrete and rock, *J. Microwave Power*, **10**(2).

Owen, B. (1972) The identification of modal resonances in ferrite loaded waveguide Y-junction and their adjustment for circulation, *Bell Syst. Tech., J.*, **51**, (3).

Owen, B. and Barnes, C. E. (1970) The compact turnstile circulator, *IEEE Trans.*, **MTT-18**, 1096–1100.

Schaug-Patterson, T. (1958) Novel design of a 3-port circulator, Norwegian Defence Research Establishment.

2

SCATTERING MATRIX OF m-PORT JUNCTION

2.1 Introduction

The entries of the scattering matrix of an m-port junction are a set of quantities that relate incident and reflected waves at its ports or terminal planes. It describes the performance of a network under any specified terminating conditions. The coefficients along the main diagonal of this matrix are reflection coefficients while those along the off-diagonal are transmission ones. The scattering matrix is modified if one or more of the terminal planes are moved. A scattering matrix exists for every linear, passive, time invariant network.

Figure 2.1 depicts a generalized junction enclosed by a surface S, which cuts the various transmission lines perpendicular to the axes and provides a definition for the ports or terminal planes of the structure. However, the text is primarily concerned with m-port junctions with the symmetry illustrated in Figures 2.2 and 2.3. It is possible to deduce important general properties of junctions containing a number of ports by invoking such properties as symmetry, reciprocity, and energy conservation. An important property of the scattering matrix is that the permissible relationships between the entries of a lossless network are readily established by having recourse to the unitary condition. The adjustment of symmetrical junctions is, however, best dealt with in terms of the eigenvalue problem in Chapter 3. In the presence of dissipation all the existing relationships between the entries of the scattering matrix based on the unitary condition become invalid. In order to cater for this effect it is usual to introduce a so-called dissipation matrix. A description of this matrix is included for completeness.

24 2 SCATTERING MATRIX OF *m*-PORT JUNCTION

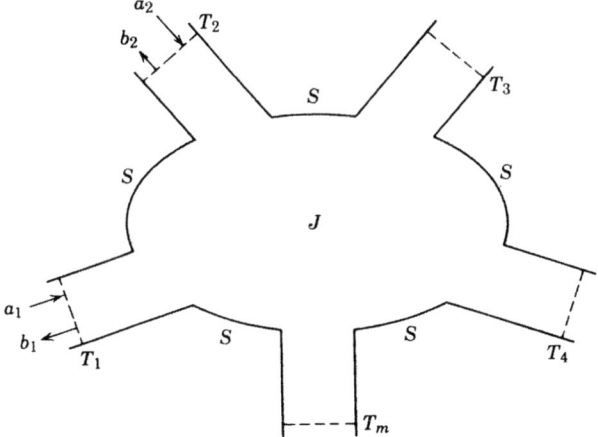

Figure 2.1 Schematic diagram of *m*-port junction showing incident and reflected waves

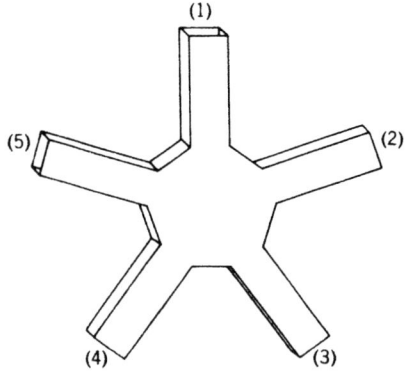

Figure 2.2 Schematic diagram of 5-port symmetrical waveguide junction

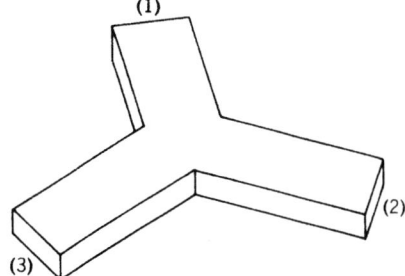

Figure 2.3 Schematic diagram of 3-port symmetrical waveguide junction

2.2 The Scattering Matrix

The scattering matrix of an m-port network is a square matrix of order m whose entries relate suitably chosen incident and reflected waves at the terminals of the network.

$$\bar{b} = \bar{S}\bar{a} \qquad (2.1)$$

$\bar{a}$ and $\bar{b}$ are incident and reflected vectors, which for a 3-port network are given by

$$\bar{a} = \begin{bmatrix} a_1 \\ a_2 \\ a_3 \end{bmatrix}, \quad \bar{b} = \begin{bmatrix} b_1 \\ b_2 \\ b_3 \end{bmatrix} \qquad (2.2)$$

The corresponding scattering matrix $\bar{S}$ is described by

$$\bar{S} = \begin{bmatrix} S_{11} & S_{12} & S_{13} \\ S_{21} & S_{22} & S_{23} \\ S_{31} & S_{32} & S_{33} \end{bmatrix} \qquad (2.3)$$

The elements along the main diagonal of this matrix are the reflection coefficients at the ports of the network, those along the off-diagonal represent the transmission coefficients between the same ports. The relationships between the incoming and outgoing waves for a 3-port network are therefore described using this nomenclature by

$$b_1 = a_1 S_{11} + a_2 S_{12} + a_3 S_{13} \qquad (2.5a)$$
$$b_2 = a_1 S_{21} + a_2 S_{22} + a_3 S_{23} \qquad (2.5b)$$
$$b_3 = a_1 S_{31} + a_2 S_{32} + a_3 S_{33} \qquad (2.5c)$$

A schematical diagram of this relationship is depicted in Figure 2.4.

A typical diagonal element of the scattering parameters of the 3-port is defined in terms of the incident and reflected waves by the preceding equations as

$$S_{11} = \left. \frac{b_1}{a_1} \right|_{a_2 = a_3 = 0} \qquad (2.6a)$$

Typical off-diagonal elements are given by

$$S_{13} = \left. \frac{b_1}{a_3} \right|_{a_1 = a_2 = 0} \qquad (2.6b)$$

2 SCATTERING MATRIX OF m-PORT JUNCTION

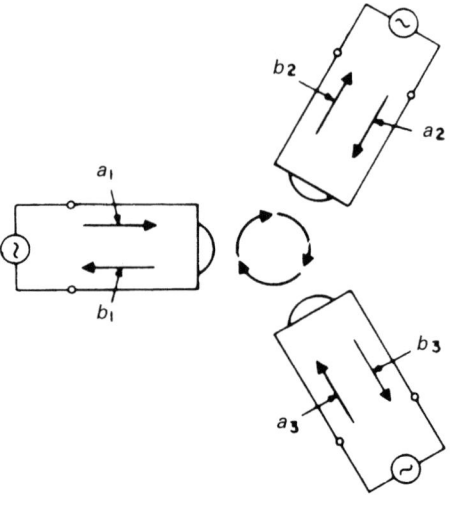

$$\begin{bmatrix} b_1 \\ b_2 \\ b_3 \end{bmatrix} = \begin{bmatrix} S_{11} & S_{21} & S_{31} \\ S_{31} & S_{11} & S_{21} \\ S_{21} & S_{31} & S_{11} \end{bmatrix} \begin{bmatrix} a_1 \\ a_2 \\ a_3 \end{bmatrix}$$

Figure 2.4 Definition of incident and reflected waves in 3-port junction

$$S_{31} = \left. \frac{b_3}{a_1} \right|_{a_2 = a_3 = 0} \tag{2.6c}$$

If it is assumed that a_i and b_i are normalized in such a way that $½a_ia_i^*$ is the available power at port i and $½b_ib_i^*$ is the emergent power at the same port then

$$a_i = \frac{1}{2}\left(\frac{V_i}{\sqrt{Z_0}} + \sqrt{Z_0}\, I_i\right), \quad i = 1,2,3 \tag{2.7}$$

$$b_i = \frac{1}{2}\left(\frac{V_i}{\sqrt{Z_0}} - \sqrt{Z_0}\, I_i\right), \quad i = 1,2,3 \tag{2.8}$$

Adopting these definitions indicates that the a's and b's at any port are linear combinations of the voltage and current variables at the same port. Such linear combinations are in fact also met in the description of a uniform transmission line for which the solution to the transmission line equations is given in terms of forward and backward travelling waves $A \exp(-\gamma z)$ and $B \exp(\gamma z)$ by

$$V = A \exp(-\gamma z) + B \exp(\gamma z) \tag{2.9}$$

$$Z_0 I = A \exp(-\gamma z) - B \exp(\gamma z) \tag{2.10}$$

A scrutiny of equations 2.6b, 2.6c indicates S_{13} connects port 3 to port 1 whereas S_{31} connects port 1 to port 3.

2.3 Circulator Definition by Means of Cyclic Substitution

In the theory of finite groups, an operation, called cyclic substitution, is defined. This operation is usually illustrated as being performed on a sequence of letters or numbers. For example, the operation of cyclic substitution (*abcd*) means that *a* be replaced by *b*, *b* by *c*, *c* by *d*, and *d* by *a*. If this operation is performed on the sequence *bdac* the result is

$$(abcd) \to bdac \to cabd$$

In the scattering relationship between incident and reflected waves the scattering matrix may be considered as indicating an operation performed on the incident waves, the result of which yields the reflected waves.

If the operator $\overline{S}$ corresponds to a cyclic substitution, then is can be seen that the device having such a scattering matrix corresponds to the intuitive notion of the circulator property. Consequently, a circulator may be defined as : 'A device, whose scattering matrix operates on the incident waves so as to produce the same result as the operation of a cyclic substitution on the incident voltages, is called a circulator'.

For a given *m*-port circulator different systems of port numbering will lead to $(m-1)$ distinct scattering matrices. In the case of a structurally symmetrical circulator the number of possibilities may be reduced, however, by specifying a standard numbering system. For an *m*-port junction having the symmetry in Figures 2.2 and 2.3, it is always possible to number the ports to represent the cyclic substitution $(1, 2, 3, \ldots, m)$.

2.4 Symmetry and Group Properties of Circulator

In the preceding section, a concept of group theory was used in formulating an exact definition of a circulator. In this section, those properties of circulators that follow directly from this definition will be examined. Primarily, it will be shown how the various symmetries that a circulator may possess are derived from its definition.

It has been shown by Dicke that the scattering matrix $\overline{S}$ of any symmetrical junction must satisfy a set of commutation relations,

$$\overline{F}\overline{S} = \overline{S}\overline{F} \tag{2.11}$$

which determine the restrictions imposed on $\overline{S}$ by the junction symmetry, where $\overline{F}$ is a 'symmetry operator'. The symmetry operators are matrices that indicate how the terminal fields transform under the operations of the symmetry group of the junction, that is, rotations and reflections that carry the junction into itself.

The problem of finding all the symmetries of $\bar{S}$ reduces to finding all the symmetry operators that commute with $\bar{S}$. Fortunately this can be done.

Since the $\bar{S}$ matrices of circulators are, by definition, equivalent to cyclic substitutions, an established theorem for cyclic substitutions will apply. This theorem states: 'The only substitutions on m letters which are commutative with a cyclic substitution are the powers of this cyclic substitution'. Since the $\bar{F}$ operators of Dicke are essentially substitutions, these $\bar{F}$ operators may be found by taking the powers of $\bar{S}$ as indicated by the theorem. Accordingly, taking a 3-port circulator as an example,

$$\bar{F}_1 = \bar{S}_0 = \begin{bmatrix} 0 & 1 & 0 \\ 0 & 0 & 1 \\ 1 & 0 & 0 \end{bmatrix} \tag{2.12}$$

$$\bar{F}_2 = \bar{S}_0^2 = \begin{bmatrix} 0 & 0 & 1 \\ 1 & 0 & 0 \\ 0 & 1 & 0 \end{bmatrix} \tag{2.13}$$

$$\bar{F}_3 = \bar{S}_3^2 = \begin{bmatrix} 1 & 0 & 0 \\ 0 & 1 & 0 \\ 0 & 0 & 1 \end{bmatrix} \tag{2.14}$$

where $\bar{S}_0$ is the scattering matrix of an ideal circulator represented by the cyclic substitution (1,2,3). Since $(\bar{S}_0)^3$ is equal to $\bar{I}$, the identity matrix, the powers of $\bar{S}_0$ are said to form a cyclic group with $\bar{S}_0$ as the 'generator' of the group.

The operators listed above would comprise a mathematically complete list of the symmetries of the 3-port circulator, except for the fact that negative $\bar{F}$ operators, which are permissible, also satisfy equation 2.11. Therefore, the complete and mathematically exhaustive list of symmetries of the 3-port circulator is contained in the following group whose order is 6:

$$F_1, F_2, I, -F_2, -I, -F_1 \tag{2.15}$$

If a structure possesses a symmetry not contained in this group, then is cannot be a circulator.

For the 3-port junction the most general form for the scattering matrix is

$$\bar{S} = \begin{bmatrix} S_{11} & S_{12} & S_{13} \\ S_{21} & S_{22} & S_{23} \\ S_{31} & S_{32} & S_{33} \end{bmatrix} \tag{2.16}$$

Applying the commutation relation

$$\bar{F}_1 \bar{S} = \bar{S} \bar{F} \tag{2.17}$$

2.4 SYMMETRY AND GROUP PROPERTIES OF CIRCULATOR

gives

$$\begin{bmatrix} S_{11} & S_{12} & S_{13} \\ S_{21} & S_{22} & S_{23} \\ S_{31} & S_{32} & S_{33} \end{bmatrix} = \begin{bmatrix} S_{13} & S_{11} & S_{12} \\ S_{23} & S_{21} & S_{22} \\ S_{33} & S_{31} & S_{32} \end{bmatrix} \quad (2.16)$$

which is possible only if

$$S_{22} = S_{33} = S_{11} \quad (2.19a)$$

$$S_{23} = S_{12} = S_{31} \quad (2.19b)$$

$$S_{32} = S_{13} = S_{21} \quad (2.19c)$$

Hence, the general form for the scattering matrix of a 3-port circulator is

$$\bar{S} = \begin{bmatrix} S_{11} & S_{21} & S_{31} \\ S_{31} & S_{11} & S_{21} \\ S_{21} & S_{31} & S_{11} \end{bmatrix} \quad (2.20)$$

Applying the commutation relation $\bar{F}_2 \bar{S} = \bar{S} \bar{F}_2$ yields a similar result.

Taking now a 4-port circulator as an example gives

$$\bar{F}_1 = \bar{S}_0 = \begin{bmatrix} 0 & 1 & 0 & 0 \\ 0 & 0 & 1 & 0 \\ 0 & 0 & 0 & 1 \\ 1 & 0 & 0 & 0 \end{bmatrix} \quad (2.21)$$

$$\bar{F}_2 = \bar{S}_0^2 = \begin{bmatrix} 0 & 0 & 1 & 0 \\ 0 & 0 & 0 & 1 \\ 1 & 0 & 0 & 0 \\ 0 & 1 & 0 & 0 \end{bmatrix} \quad (2.22)$$

$$\bar{F}_3 = \bar{S}_0^3 = \begin{bmatrix} 0 & 0 & 0 & 1 \\ 1 & 0 & 0 & 0 \\ 0 & 1 & 0 & 0 \\ 0 & 0 & 1 & 0 \end{bmatrix} \quad (2.23)$$

$$\bar{F}_4 = \bar{S}_0^4 = \begin{bmatrix} 1 & 0 & 0 & 0 \\ 0 & 1 & 0 & 0 \\ 0 & 0 & 1 & 0 \\ 0 & 0 & 0 & 1 \end{bmatrix} \quad (2.24)$$

For the 4-port junction the most general form for the scattering matrix is

$$\bar{S} = \begin{bmatrix} S_{11} & S_{12} & S_{13} & S_{14} \\ S_{21} & S_{22} & S_{23} & S_{24} \\ S_{31} & S_{32} & S_{33} & S_{34} \\ S_{41} & S_{42} & S_{43} & S_{44} \end{bmatrix} \quad (2.25)$$

Applying the commutation relation with $\bar{F}_1$ given by equation 2.21 gives

$$\begin{bmatrix} S_{11} & S_{12} & S_{13} & S_{14} \\ S_{21} & S_{22} & S_{23} & S_{24} \\ S_{31} & S_{32} & S_{33} & S_{34} \\ S_{41} & S_{42} & S_{43} & S_{44} \end{bmatrix} = \begin{bmatrix} S_{21} & S_{22} & S_{23} & S_{24} \\ S_{31} & S_{32} & S_{33} & S_{34} \\ S_{41} & S_{42} & S_{43} & S_{44} \\ S_{11} & S_{12} & S_{13} & S_{14} \end{bmatrix} \quad (2.26)$$

which is possible only if

$$S_{11} = S_{22} = S_{33} = S_{44} \quad (2.27a)$$

$$S_{12} = S_{23} = S_{34} = S_{41} \quad (2.27b)$$

$$S_{13} = S_{31} = S_{24} = S_{42} \quad (2.27c)$$

$$S_{14} = S_{43} = S_{32} = S_{21} \quad (2.27d)$$

The case associated with the first commutation relationship produces an anti-clockwise circulation solution

$$\bar{S} = \begin{bmatrix} S_{11} & S_{41} & S_{31} & S_{21} \\ S_{21} & S_{11} & S_{41} & S_{31} \\ S_{31} & S_{21} & S_{11} & S_{41} \\ S_{41} & S_{31} & S_{21} & S_{11} \end{bmatrix} \quad (2.28a)$$

The related clockwise solution which also satisfies the communication rules is

$$\bar{S} = \begin{bmatrix} S_{11} & S_{21} & S_{31} & S_{41} \\ S_{41} & S_{11} & S_{21} & S_{31} \\ S_{31} & S_{41} & S_{11} & S_{21} \\ S_{21} & S_{31} & S_{41} & S_{11} \end{bmatrix} \quad (2.28b)$$

2.5 The Unitary Condition

Since the entries of the scattering matrix have the nature of reflection and transmission parameters the amplitude of any entries is bounded by zero and unity.

2.5 THE UNITARY CONDITION

The permissible relationships between these entries will now be deduced. The derivation of the required relationships starts by recognizing that the power dissipated in an m-port network is the difference between the incident power at all ports and the reflected power at the same ports.

$$P_{diss} = \tfrac{1}{2}\left(\sum_i^m a_i a_i^* - \sum_i^m b_i b_i^*\right) \tag{2.29}$$

It may be readily demonstrated that

$$\sum_i^m a_i a_i^* = (\bar{a})^T(\bar{a}*) \tag{2.30}$$

$$\sum_i^m b_i b_i^* = (\bar{a})^T(\bar{S})^T(\bar{S}*)(\bar{a}*) \tag{2.31}$$

The power dissipated in the circuit may therefore be expressed in matrix form as

$$P_{diss} = \tfrac{1}{2}(\bar{a})^T[\bar{I} - (\bar{S})^T(\bar{S}*)](\bar{a}*) \tag{2.32}$$

I is the unit matrix, $(\bar{S})^T$ is the transpose of $\bar{S}$, $(\bar{a}*)$ is the conjugate of $\bar{a}$.

The quantity defined by the preceding equation is an Hermitian form in that the matrix inside the inner brackets

$$\bar{Q} = \bar{I} - (\bar{S})^T(\bar{S}*) \tag{2.33}$$

is its own conjugate transpose.

$$[\bar{I} - (\bar{S})^T(\bar{S}*)]^{*T} = \bar{I} - (\bar{S})^T(\bar{S}*) \tag{2.34}$$

All energy functions are in fact either Hermitian or quadratic forms.

Since the power dissipated in a network is always positive the Hermitian form in (2.32) is positive semi definite.

$$\tfrac{1}{2}(\bar{a})^T[\bar{I} - (\bar{S})^T(\bar{S}*)](\bar{a}*) \geq 0 \tag{2.35}$$

For a reactance function the Hermitian form in the preceding equation is satisfied with the equal sign. The condition for a lossless junction therefore becomes

$$\bar{I} - (\bar{S})^T(\bar{S}*) = 0 \tag{2.36}$$

This last statement indicates that the scattering matrix of a dissipationless network is unitary. It is widely used to establish the permissible relationships between the entries of the matrix $\bar{S}$ in a reactance network.

2.6 Network Definition of Junction Circulator

Many of the properties of junction circulators are best discussed in terms of the scattering matrix of the device introduced in this chapter. The required nomenclature is defined in Figure 2.3. It will now be demonstrated that a matched non-reciprocal symmetric 3-port junction is necessarily a circulator. The scattering matrix of such a junction is given by

$$\bar{S} = \begin{bmatrix} 0 & S_{21} & S_{31} \\ S_{31} & 0 & S_{21} \\ S_{21} & S_{31} & 0 \end{bmatrix} \quad (2.37)$$

Evaluating the unitary condition under this assumption indicates that

$$|S_{21}|^2 + |S_{31}|^2 = 1 \quad (2.38)$$

and

$$S_{21} S_{31}^* = 0 \quad (2.39)$$

These two equations are consistent provided

$$|S_{21}| = 1 \quad (2.40)$$

and

$$|S_{31}| = 0 \quad (2.41)$$

or

$$|S_{21}| = 0 \quad (2.42)$$

$$|S_{31}| = 0 \quad (2.43)$$

The above relationships indicate that if $S_{11} = 0$ (matched condition), then either $S_{21} = 1$ and $S_{31} = 0$ or $S_{21} = 0$ and $S_{31} = 1$. Either of these two conditions coincide with the definition of a perfect circulator.

The unitary condition may also be employed to demonstrate that it is impossible to match a lossless reciprocal 3-port junction with three-fold symmetry. If the junction is reciprocal it has the property that its scattering matrix is symmetric

$$S_{ij} = S_{ji} \quad (2.44)$$

If it is matched its diagonal elements must be identically zero.

$$S_{11} = S_{22} = S_{33} = 0 \quad (2.45)$$

2.7 SEMI-IDEAL CIRCULATOR

The scattering matrix for a matched reciprocal 3-port junction is, therefore,

$$\bar{S} = \begin{bmatrix} 0 & S_{21} & S_{21} \\ S_{21} & 0 & S_{21} \\ S_{21} & S_{21} & 0 \end{bmatrix} \quad (2.46)$$

Evaluating the unitary condition for this scattering matrix indicates that

$$2 S_{21} S_{21}^* = 1 \quad (2.47)$$

and

$$S_{21} S_{21}^* = 0 \quad (2.48)$$

Since these two equations are incompatible it is apparent that it is impossible to match such a 3-port junction. It may be separately demonstrated that the best possible match for a reciprocal 3-port symmetrical junction coincides with the condition $S_{11} = 1/3$.

2.7 Semi-Ideal Circulator

It is also possible to obtain a relationship between the scattering coefficients of a lossless but nearly matched circulator. The scattering matrix for this network is

$$S = \begin{bmatrix} S_{11} & S_{21} & S_{31} \\ S_{31} & S_{11} & S_{21} \\ S_{21} & S_{31} & S_{11} \end{bmatrix} \quad (2.49)$$

Applying the unitary condition gives

$$|S_{11}|^2 + |S_{21}|^2 + |S_{31}|^2 = 1 \quad (2.50)$$

and

$$S_{21} S_{31}^* + S_{21} S_{11}^* + S_{31} S_{11}^* = 0 \quad (2.51)$$

One solution which satisfies equation (2.50) with S_{21} close to unity and S_{11} and S_{31} small is,

$$|S_{11}| \approx |S_{31}| \quad (2.52)$$

and

$$|S_{21}| \approx 1 - 2|S_{11}|^2 \quad (2.53)$$

Thus, minimum insertion loss corresponds to both maximum isolation and minimum VSWR looking into any of the three ports.

By virtue of equation 2.51 $|S_{21}|$ can be expressed in terms of $|S_{11}|$ and $|S_{31}|$. The three terms in equation 2.50 can be interpreted as three vectors that span a triangle. Then inequality relations such as

$$|S_{21}||S_{31}| \leq |S_{11}|(|S_{21}| + |S_{31}|) \tag{2.54}$$

are valid. From these considerations Butterweck deduced that in the $|S_{11}|$, $|S_{31}|$ diagram the possible three ports are restricted to a region that is bounded by three ellipses

$$|S_{11}|^2 + |S_{11}||S_{31}| + |S_{31}|^2 - |S_{11}| - |S_{31}| = 0 \tag{2.55a}$$

$$|S_{11}|^2 - |S_{11}||S_{31}| + |S_{31}|^2 + |S_{11}| - |S_{31}| = 0 \tag{2.55b}$$

$$|S_{11}|^2 - |S_{11}||S_{31}| + |S_{31}|^2 - |S_{11}| + |S_{31}| = 0 \tag{2.55c}$$

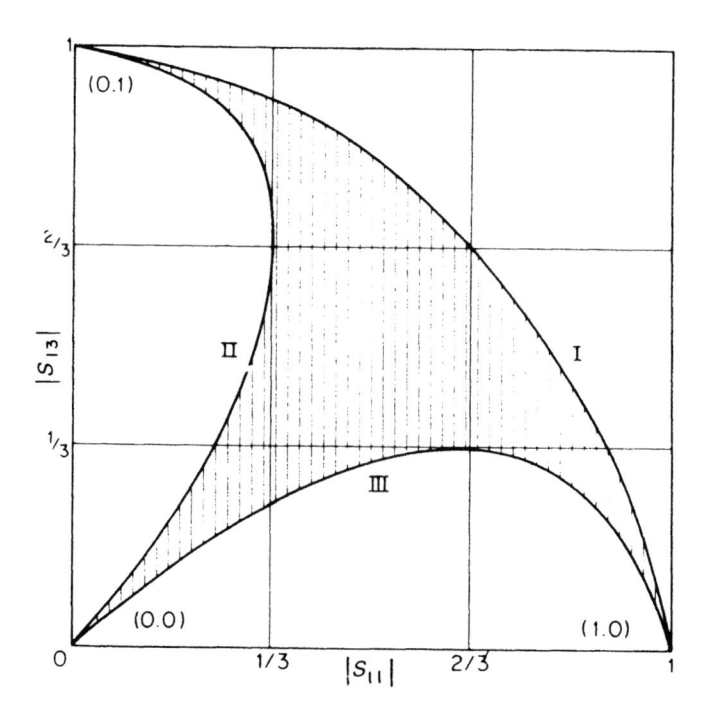

Figure 2.5 The close region of possible lossless and cyclic symmetrical 3-ports (reproduced with permission, Hagelin, S. (1966) A flow graph analysis of 3- and 4-port junction circulators, *IEEE Trans. Microwave Theory Tech.*, **MTT-14**(5), 243–249) (© 1996 IEEE)

2.8 DISSIPATION MATRIX

In Figure 2.5 this region is given by the shaded area. The origin of the diagram $|S_{11}| = 0$, $|S_{31}| = 0$, represents an ideal clockwise circulator.

2.8 Dissipation Matrix

For a lossy circulator the dissipation matrix $\bar{Q}$ must be positive real

$$\bar{Q} = \bar{I} - (\bar{S})^T(\bar{S}^*) \tag{2.56}$$

where $\bar{I}$ is a unit matrix and $\bar{S}$ is the scattering matrix. In the case of a symmetrical 3-port junction, one has

$$\bar{S} = \begin{bmatrix} S_{11} & S_{21} & S_{31} \\ S_{31} & S_{11} & S_{21} \\ S_{21} & S_{31} & S_{11} \end{bmatrix} \tag{2.57}$$

The matrix $\bar{Q}$ is given by

$$\bar{Q} = \begin{bmatrix} q_{11} & -q_{21} & -q_{31} \\ -q_{31} & q_{11} & -q_{21} \\ -q_{21} & -q_{31} & q_{11} \end{bmatrix} \tag{2.58}$$

where

$$q_{11} = -|S_{11}|^2 - |S_{21}|^2 - |S_{31}|^2 \tag{2.59a}$$

$$q_{21} = S_{11}S_{21}^* + S_{21}S_{31}^* + S_{31}S_{11}^* \tag{2.59b}$$

$$q_{31} = q_{21}^* \tag{2.59c}$$

The necessary and sufficient condition for the dissipation matrix to be positive real is that the principal minors of the dissipation matrix be non-negative.

$$q_{11} \geq 0 \tag{2.60a}$$

$$\begin{vmatrix} q_{11} & -q_{21} \\ q_{21}^* & q_{11} \end{vmatrix} \geq 0 \tag{2.60b}$$

$$\begin{vmatrix} q_{11} & -q_{21} & q_{21}^* \\ -q_{21}^* & q_{11} & -q_{21} \\ -q_{21} & -q_{21}^* & q_{11} \end{vmatrix} \geq 0 \tag{2.60c}$$

2.9 Insertion Phase Shift

A desirable quantity in the description of a 3-port junction circulator is its phase shift between ports 1 and 2. The derivation of this quantity starts with a statement of S_{21}

$$S_{21} = \frac{s_0 + \alpha s_+ + \alpha^2 s_-}{3} \tag{2.61}$$

where

$$s_0 = -1 \tag{2.61a}$$

$$s_+ = \exp -j2(\theta_1 - \Delta\theta_1 + \pi/2) \tag{2.61b}$$

$$s_- = \exp -j2(\theta_1 + \Delta\theta_1 + \pi/2) \tag{2.61c}$$

and

$$\alpha = \exp(j120) \tag{2.62a}$$

$$\alpha^2 = \exp(j240) \tag{2.62b}$$

The quantities s_0, s_+ and s_- are the so called eigenvalues of the problem region and represent the reflections coefficients of the eigen-networks of the junction. These are defined in some detail in Chapter 3.

Combining the preceding equations readily gives

$$S_{21} = \frac{-1 + \cos 2\Delta\theta_1 + \sqrt{3}\sin 2\Delta\theta_1)\exp(-j2\theta_1)}{3} \tag{2.63}$$

In the vicinity of the circulation condition $2\Delta\theta_1 = \pi/3$ and

$$S_{21} = \frac{-1 + 2\exp(-j2\theta_1)}{3} \tag{2.64}$$

At the midband frequency $\theta_1 = \pi/2$ and

$$S_{21} = -1 \tag{2.65}$$

Writing S_{21} as $|S_{21}|\exp(-j2\phi_{21})$ gives

$$|S_{21}| = \frac{[(-1 + 2\cos 2\theta_1)^2 + (2\sin 2\theta_1)^2]^{1/2}}{3} \tag{2.66}$$

and

$$\tan 2\phi_{21} = \frac{-2 \sin 2\theta_1}{(-1 + 2 \cos 2\theta_1)} \qquad (2.67)$$

In the vicinity of $\theta_1 = \pi/2$

$$\tan 2\phi_{21} = \tfrac{2}{3} \sin 2\theta_1 \qquad (2.68)$$

References and Further Reading

Auld, B. A. (1962) The synthesis of symmetrical waveguide circulators, *IRE Trans. Microwave Theory Tech.*, **MTT-7**, 137–46.

Butterweck, H. J. (1963) The Y circulator, *Arch. Elek. Ubertragung* **17**(4), 163–76.

Hagelin, S. (1966) A flow graph analysis of 3- and 4 port junction circulators, *IEEE Trans. Microwave Theory Tech.*, **MTT-14**(5), 243–249.

Humphreys, B. L. and Davies, J. B. (1962) The synthesis of *n*-port circulators, *IRE Trans. Microwave Theory Tech.*, **MTT-10**, 551–554.

Kerns, D. M. (1951) Analysis of symmetrical waveguide junctions, *J. Res. Nat. Bur. Stand. Sect.* **46**.

Montgomery, C. G., Dicke, R. H. and Purcell, E. M. (1948) *Principles of Microwave Circuits*, McGraw-Hill, New York.

Treuhaft, M. A. (1956) Network properties of circulators based on the scattering concept, *Proc. IRE*, **44**, 1394–1402.

3

EIGENVALUE ADJUSTMENT OF 3-PORT CIRCULATOR

3.1 Introduction

While the unitary condition introduced in Chapter 2 may be used to define an ideal circulator it does not give any insight into its physical adjustment. In order to do so it is necessary to establish the eigenvalues s_i and eigenvectors U_i associated with its' scattering matrix description. A study of this problem indicates that the entries of the scattering matrix may be reduced to linear combinations of the so called eigenvalues of the junction. A typical eigenvalue of a symmetrical m-port junction is a 1-port reflection coefficient or immittance at any port each corresponding to the application of a typical eigenvector of the device. The eigenvectors are the m-possible ways of exciting the junction and are determined by its symmetry only. The 1-port circuits formed in this way are known as the eigennetworks of the problem region. The adjustment of the ideal 3-port circulator requires that its reflection eigenvalues lie equally spaced on a unit circle. This condition is both necessary and sufficient and can be achieved by having recourse to only two independent variables.

There are altogether four possible ideal eigenvalue diagrams of degree-1, each of which is associated with a unique complex gyrator circuit. One eigenvalue is usually associated with a so called in-phase eigenvector and the other two by counter-rotating ones. In a reciprocal 3-port junction the latter two eigenvalues are degenerate whereas in a non-reciprocal one these are split by the gyromagnetic resonator.

3.2 Scattering Matrix, Eigenvalues, and Eigenvectors

The importance of the eigenvalue problem in network theory, to be described now, resides in the fact that the entries of a symmetrical or Hermitian square matrix may always be expressed in terms of the roots (eigenvalues) of its characteristic equation. One such matrix is the scattering matrix, met in connection with the description of incident and reflected waves in an m-port circuit.

$$\bar{b} = \bar{S}\bar{a} \tag{3.1}$$

The formulation of the eigenvalue problem starts by examining the solutions of the standard equation in (3.1) for the case for which the m-independent variables are related by a scalar quantity

$$\bar{a} = \bar{U}_i \tag{3.2}$$

$$\bar{b} = s_i \bar{U}_i \tag{3.3}$$

s_i is known as an eigenvalue and $\bar{U}_i$ as an eigenvector.

The eigenvalue equation is now obtained by substituting the two proceeding conditions into the original equation. The result is

$$\bar{S}\bar{U}_i = s_i \bar{U}_i \tag{3.4}$$

This equation has a non-vanishing value for $\bar{U}_i$ provided

$$\det |\bar{S}| - s_i |\bar{I}| = 0 \tag{3.5}$$

where $\bar{I}$ is a unit vector.

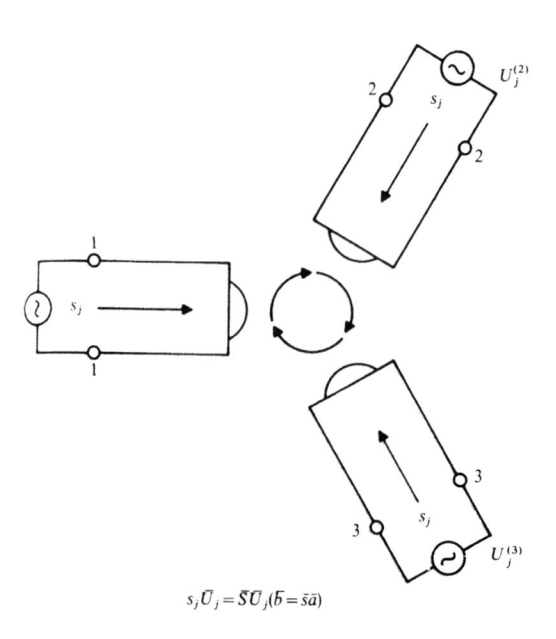

Figure 3.1 Schematic diagram of 3-port junction indicating definition of eigenvalue equation

3.3 EIGENVALUE ADJUSTMENT OF 3-PORT CIRCULATOR

The determinant defined by this equation is a polynomial of degree m and its m roots are the m-eigenvalues of the scattering matrix, some of which may be equal (degenerate). The corresponding m-eigenvectors may be evaluated by substituting each eigenvalue one at a time into the eigenvalue equation. In the special case of the symmetrical m-port network considered here the m-eigenvalues are 1-port reflection coefficients at any port of the junction. For a lossless junction, these lie in the complex plane with unit amplitude. These eigenvalues can be calculated once the coefficients of the scattering matrix are specified. The m-eigenvectors are the m possible voltage settings at the ports of the junction and are fixed by its symmetry only. The 1-port circuits associated with these reflection eigenvalues are known as the eigen-networks of the device. Since the eigenvectors are completely determined by the symmetry of the junction, a symmetrical perturbation of the junction alters the phase angles of the eigenvalues but leaves the eigenvectors unchanged. A schematic diagram of the eigenvalues equation is depicted in Figure 3.1.

3.3 Eigenvalue Adjustment of 3-port Circulator

Since the eigenvalue arrangement of a reciprocal 3-port junction for which S_{11} is a minimum coincides with the first of the two circulation conditions of an ideal 3-port circulator it will be derived first.

For a reciprocal 3-port junction with three-fold symmetry, the scattering matrix is

$$\bar{S} = \begin{bmatrix} S_{11} & S_{21} & S_{21} \\ S_{21} & S_{11} & S_{21} \\ S_{11} & S_{21} & S_{11} \end{bmatrix} \tag{3.1}$$

The characteristic equation associated with this matrix is

$$(S_{11} - s_i)^3 - 3(S_{11} - s_i)S_{21}^2 + 2S_{21}^3 = 0 \tag{3.7}$$

Its three eigenvalues or roots are

$$s_0 = S_{11} + 2S_{21} \tag{3.8}$$

$$s_+ = s_- = S_{11} - S_{21} \tag{3.9}$$

This result indicates that two of the three eigenvalues of a reciprocal 3-port junction are degenerate.

Taking linear combinations of these two equations indicates that the entries of the scattering matrix can also be written in terms of the eigenvalues as

$$S_{11} = \frac{s_0 + 2s_+}{3} \tag{3.10}$$

$$S_{21} = \frac{s_0 - s_+}{3} \tag{3.11}$$

It is observed from the first of these two relationships that the reflection coefficient S_{11} in such a junction is a minimum equal to |⅓| provided

$$s_+ = -s_0 \tag{3.12}$$

The second of these equations indicates that the condition for which the reflection coefficient S_{11} is a minimum coincides with that for which the transmission one S_{21} is a maximum. The eigenvalue diagram for this situation is illustrated in Figure 3.2.

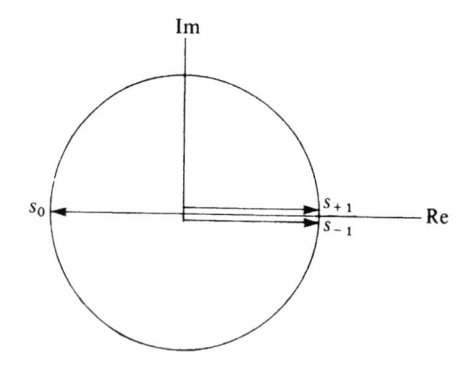

Figure 3.2 Eigenvalue diagram of reciprocal 3-port junction for maximum power transfer

One possible solution for the 1-port reflection coefficients that meets the last two conditions is

$$s_+ = \exp\left[-j2(\theta_1 + \pi/2)\right] \tag{3.13a}$$

$$s_- = \exp\left[-j2(\theta_1 + \pi/2)\right] \tag{3.13b}$$

$$s_0 = \exp\left[-j2(\theta_0)\right] \tag{3.13c}$$

provided that

$$\theta_0 = \theta_1 = \pi/2 \tag{3.13d}$$

A scrutiny of these reflection coefficients indicates that these are associated with the 1-port short- and open-circuited transmission lines illustrated in Figure 3.3. The choice of these solutions will be discussed separately in connection with the description of the eigenvectors of the junction.

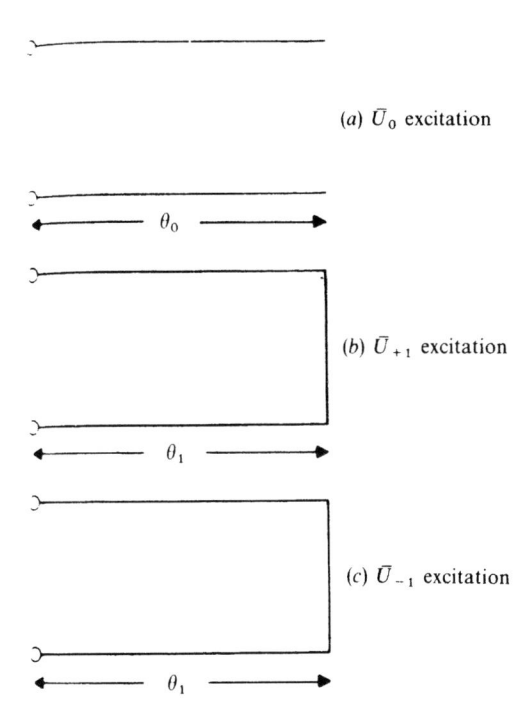

3.3 EIGENVALUE ADJUSTMENT OF 3-PORT CIRCULATOR 43

Figure 3.3 Eigen-networks of reciprocal 3-port junction

(a) $\bar{U}_0$ excitation

(b) $\bar{U}_{+1}$ excitation

(c) $\bar{U}_{-1}$ excitation

The scattering matrix for the arrangement for which S_{11} is a minimum is now obtained by combining equations (3.10) and (3.11) with equations (3.13a to 3.13c). The required result is

$$\bar{S} = \begin{bmatrix} \frac{1}{3} & -\frac{2}{3} & -\frac{2}{3} \\ -\frac{2}{3} & \frac{1}{3} & -\frac{2}{3} \\ -\frac{2}{3} & -\frac{2}{3} & \frac{1}{3} \end{bmatrix} \tag{3.14}$$

The derivation of the eigenvalue diagram of an ideal circulator likewise proceeds with the definition of its scattering matrix

$$\bar{S} = \begin{bmatrix} 0 & S_{21} & 0 \\ 0 & 0 & S_{21} \\ S_{21} & 0 & 0 \end{bmatrix} \tag{3.15}$$

where S_{21} has unit amplitude.

The characteristic equation associated with this matrix is given by

$$-s_i^3 + S_{21}^3 = 0 \tag{3.16}$$

One result for a device for which $S_{21} = -1$ is

$$s_+ = \exp[-j2(\theta_1 + \theta_+ + \pi/2)] \qquad (3.17a)$$

$$s_- = \exp[-j2(\theta_1 + \theta_- + \pi/2)] \qquad (3.17b)$$

$$s_0 = \exp[-j2(\theta_0)] \qquad (3.17c)$$

provided that

$$\theta_1 = \theta_0 = \pi/2 \qquad (3.18a)$$

$$\theta_+ = -\theta_- = \pi/6 \qquad (3.18b)$$

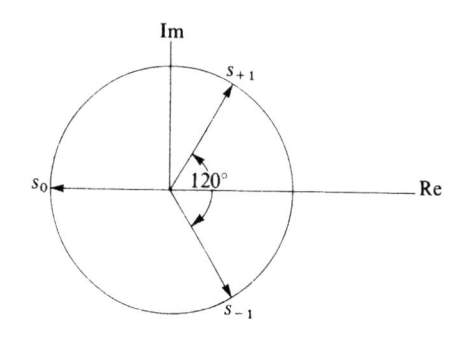

Figure 3.4 Eigenvalue diagram of ideal 3-port junction circulator

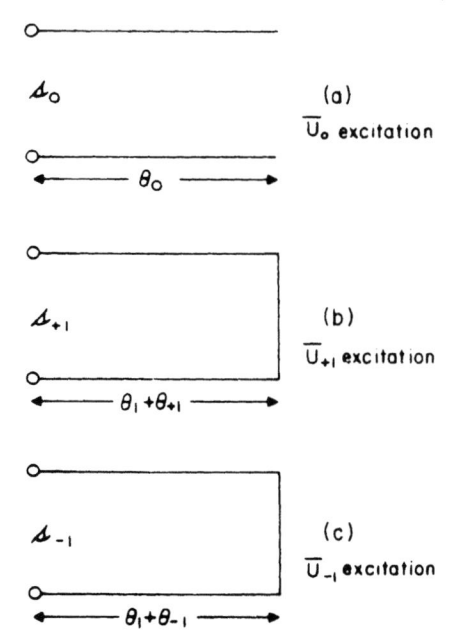

Figure 3.5 Eigen-networks of ideal 3-port junction circulator

The three reflection coefficients of an ideal circulator therefore lie equally spaced on a unit circle in the manner depicted in Figure 3.4. One possible set of 1-port eigen-networks that is compatible with this solution is indicated in Figure 3.5.

In what follows it will be demonstrated that the eigen-vectors $\bar{U}$ corresponding to the eigenvalues s_+ and s_- produce counter-rotating magnetic fields on the axis of the junction. Since such magnetic fields establish different scalar permeabilities in a gyromagnetic medium a practical means of removing the degeneracy between the reflection eigenvalues is therefore at hand.

3.4 Eigenvectors

The three eigenvectors of the problem region may be established by solving the eigenvalue equation one at a time. The equation for the eigenvector $\bar{U}_0$ and the eigenvalue s_0 of an ideal 3-port circulator is

$$\begin{bmatrix} 0 & -1 & 0 \\ 0 & 0 & -1 \\ -1 & 0 & 0 \end{bmatrix} \begin{bmatrix} U_0^{(1)} \\ U_0^{(2)} \\ U_0^{(3)} \end{bmatrix} = s_0 \begin{bmatrix} U_0^{(1)} \\ U_0^{(2)} \\ U_0^{(3)} \end{bmatrix} \quad (3.19)$$

where

$$s_0 = \exp(-j\pi) \quad (3.20)$$

Taking $U_0^{(1)}$ as 1 readily gives one solution as

$$U_0^{(1)} = 1 \quad (3.21a)$$

$$U_0^{(2)} = 1 \quad (3.21b)$$

$$U_0^{(3)} = 1 \quad (3.21c)$$

The eigenvalue equation for the eigenvector $\bar{U}_+$ and the eigenvalue s_+ is

$$\begin{bmatrix} 0 & -1 & 0 \\ 0 & 0 & -1 \\ -1 & 0 & 0 \end{bmatrix} \begin{bmatrix} U_+^{(1)} \\ U_+^{(2)} \\ U_+^{(3)} \end{bmatrix} = s_+ \begin{bmatrix} U_+^{(1)} \\ U_+^{(2)} \\ U_+^{(3)} \end{bmatrix} \quad (3.22)$$

where

$$s_+ = \exp(-j\pi/3) \quad (3.23)$$

Taking $U_+^{(1)}$ as 1 indicates that the solution is in this instance given by

$$U_+^{(1)} = 1 \quad (3.24a)$$

$$U_+^{(2)} = \alpha \tag{3.24b}$$

$$U_+^{(3)} = \alpha^2 \tag{3.24c}$$

where

$$\alpha = \exp(j120) \tag{3.25a}$$

$$\alpha^2 = \exp(j240) \tag{3.25b}$$

The solution of the eigenvalue equation for the eigenvector $\bar{U}_-$ of the junction is given without ado by

$$U_-^{(1)} = 1 \tag{3.26a}$$

$$U_-^{(2)} = \alpha^2 \tag{3.26b}$$

$$U_-^{(3)} = \alpha \tag{3.26c}$$

The eigen-vectors are normalized in such a way that

$$(\bar{U}_\pm)^T \bar{U}_\pm^* = 1 \tag{3.27}$$

One solution is therefore described by

$$\bar{U}_0 = \frac{1}{\sqrt{3}}\begin{bmatrix}1\\1\\1\end{bmatrix}, \quad \bar{U}_+ = \frac{1}{\sqrt{3}}\begin{bmatrix}1\\\alpha\\\alpha^2\end{bmatrix}, \quad \bar{U}_- = \frac{1}{\sqrt{3}}\begin{bmatrix}1\\\alpha^2\\\alpha\end{bmatrix} \tag{3.28}$$

The three eigensolutions of the problem are illustrated in Figure 3.6. Application of each eigenvector one at a time reveals the variables s_0, s_+, and s_- at any port.

Scrutiny of the magnetic field pattern in the transverse plane one at a time for the $U_\pm$ eigenvectors indicates that each is circularly polarized on the axis of the junction with a different hand. Since a suitably magnetized magnetic insulator displays different scalar permeabilities $\mu \mp \kappa$, for such rotating fields such a magnetized junction may be employed to split the degeneracy between the $s_\pm$ reflection coefficients. This arrangement therefore provides a practical means for the construction of the ideal circulator.

A property of a symmetric junction that will now be verified is that a single signal at any port establishes all the eigenvectors of the problem region with equal amplitudes. This statement may be demonstrated without difficulty by constructing a linear combination of the eigenvectors associated with the problem under consideration.

3.4 EIGENVECTORS

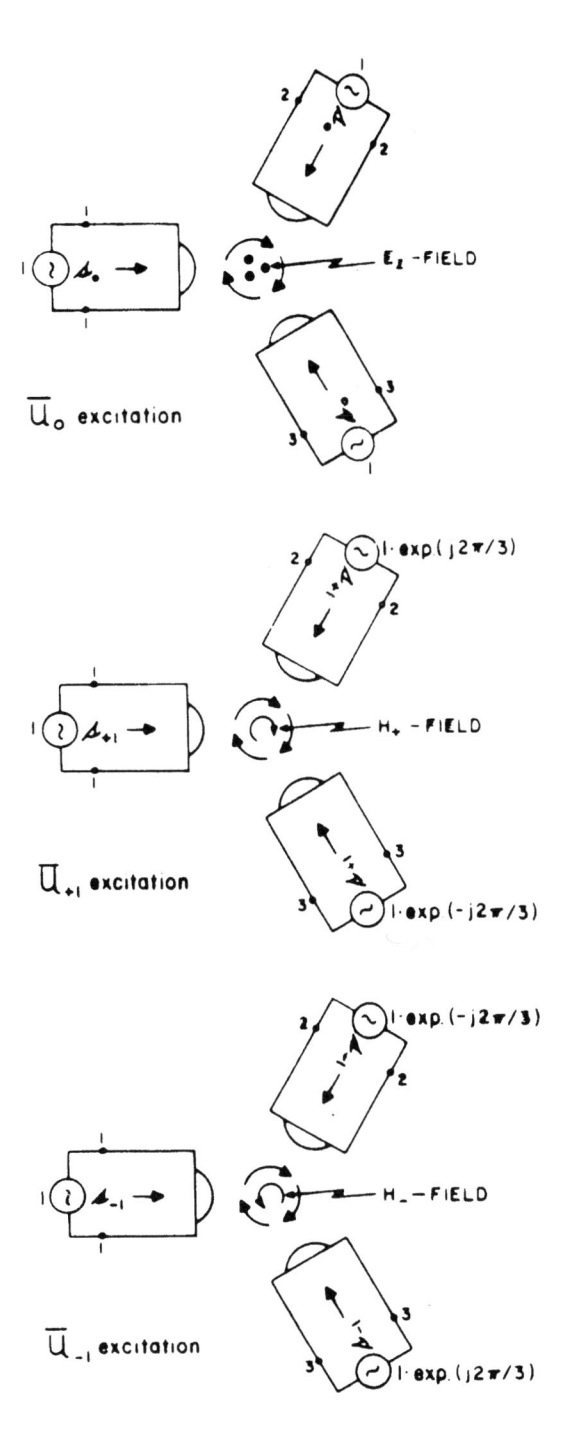

Figure 3.6 Eigensolutions of 3-port circulator

$$\begin{bmatrix} 1 \\ 0 \\ 0 \end{bmatrix} = \frac{1}{3}\begin{bmatrix} 1 \\ 1 \\ 1 \end{bmatrix} + \frac{1}{3}\begin{bmatrix} 1 \\ \alpha \\ \alpha^2 \end{bmatrix} + \frac{1}{3}\begin{bmatrix} 1 \\ \alpha^2 \\ \alpha \end{bmatrix} \qquad (3.29)$$

The equality is met by provided the following identity is satisfied

$$1 + \alpha + \alpha^2 = 0 \qquad (3.30)$$

Since the eigenvectors are fixed by the junction symmetry, a symmetric perturbation of the junction alters the phases of the eigenvalues but leaves the eigenvectors unchanged.

3.5 Scattering Matrix of 3-port Junction Circulator

One technique, that is often employed in the analysis of a symmetrical m-port network is to express the entries of its matrix description in terms of the solutions of more simple (usually 1-port) circuits that reflect the symmetry of the problem region. One unique decomposition may be achieved by adopting the eigenvectors of the network for this purpose. The 1-port circuits established in this way are known as the eigen-networks of the problem region. The eigenvalues of the scattering matrix are then the reflection coefficients associated with these eigen-networks. Furthermore, the entries of the scattering matrix are now reduced to simple linear combinations of these same eigenvalues. This procedure will now be demonstrated in connection with the description of the 3-port junction circulator by considering each eigenvector one at a time as a preamble to laying out a more universal mathematical technique.

The required relationship between the scattering parameters and the eigenvalues of the problem region starts by constructing the incident and reflected waves for the first triplet of voltage settings or eigenvector.

$$\begin{bmatrix} b_1 \\ b_2 \\ b_3 \end{bmatrix} = \begin{bmatrix} S_{11} & S_{21} & S_{31} \\ S_{31} & S_{11} & S_{21} \\ S_{21} & S_{31} & S_{11} \end{bmatrix} \begin{bmatrix} \frac{1}{3} \\ \frac{1}{3} \\ \frac{1}{3} \end{bmatrix} \qquad (3.31)$$

This gives

$$\frac{b_1}{(\frac{1}{3})} = S_{11} + S_{21} + S_{31} \qquad (3.32a)$$

$$\frac{b_2}{(\frac{1}{3})} = S_{31} + S_{11} + S_{21} \qquad (3.32b)$$

$$\frac{b_3}{(\frac{1}{3})} = S_{21} + S_{31} + S_{11} \qquad (3.32c)$$

3.5 SCATTERING MATRIX OF 3-PORT JUNCTION CIRCULATOR

The reflection coefficient at each port is therefore identical and is in this instance denoted by ρ_0.

$$\frac{b_1}{\left(\frac{1}{3}\right)} = \frac{b_2}{\left(\frac{1}{3}\right)} = \frac{b_3}{\left(\frac{1}{3}\right)} = \rho_0 \tag{3.33}$$

where

$$\rho_0 = S_{11} + S_{21} + S_{31} \tag{3.34}$$

The factor ⅓ in the definition of the incident waves stems from that fact that each port is simultaneously excited.

Taking the second triplet of voltage settings or eigenvector gives

$$\begin{bmatrix} b_1 \\ b_2 \\ b_3 \end{bmatrix} = \begin{bmatrix} S_{11} & S_{21} & S_{31} \\ S_{31} & S_{11} & S_{21} \\ S_{21} & S_{31} & S_{11} \end{bmatrix} \begin{bmatrix} \frac{1}{3} \\ \frac{\alpha}{3} \\ \frac{\alpha^2}{3} \end{bmatrix} \tag{3.35}$$

or

$$\frac{b_1}{\left(\frac{1}{3}\right)} = S_{11} + \alpha S_{21} + \alpha^2 S_{31} \tag{3.36a}$$

$$\frac{b_2}{\left(\frac{\alpha}{3}\right)} = S_{11} + \alpha S_{21} + \alpha^2 S_{31} \tag{3.36b}$$

$$\frac{b_3}{\left(\frac{\alpha^2}{3}\right)} = S_{11} + \alpha S_{21} + \alpha^2 S_{31} \tag{3.36c}$$

provided $\alpha^3 = 1$.

The reflection coefficient at each port is again the same. Denoting it by ρ_+ gives

$$\rho_+ = S_{11} + \alpha S_{21} + \alpha^2 S_{31} \tag{3.37}$$

Likewise

$$\rho_- = S_{11} + \alpha^2 S_{21} + \alpha S_{31} \tag{3.38}$$

The required result is now established by taking suitable combinations of ρ_0, ρ_+ and ρ_-. Taking S_{11} by way of an example indicates that

$$S_{11} = \frac{\rho_0 + \rho_+ + \rho_-}{3} \tag{3.39a}$$

as asserted.

The other two relationships are similarly obtained

$$S_{21} = \frac{\rho_0 + \alpha\rho_+ + \alpha^2\rho_-}{3} \qquad (3.39b)$$

$$S_{31} = \frac{\rho_0 + \alpha^2\rho_+ + \alpha\rho_-}{3} \qquad (3.39c)$$

The entries of the scattering matrix are therefore simple linear combinations of the 1-port variables or eigenvalues revealed by the different eigenvectors of the problem region as asserted,

If the network is reciprocal then

$$\rho_+ = \rho_- \qquad (3.40)$$

and

$$S_{11} = \frac{\rho_0 + 2\rho_+}{3} \qquad (3.41a)$$

$$S_{21} = S_{31} = \frac{\rho_0 - \rho_+}{3} \qquad (3.42b)$$

in keeping with equations (3.10) and (3.11).

3.6 Diagonalization

If the eigenvectors and eigenvalues are known it is possible to expand the coefficients of the scattering matrix using a simple mathematical technique. This may be done by using the following similarity transformation

$$\bar{S} = \bar{U}\bar{\lambda}\bar{U}^{-1} \qquad (3.42)$$

The columns of the matrix $\bar{U}$ are constructed in terms of the eigenvectors $\bar{U}_i$ of the problem region, and $\bar{U}^{-1}$ is the inverse of $\bar{U}$. The matrix $\bar{\lambda}$ is a diagonal matrix with the eigenvalues of $\bar{S}$ along its main diagonal. If the eigenvectors of $\bar{S}$ are linearly independent and $\bar{S}$ is a real symmetric or Hermitian matrix it may be demonstrated that

$$\bar{U}^{-1} = (\bar{U}*)^T \qquad (3.43)$$

where $(\bar{U}*)^T$ is the transpose of the complex conjugate of $\bar{U}$. The relationship between the eigenvalue and the coefficients of the scattering matrix is then obtained by multiplying out the similarity identity.

3.7 THE 4-PORT JUNCTION

The diagonalization procedure will now be developed in the case of an ideal circulator circuit by constructing the square matrices $\bar{U}$ and $(\bar{U}^*)^T$ in terms of the eigenvectors of the problem region.

$$\bar{U} = \frac{1}{\sqrt{3}} \begin{bmatrix} 1 & 1 & 1 \\ 1 & \alpha & \alpha^2 \\ 1 & \alpha^2 & \alpha \end{bmatrix} \tag{3.44}$$

$$(\bar{U}^*)^T = \frac{1}{\sqrt{3}} \begin{bmatrix} 1 & 1 & 1 \\ 1 & \alpha^* & (\alpha^*)^2 \\ 1 & (\alpha^*)^2 & \alpha^* \end{bmatrix} \tag{3.45}$$

In forming the matrix $\bar{U}$ in terms of the m eigenvectors of the problem region care has been taken to ensure that the resultant matrix is symmetrical.

The diagonal matrix $\bar{\lambda}$ is constructed using the eigenvalues of $\bar{S}$ as

$$\bar{\lambda} = \begin{bmatrix} s_0 & 0 & 0 \\ 0 & s_+ & 0 \\ 0 & 0 & s_- \end{bmatrix} \tag{3.46}$$

Diagonalizing the matrix $\bar{S}$ gives

$$\begin{bmatrix} S_{11} & S_{21} & S_{31} \\ S_{31} & S_{11} & S_{21} \\ S_{21} & S_{31} & S_{11} \end{bmatrix} = \frac{1}{3} \begin{bmatrix} 1 & 1 & 1 \\ 1 & \alpha & \alpha^2 \\ 1 & \alpha^2 & \alpha \end{bmatrix} \begin{bmatrix} s_0 & 0 & 0 \\ 0 & s_+ & 0 \\ 0 & 0 & s_- \end{bmatrix} \begin{bmatrix} 1 & 1 & 1 \\ 1 & \alpha^* & (\alpha^*)^2 \\ 1 & (\alpha^*)^2 & \alpha^* \end{bmatrix} \tag{3.47}$$

The required result is in accord with that described by equations (3.39), (3.40) and (3.41).

The eigenvalues appearing in the scattering description of the junction correspond to 1-port reflection variables. Such variables are displayed at any port provided the generator settings coincides with the corresponding eigenvectors.

This result again indicates that the reflection eigenvalues of an ideal 3-port junction circulators are displaced by 120° on a unit circle.

3.7 The 4-port Junction

Another interesting and practical junction circulator is the symmetrical 4-port one illustrated in Figure 3.7. Its description is briefly summarized in this section. The scattering matrix is defined by

$$\bar{S} = \begin{bmatrix} S_{11} & S_{21} & S_{31} & S_{41} \\ S_{41} & S_{11} & S_{21} & S_{31} \\ S_{31} & S_{41} & S_{11} & S_{21} \\ S_{21} & S_{31} & S_{41} & S_{11} \end{bmatrix} \tag{3.48}$$

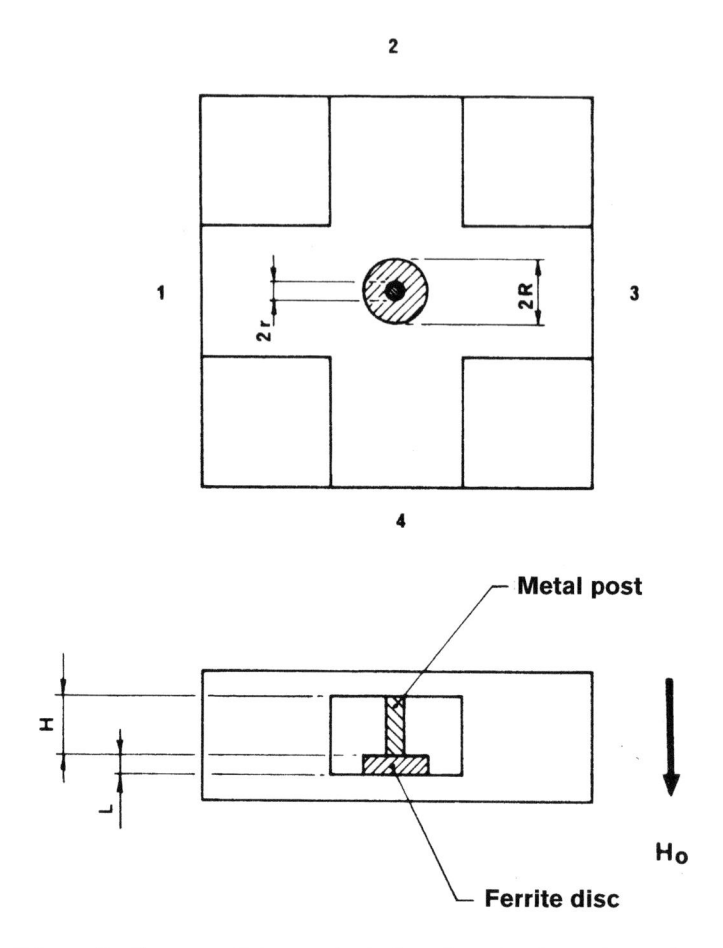

Figure 3.7 Schematic diagram of 4-port single junction circulator (reproduced with permission, Helszajn, J. and Buffler, C.R. (1968) Adjustment of the 4-port single junction circulator, *Radio Electronic Eng.* **35**, 357–360)

One set of eigenvectors which satisfy the symmetry of the network is

$$\bar{U}_1 = \frac{1}{\sqrt{4}}\begin{bmatrix}1\\1\\1\\1\end{bmatrix}, \quad \bar{U}_2 = \frac{1}{\sqrt{4}}\begin{bmatrix}1\\ \beta\\ \beta^2\\ \beta^3\end{bmatrix}, \quad \bar{U}_3 = \frac{1}{\sqrt{4}}\begin{bmatrix}1\\ \beta^2\\ \beta^3\\ \beta\end{bmatrix}, \quad \bar{U}_4 = \frac{1}{\sqrt{4}}\begin{bmatrix}1\\ \beta^3\\ \beta\\ \beta^2\end{bmatrix} \quad (3.49)$$

where

$$\beta = \exp(-j\,\pi/2) \quad (3.50)$$

3.7 THE 4-PORT JUNCTION

These eigenvectors have again the property that a single input at any one port may be decomposed into a linear combination of the individual eigenvectors with equal amplitude.

The eigenvalues appearing in the description of the scattering coefficients may again be revealed by having recourse to the diagonalization procedure utilized in the description of the 3-port circulator. The matrix $\overline{U}$ is in this instance given by

$$\overline{U} = \begin{bmatrix} 1 & 1 & 1 & 1 \\ 1 & \beta & \beta^2 & \beta^3 \\ 1 & \beta^2 & \beta^3 & \beta \\ 1 & \beta^3 & \beta & \beta^2 \end{bmatrix} \quad (3.51)$$

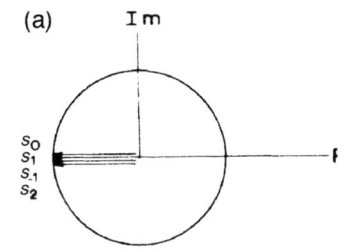

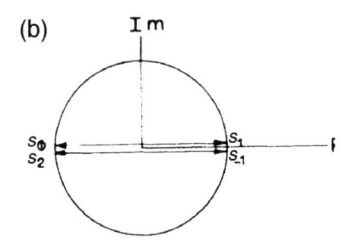

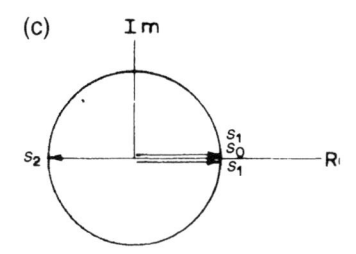

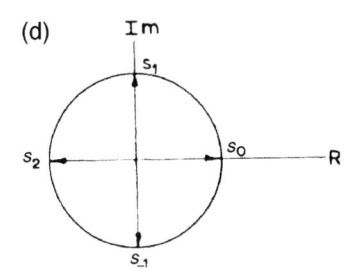

Figure 3.8 Eigenvalue adjustment of ideal 4-port junction circulator (reproduced with permission, Helszajn (1970) The adjustment of the m-port single junction circulator, *IEEE Trans.*, **MTT-18**, 705–711) (© 1970 IEEE)

The diagonal matrix $\bar{\lambda}$ has one degenerate pair of eigenvalues and two non-degenerate ones

$$\bar{\lambda} = \begin{bmatrix} s_0 & 0 & 0 & 0 \\ 0 & s_{+1} & 0 & 0 \\ 0 & 0 & s_{-1} & 0 \\ 0 & 0 & 0 & s_2 \end{bmatrix} \qquad (3.52)$$

The required result is

$$S_{11} = \frac{s_0 + s_{+1} + s_{-1} + s_2}{4} \qquad (3.53a)$$

$$S_{21} = \frac{s_0 + js_{+1} - js_{-1} - s_2}{4} \qquad (3.53b)$$

$$S_{31} = \frac{s_0 - s_{+1} - s_{-1} - s_2}{4} \qquad (3.53c)$$

$$S_{41} = \frac{s_0 - js_{+1} + js_{-1} - s_2}{4} \qquad (3.53d)$$

Scrutiny of the first of these four equations indicates that the spur or trace of the scattering matrix is equal to the sum of the eigenvalues. This is a general result.

One property of a single 4-port junction circulator is that S_{11} is equal to zero. Scrutiny of this condition suggests that the eigenvalues of an ideal 4-port single junction circulator lie equally spaced on a unit circle. This solution is illustrated in Figure 3.8.

3.8 Dissipation Eigenvalues

If the scattering and dissipation matrices have common eigenvectors, the respective eigenvalues are related by the following theorem:

If

$$\bar{Q}\bar{U}_i = q_i \bar{U}_i$$

then

$$f(\bar{Q})\bar{U}_i = f(q_i)\bar{U}_i$$

where $\bar{U}_i$ is an eigenvector.

This theorem may be used to deduce the eigenvalues of the dissipation matrix in terms of those of the scattering one.

Making use of the relationship between the two in Chapter 2 gives in the case of the 3-port junction

$$q_0 = 1 - s_0 s_0^* \quad (3.54a)$$

$$q_{+1} = 1 - s_{+1} s_{+1}^* \quad (3.54b)$$

$$q_{-1} = 1 - s_{-1} s_{-1}^* \quad (3.54c)$$

These last three equations may be used to construct the scattering matrix in terms of the eigenvalues of the dissipation one. In a lossless junction, the amplitudes of the scattering matrix eigenvalues are unity, while if the junction is lossy the amplitudes will depart from unity.

The above discussion indicates that the eigenvalues of the dissipation matrix represent the dissipation associated with each possible way of exciting the junction. These eigenvalues are real quantities that become zero when those of the scattering matrix become unity.

The entries of the dissipation matrix are

$$q_{11} = \frac{q_0 + q_+ + q_{-1}}{3} \quad (3.55a)$$

$$q_{12} = \frac{q_0 + \alpha q_{+1} + \alpha^2 q_{-1}}{3} \quad (3.55b)$$

$$q_{13} = \frac{q_0 + \alpha^2 q_{+1} + \alpha q_{-1}}{3} \quad (3.55c)$$

Here, q_{11} represents the total dissipation of the junction, and q_{12} is a complex quantity that determines the allowable relationships between the scattering parameters.

The entries of the dissipation and scattering matrices may in practice be directly evaluated by relating their eigenvalues to the loaded and unloaded Q-factors of the junction eigen-networks.

References and Further Reading

Auld, B. A. (1959) The synthesis of symmetrical waveguide circulators, *Trans. IRE Microwave Theory Tech.*, **MTT-7**, 238–26.

Bex, H. and Schwartz, E. Inequalities for scattering coefficients of lossy circulators, *IEEE Trans. Microwave Theory Tech.*, **MTT-19**.

Bosma, H. (1966) Performance of lossy H-plane Y-circulators, *IEEE Trans. Magnetics*, **MAG-2**, 273–277.

Carlin, H. J. and Giordano, A. B. *Network Theory. An Introduction to Reciprocal and Nonreciprocal Circuits*, Prentice-Hall, Englewood Cliffs, NJ.

Davies, J. B. (1962) An analysis of the m-port symmetrical H-plane waveguide junction with central ferrite post, *IRE Trans.*, **MTT-10**, 596–604.

Hagelin, S. (1969) Analysis of lossy symmetrical three-port networks with circulator properties, *IEE Trans. Microwave Theory Tech.*, **MTT-17**, (6) 328.

Helszajn, J. (1970) Wideband circulator adjustment using $n = \pm 1$ and $n = 0$ electromagnetic-field patterns, *Electronic Lett.*, **6**, pp. 729–731.

Helszajn, J. (1970) The adjustment of the m-port single junction circulator', *IEEE Trans.*, **MTT-18**, 705–711.

Helszajn, J. and Buffler, C. R. (1968) Adjustment of the 4-port single junction circulator, *Radio and Electronic Eng.*, **35**, 357–360.

Knerr, R. H., Barnes, C. E. and Bosch, F. (1970) A compact broadband thin-film limped element L-band circulator, *IEEE Trans.*, **MTT-18**, 1100–1108.

Konishi, Y. (1965) Lumped Element Circulator, *IEEE Trans. Microwave Theory Tech.* **MTT-13**, 852–864.

Milano, U., Saunders, J. H. and Davies, L. Jr. (1960) A Y-junction stripline circulator, *IRE Trans.* **MTT-9**, 346–351.

Milano, U., Saunders, J. H. and Davies, J. R. L. (1960) A Y-junction stripline circulator, *IRE Trans. Microwave Theory Tech.* **MTT-8**, 346–350.

Montgomery, C. G., Dicke, R. H. and Purcell, E. M. (1948) *Principles of Microwave Circuits.* McGraw-Hill, New York.

4

IMPEDANCE MATRIX OF JUNCTION CIRCULATOR

4.1 Introduction

There are in general six different ways in which the boundary conditions of a circulator can be applied. Its scattering matrix and its immittance matrices provide three possible descriptions and the corresponding eigenvalues provide three more. The actual choice between the different possibilities is usually determined by the physical problem. It is of note that while it is always possible to construct the scattering matrix of a junction its impedance or admittance matrix need not be realizable. While the definition of an ideal circulator is best established by having recourse to its scattering matrix its adjustment requires a knowledge of its eigenvalues. The 1-port equivalent circuits separately require a definition of one of its immittance matrices. There are altogether four degree-1 circuits and four degree-2 arrangements. The derivations of the most commonly encountered degree-1 and 2 circuits are the main endeavour of this chapter.

Since the impedance eigenvalues are related to the scattering ones, an impedance matrix can be constructed after each symmetrical adjustment of the junction. This means that it is possible, in principle, to form $(m-1)$ equivalent networks each of which corresponds to the $(m-1)$ adjustments of the matrix eigenvalues.

4.2 Impedance Matrix of Junction Circulators

The derivation of equivalent circuits requires a knowledge of an immittance rather than the scattering description of the network. The impedance matrix of a 3-port symmetrical but non-reciprocal network is

4 IMPEDANCE MATRIX OF JUNCTION CIRCULATOR

$$\bar{Z} = \begin{bmatrix} Z_{11} & Z_{12} & Z_{13} \\ Z_{13} & Z_{11} & Z_{12} \\ Z_{12} & Z_{13} & Z_{11} \end{bmatrix} \quad (4.1)$$

The impedance matrix can be diagonalized to reveal the eigenvalues in a similar way to that of the scattering one. This may be done by noting that the eigenvectors of $\bar{S}$ are also those of $\bar{Z}$ since the two commute. Hence

$$\bar{Z} = \bar{U} \bar{z} \bar{U}^{-1} \quad (4.2)$$

where $\bar{z}$ is a diagonal matrix with the eigenvalues of the impedance matrix $\bar{Z}$. It is here assumed that the matrix $\bar{Z}$ exists. This statement requires that the matrix $\bar{I} - \bar{S}$ is non-singular. The voltages and currents on this sort of circuit are indicated in Figure 4.1.

Following the procedure outlined in Chapter 3 to derive the relationships between the scattering coefficients and its eigenvalues one obtains in the case of the $\bar{Z}$ matrix

$$Z_{11} = \frac{z_0 + z_+ + z_-}{3} \quad (4.3a)$$

$$Z_{12} = \frac{z_0 + \alpha z_+ + \alpha^2 z_-}{3} \quad (4.3b)$$

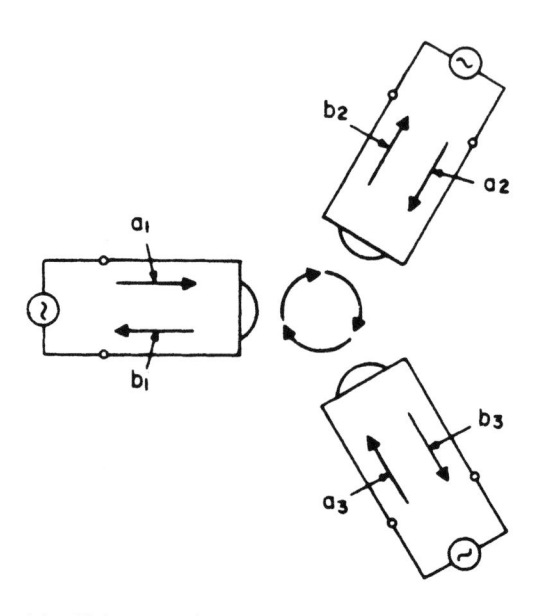

Figure 4.1 Voltage and current variables on 3-port circulator

4.2 IMPEDANCE MATRIX OF JUNCTION CIRCULATORS

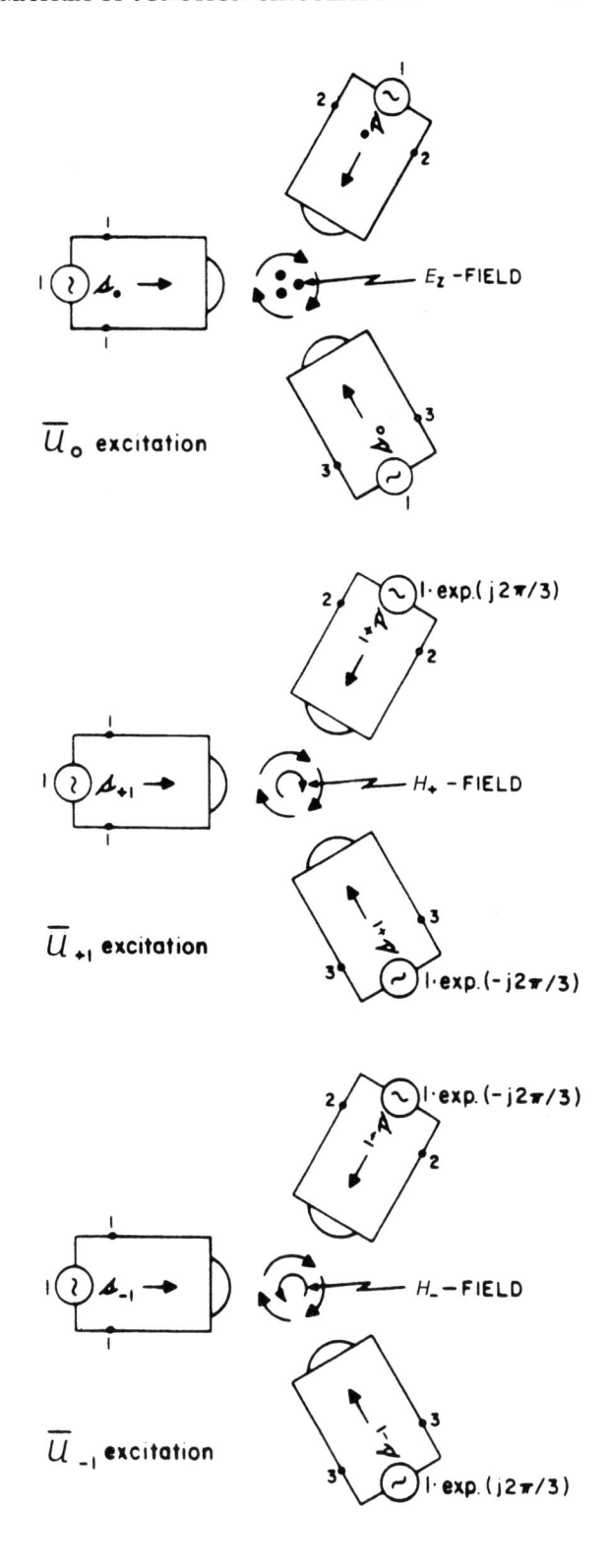

Figure 4.2 Eigensolutions of 3-port circulator

$$Z_{13} = \frac{z_0 + \alpha^2 z_+ + \alpha z_-}{3} \tag{4.3c}$$

The magnetic field in the transverse plane for the in-phase set of generator settings ($\bar{U}_0$) for one solution is zero on the axis of the junction while the electric field along the axis is a maximum there. A possible 1-port equivalent circuit is in this instance an open-circuited one. The solution for the counter-rotating generator settings ($\bar{U}_+$ and $\bar{U}_-$) produces counter-rotating circularly polarized magnetic fields in the transverse plane on the axis of the junction and produces an electric field which is zero there. One-port short-circuited transmission lines are therefore possibilities in this instance. The eigen solutions of a 3-port junction are illustrated in Figure 4.2. The three eigen-networks of the circuit are reproduced in Figure 4.3. The open-circuit parameters of the single 4-junction are likewise related to the impedance eigenvalues of the network by

$$Z_{11} = \frac{z_0 + z_{+1} + z_{-1} + z_2}{4} \tag{4.4a}$$

$$Z_{12} = \frac{z_0 + jz_{+1} - jz_{-1} - z_2}{4} \tag{4.4b}$$

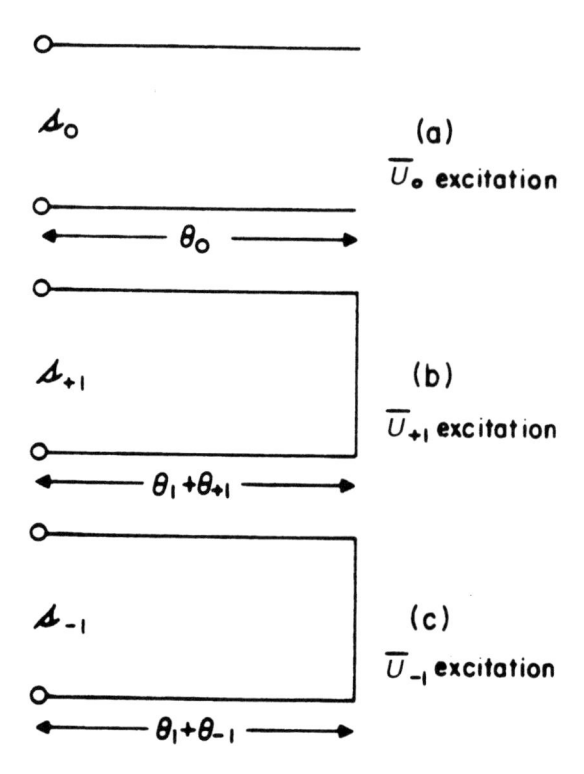

Figure 4.3 Eigen-networks of 3-port junction circulator

4.3 EIGENVALUES OF IMMITTANCE MATRICES

$$Z_{13} = \frac{z_0 - z_{+1} - z_{-1} + z_2}{4} \tag{4.4c}$$

$$Z_{14} = \frac{z_0 - jz_{+1} + jz_{-1} - z_2}{4} \tag{4.4d}$$

4.3 Eigenvalues of Immittance Matrices

The eigenvalues of the $\bar{Z}$ or $\bar{Y}$ matrices can be found in the usual way by solving the eigenvalues equation. If the immittance matrices are constructed in terms of the scattering one, it is possible to avoid the need to invert matrices by first finding the eigenvalues of $\bar{S}$. These eigenvalues are then used to construct the corresponding immittance ones and thereafter the immittance matrices.

The eigenvalues of the $\bar{S}$, $\bar{Z}$, and $\bar{Y}$ matrices, which have common eigenvectors, may be related by making use of the following theorem which has already been introduced in Chapter 3:

If

$$\bar{S}\bar{U}_i = s_i \bar{U}_i \tag{4.5}$$

then

$$f(\bar{S})\bar{U}_i = f(s_i)\bar{U}_i \tag{4.6}$$

This theorem will first be applied to obtain the relation between the eigenvalues of the $\bar{S}$ and $\bar{Z}$ matrices. The bilinear transformation between the two matrices is

$$\bar{S} = (\bar{Z} - I)(\bar{Z} + I)^{-1} \tag{4.7}$$

Making use of the theorem in equation (4.6) immediately produces the relationship between s_i and z_i as

$$s_i = \frac{z_i - 1}{z_i + 1} \tag{4.8}$$

Writing z_i in terms of s_i gives

$$z_i = \frac{1 + s_i}{1 - s_i} \tag{4.9}$$

where z_i is a normalized eigenvalue that satisfies the eigenvalue equation given by

$$\bar{Z}\bar{U}_i = z_i\bar{U}_i \qquad (4.10)$$

The derivation of the relationship between the eigenvalues of the scattering and admittance matrices, again starts with the bilinear relationship between the two matrix quantities

$$\bar{S} = (I - \bar{Y})(I + \bar{Y})^{-1} \qquad (4.11)$$

The relationship between the eigenvalues is now given as

$$s_i = \frac{1 - y_i}{1 + y_i} \qquad (4.12)$$

Writing y_i in terms of s_i indicates that the two are related by

$$y_i = \frac{1 - s_i}{1 + s_i} \qquad (4.13)$$

where y_i is a normalized eigenvalue that satisfies the eigenvalues equation defined by

$$\bar{Y}\bar{U}_i = y_i\bar{U}_i \qquad (4.14)$$

4.4 Complex Gyrator Immittance of 3-port Circulator

The complex gyrator immittance of the 3-port junction circulator is perhaps the most important quantity in the description of this class of device. It is defined by

$$Z_{in} = \frac{V_1}{I_1} \qquad (4.15)$$

with

$$V_3 = I_3 = 0 \qquad (4.16)$$

The origin of this definition may be understood by writing down the voltage current relationships of the network

$$\begin{bmatrix} V_1 \\ V_2 \\ 0 \end{bmatrix} = \begin{bmatrix} Z_{11} & Z_{21} & Z_{31} \\ Z_{31} & Z_{11} & Z_{21} \\ Z_{21} & Z_{31} & Z_{11} \end{bmatrix} \begin{bmatrix} I_1 \\ I_2 \\ 0 \end{bmatrix} \qquad (4.17)$$

4.4 COMPLEX GYRATOR IMMITTANCE OF 3-PORT CIRCULATOR

The required result is

$$Z_{in} = Z_{11} - \frac{Z_{21}^2}{Z_{31}} \tag{4.18}$$

The condition at port 2 is then given by

$$Z_{out} = \frac{V_2}{-I_2} = Z_{in}^* \tag{4.19}$$

This relationship indicates that terminating each port by Z_{in}^* in a cyclic manner is sufficient to match the device. Figure 4.4 illustrates the schematic diagram of this arrangement.

If the frequency variation of the in-phase impedance eigenvalue z_0 may be neglected compared to those of the degenerate or split ones then it is possible to deduce an especially simple model for this class of device. The required realization starts by writing

$$z_0 = 0 \tag{4.20}$$

It continues by developing the open-circuit parameters under this condition

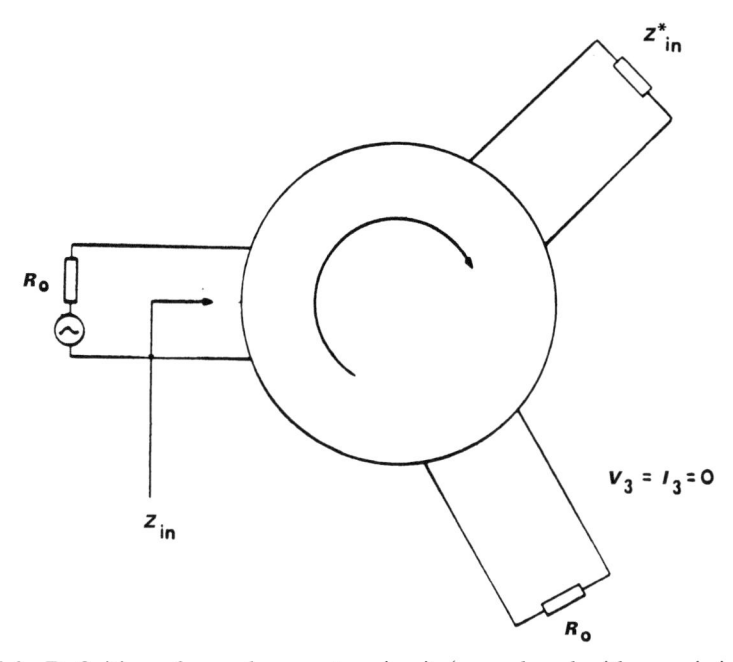

Figure 4.4 Definition of complex gyrator circuit (reproduced with permission, Simon, J. (1965) Broadband Strip-Transmission Line Y junction circulators, *IEEE Trans. Microwave Theory Tech.*, **MTT-13**, 335–345) (© 1965 IEEE)

$$Z_{11} \approx \frac{z_+ + z_-}{3} \qquad (4.21a)$$

$$Z_{21} \approx -\left(\frac{z_+ + z_-}{6}\right) + j\sqrt{3}\left(\frac{z_+ - z_-}{6}\right) \qquad (4.21b)$$

$$Z_{31} \approx -\left(\frac{z_+ + z_-}{6}\right) - j\sqrt{3}\left(\frac{z_+ - z_-}{6}\right) \qquad (4.21c)$$

Forming Y_{in} instead of Z_{in} gives

$$Y_{in} = \frac{1}{Z_{in}} = \frac{Z_{31}}{Z_{11}Z_{31} - Z_{21}^2} \qquad (4.22)$$

Noting that

$$Z_{11}Z_{31} - Z_{21}^2 = \frac{-z_+ z_-}{3}$$

gives

$$Y_{in} = \frac{(z_+ + z_-) + j\sqrt{3}(z_+ - z_-)}{2 z_+ z_-}$$

or

$$Y_{in} = \left(\frac{y_+ + y_-}{2}\right) - j\sqrt{3}\left(\frac{y_+ - y_-}{2}\right) \qquad (4.24)$$

The imaginary and real parts of Y_{in} are therefore related to the sum and difference of the split admittance eigenvalues in a particularly simple way.

$$j B_{in} = \left(\frac{y_+ + y_-}{2}\right) \qquad (4.24)$$

$$G_{in} = -j\sqrt{3}\left(\frac{y_+ - y_-}{2}\right) \qquad (4.25)$$

One equivalent circuit of the 3-port junction circulator is therefore a simple 1-port LCR network. This classic result is illustrated in Figure 4.5. Furthermore, a knowledge of y_+ and y_- is sufficient to describe this class of device.

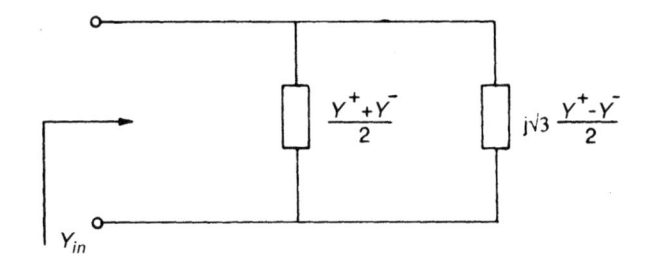

Figure 4.5 Schematic diagram of degree-1 complex gyrator circuit

4.5 Synthesis of Junction Circulators using Resonant In-Phase Eigen-network

While the in-phase eigen-networks of a 3-port junction circulator may often be idealized by a frequency-independent short-circuit boundary condition at its input terminals, it may also be adjusted to exhibit a series resonance there. The complex gyrator impedance is in this instance approximately given by

$$Z_{in} \approx \frac{8z^0 - (z^+ + z^-)}{6} + j\frac{(z^+ - z^-)}{2\sqrt{3}} \qquad (4.26)$$

In obtaining this result, the in-phase eigenvalue z^0 has been idealized by a short-circuit boundary condition in forming the real part of the gyrator immittance, but has been retained in describing its imaginary part. This impedance is readily realized in the form indicated in Figure 4.6 by expanding Z_{in} as

$$Z_{in} = Z_1 + \frac{1}{Y_1} \qquad (4.27)$$

where

$$Z_1 \approx \frac{4z^0}{3} \qquad (4.28a)$$

$$Y_1 \approx \frac{(y_+ + y_-)}{2} - j\sqrt{3}\frac{(y^+ - y^-)}{2} \qquad (4.28b)$$

The equivalent circuit in Figure 4.6 reduces to the usual approximation by omitting the z^0 term in the derivation. This circuit has the nature of a bandpass filter which may be adjusted to display a reflection or transmission characterisitic akin to that of a quarter-wave coupled junction with its in-phase eigen-networks idealized by a short-circuit boundary condition.

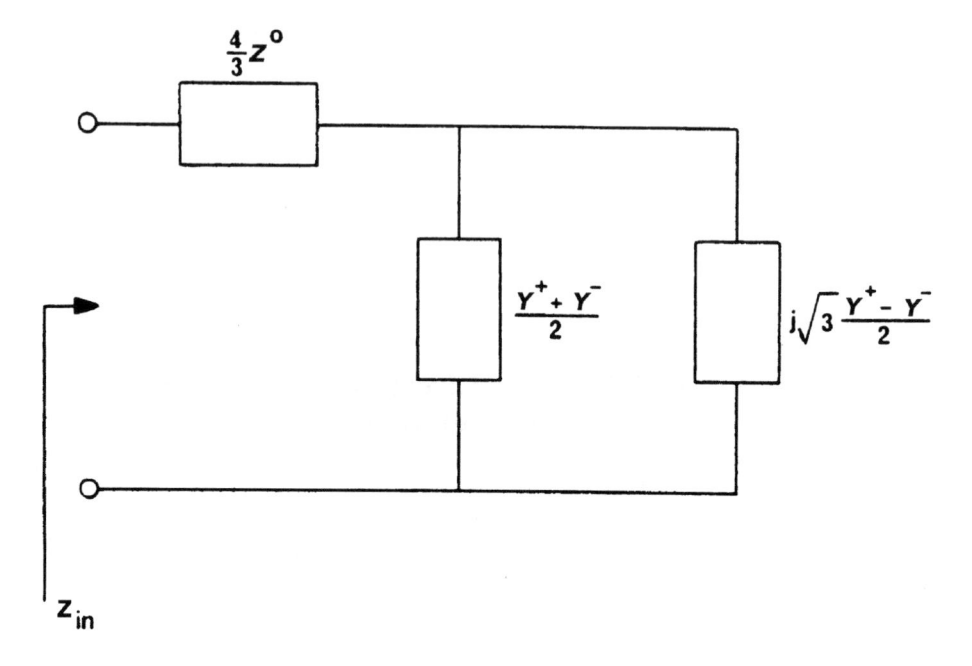

Figure 4.6 Schematic diagram of degree-2 complex gyrator circuit

4.6 Equivalent Circuit

An equivalent 3-port circuit based on ideal gyrators can be synthesised from the impedance matrix by having recourse to simple matrix addition. Making use of the bilinear mapping between the reflection eigenvalues of an ideal circulator and the impedance ones gives

$$z^0 = 0 \tag{4.29a}$$

$$z^+ = -jZ_0/\sqrt{3} \tag{4.29b}$$

$$z^- = jZ_0/\sqrt{3} \tag{4.29c}$$

The open circuit parameters of an ideal circulator are therefore

$$Z_{11} = 0 \tag{4.30a}$$

$$Z_{12} = R_0 \tag{4.30b}$$

$$Z_{13} = -R_0 \tag{4.30c}$$

4.6 EQUIVALENT CIRCUIT

The impedance matrix of an ideal 3-port circulator is therefore given by

$$\overline{Z} = \begin{bmatrix} 0 & R_0 & -R_0 \\ -R_0 & 0 & R_0 \\ R_0 & -R_0 & 0 \end{bmatrix} \qquad (4.31)$$

Making use of the rule of matrix addition indicates that

$$\begin{bmatrix} 0 & R_0 & -R_0 \\ -R_0 & 0 & R_0 \\ R_0 & -R_0 & 0 \end{bmatrix} = \begin{bmatrix} 0 & R_0 & 0 \\ -R_0 & 0 & 0 \\ 0 & 0 & 0 \end{bmatrix} + \begin{bmatrix} 0 & 0 & -R_0 \\ 0 & 0 & 0 \\ R_0 & 0 & 0 \end{bmatrix} + \begin{bmatrix} 0 & 0 & 0 \\ 0 & 0 & R_0 \\ 0 & -R_0 & 0 \end{bmatrix} \qquad (4.32)$$

The first impedance matrix represents an ideal gyrator circuit connected between ports 1 and 2 of the junction. The other two matrices represent similar gyrator circuits between ports 2 and 3 and 3 and 1. The equivalent circuit is shown in Figure 4.7.

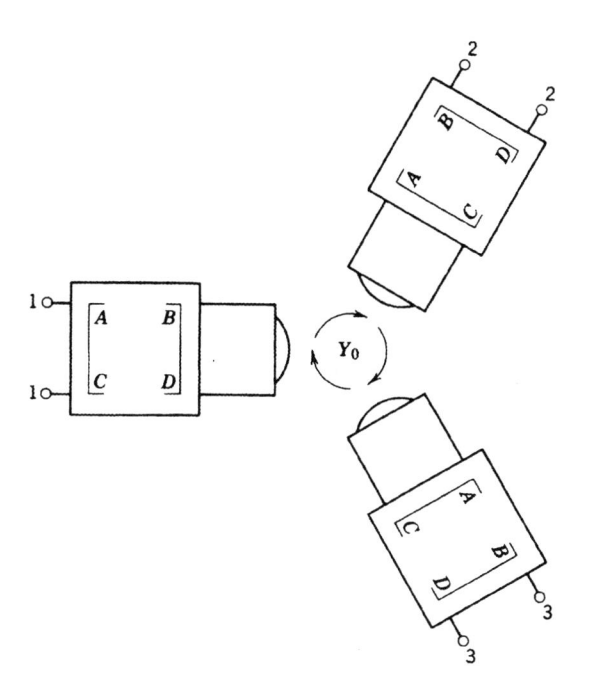

Figure 4.7 Topology of 3-port junction circulator in terms of 2-port gyrator circuits

4.7 Quality Factor of Junction Circulator

The real part of the complex gyrator admittance is sometimes written in terms of the susceptance slope parameter of the degenerate counter-rotating eigen-networks and the resonant frequencies of the split eigen-networks. This approach is of special interest in the network problem of the device. The derivation of this result begins by writing

$$y_+ \approx j(y_1 + \Delta y_1)\cot(\theta_1 - \Delta\theta_1) \qquad (4.33a)$$

$$y_- \approx j(y_1 - \Delta y_1)\cot(\theta_1 + \Delta\theta_1) \qquad (4.33b)$$

$\theta_1 \mp \Delta\theta_1$ are the electrical lengths of the split eigen-networks

$$\theta_1 \mp \Delta\theta_1 = (\pi/2)(1 + \delta_\pm) \qquad (4.34)$$

The normalized frequency variables are expanded about the frequencies $\omega_\pm$ of each eigen-network.

$$\delta_\pm = \frac{\omega - \omega_\pm}{\omega_\pm} \qquad (4.35)$$

At $\theta_1 + \pi/2$

$$y_+ = -j(y_1 + \Delta y_1)\tan(\pi\delta_+/2) \qquad (4.36a)$$

$$y_- = -j(y_1 - \Delta y_1)\tan(\pi\delta_-/2) \qquad (4.36b)$$

Introducing these relationships into the real part of the complex gyrator admittance and assuming that

$$\tan(\pi\delta_+/2) \approx (\pi\delta_+/2)$$

readily gives the desired result

$$g \approx \sqrt{3}\, b' \left(\frac{\omega_+ - \omega_-}{\omega_0}\right) \qquad (4.37)$$

b' is the susceptance slope parameter of the degenerate counter-rotating eigen-network.

$$b' = \frac{\pi y_1}{4} \qquad (4.38)$$

4.8 DEGENERATE COUNTER-ROTATING EIGEN-NETWORK (s_1)

Equation (4.37) is also sometimes written as

$$\frac{1}{Q_L} = \frac{g}{b'} = \sqrt{3}\left(\frac{\omega_+ - \omega_-}{\omega_0}\right) \qquad (4.39)$$

This relationship permits any one of the above variables to be deduced from a knowledge of the other two.

4.8 Degenerate Counter-Rotating Eigen-network (s_1)

While a typical degenerate eigenvalue may be represented by an equivalent transmission line a knowledge of its topology is often desirable. One possible way to deduce its geometry will now be derived. The derivation of its eigen-network starts by placing a magnetic wall at port 3 of the junction and constructing the relationship between the other two ports. This gives

$$I_3 = 0 \qquad (4.40)$$

and

$$\begin{bmatrix} V_1 \\ V_2 \\ V_3 \end{bmatrix} = \begin{bmatrix} Z_{11} & Z_{12} & Z_{12} \\ Z_{12} & Z_{11} & Z_{12} \\ Z_{12} & Z_{12} & Z_{11} \end{bmatrix} \begin{bmatrix} I_1 \\ I_2 \\ 0 \end{bmatrix} \qquad (4.41)$$

Z_{11} and Z_{12} are linear combinations of the eigenvalues of the 3-port circuit in the usual way.

$$Z_{11} = \frac{z_0 + 2z_1}{3} \qquad (4.42a)$$

$$Z_{12} = \frac{z_0 - z_1}{3} \qquad (4.42b)$$

z_0 is the in-phase eigenvalue, z_1 is the degenerate counter-rotating one.

The relationship between ports 1 and 2 of the reduced symmetrical network in Figure 4.8 is then given by

$$\begin{bmatrix} V_1 \\ V_2 \end{bmatrix} = \begin{bmatrix} Z_{11} & Z_{12} \\ Z_{12} & Z_{11} \end{bmatrix} \begin{bmatrix} I_1 \\ I_2 \end{bmatrix} \qquad (4.43)$$

The in-phase and out-of-phase eigenvalues z'_1 and z'_2 of the new circuit indicated in Figure 4.8 are now deduced by taking the sum and difference of the open-circuit parameters.

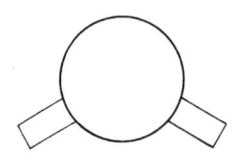

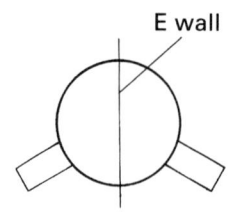

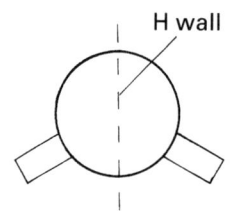

[Figure 4.8: Construction of degenerate eigen-network — three diagrams showing a circular junction with two ports, labeled "E wall" and "H wall"]

Figure 4.8 Construction of degenerate eigen-network

$$z'_1 = Z_{11} + Z_{12} \quad (4.44a)$$

$$z'_2 = Z_{11} - Z_{12} \quad (4.44b)$$

Evaluating these quantities in terms of the eigenvalues of the original circuit indicates that

$$z'_1 = (z_0 - z_1)/3 \quad (4.45a)$$

$$z'_2 = z_1 \quad (4.45b)$$

This result indicate that the out-of-phase eigen-network of the reduced 2-port circuit coincides with that of the degenerate counter-rotating one of the original 3-port circuit.

It is also of note that a knowledge of z'_1 and the in-phase eigen-network of the reduced 2-port circuit may be separately employed to evaluate the in-phase eigen-network of the original 3-port junction.

It is of separate note that for the out-of-phase eigen-network

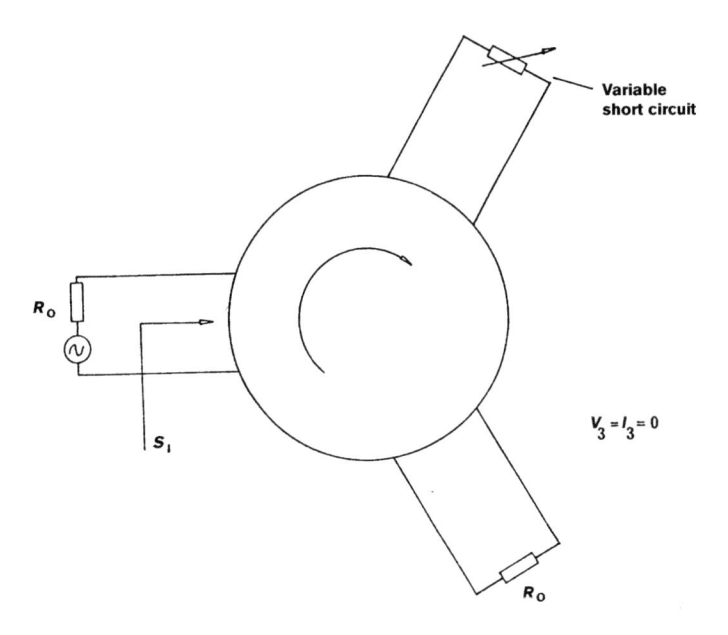

Figure 4.9 Experimental arrangement for measurement of degenerate reflection eigenvalue S_1

$$I_1 = -I_2$$

and the voltage at port 3 is $V_3 = 0$ in addition to the condition $I_3 = 0$ at the same port. If $I_1 = -I_2$ then

$$\frac{V_1}{I_1} = \frac{V_2}{I_2} = Z_{11} - Z_{12} \qquad (4.46)$$

in keeping with the eigenvalue problem.

One possible way in which the eigenvalue s_1 of a reciprocal 3-port junction can be experimentally determined without the need to fabricate its eigen-network consists of placing a sliding short-circuit at port 2 and a matched load at port 3. The variable short-circuit is varied until there is total reflection at port 1. The reflection coefficient at port 1 is then the eigenvalue s_1. This technique is especially appropriate in the case of a waveguide junction for which the reference terminals are usually ill defined.

The experimental arrangement is shown in Figure 4.9.

4.9 In-Phase Eigen-Network

The 1-port in-phase eigen-network is also a classic circuit which may be separately fabricated. Figure 4.10 indicates this circuit in the case of a junction using a planar disk resonator. It is obtained by partitioning the circuit with appropriate magnetic

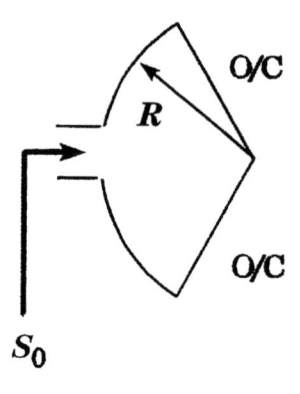

Figure 4.10 Topology of in-phase eigen-network

walls. This 1-port circuit accurately displays the admittance and reflection eigenvalues of the circuit provided its boundaries are idealised by perfect magnetic walls. It therefore represents the simplest test fixture for the characterization of this eigen-network. A shortcoming of this circuit, however, is that the open walls exposed by this boundary condition are ill defined in practical circuits.

The phase angle of the in-phase eigen-network of a demagnetized junction may be directly measured using the eigenvalue approach by applying equal in-phase signals at the three ports of the junction or by making use of the relationship between the scattering variable S_{11} and the in-phase and degenerate counter-rotating eigenvalues s_0 and s_1.

$$S_{11} = \frac{s_0 + 2s_1}{3} \tag{4.47}$$

4.10 The Gyrator Network

The basic 2-port circuit met in connection with nonreciprocal networks is the gyrator one introduced in Chapter 2.

It will now be demonstrated that both the impedance and admittance matrices exist for the gyrator network. The impedance eigenvalues are

$$\frac{Z_+}{R_0} = \frac{1 + s_+}{1 - s_+} = \frac{1 + j}{1 - j} = j \tag{4.48a}$$

$$\frac{Z_-}{R_0} = \frac{1 + s_-}{1 - s_-} = \frac{1 - j}{1 + j} = -j \tag{4.48b}$$

The admittance eigenvalues are the reciprocal of the impedance ones.

4.10 THE GYRATOR NETWORK

$$\frac{Y_+}{Y_0} = \frac{R_0}{Z_+} = -j \tag{4.49a}$$

$$\frac{Y_-}{Y_0} = \frac{R_0}{Z_-} = j \tag{4.49b}$$

The admittance matrix is now diagonalized using equation (4.2)

$$\bar{Y} = \frac{1}{2}\begin{bmatrix} (Y_+ + Y_-), & j(Y_+ - Y_-) \\ -j(Y_+ - Y_-), & (Y_+ + Y_-) \end{bmatrix} \tag{4.50}$$

In terms of the original variables the results is

$$\bar{Y} = \begin{bmatrix} 0 & Y_0 \\ -Y_0 & 0 \end{bmatrix} \tag{4.51}$$

The $\bar{Y}$-matrix schematic of the 2-port gyrator is illustrated in Figure 4.11. The result for the $\bar{Z}$-matrix is

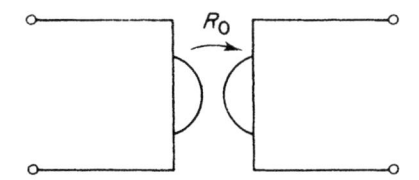

Figure 4.11 Schematic diagram of 2-port gyrator network using impedance gyrator

$$\bar{Z} = \begin{bmatrix} 0 & -R_0 \\ R_0 & 0 \end{bmatrix} \tag{4.52}$$

The $\bar{Z}$-matrix schematic of the 2-port gyrator is depicted in Figure 4.12.

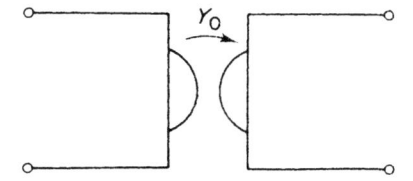

Figure 4.12 Schematic diagram of 2-port gyrator network using admittance gyrator

4.11 Input Impedance of Waveguide Circulator

A knowledge of the input impedance at port 1 of a symmetrical 3-port circulator with ports 2 and 3 terminated in matched loads is also in some instances desirable. The derivation of the required result starts with a statement of the voltage–current relationships of the junction

$$\begin{bmatrix} V_1 \\ V_2 \\ V_3 \end{bmatrix} = \begin{bmatrix} Z_{11} & Z_{12} & Z_{13} \\ Z_{13} & Z_{11} & Z_{12} \\ Z_{12} & Z_{13} & Z_{11} \end{bmatrix} \begin{bmatrix} I_1 \\ I_2 \\ I_3 \end{bmatrix} \quad (4.56)$$

It continues by defining the boundary conditions at ports 2 and 3

$$V_2 = -I_2 Z_0 \quad (4.57a)$$

$$V_3 = -I_3 Z_0 \quad (4.57b)$$

as a preamble to forming the impedance Z_{in} at port 1

$$Z_{in} = \frac{V_1}{I_1}$$

This gives

$$\left(\frac{V_1}{I_1}\right) = Z_{11} + Z_{12}\left(\frac{I_2}{I_1}\right) + Z_{13}\left(\frac{I_3}{I_1}\right) \quad (4.58a)$$

$$\left(\frac{-I_2}{I_1}\right) Z_0 = Z_{13} + Z_{11}\left(\frac{I_2}{I_1}\right) + Z_{12}\left(\frac{I_3}{I_1}\right) \quad (4.58b)$$

$$\left(\frac{-I_3}{I_1}\right) Z_0 = Z_{12} + Z_{13}\left(\frac{I_2}{I_1}\right) + Z_{11}\left(\frac{I_3}{I_1}\right) \quad (4.58c)$$

The required result is readily deduced by rearranging the proceeding equations

$$Z_{in} = Z_{11} + \frac{Z_{12}^3 - Z_{12}^{*3} + 2Z_{12}Z_{12}^*(Z_{11} + Z_0)}{(Z_{11} + Z_0)^2 + Z_{12}Z_{12}^*} \quad (4.59)$$

In obtaining this result use has also been made of the fact that

$$Z_{13} = -Z_{12}^* \quad (4.60)$$

This important identity is left to the reader to verify at his leisure.

Further Reading

Aitken, F. M. and McLean, R. (1963) Some properties of the waveguide Y circulator, *Proc. IEE*, **110**(2), 256–260.

Auld, B. A. (1959) The synthesis of symmetrical waveguide circulators, *IRE Trans. Microwave Theory Tech.*, **MTT-7**, 238–246.

Bergman, J. O. and Christensen, C. (1968) Equivalent circuit for a lumped element Y circulator, *IEEE Trans. Microwave Theory Tech.*, **MTT-16**, 308–310.

Bittar, J. and Verszely, J. (1980) A general equivalent network of the input impedance of symmetric three-port circulators, *IEEE Trans. Microwave Theory Tech.*, **MTT-28**, 807–808.

Carlin, H. J. and Giodano, Q. B. (1964) *Network Theory, an Introduction to Reciprocal and Nonreciprocal Circuits*, Prentice-Hall, Inc., Englewood Cliffs, NJ.

Chait, H. N. and Curry, T. R. (1959) Y-Circulator, *J. Appl. Phys.*, **30**, 152.

Feoktistov, V. G. (1968) Design of a strip Y-circulator, *Radio Eng. Electronic Physics*, **13**(7).

Fowler, H. (1950) Paper presented at the 1950 symposium on microwave properties and applications of ferrites, Harvard University, Cambridge, MA.

Hagelin, S. (1966) A flow graph analysis of 3 and 4-port junction circulators, *IEEE Trans. Microwave Theory Tech.*, **MTT-14**(5), 243–249.

Helszajn, J. (1972) Three resonant mode adjustment of the waveguide circulator, *Radio Electron Eng.*, **42**(5), 213–216.

Helszajn, J. (1973) Frequency response of quarter-wave coupled reciprocal junctions, *IEEE Trans. Microwave Theory Tech.*, **MTT-21**, 533–538.

Helszajn, J. (1976) *Non-reciprocal microwave junctions and circulators*. John Wiley & Sons, NY.

Helszajn, J. (1985) Synthesis of quarter-wave coupled junction circulators with degree 1 and 2 complex gyrator circuits, *IEEE Trans. Microwave Theory Tech.*, **MTT-32**, 382–390.

Helszajn, J. (1987) Complex gyrator of an evanescent mode E-plane junction circulator using H-plane turnstile resonators, *IEEE Trans.*, **MTT-35**, 797–806.

Hogan, C. L. (1952) The microwave gyrator, *Bell Sys. Tech. J.*, **31**(1).

Konishi, Y. (1965) Lumped element Y circulator, *IEEE Trans. Microwave Theory Tech.*, **MTT-13**(6), 852–864.

Richardson, J. K. (1967) An approximate method of calculating Z_0 of a symmetrical strip line, *IEEE Trans.*, **MTT-15**, 130.

Salay, S. J. and Peppiatt, H. J. (1972) An accurate junction circulator design procedure, *IEEE Trans. Microwave Theory Tech.*, **MTT-20**, 192–193.

Solbach, K. (1982) Equivalent circuit representation for the E-plane circulator, *IEEE Trans. Microwave Theory Tech.*, **MTT-30**, 806–809.

Tellegen, (1948) The gyrator, a new electric network element, *Philips Res. Repts.*, **3**, 81–101.

5

THE POST GYROMAGNETIC RESONATOR

5.1 Introduction

The two families of gyromagnetic resonators encountered in the design of waveguide circulators are the planar or post ones for which the alternating radio frequency fields do not vary along the axis of the resonator and the cavity ones for which the fields display a standing wave there. The purpose of this chapter is to deal with the post arrangement with top and bottom electric walls and with magnetic sidewalls. The two most important shapes met in the construction of waveguide circulators which have the required symmetry of the problem region are the disk and equilateral triangular ones. Each shape supports an infinite number of modes which are characterized by a unique resonant frequency and a unique standing wave pattern in its plane. The gyromagnetic problem of the disk geometry is dealt with in some detail whereas that of the equilateral triangle is tackled phenomenologically. The application of a direct magnetic field along the axis of such a resonator provides a means of rotating its standing wave pattern and therefore provides one means of decoupling one port from the other two in the construction of the classic circulator. The possibility of utilizing a square gyromagnetic resonator in the construction of 4-port single junction circulators is separately understood.

5.2 Mode Patterns in Planar Circular Disk

An important microwave resonator is the planar one with a magnetic sidewall and top and bottom electric ones. The field patterns and frequencies of the disk resonator outlined in Figure 5.1 will now be derived.

5 THE POST GYROMAGNETIC RESONATOR

Planar resonators

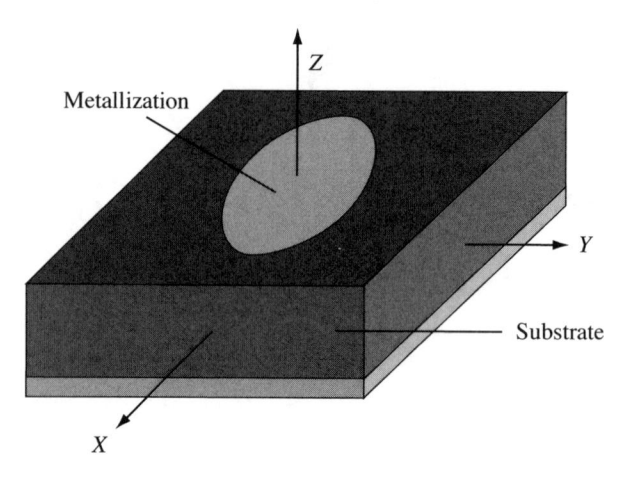

Figure 5.1 Schematic diagram of planar disc resonator

In what follows it is assumed that the substrate thickness and relative dielectric constant are such that no integral half wave-long value can be accommodated in the z-direction. This means that

$$\frac{\partial}{\partial z} = 0 \tag{5.1}$$

and there is only a z-component of the electric field

$$\bar{E} = \bar{i}_z E_z \tag{5.2}$$

where $\bar{i}_z$ is a unit vector. The other field components are then expressed in terms of E_z, by having recourse to Maxwell's first curl equation, as

$$H_r = \frac{j}{\omega \mu_0} \left[\frac{1}{r} \frac{\partial E_z}{\partial \phi} \right] \tag{5.3a}$$

$$H_\phi = \frac{-j}{\omega \mu_0} \left[\frac{\partial E_z}{\partial r} \right] \tag{5.3b}$$

$$E_\phi = E_r = H_z = 0 \tag{5.3c}$$

The derivation is completed once a solution for E_z which satisfies both the wave equation

$$\left[\frac{\partial}{\partial r^2} + \frac{1}{r} \frac{\partial}{\partial r} + \frac{1}{r^2} \frac{\partial}{\partial \phi^2} + k^2 \right] E_z = 0 \tag{5.4}$$

and the boundary conditions has been determined.

5.2 MODE PATTERNS IN PLANAR CIRCULAR DISK

The quantity k in the wave equation is defined in the usual way by

$$k = \omega \sqrt{\varepsilon_0 \varepsilon_r \mu_0 \mu_r}$$

The wave equation met here is in fact identical to that encountered in connection with the description of a circular waveguide except that the derivatives with respect to the z coordinate are omitted. Its solution may be described by the product of a pure function of R and a pure function of Φ.

$$E_z = R \cdot \Phi$$

Substituting the trial function into the wave equation gives

$$r^2 \left(\frac{R''}{R}\right) + r\left(\frac{R'}{R}\right) + k^2 r^2 = -\frac{\phi''}{\phi} \tag{5.5}$$

This equation is satisfied provided

$$-\frac{\phi''}{\phi} = n^2 \tag{5.6a}$$

$$r^2 \left(\frac{R''}{R}\right) + r\left(\frac{R'}{R}\right) + (k^2 r^2 - n^2) = 0 \tag{5.6b}$$

One solution to the first differential equation is given in terms of a linear combination of exponential functions

$$\phi = A_n[\exp(jn\phi) + \exp(-jn\phi)] \tag{5.7a}$$

The solution to the second one involves Bessel functions of the first kind of order n and argument kr

$$R = J_n(kr) \tag{5.7b}$$

The required product solution for E_z is therefore given by

$$E_z = A_n J_n(kr)[\exp(jn\phi) + \exp(-jn\phi)] \tag{5.8}$$

It satisfies

$$E_z = A_n J_n(kr) \tag{5.9}$$

at $\phi = 0$.

The complete standing wave solution in this type of planar resonator is therefore described by

5 THE POST GYROMAGNETIC RESONATOR

$$E_z = A_n J_n(kr) \cos n\phi \tag{5.10a}$$

$$H_r = -jn\zeta_r A_n \frac{J_n(kr)}{kr} \sin n\phi \tag{5.10b}$$

$$H_\phi = -j\zeta_r A_n J'_n(kr) \cos n\phi \tag{5.10c}$$

and TM conditions prevail.

The free space wave admittance ζ_r is

$$\zeta_r = \sqrt{\frac{\varepsilon_0 \varepsilon_r}{\mu_0}} \tag{5.11}$$

At the edge of the disk it is assumed that the radial component of the surface current must vanish, and consequently

$$H_\phi(R) = 0 \quad \text{at } r = R \tag{5.12a}$$

This relationship is satisfied whenever

$$J'_n(kR) = 0 \tag{5.12b}$$

This last equation fixes the radius of the ferrite disk for a particular mode. Its solution is denoted by

$$(kR)_{nj} \tag{5.12c}$$

nj implies the jth root of the nth order Bessel function. For $n = 0$, the radius of the disk for the first root is

$$(kR)_{0,1} = 3.83171 \tag{5.13}$$

and the field components of the TM_{01} mode are described by

$$E_z = J_0(kr) \tag{5.14a}$$

$$H_\phi = j\zeta_r J'_0(kr) \tag{5.14b}$$

$$H_r = E_r = E_\phi = H_z = 0 \tag{5.14c}$$

In order to evaluate the field pattern of this mode, it is necessary to re-introduce the time variation $\exp(j\omega t)$ and evaluate the real part of each component.

5.2 MODE PATTERNS IN PLANAR CIRCULAR DISK

$$E_z = J_0(kr) \cos \omega t \qquad (5.15a)$$

$$H_\phi = -\zeta_r J'_0(kr) \sin \omega t \qquad (5.15b)$$

$$H_r = E_r = E_\phi = H_z = 0 \qquad (5.15c)$$

The result at $\omega t = 0$, $\pi/2$, π and $3\pi/2$ is illustrated in Figure 5.2. For $n = 1$, the radius of the disk is fixed by

$$(kR)_{1,1} = 1.84118 \qquad (5.16)$$

and the field components are

$$E_z = J_1(kr) \cos \phi \qquad (5.17a)$$

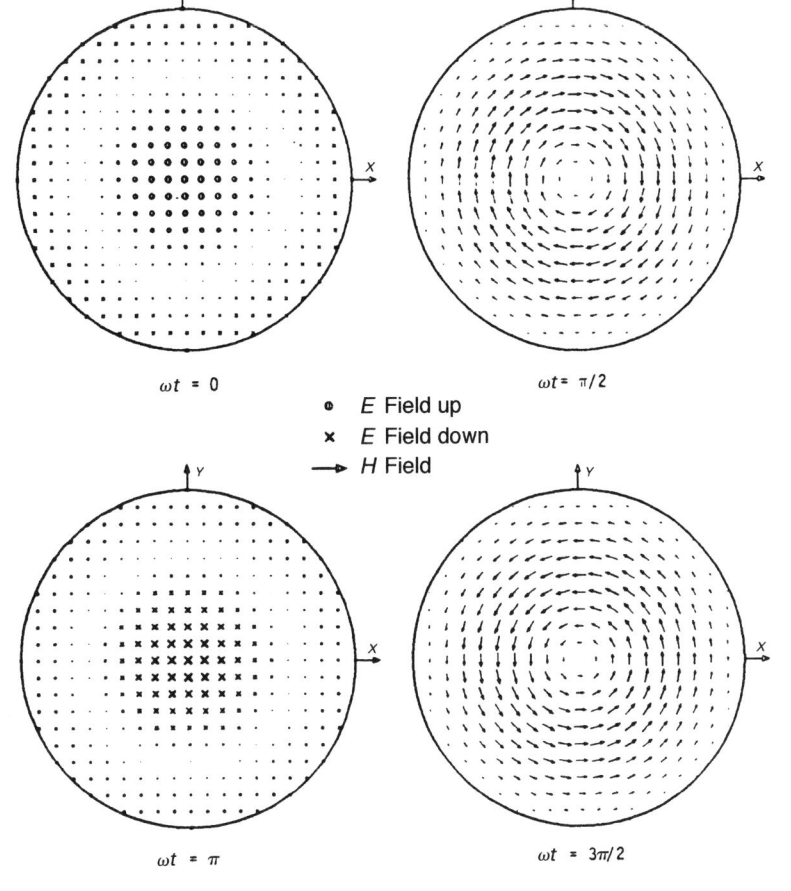

Figure 5.2 Electromagnetic field patterns for $TM_{0,1,0}$, mode (reproduced with permission, Watkins, J. (1969) Circular resonant structures in microstrip, *Electronics Lett.*, **5**, 524–525)

$$H_r = -j\zeta_r \frac{J_1(kr)}{kr} \sin\phi \tag{5.17b}$$

$$H_\phi = -j\zeta_r J'_1(kr) \cos\phi \tag{5.17c}$$

$$H_z = E_r = E_\phi = 0 \tag{5.17d}$$

The field components at $\omega t = 0$, $\pi/2$, π and $3\pi/2$ are depicted in Figure 5.3.

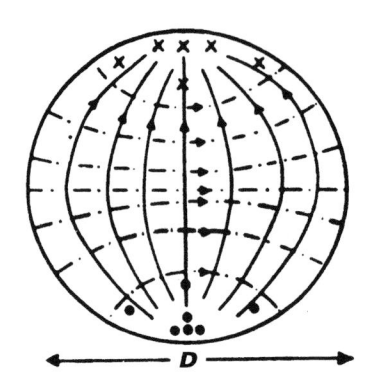

Figure 5.3 Electromagnetic field patterns for $TM_{1,1,0}$ mode (reproduced with permission, Watkins, J. (1969) Circular resonant structures in microstrip, *Electronics Lett.*, **5**, 524–525)

For $n = 2$, the radius of the ferrite disk is determined by

$$(kR)_{2,1} = 3.05424 \tag{5.18}$$

The radii for the $n = 1$ and 2 cases are smaller than for the $n = 0$ one. The field components are

$$E_z = J_2(kr) \cos 2\phi \tag{5.18a}$$

$$H_r = -j2\zeta_r \frac{J_2(kr)}{kr} \sin 2\phi \tag{5.18b}$$

$$H_\phi = -j\zeta_r J'_2(kr) \cos 2\phi \tag{5.18c}$$

$$H_z = E_r = E_\phi = 0 \tag{5.18d}$$

The field patterns for this mode are illustrated in Figure 5.4. The solution for $n = 3$ is given by

$$(kR)_{3,1} = 4.20119 \tag{5.19}$$

$$E_z = J_3(kr) \cos 3\phi \tag{5.20a}$$

$$H_r = -j3\zeta_r \frac{J_3(kr)}{kr} \sin 3\phi \tag{5.20b}$$

5.2 MODE PATTERNS IN PLANAR CIRCULAR DISK

Figure 5.4 Electromagnetic field patterns for $TM_{2,1,0}$ mode (reproduced with permission, Watkins, J. (1969) Circular resonant structures in microstrip, *Electronics Lett.*, **5**, 524–525)

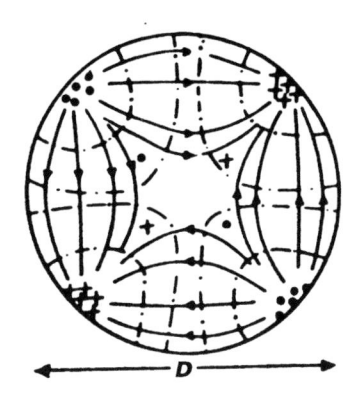

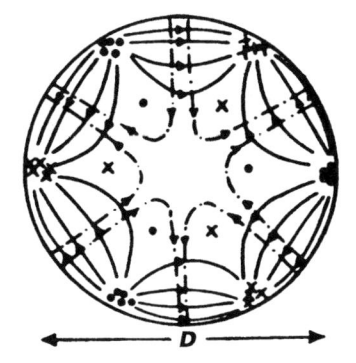

Figure 5.5 Electromagnetic field patterns for $TM_{3,1,0}$ mode (reproduced with permission, Watkins, J. (1969) Circular resonant structures in microstrip, *Electronics Lett.*, **5**, 524–525)

$$H_\phi = -j\zeta_r J_3'(kr) \cos 3\phi \qquad (5.20c)$$

$$H_z = E_r = E_\phi = 0 \qquad (5.20d)$$

Its field pattern is indicated in Figure 5.5.

A knowledge of the series form for $J_0(x)$ and $J_1(x)$

$$J_0(x) = 1 - 2.2499997 \left(\frac{x}{3}\right)^2 + 1.2656208 \left(\frac{x}{3}\right)^4 - 0.3163866 \left(\frac{x}{3}\right)^6 +$$

$$0.0444479 \left(\frac{x}{3}\right)^8 - 0.0039444 \left(\frac{x}{3}\right)^{10} + 0.0002100 \left(\frac{x}{3}\right)^{12}$$

$$J_1(x) = x \left[0.50 - 0.56249985 \left(\frac{x}{3}\right)^2 + 0.21093573 \left(\frac{x}{3}\right)^4 - 0.03954289 \left(\frac{x}{3}\right)^6 + \right.$$

$$\left. 0.00443319 \left(\frac{x}{3}\right)^8 - 0.00031761 \left(\frac{x}{3}\right)^{10} + 0.00001109 \left(\frac{x}{3}\right)^{12} \right]$$

and the recurrence formula

$$J_{n+1}(x) = J_{n-1}(x) + \frac{2n}{x} J_n(x)$$

is sufficient for computational purposes.

5.3 Mode Chart of Gyromagnetic Resonator

A knowledge of the modes in a gyromagnetic resonator provides some valuable insight into the operation of junction circulators. If a magnetic field is applied along the axis of the disk the two counter-rotating modes are no longer resonant at the same frequency. In addition the standing pattern formed by the two counter-rotating field patterns is rotated within the disk.

If the dielectric substrate is replaced by a magnetized ferrite substrate the derivation proceeds in essentially the same way as for the isotropic problem except that μ_r now takes on a tensor form

$$[\mu_r] = \begin{bmatrix} \mu & -j\kappa & 0 \\ j\kappa & \mu & 0 \\ 0 & 0 & 1 \end{bmatrix} \tag{5.21}$$

If it is assumed that the alternating fields do not vary in the z-direction the field components H_r and H_ϕ are readily expressed in terms of E_z by having recourse to Maxwell's first curl equation

$$H_r = j \left[\frac{1}{r} \frac{\partial E_z}{\partial \phi} - j \frac{\kappa}{\mu} \frac{\partial E_z}{\partial r} \right] / \omega \mu_0 \mu_e \tag{5.22}$$

$$H_\phi = -j \left[\frac{\partial E_z}{\partial r} + j \frac{\kappa}{\mu} \frac{1}{r} \frac{\partial E_z}{\partial \phi} \right] / \omega \mu_0 \mu_e \tag{5.23}$$

where

$$\mu_e = \frac{\mu^2 - \kappa^2}{\mu} \tag{5.24}$$

One solution for E_z which satisfies the wave equation is

$$E_z = A_n J_n(k_e r) \exp jn\phi \tag{5.25}$$

except that A_n is now different for n positive and negative. Combining the preceding relationships gives

5.3 MODE CHART OF GYROMAGNETIC RESONATOR

$$H_r = -A_n \zeta_e \left[\frac{n J_n(k_e r)}{k_e r} - \frac{\kappa}{\mu} J'_n(k_e r) \right] \exp(jn\phi) \qquad (5.26)$$

$$H_\phi = -jA_n \zeta_e \left[J'_n(k_e r) - \frac{\kappa n}{\mu} \frac{J_n(k_e r)}{k_e r} \right] \exp(jn\phi) \qquad (5.27)$$

The wave number is defined by

$$k_e = \omega \sqrt{\varepsilon_0 \varepsilon_r \mu_0 \mu_e} \qquad (5.28)$$

and the wave admittance is given by

$$\zeta_e = \frac{1}{\eta_e} = \sqrt{\frac{\varepsilon_0 \varepsilon_r}{\mu_0 \mu_e}} \qquad (5.29)$$

Imposing a magnetic wall boundary condition at $r = R$ gives

$$H_\phi(R) = 0 \qquad (5.30)$$

or

$$J'_n(k_e R) - \frac{\kappa n}{\mu} \frac{J_n(k_e R)}{k_e R} = 0 \qquad (5.31)$$

There are, therefore, two roots for the planar magnetized disk resonator:

$$(k_e R)_{-nj} \qquad (5.32a)$$

$$(k_e R)_{+nj} \qquad (5.32b)$$

The resonant frequencies defined by these two roots are

$$\omega_{+nj} \sqrt{\varepsilon_0 \varepsilon_r \mu_0 \mu_e} \, R = (k_e R)_{+nj} \qquad (5.33a)$$

$$\omega_{-nj} \sqrt{\varepsilon_0 \varepsilon_r \mu_0 \mu_e} \, R = (k_e R)_{-nj} \qquad (5.33b)$$

The resonant frequency of the unmagnetized disk, for which κ/μ is zero, lies between the two split frequencies.

$$\omega_{nj} \sqrt{\varepsilon_0 \varepsilon_r \mu_0 \mu_e} \, R = (kR)_{nj} \qquad (5.34)$$

If κ/μ is small the difference in the $(k_e R)_{nj}$ values of a pair of resonances is determined by

$$(\Delta k_e R)_{nj} \approx \frac{2n(k_e R)_{nj}}{(k_e R)_{nj}^2 - n^2} \cdot \left(\frac{\kappa}{\mu}\right) \tag{5.35}$$

This last equation indicates that the splitting of a pair of resonant frequencies is proportional to κ/μ for κ/μ small. This result is obtained by making use of the identity below

$$\frac{xJ_n'(x)}{J_n(x)} = \left(\frac{x - x_{mn}}{x_{mn}}\right)(x_{mn}^2 - n^2)$$

Introducing this relationship into equation (5.31) gives

$$\left(\frac{x - x_{mn}}{x_{mn}}\right)(x_{mn}^2 - n^2) = \pm n \left(\frac{\kappa}{\mu}\right) \tag{5.36}$$

The required result in equation (5.35) is obtained by solving for the values of x.

Figure 5.6 indicates the mode chart of such a gyromagnetic resonator. Another important result of this sort of problem region is that the field pattern in the ferrite disc is rotated as the resonator is magnetized. The direction in which it is

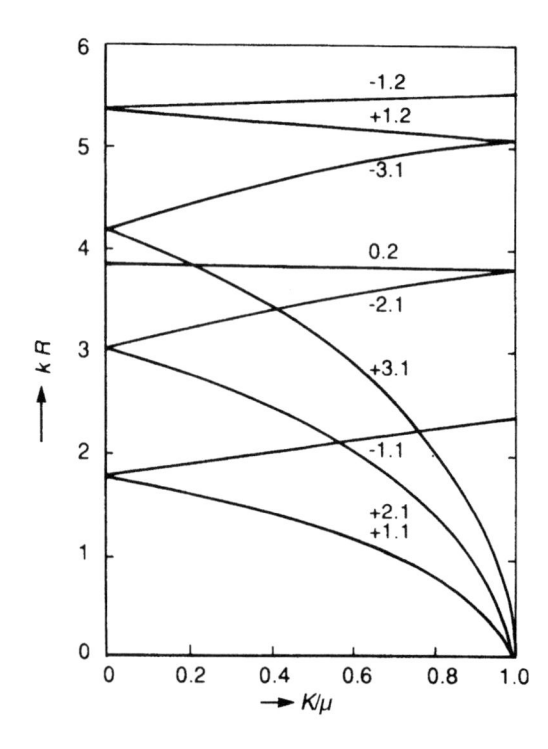

Figure 5.6 Mode chart of gyromagnetic disc resonator (reproduced with permission, Bosma, H. (1964) On stripline Y-circulation at UHF, **MTT-12**, 61–72)

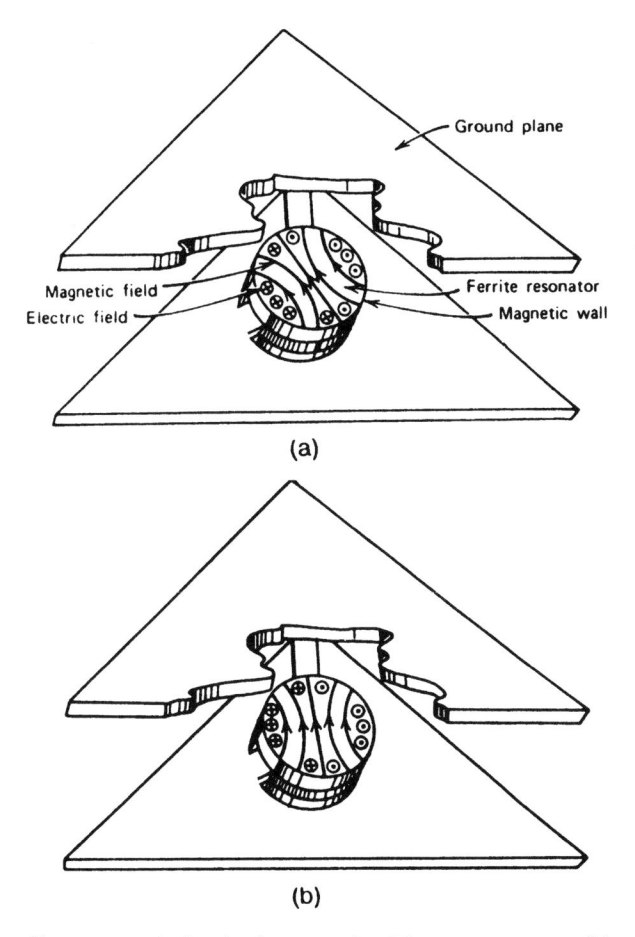

Figure 5.7 (a) Standing wave solution in demagnetized 3-port resonator; (b) standing wave solution in magnetized resonator

rotated is determined by that of the direct magnetic field. If it is rotated by 30° one of the ports is completely decoupled, and transmission occurs between the other two ports. In this configuration the 3-port junction behaves as a circulator. This is indicated in Figure 5.7.

References and Further Reading

Bosma, H. (1964) On stripline Y-circulation at UHF, *IEEE Trans. Microwave Theory Tech.*, **MTT-12**, 61–72.

Chew, W. C. and Kong, J. A. (1980) Resonance of the axial-symmetric modes in microstrip disk resonators, *J. Math. Phys.*, **21**(3), 582–591.

Helszajn, J. and James, D. S. (1978) Planar triangular resonators with magnetic walls, *IEEE Trans. Microwave Theory Tech.*, **MTT-26**(2), 95–100.

Irish, T. R. (1972) Annular resonant structures and their users as microwave filters, *Radio and Electronic Eng.*, **42**(2), 85–90.

Itoh, T. (1974) Analysis of microstrip resonators, *IEEE Trans.*, **MTT-22**, 946–952.

Watkins, J. (1969) Circular resonant structures in microstrip, *Electronics Lett.*, **5**, 524–525.

Wolff, I. and Knoppik, N. (1974) Rectangular and circular microstrip disk capacitors and resonators, *IEEE Trans.*, **MTT-22**, 857–864.

6

OKADA RESONATOR

6.1 Introduction

Conventional waveguide junction circulators either employ a single quarter-wave long gyromagnetic resonator or a pair of such resonators in a turnstile arrangement or a planar post one in contact with the top and bottom walls at the junction of three H-plane waveguides. Another geometry that has much to comment it is that obtained by suppressing the axial resonance in the geometry using either the single or the pair of coupled quarter-wave long resonators. The schematic diagrams of two possible arrangements are indicated in Figures 6.1 and 6.2. The quasi-planar resonator formed in this way is particularly attractive for the design of high-mean power devices in that its surface area is larger than that of the original prototype and that the temperature drop across its thickness is reduced. It also has the merit that its insertion loss is if anything less than that associated with the corresponding homogeneous arrangement. Some of the considerations entering into the description of the resonator are average and peak power ratings, cooling provisions, magnetic and dielectric losses, gain-bandwidth or quality factor. In this sort of structure the filling factor is fixed by the quality factor of the resonator, the magnetic parameters of the ferrite material by the magnetic losses, the gap between the ferrite layers by the peak power specification. Circulators using such resonators have been tested at 6000 W-CW at S-band, 1500 W-CW at X-band and 400 W-CW at J-band.

One way to increase the power rating of a planar resonator still further is to have recourse to the Okada geometry. It consists of stacked circular metal plates on which are mounted ferrite disks separated by a dielectric or free space region. The number of chambers in this arrangement is fixed by the average power rating of the device. Such resonators have been designed to handled 1.3 MW-CW at 350 MHz.

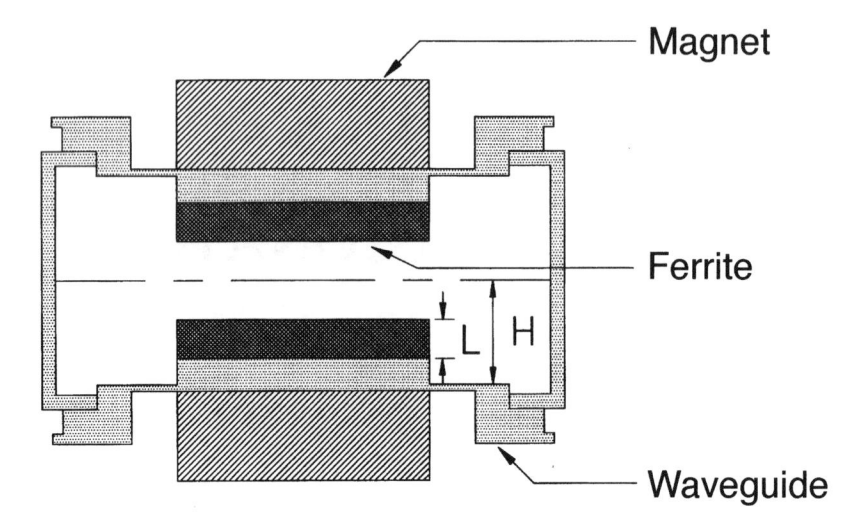

Figure 6.1 Schematic diagram of symmetric planar resonator

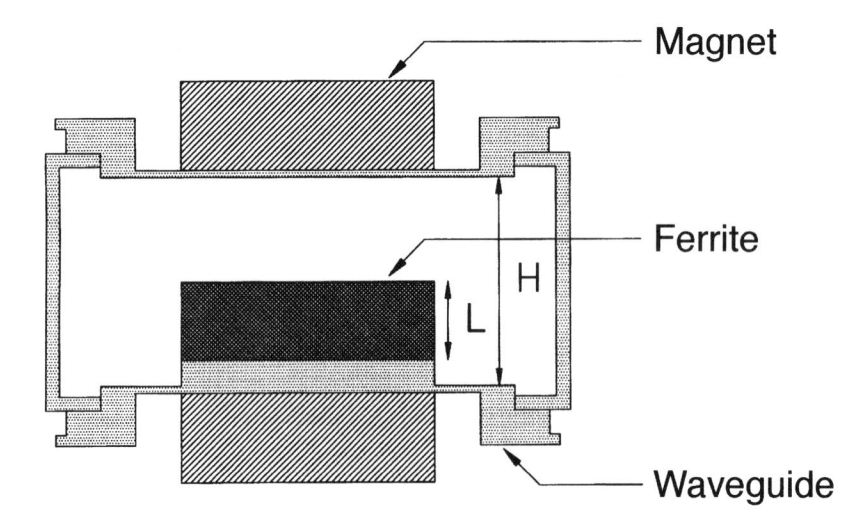

Figure 6.2 Physical details of single chamber planar resonator

6.2 Effective Constitutive Dielectric Constant

The description of an inhomogeneous quasi planar resonator usually proceeds as in the case of the related homogeneous problem region once effective constitutive parameters are introduced. The derivation of the effective dielectric constant of a resonator composed of two thin planar regions between circular metal plates begins by either satisfying the conservation of capacitance of the overall assembly or by having recourse to a transverse resonance formulation. Adopting the usual

6.2 EFFECTIVE CONSTITUTIVE DIELECTRIC CONSTANT

approximation to the latter problem, but replacing ε_f by ε_{eff} gives the standard result.

$$\frac{\varepsilon_{eff} k_0}{\beta_0} \cot(\beta_0 L) - \frac{\varepsilon_d k_0}{\alpha_0} \coth \alpha_0 (H - L) = 0 \tag{6.1}$$

where

$$\left(\frac{\beta_0}{k_0}\right)^2 = \varepsilon_{eff} - \left(\frac{k_{mn}}{k_0}\right)^2 \tag{6.2}$$

$$\left(\frac{\alpha_0}{k_0}\right)^2 = \left(\frac{k_{mn}}{k_0}\right)^2 - \varepsilon_d \tag{6.3}$$

ε_{eff} is the effective dielectric constant of the dielectric or gyromagnetic waveguide; it is obtained from either measurement or calculation. ε_d is that of the relative dielectric constant of the spacers. For the dominant mode considered here

$$k_{mn} = \left(\frac{1.84}{R}\right) \tag{6.4}$$

The derivation of the planar solution begins by approximating the ordinary and hyperbolic trigonometric tangent quantities in the transverse resonance condition by the appropriate small angle descriptions. This gives

$$\left(\frac{\varepsilon_{eff}}{\beta_0}\right)\left(\frac{\alpha_0(1-k)}{k}\right) = \left(\frac{\varepsilon_d}{\alpha_0}\right)(\beta_0) \tag{6.5}$$

The axial electric filling factor is related to the dimensions of the gyromagnetic and dielectric layers of the half-space bounded by the top and bottom circular plates by

$$k = \frac{L}{H} \tag{6.6}$$

This equation may be readily solved for $k_0 R$ by expressing α_0 and β_0 in terms of the original variables. The required result is

$$(k_0 R)^2 \varepsilon_e(\text{eff}) = (1.84)^2 \tag{6.7}$$

where

$$\varepsilon_e(\text{eff}) = \frac{\varepsilon_{\text{eff}}\varepsilon_d}{\varepsilon_{\text{eff}} - k(\varepsilon_{\text{eff}} - \varepsilon_d)} \quad (6.8)$$

The former equation is recognized as the cut-off number of the dominant mode of a disk planar circuit and the latter one as the effective dielectric constant of a two-dielectric planar capacitor. The latter quantity may of course be directly constructed by forming the overall capacitance of a two-layer dielectric structure and satisfying the conservation in capacitance in the assembly

$$\frac{H}{\varepsilon_0\varepsilon_e(\text{eff})A} = \frac{L}{\varepsilon_0\varepsilon_{\text{eff}}A} + \frac{(H-L)}{\varepsilon_0 A} \quad (6.9)$$

The radial wave-number is also sometimes written in terms of an effective wave-number

$$k_{\text{eff}} = k_0 \sqrt{\varepsilon_e(\text{eff})} \quad (6.10)$$

6.3 Effective Gyrotropy

A knowledge of the effective diagonal and off-diagonal entries of the tensor permeability are also necessary for the description of an inhomogenous gyromagnetic planar circuit. These quantities may again be deduced by having recourse to the transverse resonance condition without ado. The derivation of the required result starts by constructing the small angle condition of the problem region with β_0 replaced by $\beta_\pm$. Adopting the perturbation formulation of the split phase constants gives

$$\varepsilon_{\text{eff}}\left[\left(\frac{1.84}{k_\pm R}\right)^2 - \varepsilon_d\right]\left(\frac{1-k}{k}\right) = \varepsilon_d\left[\varepsilon_{\text{eff}}(\mu \mp C_{11}\kappa) - \left(\frac{1.84}{k_\pm R}\right)^2\right] \quad (6.11)$$

where

$$C_{11} = \frac{2}{(k_{\text{eff}}R)^2 - 1} \quad (6.12)$$

Rearranging the preceding relationship readily give the desired result

$$(k_\pm R)^2[\mu(\text{eff}) \mp C_{11}\kappa(\text{eff})]\varepsilon_e(\text{eff}) = (1.84)^2 \quad (6.13)$$

where

$$\mu(\text{eff}) = 1 + k(1-\mu) \quad (6.14)$$

$$\kappa(\text{eff}) = k\kappa \quad (6.15)$$

and $\varepsilon_e(\text{eff})$ has the meaning met in connection with the dielectric problem region.

6.3 EFFECTIVE GYROTROPY

The nature of the effective diagonal element of the tensor permeability may be derived from first principles, or it may be obtained by replacing $\varepsilon_e(\text{eff})$ by $1/\mu_e(\text{eff})$ and $\varepsilon(\text{eff})$ by $1/\mu$ in the description of $\varepsilon_e(\text{eff})$.

When the isotropic resonator consists of a demagnetized ferrite one instead of a dielectric one then the radial wavenumber in equation (6.7) must be replaced by

$$(k_0 R)^2 \, \mu_e(\text{eff}) \, \varepsilon_e(\text{eff}) = (1.84)^2 \tag{6.16}$$

where

$$\mu_e(\text{eff}) = \frac{\mu(\text{eff})^2 - C_{11}^2 \kappa(\text{eff})^2}{\mu(\text{eff})} \tag{6.17}$$

Another quantity that enters into the description of this sort of problem region is its effective gyrotropy. It is related to the actual one and the filling factor of the planar resonator by

$$\left(\frac{\kappa}{\mu}\right)_{\text{eff}} = \frac{k\mu}{1 + k(\mu - 1)} \left(\frac{\kappa}{\mu}\right) \tag{6.18}$$

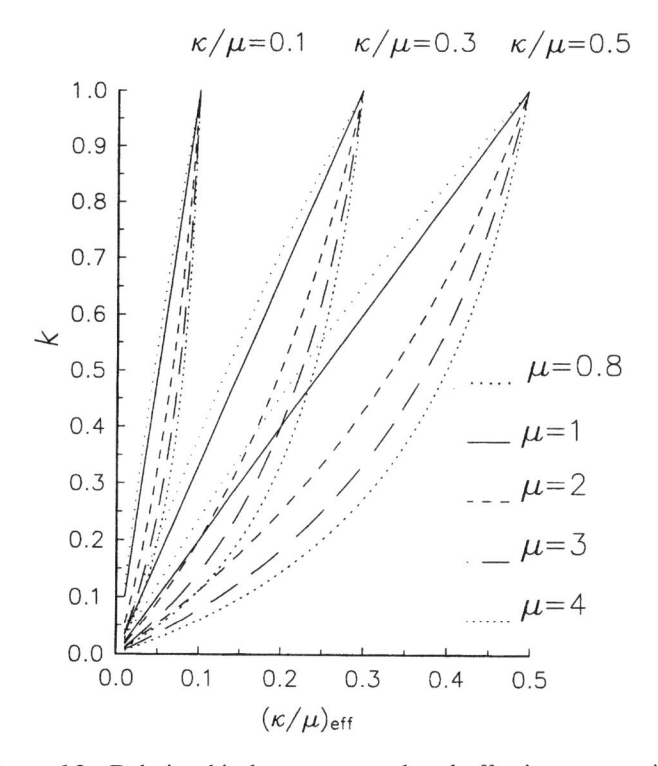

Figure 6.3 Relationship between actual and effective gyrotropies

μ and κ are the actual diagonal and off diagonal elements of the tensor permeability and k is the filling factor.

In a below resonance device μ is usually not very different from unity and

$$\left(\frac{\kappa}{\mu}\right)_{\text{eff}} = k\left(\frac{\kappa}{\mu}\right) \tag{6.19}$$

Values of κ/μ between $0.10 \leq \kappa/\mu \leq 0.70$ are realizable in below resonance devices.

Figure 6.3 indicates the relationship between the effective gyrotropy and the axial filling factor (k) of the resonator for parametric values of the diagonal element of the tensor permeability with the radial filling factor (q) equal to unity. μ is equal to or less than unity below the Kittle resonance and it is equal to or larger than unity above it.

6.4 Quality Factor

In practice, the radius of the device cannot be fixed until the filling factor of the resonator has been determined in terms of the quality factor of the device. This quantity determines its gain bandwidth product. It is related to the split frequencies of the resonator by

$$\frac{1}{Q_L} = \sqrt{3}\left(\frac{\omega_+ - \omega_-}{\omega_0}\right) \tag{6.20}$$

For a weakly magnetized planar disk resonator

$$\frac{\omega_+ - \omega_-}{\omega_0} = \frac{2}{(k_{\text{eff}}R)^2 - 1}\left(\frac{\kappa}{\mu}\right)_{\text{eff}}$$

and

$$\frac{1}{Q_L} = \frac{2\sqrt{3}}{(k_{\text{eff}}R)^2 - 1}\left(\frac{\kappa}{\mu}\right)_{\text{eff}} \tag{6.21}$$

$k_{\text{eff}}R$ is 1.84 in keeping with equation (6.7).

Values of Q_L between 2 and 2½ are suitable for the design of quarter-wave coupled devices with modest specifications. This gives

$$0.27 \leq \left(\frac{\kappa}{\mu}\right)_{\text{eff}} \leq 0.34$$

The resonant frequencies of the split counter-rotating modes of a magnetized resonator may be shown to coincide with those at which the VSWR at one port with the other two ports terminated are defined by

$$VSWR = 2 \tag{6.22}$$

6.5 Effective Dielectric and Magnetic Loss Tangents of Planar Resonators

A complete representation of a planar composite resonator not only requires a definition of the constitutive parameters of the circuit but also a knowledge of either its effective dielectric and magnetic loss tangents or unloaded Q-factors. The derivation of the effective dielectric loss tangent commences with the definition of the effective dielectric constant in equation (6.8) with ε_d equal to unity.

$$\frac{1}{\varepsilon_{\text{eff}}} = \frac{k}{\varepsilon_r} + (1-k) \tag{6.23}$$

It proceeds by introducing the usual complex quantities for ε_{eff} and ε_r to represent dissipation effects

$$\varepsilon_{\text{eff}} = \varepsilon'_{\text{eff}} - j\varepsilon''_{\text{eff}} \tag{6.24a}$$

$$\varepsilon_r = \varepsilon'_r - j\varepsilon''_r \tag{6.24b}$$

Introducing these relationships into the definition of ε_{eff} indicates that

$$\frac{\varepsilon'_{\text{eff}} + j\varepsilon''_{\text{eff}}}{(\varepsilon'_{\text{eff}})^2 + (\varepsilon''_{\text{eff}})^2} = \frac{k(\varepsilon'_r + j\varepsilon''_r)}{(\varepsilon'_r)^2 + (\varepsilon''_r)^2} + (1-k)$$

The real and imaginary parts of the preceding equation are now given by

$$\frac{\varepsilon'_{\text{eff}}}{(\varepsilon'_{\text{eff}})^2 + (\varepsilon''_{\text{eff}})^2} \approx \frac{1}{\varepsilon'_{\text{eff}}}$$

$$\frac{\varepsilon''_{\text{eff}}}{(\varepsilon'_{\text{eff}})^2 + (\varepsilon''_{\text{eff}})^2} \approx \frac{k\varepsilon''_r}{(\varepsilon'_{\text{eff}})^2}$$

The dielectric loss tangent is then defined by

$$\tan \delta'_d = \frac{\varepsilon''_{\text{eff}}}{\varepsilon'_{\text{eff}}} = k \left(\frac{\varepsilon''_r}{\varepsilon'_r}\right)\left(\frac{\varepsilon'_r}{\varepsilon'_{\text{eff}}}\right) \tag{6.25}$$

The effective dielectric loss tangent of a quasi-planar inhomogeneous resonator is therefore reduced by a factor of

$$k\left(\frac{\varepsilon'_r}{\varepsilon'_{\text{eff}}}\right)$$

compared to that of an homogeneous one.

The derivation of the effective magnetic loss tangent in a magnetic material also starts with the definition of its permeability in equation (6.14). It again proceeds by introducing complex constitutive parameters to represent dissipation effects

$$\mu_e = \mu'_e - j\mu''_e \tag{6.26a}$$

$$\mu_{eff} = \mu'_{eff} - j\mu''_{eff} \tag{6.26b}$$

The real and imaginary parts of the complex effective permeability can therefore be written in terms of the complex constitutive ones as

$$\mu'_{eff} - j\mu''_{eff} - 1 + k(\mu'_e - j\mu''_e - 1) \tag{6.27}$$

The derivation of the required result now continues by equating the real and imaginary parts of this equation, This gives

$$\mu'_{eff} = 1 + k(\mu'_e - 1) \tag{6.28a}$$

$$\mu''_{eff} = k\mu''_e \tag{6.28b}$$

The magnetic loss tangent is in this instance given by

$$\tan \delta'_m = \frac{\mu''_{eff}}{\mu'_{eff}} = k\left(\frac{\mu'_e}{\mu'_{eff}}\right)\tan \delta_m \tag{6.29}$$

The effective magnetic loss tangent of a composite resonator is therefore reduced by a factor

$$k\left(\frac{\mu'_e}{\mu'_{eff}}\right)$$

compared to that of an homogeneous arrangement.

6.6 Effective Unloaded Quality Factor of Planar Resonator

The unloaded Q-factor of such a composite resonator may be obtained once the effective loss tangents are at hand, it is defined in terms of the effective dielectric and magnetic loss tangents in the usual way by

$$\frac{1}{Q'_U} = \frac{1}{Q'_m} + \frac{1}{Q'_d} \tag{6.30}$$

where

$$Q'_m = \frac{1}{\tan \delta'_m} \qquad (6.31)$$

$$Q'_d = \frac{1}{\tan \delta'_d} \qquad (6.32)$$

$\tan \delta'_d$ and $\tan \delta'_m$ are defined by equations (6.25) and (6.29)

6.7 Voltage Breakdown

An estimate of the peak power rating of a planar disk resonator containing two different dielectric regions may be established under the simplifying condition that its electric field resembles that of a two-region parallel-plate capacitor. Assuming that the gap region breaks down first, the derivation of the required result starts by forming the voltage in the gap in terms of the applied one.

$$V_{gap} = \frac{C_d}{C_g + C_d} V_{applied} \qquad (6.33)$$

where

$$C_d = \frac{\varepsilon_r \varepsilon_0 A}{L_f}, \quad \varepsilon_r \neq 1 \qquad (6.34)$$

$$C_{gap} = \frac{\varepsilon_0 A}{H_r - L_f}, \quad \varepsilon_r = 1 \qquad (6.35)$$

and A is the cross-sectional area of the resonator.
The relationship between the applied and gap voltages is therefore given by

$$V_{gap} = \frac{\varepsilon_r (1 - k)}{\varepsilon_r + k(1 - \varepsilon_r)} \cdot V_{applied} \qquad (6.36)$$

k is the usual filling factor of the resonator

$$k = \frac{L_f}{H_r} \qquad (6.37)$$

In problems of this sort it is usual to deal with the electric field strength in the gap instead of the voltage there. The relationship between the two is

$$E_{gap} = \frac{V_{gap}}{H_r - L_f}, \quad \text{V/m} \tag{6.38}$$

The average electric field strength between the top and bottom plates of a two region arrangement is separately specified by

$$E_{applied} = \frac{V_{applied}}{H_r}, \quad \text{V/m} \tag{6.39}$$

The electric field strength in the gap is now constructed by replacing the voltage variables by the corresponding fields quantities. This gives

$$(H_r - L_f) E_{gap} = \frac{\varepsilon_r(1-k)}{\varepsilon_r + k(1-\varepsilon_r)} (H_r) E_{app} \tag{6.40}$$

or

$$E_{gap} = \varepsilon_{eff} E_{app} \tag{6.41}$$

where ε_{eff} is the effective dielectric constant of the arrangement given by equation (6.1).

When the gap is very small $L_f \to H_r$, $H_r - L_f \to 0$, $k \to 1$ and $\varepsilon_{eff} \approx \varepsilon_r$, so that

$$E_{gap} \to \varepsilon_r E_{app} \tag{6.42}$$

For $k = 0.35$ (say)

$$\varepsilon_{eff} \approx 1$$

and

$$E_{gap} \approx E_{app} \tag{6.43}$$

In order to proceed with a calculation of the actual gap it is necessary to have the value at which the electric field breakdowns in air. It is given by

$$E_{rms} = 2.5 \text{ kV/mm}$$

If, for example, E_{gap} should not reach 3.0 kV/mm for a material with a relative dielectric constant of $\varepsilon_r = 15$, E_{app} must not exceed 2 kV/mm.

The breakdown strength of air under uniform field condition does not apply at the edges of electrodes. One way to cater for this difficulty is to make the radius of any curvature at least twice the gap distance, but this is a rough engineering guide. A more exact guide may be obtained by having recourse to Paschen's Law curve.

The development of the peak power rating of a resonator arrangement consisting of two ferrite regions separated by a dielectric one follows without ado.

6.8 Power Rating of Multilayer Resonators

The design of any device is not completed until its average and peak power ratings are established. The peak power rating is determined by breakdown or arcing and by the onset of spinwave instability. The average power rating by the permissible temperature drop ΔT between the open face of each ferrite element and the heat sink. It is given by

$$\Delta T = \frac{P_{abs} L}{2nk_t A} \qquad (6.44)$$

L is the thickness of each ferrite element (calculated using a two layer prototype), A is the surface area, defined by the region within which the dissipation in the resonator takes place, k_t is the thermal conductivity, n is the number of ferrite doublets. The factor 2 is an empirical constant to cater for the fact that all the power is not necessarily dissipated at the open face, P_{abs} is the power dissipated, in each ferrite element

$$P_{abs} = \frac{P_{in}}{2n}\left[1 - 10^{L(dB)/10}\right], \quad \text{Watts} \qquad (6.45)$$

$L(dB)$ is the insertion loss in decibels which is here assumed a negative quantity. A typical value of the thermal conductivity of a ferrite material is

$$k_t = 0.0063 \text{ Wmm}^{-1} \, °\text{C}^{-1}$$

Combining the two proceeding relationships indicates that the power rating of such a structure is related to its insertion loss by

$$P_{in} = \frac{4n^2 k \Delta T}{[1 - 10^{L(dB)/10}]} \left(\frac{A}{L}\right), \quad \text{Watts} \qquad (6.46)$$

The power rating of the device is therefore fixed by ΔT, n, L (dB) and A/L. Taking

$$\Delta T = 35°\text{C}$$

$$k_t = 0.0063 \text{ W mm}^{-1} \, °\text{C}^{-1}$$

$$A/L = 1500 \text{ mm}$$

and a worst case condition for L (dB) of

$$L \text{ (dB)} = -0.19 \text{ (say)}$$

gives P_{in} as $P_{in} = 34\,815 \, n^2$, Watts

6.9 Okada Resonator

One way to increase the power rating of a planar resonator is to have recourse to the Okada resonator. This resonator consists of stacked circular metal plates on which are mounted ferrite disks separated by a dielectric or free space region. Figure 6.4 illustrates one structure using four double chambers. The equivalence between this resonator and one using top and bottom plates only may be accomplished by constructing a one-to-one equivalence between the capacitance's of the two arrangements. Figure 6.5 indicates the equivalence between a planar resonator consisting of a pair of dielectric disks and one using two pairs. The equivalence between the two is met provided the effective dielectric constants of the two arrangements are equal. This condition is satisfied if the thickness of each ferrite layer in the reorganised arrangement is half that of the original structure ($L/2$). Reorganizing the parallel plate capacitance in this manner doubles the surface area of the resonator which is in contact with the heat sink and halves its thickness. The power rating of the device is therefore increased by a factor of four. The required equivalence between the two structures will now be demonstrated for a circuit with a filling factor of 0.50 by way of an example. The mapping between the two arrangements starts by introducing an electric wall at the symmetry plane of the original circuit. Since the electric field is perpendicular to

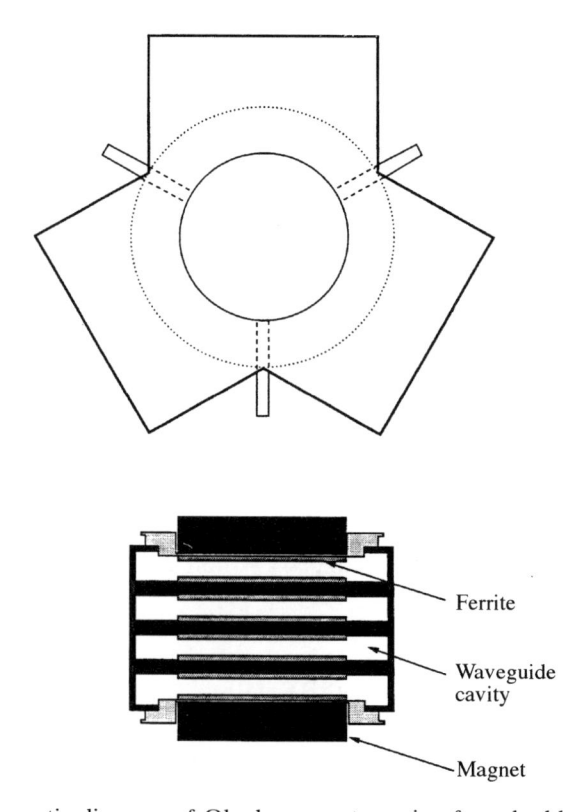

Figure 6.4 Schematic diagram of Okada resonator using four double chambers

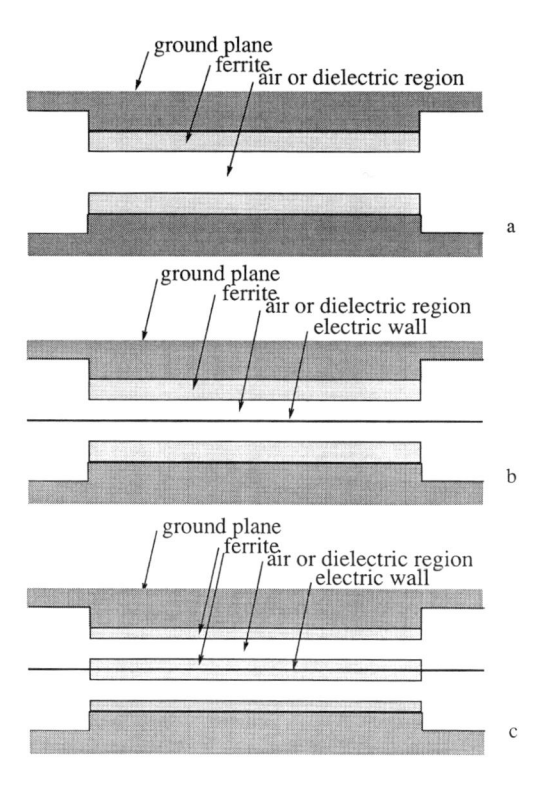

Figure 6.5 Equivalence between one and two Okada chambers

this wall it is unaffected by it. (The magnetic field in such a resonator is parallel to it and so it is also unaffected by the introduction of such a wall.) The derivation is completed by noting that the effective dielectric constant and consequently the capacitance of the circuit is unaltered if the dielectric disks on the top and bottom of the original circuit are equally divided between the original walls and the surfaces revealed by the electric wall. Figure 6.6 depicts the mapping between a two-layered structure and a four-layered one. The power rating of the device is in this instance increased by a factor of sixteen compared to that of the original prototype. The power rating of this sort of circuit is therefore in general increased by a factor of n^2. In a practical arrangement, the infinitely thin dielectric walls are replaced by conducting sheets which are water cooled. Figure 6.7 depicts one possibility.

6.10 Okada Resonator using Composite Resonators

A means of increasing the power rating of this gyromagnetic resonator still further is to use a composite structure consisting of a ferrite disk and a dielectric ring. The principle of this geometry relies on the fact that the outer rim of the resonator

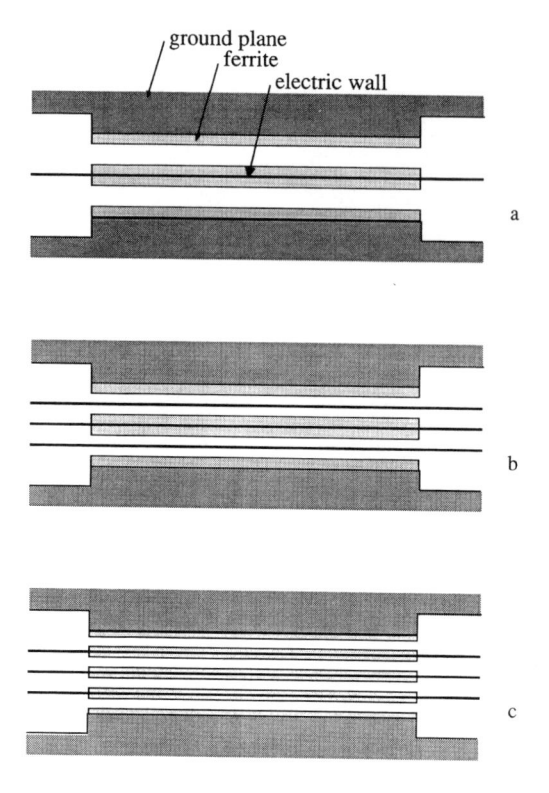

Figure 6.6 Equivalence between two and four Okada chambers

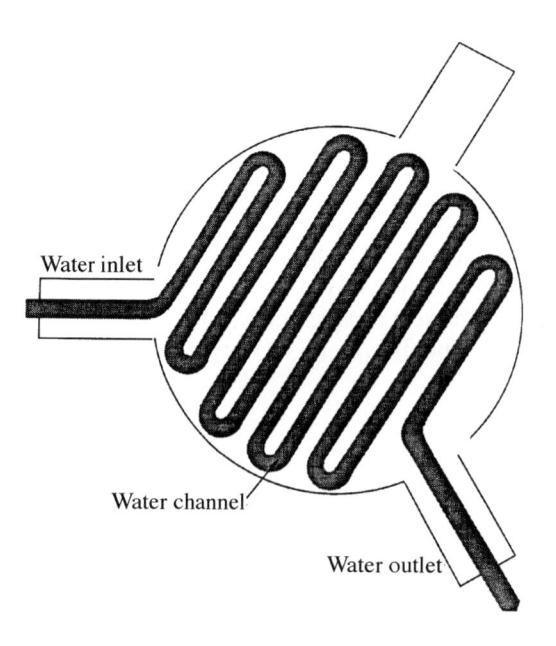

Figure 6.7 Cooling geometry of struts

6.10 OKADA RESONATOR USING COMPOSITE RESONATORS

does not to first order contribute to the gyrotropy of the junction. It may therefore be replaced by a dielectric material with a similar value of dielectric constant as that of the ferrite without unduly perturbing its operation. The basis of this technique is the fact that the thermal conductivity of a dielectric material such as alumina or berylia oxide is much higher than that of a ferrite one. It also has the merit of reducing the cross-sectional area of the magnetic circuit. If the gyromagnetic region of the resonator is assumed to reside within 0.707 of the outside radius then the number of ferrite tiles is reduced by a factor of 2. This arrangement is depicted in Figure 6.8. Still another possibility that may have some value is to embody magnetic tiles between the inner gyromagnetic ones and the outer dielectric ones in order to focus the direct magnetic field. These tiles may be constructed from a similar material to that of the gyromagnetic ones except for its magnetization. The value of this quantity is then chosen such that its Kittel

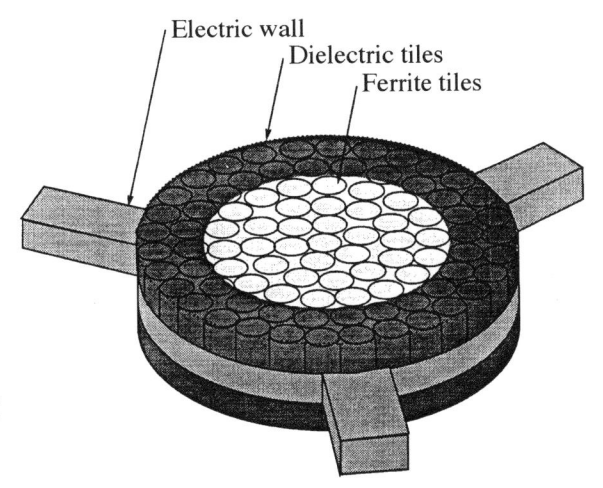

Figure 6.8 Schematic diagram of composite resonator using ferrite and dielectric tiles

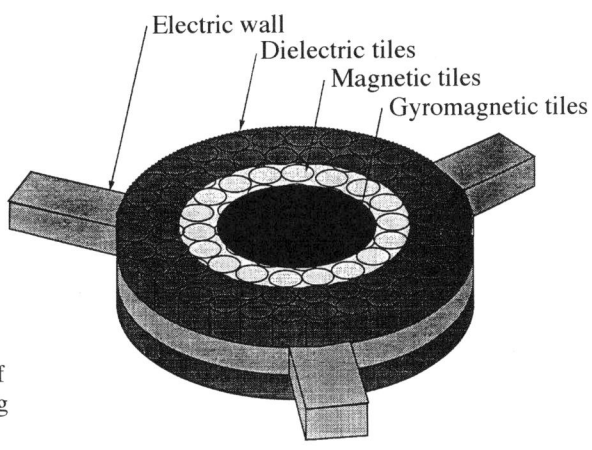

Figure 6.9 Schematic diagram of composite resonator with focusing ferrite tiles

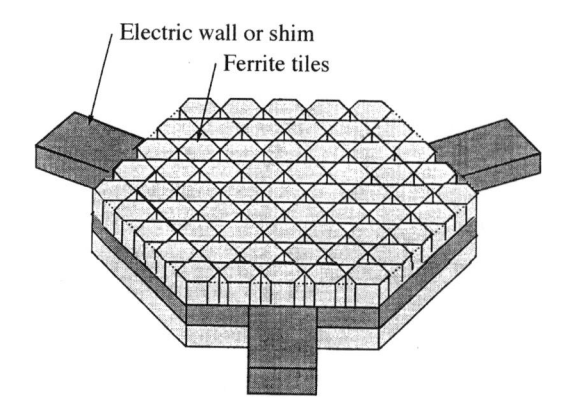

Figure 6.10 Schematic diagram of hexagonal resonator using ferrite tiles (reprinted with permission, Helszajn, J., Guixa, R., Girones, J., Hoyos, E. and Garca-Taheno, J. (1995) Adjustment of Okada resonator using composite chambers, *IEEE Trans. MTT*, **MMT-43**, 2524–2531)

line resides at a lower direct magnetic field at the operating frequency of the resonator from that of the gyromagnetic region. This geometry is illustrated in Figure 6.9.

The size of the planar circular or hexagonal plates required for the fabrication of planar resonators at UHF frequencies poses, in practice, a manufacturing problem in that it is not possible to fire such large ferrite assemblies. Another difficulty, in the design of high mean power devices is that large ferrite surfaces bounded to ground planes are likely to shutter under thermal shock. One way to avoid both difficulties is to replace the ferrite substrate by a number of circular or hexagonal tiles. This approach has also the merit that tiles may be replaced if damaged. The use of a planar disk resonator in the design of a junction circulator is of course not unique. Another possibility is a regular hexagonal shape, still other possibilities are irregular hexagonal and triangular ones. Figure 6.10 illustrates a regular hexagonal arrangement.

Rubber based adhesive is usually employed in the design of high power devices to bond the ferrite tiles to metal surfaces.

6.11 Experimental Okada Resonator Mode Chart

The experimental mode chart of one Okada resonator at the junction of three WR2300 waveguides is illustrated in Figure 6.11. It is obtained by recording the frequencies of the split absorption lines at one port in the return loss of the device with the other two ports terminated in matched loads. The geometry considered here consisted of five inner chambers using pairs of ferrite layers and two outer ones using single ones. Of note is that each branch displays four split frequencies in keeping with the four symmetry planes of the direct magnetic field along the axis if the resonator. These symmetry planes may be understood in connection with the constant direct magnetic field contours of the electromagnet illustrated in Figure 6.12. This suggests that the middle of the chamber resonates at f_0, the inner pair of chambers at $f_0 + \Delta f_0$, the next pair of chambers at $f_0 + 2\Delta f_0$ and the outer two chambers at $f_0 + 3\Delta f_0$.

6.11 EXPERIMENTAL OKADA RESONATOR MODE CHART

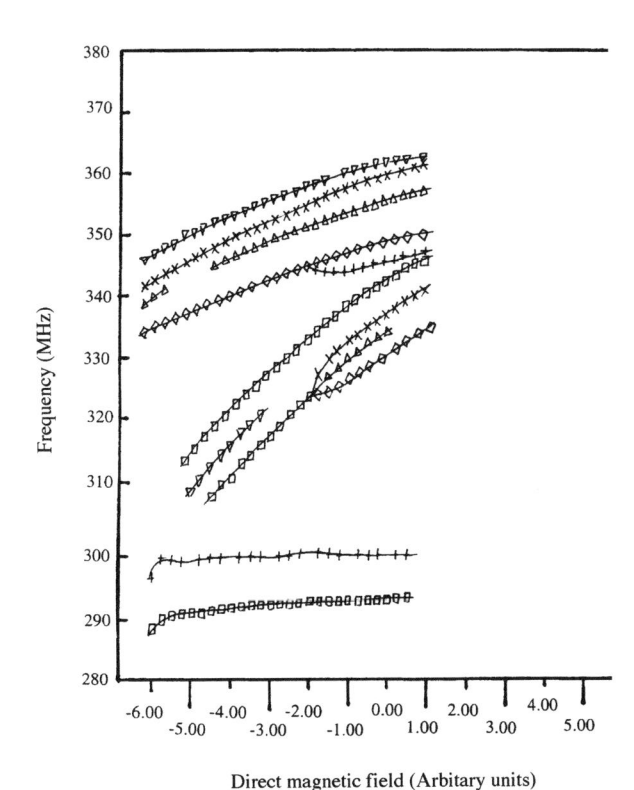

Figure 6.11 Experimental split frequencies of Okada composite resonator using five inner double chambers and two outer single ones (reprinted with permission, Helszajn, J., Guixa, R., Girones, J., Hoyos, E. and Garca-Taheno, J. (1995) Adjustment of Okada resonator using composite chambers, *IEEE Trans. MTT*, **MTT-43**, 2524–2531)

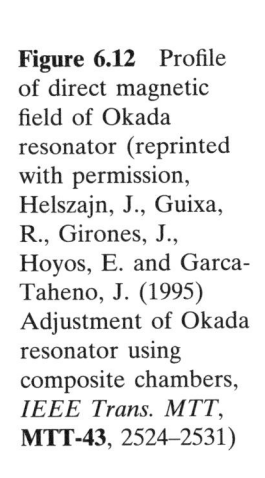

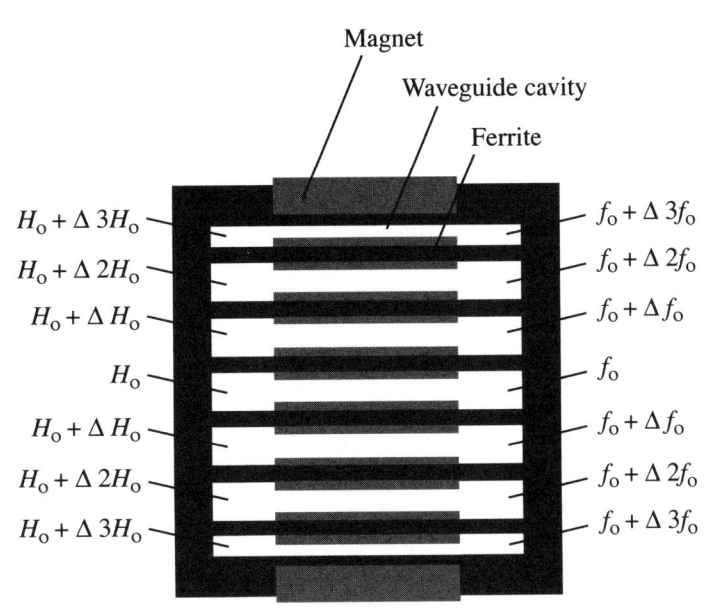

Figure 6.12 Profile of direct magnetic field of Okada resonator (reprinted with permission, Helszajn, J., Guixa, R., Girones, J., Hoyos, E. and Garca-Taheno, J. (1995) Adjustment of Okada resonator using composite chambers, *IEEE Trans. MTT*, **MTT-43**, 2524–2531)

$H_o + \Delta 3H_o$ — $f_o + \Delta 3f_o$
$H_o + \Delta 2H_o$ — $f_o + \Delta 2f_o$
$H_o + \Delta H_o$ — $f_o + \Delta f_o$
H_o — f_o
$H_o + \Delta H_o$ — $f_o + \Delta f_o$
$H_o + \Delta 2H_o$ — $f_o + \Delta 2f_o$
$H_o + \Delta 3H_o$ — $f_o + \Delta 3f_o$

Since the mode chart has been split by the direct field profile of the resonator the average upper and lower split frequencies must be used for calculations.

Scrutiny of the mode chart in Figure 6.11 gives one possible solution between any pair of split modes

$$(f_0) \text{ average} \approx 325 \text{ MHz}$$

$$(f_+) \text{ average} \approx 335 \text{ MHz}$$

$$(f_-) \text{ average} \approx 307 \text{ MHz}$$

The applied direct magnetic field was established by superimposing a small direct magnetic field using a coil arrangement on a re-entrant magnetic structure.

The quality factor is

$$\frac{1}{Q_L} = \sqrt{3}\left(\frac{\omega_+ - \omega_-}{\omega_0}\right)$$

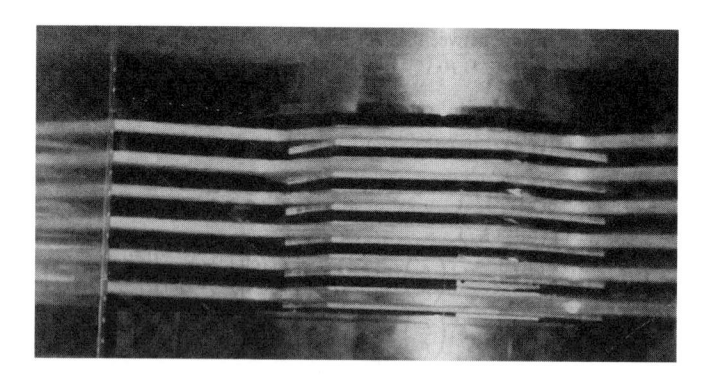

Figure 6.13 Photo of Okada 3-port junction circulator (reprinted with permission, Helszajn, J., Guixa, R., Girones, J., Hoyos, E. and Garca-Taheno, J. (1995) Adjustment of Okada resonator using composite chambers, *IEEE Trans. MTT*, **MTT-43**, 2524–2531)

Figure 6.14 Photo of Okada resonator (reprinted with permission, Helszajn, J., Guixa, R., Girones, J., Hoyos, E. and Garca-Taheno, J. (1995) Adjustment of Okada resonator using composite chambers, *IEEE Trans. MTT*, **MTT-43**, 2524–2531)

or

$$Q_L = 6.25$$

This compares with a calculated value of $Q_L = 5.14$ at the same value of direct magnetic field.

Figure 6.13 illustrates a photo of a practical 320 MHz Okada junction in WR2300 waveguide using five inner double chambers and two single outer ones. Figure 6.14 indicates the resonator in some more detail.

References and Further Reading

Buffler, C. R. and Helszajn, J. (1968) The use of composite junctions in the design of high power circulators, *IEEE Intl., Microwave Symp.* **MTT, 20–20**, 239–249.

Fay, C. E. and Comstock, R. L. (1965) Operation of the ferrite junction circulator, *IEEE Trans. Microwave Theory Tech.*, **MTT-13**, 15–27.

Hauth, W., Lenz, S. and Pivit, E. (1986) Design and realization of very-high-power waveguide junction circulators, *Frequency*, **40**, 2–11.

Helszajn, J. (1967) An H-plane high-power TEM ferrite junction circulator, *Radio and Electron. Eng.*, **33**, 257–261.

Helszajn, J. (1970) Frequency and bandwidth of *H*-plane TEM junction circulator, *Proc. IEE*, **117**(7), 1235–1238.

Helszajn, J. (1973) Microwave measurement of circulator network, *IEEE Trans. Microwave Theory Tech.*, **MTT-21**, 347–53.

Helszajn, J. (1974) Composite stripline junctions using ferrite disks and dielectric rings, *IEEE Trans. Microwave Theory Tech.*, **MTT-22**, 400–410.

Helszajn, J., Walker, P. N. and Davidson, E. (1980) Design of low loss waveguide circulators at large peak and mean power levels, *Military Microwave*, 1–7.

Helszajn, J., Guixa, R., Girones, J., Hoyos, E. and Garca-Taheno, J. (1995) Adjustment of Okada resonator using composite chambers *IEEE Trans. Microwave Theory Tech.*, **MTT-43**, 2524–2531.

Konishi, Y. (1969) A high power uhf circulator, *IEEE Trans. Microwave Theory Tech.*, **MTT-15**, 700–708.

Okada, F. and Ohwit, K. (1975) The development of a high power microwave circulator for use in breaking of concrete and rock, *J. of Microwave Power*, **10**(2).

Okada, F. and Ohwit, K. (1978) Design of a high power CW Y-junction waveguide circulator, *IEEE Trans. Microwave Theory Tech.*, **MTT-26**, 364–369.

Okada, F. and Ohwit, K. (1981) High-power circulators for industrial processing systems, *IEE Trans. Magnetics*, **MAG-17**, 2957–2960.

Okada, F., Ohwit, K., Mori, M. and Yasude, M. (1977) A 100 kW waveguide Y-junction circulator for microwave power systems at 915 MHz, *J. of Microwave Power*, **12**(3).

Okada, F., Ohwit, K. and Yasude, M. (1977) Design procedure of high power CW Y-junction waveguide circulators', *IMPI*, 118–120.

Pucel R. Q. and Masse, D. (1972) Microstrip propagation on magnetic substrates – Part 1: design theory, *IEEE Trans. Microwave Theory Tech.*, **MTT-20**, 304–308.

7

ISOTROPIC, ANISOTROPIC AND GYROMAGNETIC CIRCULAR WAVEGUIDES

7.1 Introduction

A waveguide that exhibits degenerate counter-rotating circularly polarized magnetic fields along it axis is the dielectric-filled circular waveguide with electric and magnetic walls propagating the TE_{11} and TM_{11} modes respectively. These two solutions are first examined as a preamble to treating the problems of propagation in open dielectric waveguides in Chapter 8. It is now understood that this type of degeneracy may be split by perturbing the waveguide by a suitably magnetized ferrite medium. These types of waveguides therefore exhibit so-called Faraday rotation. Such gyromagnetic waveguides may also be employed in conjunction with quarter-wave plates to fabricate ferrite phase shifters and other non-reciprocal devices. Propagation in round waveguides with magnetic and electric walls filled with an anisotropic ferrite medium are also described. Scrutiny of these two problem regions indicate that the cut-off number of the dominant TE_{11} mode in a round wave-guide with an electric wall is unaltered if the transverse permeability is made unequal to that along the longitudinal co-ordinate; the cut-off number of the dominant TM_{11} mode in a similar waveguide, with a magnetic wall is, however, modified under the same circumstances. The phase constants of both types of waveguides are, on the other hand, split. A comparison between the cut-off numbers and propagation constants of the dominant mode of ansotropic and gyromagnetic waveguides indicates that the two solutions are qualitatively alike.

7.2 Maxwell's Equations in Cylindrical Coordinates

A property of an open or inhomogenous waveguide is that the transverse fields in each region may be constructed from a knowledge of E_z and H_z in the appropriate regions. This is a general result that will now be demonstrated as a preamble

7 ISOTROPIC, ANISOTROPIC, GYROMAGNETIC CIRCULAR WAVEGUIDES

to studying the closed and open circular waveguides. The derivation begins by splitting the alternating fields and vector operators into longitudinal and transverse components with γ replaced by $j\beta$:

$$\bar{E} = (\bar{E}_t + \bar{a}_z E_z) \exp(-j\beta_r) \tag{7.1}$$

$$\bar{H} = (\bar{H}_t + \bar{a}_z H_z) \exp(-j\beta_r) \tag{7.2}$$

$$\nabla = \nabla_t - j\bar{a}_z \beta_r \tag{7.3}$$

It proceeds by substituting these identities into Maxwell's first and second curl and divergence equations. The result is

$$(\nabla_t \times \bar{H}_t) - j\beta(\bar{a}_z \times \bar{H}_t) + (\nabla_t \times \bar{a}_z) H_z = j\omega\varepsilon_r\varepsilon_0 (\bar{E}_t + \bar{a}_z E_z) \tag{7.4}$$

$$(\nabla_t \times \bar{E}_t) - j\beta(\bar{a}_z \times \bar{E}_t) + (\nabla_t \times \bar{a}_z) E_z = -j\omega\mu_0 (\bar{H}_t + \bar{a}_z H_z) \tag{7.5}$$

$$\nabla_t \bar{E}_t - j\beta E_z = 0 \tag{7.6}$$

$$\nabla_t \bar{H}_t - j\beta H_z = 0 \tag{7.7}$$

$\bar{E}_t$ and $\bar{H}_t$ and now expressed in terms of linear combinations of E_z and H_z by eliminating $\bar{E}_t$ and $\bar{H}_t$ in turn in the two curl equations:

$$k_c^2 \bar{H}_t + j\beta \nabla_t H_z + j\omega\varepsilon_0 \varepsilon_r (\bar{a}_z \times \nabla_t) E_z = 0 \tag{7.8}$$

and

$$k_c^2 \bar{E}_t + j\beta \nabla_t E_z - j\omega\mu_0 \varepsilon_r (\bar{a}_z \times \nabla_t) H_z = 0 \tag{7.9}$$

The separation constant k_c is given by

$$k_c^2 = -\beta^2 + \omega^2 \mu_0 \varepsilon_0 \varepsilon_r \tag{7.10}$$

The radial and azimuthal field components are then evaluated by noting the following vector identities:

$$\nabla_t = \bar{a}_r \left(\frac{\partial}{\partial r}\right) + \bar{a}_\theta \left(\frac{1}{r}\frac{\partial}{\partial \theta}\right) \tag{7.11}$$

$$\bar{a}_z \times \nabla_t = \bar{a}_r \left(\frac{-1}{r}\frac{\partial}{\partial \theta}\right) + \bar{a}_\theta \left(\frac{\partial}{\partial r}\right) \tag{7.12}$$

The required result

7.2 MAXWELL'S EQUATIONS IN CYLINDRICAL COORDINATES

$$k_c^2 H_r = j\frac{\omega\varepsilon_0\varepsilon_r}{r}\left(\frac{\partial E_z}{\partial \theta}\right) - j\beta\left(\frac{\partial H_z}{\partial r}\right) \qquad (7.13)$$

$$k_c^2 H_\theta = -j\omega\varepsilon_0\varepsilon_r\left(\frac{\partial E_z}{\partial r}\right) - j\frac{\beta}{r}\left(\frac{\partial H_z}{\partial \theta}\right) \qquad (7.14)$$

$$k_c^2 E_r = -j\beta\left(\frac{\partial E_z}{\partial r}\right) - j\frac{\omega\mu_0}{r}\left(\frac{\partial H_z}{\partial \theta}\right) \qquad (7.15)$$

$$k_c^2 E_\theta = -j\frac{\beta}{r}\left(\frac{\partial E_z}{\partial \theta}\right) + j\omega\mu_0\left(\frac{\partial H_z}{\partial r}\right) \qquad (7.16)$$

is recognized as a simple linear combination of pure TM and TE solutions.

The derivation is complete once solutions to E_z and H_z which satisfy the respective wave equations

$$(\nabla_t^2 + k_c^2)E_z = 0 \qquad (7.17)$$

$$(\nabla_t^2 + k_c^2)H_z = 0 \qquad (7.18)$$

and the boundary conditions at the open or closed waveguide wall are deduced. The ∇_t^2 operator is defined in terms of circular variables by

$$\nabla_t^2 = \frac{1}{r}\frac{\partial}{\partial r} + \frac{\partial^2}{\partial r^2} + \frac{1}{r^2}\frac{\partial^2}{\partial \theta^2} \qquad (7.19)$$

The solutions are catalogued as TE, TM, HE, EH modes according to whether

$$E_z = 0, H_z \neq 0, \quad \text{TE} \qquad (7.20a)$$

$$E_z = 0, H_z = 0, \quad \text{TM} \qquad (7.20b)$$

$$E_z = 0, H_z \neq 0, \quad \text{HE or EH} \qquad (7.20c)$$

The first two solutions are met in the description of the round waveguide with an electric or magnetic wall; the third possibility is encountered in the development of the dielectric waveguide.

7.3 TM$_{nm}$ Modes in Circular Waveguide with Electric Wall

In the case of the TM$_{nm}$ family of modes in the round waveguide in Figure 7.1, $H_z = 0$ and equations (7.13) to (7.16) become

$$H_r = \frac{j\omega\varepsilon_0\varepsilon_r/r}{k_c^2}\left(\frac{\partial E_z}{\partial \theta}\right) \tag{7.21}$$

$$H_\theta = \frac{-j\omega\varepsilon_0\varepsilon_r}{k_c^2}\left(\frac{\partial E_z}{\partial r}\right) \tag{7.22}$$

$$E_r = \frac{-j\beta}{k_c^2}\left(\frac{\partial E_z}{\partial r}\right) \tag{7.23}$$

$$E_\theta = \frac{-j\beta/r}{k_c^2}\left(\frac{\partial E_z}{\partial \theta}\right) \tag{7.24}$$

The solution for E_z which satisfies the wave equation in (7.17) is completed by having recourse to the method of separation of variables by writing E_z as a product of a pure function of r and a pure function of θ: it is described with time variation taken as $\exp(j\omega t)$ and the z variation given by $\exp(-j\beta z)$ by

$$E_z = A_n J_n(k_c r) \exp j(\omega t - \beta z - n\theta) \tag{7.25}$$

The derivation of the required result now proceeds by satisfying the electric wall boundary conditions at the wall of the waveguide:

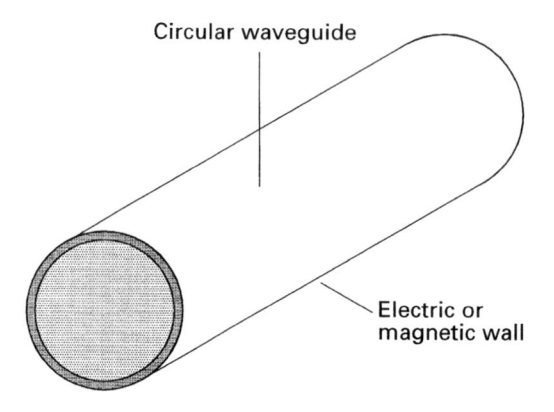

Figure 7.1 Schematic diagram of circular waveguide with electric wall

7.3 TM$_{nm}$ MODES IN CIRCULAR WAVEGUIDE

$$E_z = 0 \quad \text{at } r = R \tag{7.26}$$

or

$$J_n(k_c R) = 0 \tag{7.27}$$

Since the Bessel function $J_n(x)$ of order n has an infinite number of values x for which it becomes zero, this condition may be met by any one of them; that is if U_{nm} is the mth root of equation (7.27) it is satisfied if

$$(k_c)_{nm} = \frac{U_{nm}}{R} \tag{7.28}$$

This equation defines a doubly infinite set of possible values for k_c in equation (7.10), one for each combination of the integers n and m, each of which defines a particular mode TM$_{nm}$. The lowest mode is given by the first root of equation (7.27). This solution coincides with $n = 0$ and $m = 1$:

$$U_{01} = 2.40 \tag{7.29}$$

The next roots are given by $U_{11} = 3.85$, $U_{02} = 5.52$, etc.

The complete description of the first TM$_{nm}$ mode in a lossless circular waveguide with an electric wall is now obtained by evaluating equations (7.21) to (7.24) and (7.25) with appropriate values of n and m. Taking $n = 0$ and $m = 1$ as an example gives

$$\beta^2 = \omega^2 \mu_0 \varepsilon_0 \varepsilon_r - \left(\frac{2.40}{R}\right)^2 \tag{7.30a}$$

and

$$H_r = 0 \tag{7.30b}$$

$$H_\theta = \frac{-j\omega\varepsilon_0\varepsilon_r}{k_c^2} J_0'(k_c r) \exp j(\omega t - \beta z) \tag{7.30c}$$

$$H_z = 0 \tag{7.30d}$$

$$E_r = \frac{-j\beta}{k_c^2} J_0'(k_c r) \exp j(\omega t - \beta z) \tag{7.30e}$$

$$E_\theta = 0 \tag{7.30f}$$

$$E_z = J_0(k_c r) \exp j(\omega t - \beta z) \tag{7.30g}$$

7.4 TE$_{nm}$ Modes in Circular Waveguide with Electric Wall

The derivation of the TE$_{nm}$ field of solutions in a circular waveguide proceeds in a similar fashion to that outlined in the case of the TM$_{nm}$ problem region. The result is summarized below with the aid of equations (7.13) to (7.16).

$$H_r = \frac{-j\beta}{k_c^2}\left(\frac{\partial H_z}{\partial r}\right) \tag{7.31}$$

$$H_\theta = \frac{-j\beta/r}{k_c^2}\left(\frac{\partial H_z}{\partial \theta}\right) \tag{7.32}$$

$$E_r = \frac{-j\omega\mu_0/r}{k_c^2}\left(\frac{\partial H_z}{\partial \theta}\right) \tag{7.33}$$

$$E_\theta = \frac{j\omega\mu_0}{k_c^2}\left(\frac{\partial H_z}{\partial r}\right) \tag{7.34}$$

with

$$E_z = 0 \tag{7.35}$$

and

$$H_z = A_n J_n(k_c r)\exp j(\omega t - \beta z - n\theta) \tag{7.36}$$

The boundary condition in the case of the electric wall situation is here met by

$$E_\theta = 0 \quad \text{at } r = R \tag{7.37}$$

or equivalently

$$\frac{\partial H_z}{\partial r} = 0 \quad \text{at } r = R \tag{7.38}$$

which coincides with

$$J'_n(kR) = 0 \tag{7.39}$$

The mth root of this nth order equation is defined by

$$(k_c)_{nm} = \frac{U'_{nm}}{R} \tag{7.40b}$$

7.4 TE$_{nm}$ MODES IN CIRCULAR WAVEGUIDE

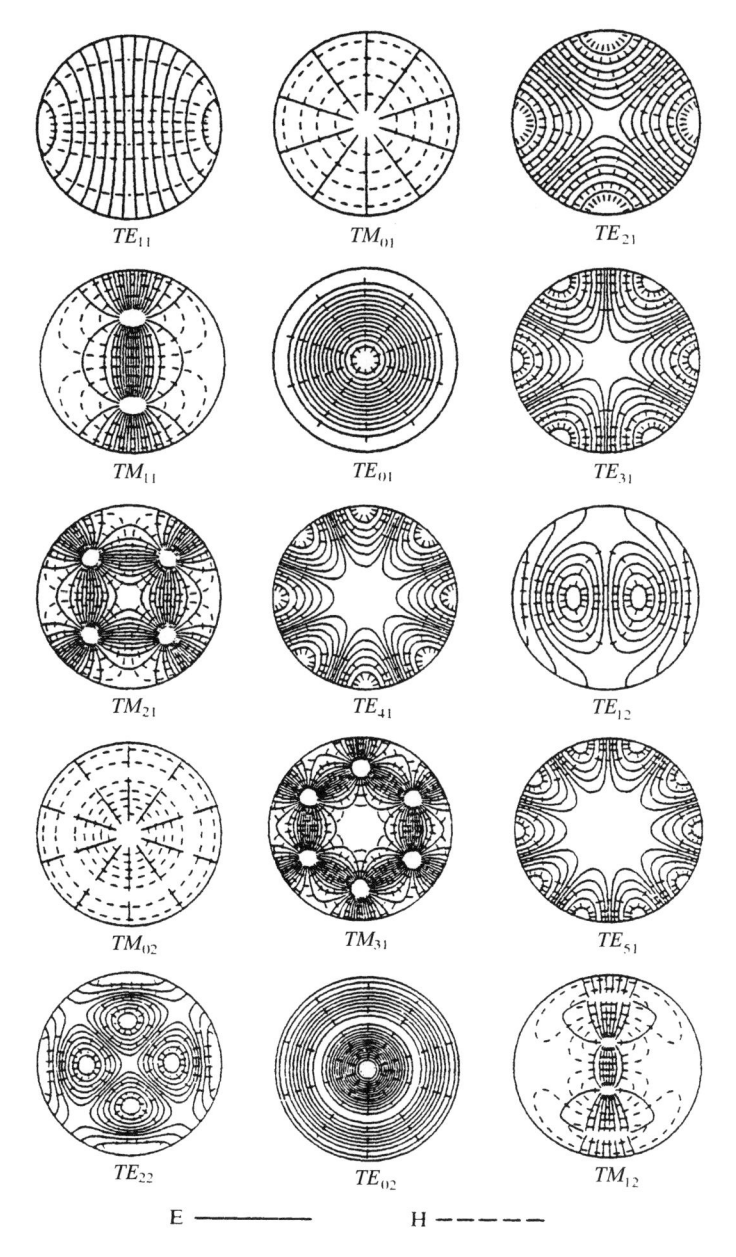

Figure 7.2 Transverse field distribution for a circular waveguide with an electric wall (Source: Lee, C. S., Lee, S. W. and Chuang, S. L. (1985) Plot of modal field distribution in rectangular and circular waveguides, *IEEE Trans. Microwave Theory Tech.*, **MTT-33**, 271–274, March 1985, first 15 modes) (© 1985 IEEE)

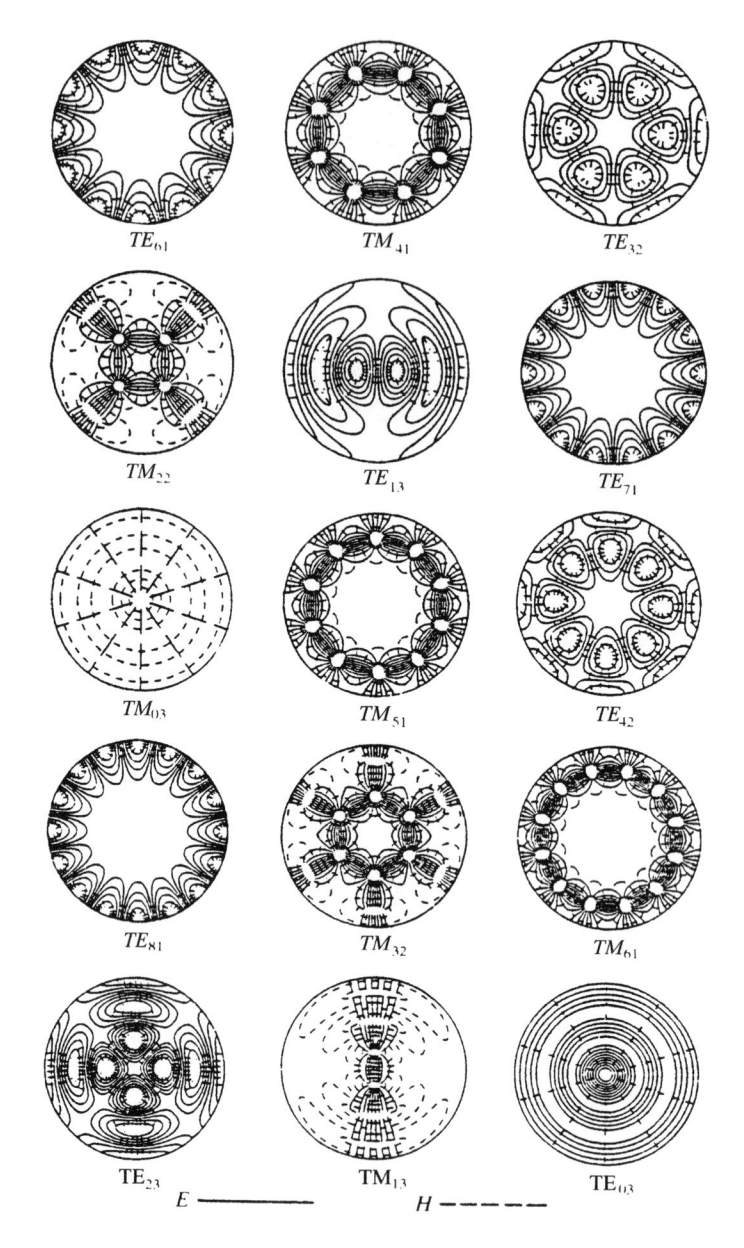

Figure 7.3 Transverse field distribution for a circular waveguide with an electric wall (Source: Lee, C. S., Lee, S. W. and Chuang, S. L. (1985) Plot of modal field distribution in rectangular and circular waveguides, *IEEE Trans. Microwave Theory Tech.*, **MTT-33**, 271–274, March 1985, first 15 modes)

7.5 TM$_{nm}$ and TE$_{nm}$ MODES IN CIRCULAR WAVEGUIDE   117

and the lowest root is realized with $n = 1$ and $m = 1$:

$$U'_{11} = 1.84 \qquad (7.40c)$$

The next roots are given by $U'_{21} = 3.05$, $U'_{01} = 3.83$, etc.

The TE$_{11}$ field of solutions in the electric wall case is therefore described by equations (7.31) to (7.36) with k_c and β set by U'_{11}:

$$H_r = \left(\frac{-j\beta}{k_c}\right) J'_1(k_c r) \qquad (7.41a)$$

$$H_\theta = \left(\frac{-j\beta}{k_c^2 r}\right) J_1(k_c r) \qquad (7.41b)$$

$$H_z = 0 \qquad (7.41c)$$

$$E_r = \left(\frac{-\omega\mu_0}{k_c^2 r}\right) J_1(k_c r) \qquad (7.41d)$$

$$E_\theta = \left(\frac{j\omega\mu_0}{k_c}\right) J'_1(k_c r) \qquad (7.41e)$$

$$E_z = A_n J_n(k_c r) \qquad (7.41f)$$

where the exp $[j(\omega t - \beta z - \theta)]$ variation is understood.

In order to plot the field patterns it is necessary to take the real parts of the exponential form of the fields employed so far and evaluate them at say, $\omega t = 0$ and $\beta z = 0, \pi/2, \pi, 3\pi/2$.

Figures 7.2 and 7.3 indicate the transverse fields of some such solutions for a circular waveguide with an electric wall.

7.5 TM$_{nm}$ and TE$_{nm}$ Modes in Circular Waveguide with a Magnetic Wall

The circular waveguide with a magnetic wall boundary condition is of special interest in the design of dielectric and gyromagnetic resonators in that the solution of the open wall dielectric waveguide is asymptotic to it as its relative dielectric constant becomes large. The derivation of the cut-off space of either the TM$_{nm}$ or TE$_{nm}$ modes in this type waveguide starts with a statement of the appropriate boundary condition at $r = R$ and a statement of the TM$_{nm}$ or TE$_{nm}$ condition. A comparison between the TM$_{nm}$ solution with a magnetic wall boundary condition with that of the TE$_{nm}$ one with an electric wall one indicates that the two have the same cut-off numbers. It also indicates that the electric field of the latter maps into the magnetic field of the former with the exception of a reversal in the sign of its components and that the magnetic field maps into electric one without ado. The field plot of the TM_{nm} solution discussed here may therefore be obtained from the TE_{nm} one in an electric wall enclosure by exchanging the electric lines by magnetic ones except for the reversal of the signs, the magnetic ones by electric ones. The TE_{nm} solution in this type of waveguide has the same cut-off number as the TM_{nm} one with an electric wall and the fields of the one is mapped into the other in a dual fashion. The results obtained in this way, except for the change in the signs of the magnetic or electric fields may be visualized by replacing the electric fields by magnetic ones and the magnetic ones by electric ones in Figures 7.2 and 7.3. Taking the TM_{nm} solution by way of an example gives

$$H_z = 0 \tag{7.42a}$$

$$E_z = A_n J_n(k_c r) \exp j(\omega t - \beta z - n\theta) \tag{7.42b}$$

$$E_r = \frac{-j\beta}{k_c^2}\left(\frac{\partial E_z}{\partial r}\right) \tag{7.42c}$$

$$E_\theta = \frac{-j\beta/r}{k_c^2}\left(\frac{\partial E_z}{\partial \theta}\right) \tag{7.42d}$$

$$H_r = \frac{-j\omega\varepsilon_0\varepsilon_r}{k_c^2}\left(\frac{\partial E_z}{\partial \theta}\right) \tag{7.42e}$$

$$H_\theta = \frac{-j\omega\varepsilon_0\varepsilon_r}{k_c^2}\left(\frac{\partial E_z}{\partial r}\right) \tag{7.42f}$$

The required boundary condition on the magnetic wall is

$$H_\theta = 0 \quad \text{at} \quad r = R \tag{7.43a}$$

or

$$\frac{\partial E_z}{\partial r} = 0 \quad \text{at} \quad r = R \tag{7.44}$$

This condition is readily satisfied provided

$$J'_n(k_c R) = 0 \tag{7.45}$$

or

$$(k_c)_{nm} = \frac{U'_{nm}}{R} \tag{7.46}$$

The lowest root is obtained with $n = m = 1$

or
$$U'_{11} = 1.84 \tag{7.47}$$

A comparison between this problem region and that of the TE$_{nm}$ one associated with an electric wall boundary condition indicates that the two have the same cut-spaces.

The following duality between the TM$_{11}$ solution in a round waveguide with a magnetic wall and the TE$_{11}$ one in a similar waveguide with an electric wall is also noted:

$$\text{TM}_{11} \rightarrow \text{TE}_{11}$$

$$E_r \rightarrow H_r$$

$$E_\theta \rightarrow H_\theta$$

$$E_z \rightarrow H_z$$

$$H_r \rightarrow -E_r \quad \text{with } \varepsilon_0 \text{ replaced by } \mu_0$$

$$H_\theta \rightarrow -E_\theta \quad \text{with } \varepsilon_0 \text{ replaced by } \mu_0$$

$$H_z \rightarrow E_z \rightarrow 0$$

The field patterns of this problem region are therefore obtained from those of the dual problem by replacing the electric field points by magnetic ones and the magnetic field points by electric ones.

The duality between the TE$_{nm}$ in the round waveguide with a magnetic wall and the TM$_{nm}$ one with an electric wall proceeds in a like manner.

7.6 Anisotropic Waveguides with Electric and Magnetic Walls

In a demagnetized ferrite insulator the diagonal elements μ_x, μ_y and μ_z, of the tensor permeability are different from unity. The nature of the permeability may be summarized in terms of a diagonal permeability:

$$[\mu] = \begin{bmatrix} \mu_x & 0 & 0 \\ 0 & \mu_y & 0 \\ 0 & 0 & \mu_z \end{bmatrix} \tag{7.48}$$

The dielectric case already treated is given by

$$\mu_x = \mu_y = \mu_z = 1 \tag{7.49}$$

7 ISOTROPIC, ANISOTROPIC, GYROMAGNETIC CIRCULAR WAVEGUIDES

For a demagnetized polycrystalline ferrite material

$$\mu_x = \mu_y = \mu_z = \mu_d \qquad (7.50)$$

In a material such as a garnet with cubic anisotropy

$$\mu_x \neq \mu_y \neq \mu_z \qquad (7.51)$$

In some situations the tensor permeability associated with a partially magnetized ferrite is also approximated by a diagonal one with

$$\mu_x = \mu_y = \mu_t \qquad (7.52a)$$

$$\mu_z = 1 \qquad (7.52b)$$

and in a saturated material with

$$\mu_x = \mu_y = \mu_t \qquad (7.53a)$$

$$\mu_z = 1 \qquad (7.53b)$$

where

$$\mu_t = \mu \pm \kappa \qquad (7.54a)$$

and

$$\mu_t = 1 \pm \kappa \qquad (7.54b)$$

respectively.

The corresponding diagonal tensors quantities are

$$[\mu] = \begin{bmatrix} \mu \pm \kappa & 0 & 0 \\ 0 & \mu \pm \kappa & 0 \\ 0 & 0 & \mu_z \end{bmatrix} \qquad (7.55)$$

where $\mu_z = 1$ in a saturated material.

Some or all of these different forms have been incorporated at one time or another in the description of propagation and power flow in fully ferrite filled round waveguides with electric walls or in open ferrite waveguides.

The dispersion relationship of the TE_{11} mode in a waveguide with an electric wall filled with an anisotropic medium is

$$\beta^2 = k_0^2 \varepsilon_f \mu_t - k_c^2 \left(\frac{\mu_t}{\mu_z} \right) \qquad (7.56)$$

7.6 ANISOTROPIC WAVEGUIDES

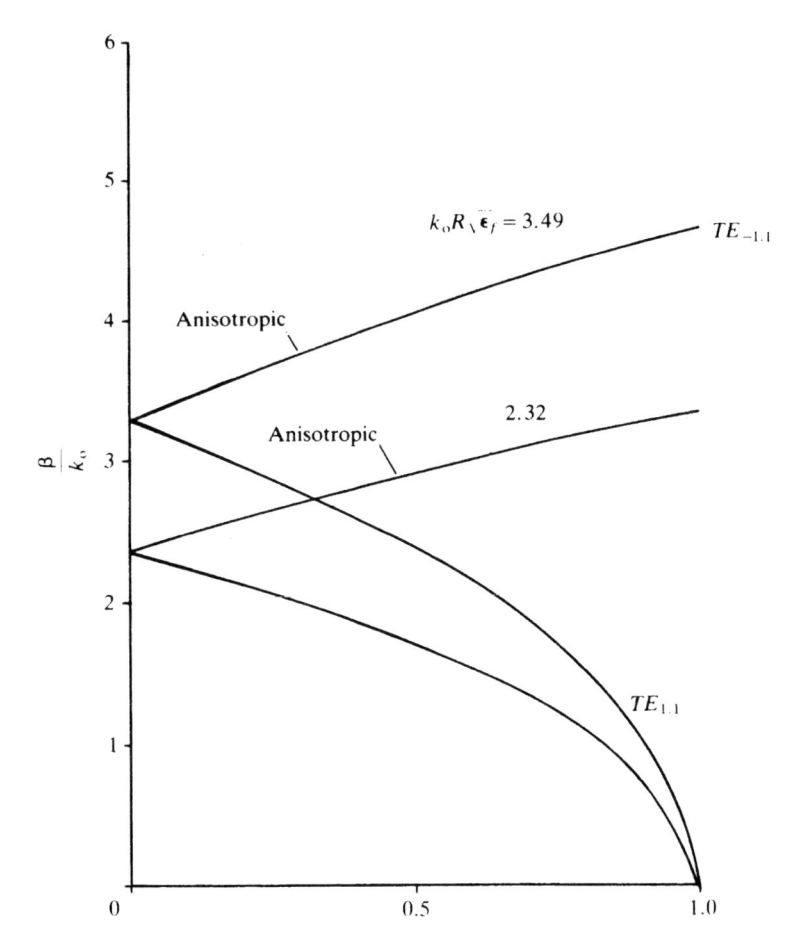

Figure 7.4 Split phase constants of TE$_{11}$ mode in closed anisotropic waveguide with electric wall (Source: Helszajn, J. and Gibson, A. A. P. (1987) Mode nomenclature of circular gyromagnetic and anisotropic waveguides with magnetic and open walls, *Proc. IEE*, **134**, Part H, No. 6, December)

The corresponding result for the TM$_{11}$ mode in round waveguide with a magnetic wall is

$$\beta^2 = k_0^2 \varepsilon_f \mu_t - k_c^2 \tag{7.57}$$

The latter waveguide displays split cut-off numbers and propagation constants; the former one is characterized by degenerate cut-off numbers and split propagation constants. The phase constants of these two situations are indicated in Figures 7.4 and 7.5.

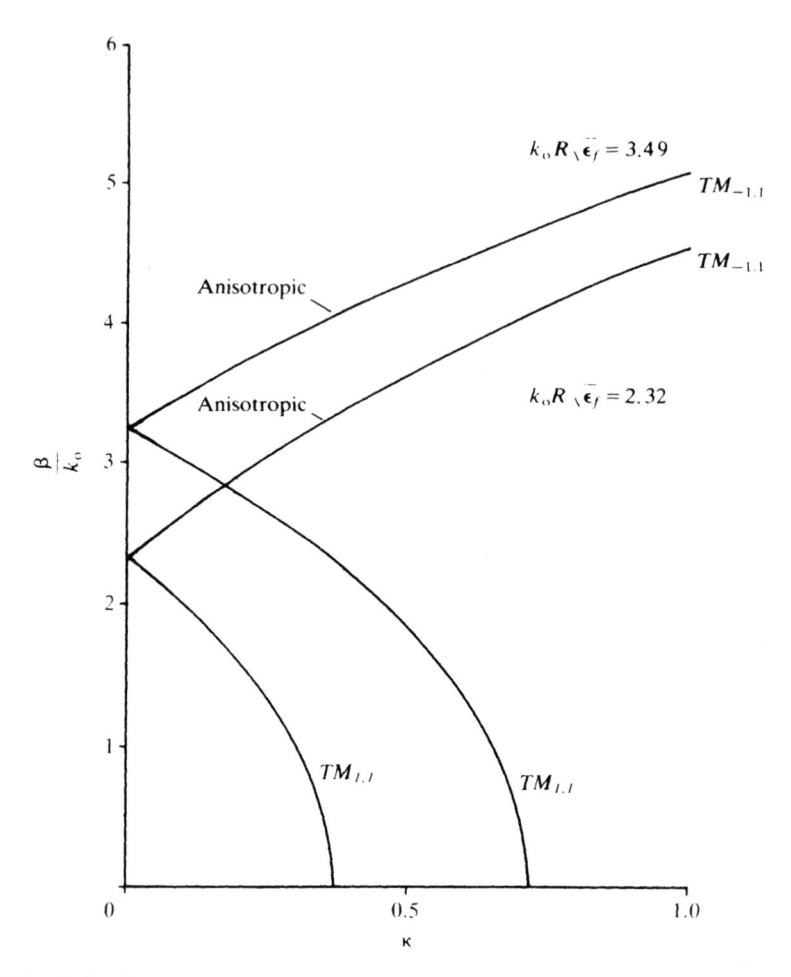

Figure 7.5 Split phase constants TM_{11} mode in closed anisotropic waveguide with magnetic wall (Source: Helszajn, J. and Gibson, A. A. P., Mode nomenclature of circular gyromagnetic and anisotropic waveguides with magnetic and open walls, *Proc. IEE*, **134**, Part H, No. 6, December 1987)

7.7 The Gyromagnetic Waveguide

The gyromagnetic waveguide with an electric or magnetic wall is the fundamental building block in the construction of Faraday rotation devices and gyromagnetic resonators in the design of junction circulators. The detailed solutions of the former two problems are well understood but are outside the remit of this work. These sort of waveguides are characterized by a permeability of the form

7.7 THE GYROMAGNETIC WAVEGUIDE

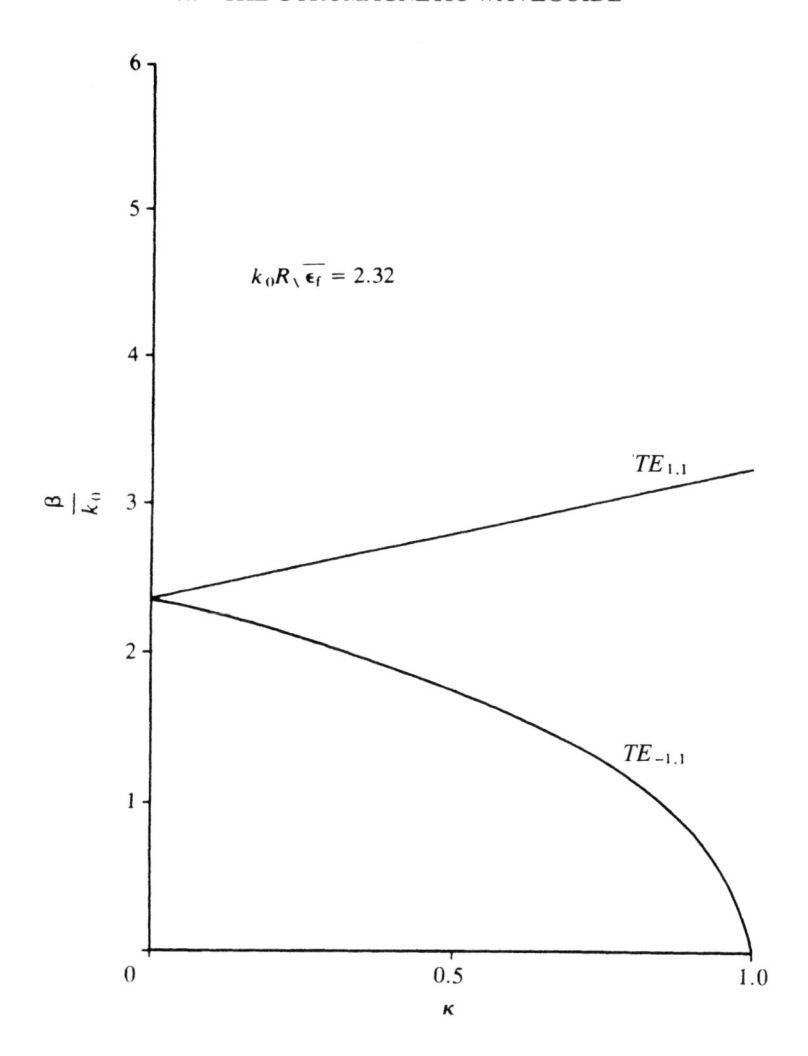

[Figure: Plot of β/k_0 vs κ showing split phase constants, labeled $k_0 R \sqrt{\epsilon_f} = 2.32$, with curves $TE_{1,1}$ and $TE_{-1,1}$ branching from approximately 2.3 at $\kappa = 0$.]

Figure 7.6 Split phase constants for the dominant quasi-TE_{11} mode for ferrite-filled circular waveguide with an electric wall ($\mu = 1$, $\mu_z = 1$, $\varepsilon_f = 15$) (Source: Helszajn, J. and Gibson, A. A. P. (1987) Mode nomenclature of circular gyromagnetic and anisotropic waveguides with magnetic and open walls, *Proc. IEE*, **134**, Part H, No. 6, December)

$$\begin{vmatrix} \mu & -j\kappa & 0 \\ j\kappa & \mu & 0 \\ 0 & 0 & \mu_z \end{vmatrix} \tag{7.58}$$

Figures 7.6 and 7.7 illustrate the split phase constants in each of these waveguides for one typical value of $k_0 R \sqrt{\varepsilon_f}$.

In order to avoid having to solve the exact characteristic equations of these sorts of problem regions polynomial representations of the split phase constants

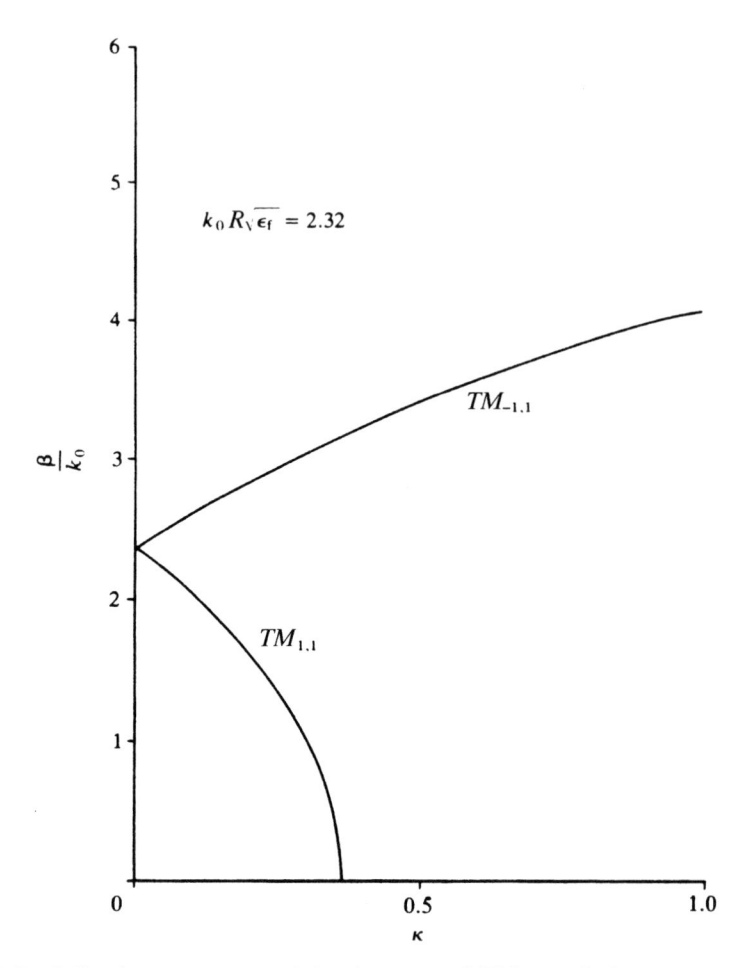

Figure 7.7 Split phase constants of dominant quasi-TM$_{11}$ mode in a gyromagnetic waveguide with a magnetic wall ($\mu = 1$, $\mu_z = 1$, $\varepsilon_f = 15$) (Source: Helszajn, J. and Gibson, A. A. P. (1987) Mode nomenclature of circular gyromagnetic and anisotropic waveguides with magnetic and open walls, *Proc. IEE*, **134**, Part H, No. 6, December)

of the dominant TE$_{11}$ mode met in the electric wall problem and the TM$_{11}$ mode met in connection with the magnetic wall problem have been constructed. Polynomial representations for $\beta_\pm/k_0$ of the surface defined by k_0R and κ/μ between ±1 based on the exact solutions of the characteristic equations of the magnetic and electric wall problem regions are

$$\beta_\pm/k_0 = -11.668xy^2 + 24.657y^3 - 66.87y_2 + 61.548y + 21.809xy$$

$$+ 3.361yx^2 - 15.874 - 12.004x - 4.599x^2 - 1.181x^3,$$

$$0.60 \leq k_0R \leq 0.90 \tag{7.59}$$

7.7 THE GYROMAGNETIC WAVEGUIDE

$$\beta_\pm/k_0 = -4.328xy^2 + 17.08y^3 - 45.504y^2 + 42.222y + 8.194xy$$
$$- 0.647yx^2 - 10.231 - 2.635x - 0.242x^2 + 0.861x^3,$$
$$0.60 \leq k_0R \leq 0.90 \quad (7.60)$$

respectively.

The variables appearing in the polynomials are defined by

$$x = \frac{\kappa}{\mu} \quad y = k_0R$$

A scrutiny of these polynomials suggests that the opening between the split phase constant curves is larger in a gyromagnetic waveguide with a magnetic wall than in one with an electric wall.

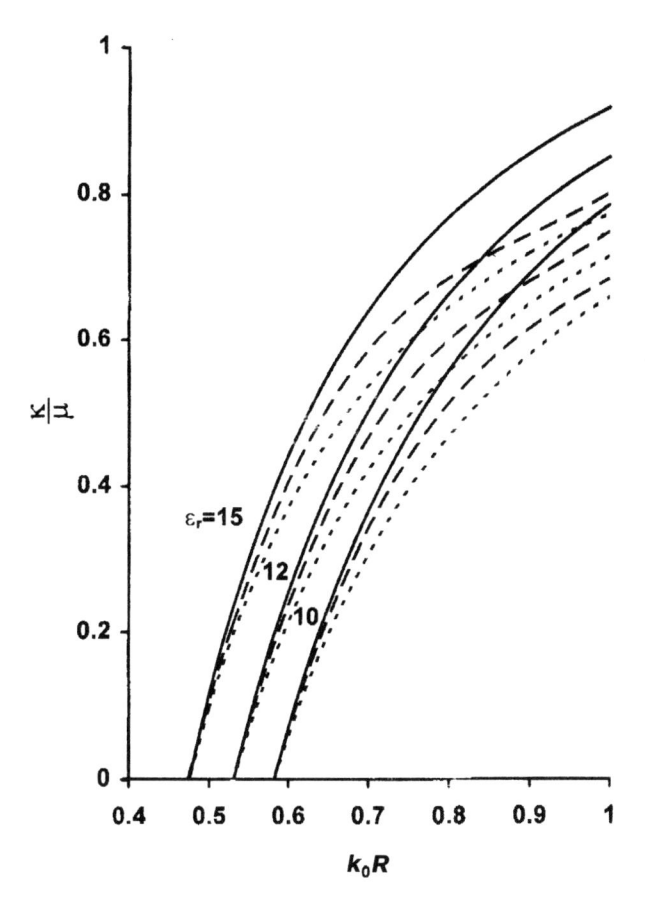

Figure 7.8 Comparison between cut-off numbers of clockwise circularly polarized mode in circular waveguide with ideal magnetic wall (- - - - Gyromagnetic solution, · · · · Anisotropic solution, ――― Perturbation solution) (reproduced with permission, Helszajn, J. and Sharp, J. (1997) Polynomial representation of split-phase constants of gyromagnetic waveguides with electric and magnetic walls, *IEEE Microwave and Guided Wave Letters*, **7**, 291)

126　7 ISOTROPIC, ANISOTROPIC, GYROMAGNETIC CIRCULAR WAVEGUIDES

The cut-off space of the lower split phase constant branch of the magnetic wall problem region is as is understood dependent upon the gyrotropy. One polynomial representation for this feature which fixes the gyrotropy of the problem region in this type of waveguide is

$$\kappa/\mu = 6.694(k_0R)^3 - 18.105(k_0R)^2 + 16.841(k_0R) - 4.630,$$

$$\varepsilon_r = 15, \quad 0.475 < k_0R < 1.0 \tag{7.61a}$$

$$\kappa/\mu = 7.01(k_0R)^3 - 19{,}509(k_0R)^2 + 18.755(k_0R) - 5.507,$$

$$\varepsilon_r = 12, \quad 0.531 < k_0R < 1.0 \tag{7.61b}$$

$$\kappa/\mu = 6.06(k_0R)^3 - 18.022(k_0R)^2 + 18.510(k_0R) - 5.861,$$

$$\varepsilon_r = 10, \quad 0.582 < k_0R < 1.0 \tag{7.61c}$$

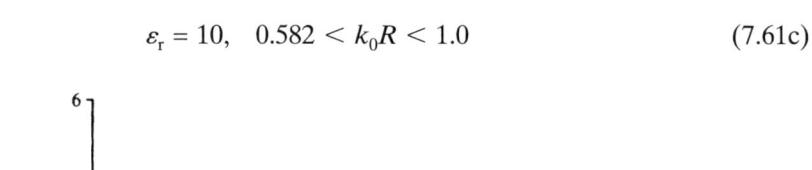

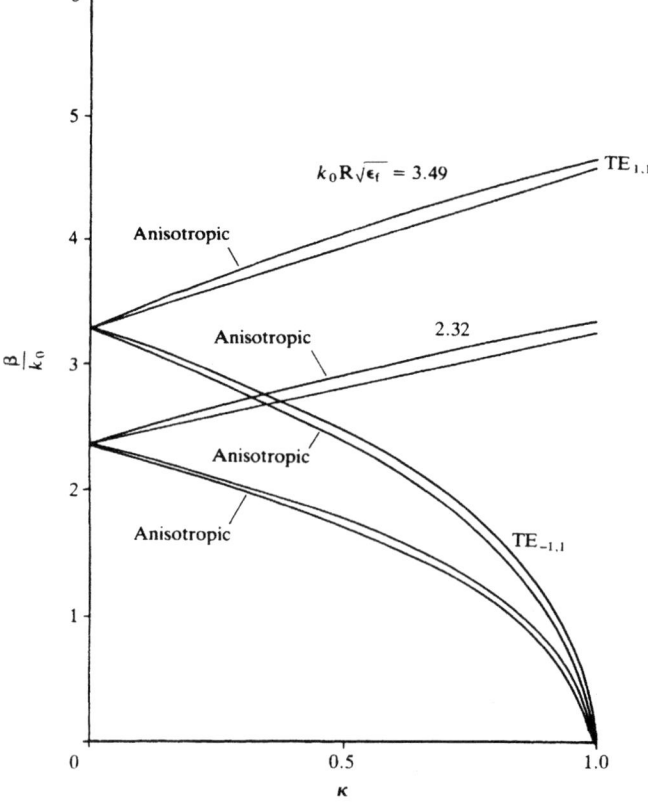

Figure 7.9 Split phase constants of (quasi-TE$_{11}$ modes) inanisotropic and gyromagnetic waveguides with electric walls (Source: Helszajn, J. and Gibson, A. A. P. (1987) Mode nomenclature of circular gyromagnetic and anisotropic waveguides with magnetic and open walls, *Proc. IEE*, **134**, Part H, No. 6, December)

7.7 THE GYROMAGNETIC WAVEGUIDE

This result is compared in Figure 7.8 with those obtained from the corresponding perturbation calculation to be dealt with in the next section and the anisotropic formulation for β_+ in this one.

A scrutiny of the cut-off numbers and split phase constants of such waveguides and those of the gyromagnetic ones indicates that the two exhibit solutions that are qualitatively alike. The dispersion relationships for the TE_{11} and TM_{11} modes in round anisotropic waveguides with electric and magnetic walls are separately compared in Figures 7.9 and 7.10. The anisotropic problem for which the entries along the main diagonal in equation (7.55) corresponds to the eigenvalues of the tensor permeability in equation (7.58) of the gyromagnetic medium is therefore of special interest in the theory of round waveguides.

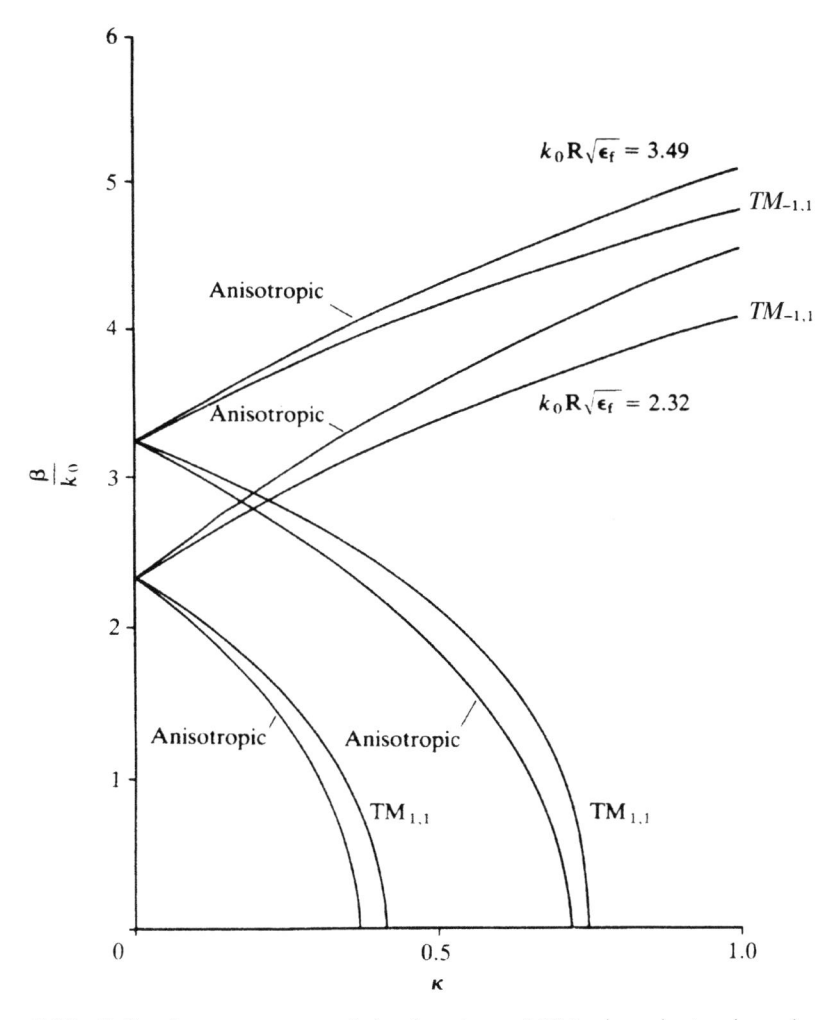

Figure 7.10 Split phase constants of dominant quasi-TM_{11} in anisotropic and gyromagnetic waveguides with magnetic wall (Source: Helszajn, J. and Gibson, A. A. P. (1987) Mode nomenclature of circular gyromagnetic and anisotropic waveguides with magnetic and open walls, *Proc. IEE*, **134**, Part H, No. 6, December)

7.8 Perturbation Theory of Gyromagnetic Circular Waveguide with Electric and Magnetic Walls

The propagation constants of the ferrite-filled circular waveguide with either an electric or a magnetic wall may also be formed by having recourse to perturbation theory. The pertubation formulation is:

$$\beta_\pm + \beta_0^* = \frac{\int_S \{\mu_0[\Delta\mu]\cdot\overline{h}_0\}\cdot\overline{h}_0^* \, ds}{\int_S a_z\cdot(\overline{E}_0 \times \overline{h}_0^* + \overline{E}_0^* \times \overline{h}_0) \, ds} \tag{7.62}$$

In this treatment the gyromagnetic filler is first considered as a dielectric one with a propagation constant

$$\beta_0^2 = k_0^2 \varepsilon_f - \left(\frac{1.84}{R}\right)^2 \tag{7.63}$$

$[\Delta\mu]$ is given by

$$[\Delta\mu] = \begin{bmatrix} 1 & 0 & 0 \\ 0 & 1 & 0 \\ 0 & 0 & 1 \end{bmatrix} - \begin{bmatrix} \mu & -j\kappa & 0 \\ j\kappa & \mu & 0 \\ 0 & 0 & 1 \end{bmatrix} \tag{7.64}$$

and the other parameters have the usual meaning.

Adopting circular variables for the TE_{11} fields of the unperturbed round waveguide with an electric wall gives

$$\beta_\pm^2 \approx \beta_0^2(1 \pm C_{11}\kappa) \tag{7.65}$$

where

$$C_{11} = \frac{2}{(U'_{11})^2 - 1} \tag{7.66}$$

Evaluating C_{11} with $U' = 1.84$ gives

$$C_1 = 0.84 \tag{7.67}$$

The split phase constants obtained using perturbation theory are compared to the exact solution in Figure 7.11. The dispersion relationships again display split phase constants but no split cut-off numbers in keeping with the exact solution.

For the dominant TM_{11} mode in the magnetic wall problem

$$\beta_\pm^2 \approx k_0^2 \varepsilon_f (1 \pm C_{11}\kappa) - \left(\frac{1.84}{R}\right)^2 \tag{7.68}$$

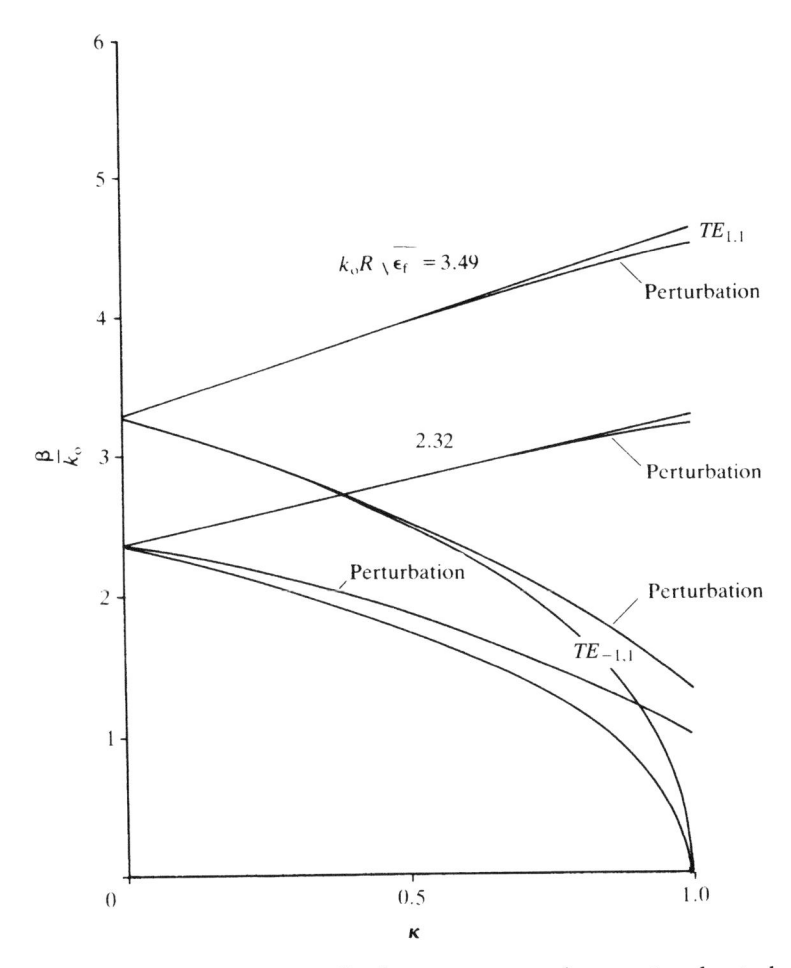

Figure 7.11 Comparison between split phase constants using exact and peturbation for a gyromagnetic waveguide with an electric wall (Source: Helszajn, J. and Gibson, A. P. (1989) Cutoff spaces of elliptical gyromagnetic planar circuits and waveguides using finite elements, *IEEE Trans. Microwave Theory Tech.*, **37**(1), January)

This result displays both split cut-off numbers and phase constants, again in keeping with the properties of the exact problem. Figure 7.12 compares this solution with the exact one.

The factor C_{11} in the preceding equations coincides with the first pair of split cut-off numbers of the planar disc resonator with magnetic sidewalls met in Chapter 5.

A different and perhaps more appropriate from for $[\Delta\mu]$ is

$$[\Delta\mu] = \begin{bmatrix} \mu & 0 & 0 \\ 0 & \mu & 0 \\ 0 & 0 & \mu_z \end{bmatrix} - \begin{bmatrix} \mu & -j\kappa & 0 \\ j\kappa & \mu & 0 \\ 0 & 0 & 1 \end{bmatrix} \quad (7.69)$$

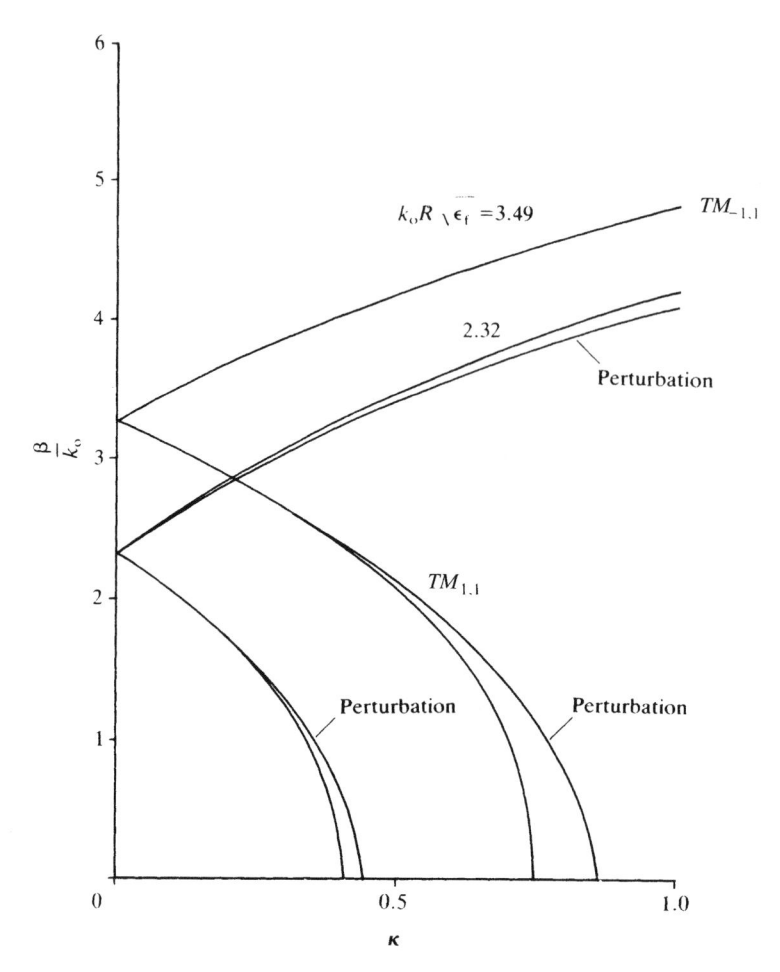

Figure 7.12 Comparison between split phase constants using exact peturbation theories for a gyromagnetic waveguide with a magnetic wall (Source: Helszajn, J. and Gibson, A. P. (1989) Cutoff spaces of elliptical gyromagnetic planar circuits and waveguides using finite elements, *IEEE Trans. Microwave Theory Tech.*, **37**(1), January)

with

$$\beta_0^2 = k_0^2 \varepsilon_f \mu - \left(\frac{1.84}{R}\right)^2 \left(\frac{\mu}{\mu_z}\right) \tag{7.70}$$

for the TE$_{11}$ mode in a circular waveguide with an electric wall and

$$\beta_0^2 = k_0^2 \varepsilon_f \mu - \left(\frac{1.84}{R}\right)^2 \tag{7.71}$$

for the TM$_{11}$ mode in a circular waveguide with a magnetic wall.

Scrutiny of the approximate dispersion relationships for the gyromagnetic waveguides in this section and the exact ones using the appropriate characteristic equations indicate that the former descriptions are adequate for engineering purposes in the interval

$$0 \leq \kappa \leq 0.50 \tag{7.72}$$

7.9 Circular Polarization in Free Space

A property of a round waveguide propagating the dominant TM$_{nm}$ is the existence of clockwise and anticlockwise circular polarization on its axis. Since such polarization plays such an important role in the operation of ferrite devices it is helpful to recall its definitions in some detail.

The two possibilities are defined in cartesian coordinates by

$$\bar{E}^-(\omega t, \beta z) = (\bar{a}_x E_0 + j\bar{a}_y E_0) \exp j(\omega t - \beta z) \tag{7.73a}$$

$$\bar{E}^+(\omega t, \beta z) = (\bar{a}_x E_0 - j\bar{a}_y E_0) \exp j(\omega t - \beta z) \tag{7.73b}$$

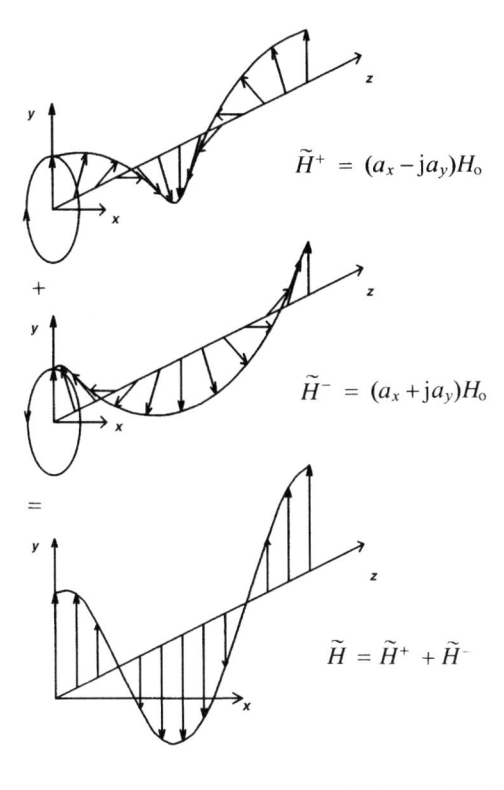

Figure 7.13 Linear polarization, clockwise, counter-clockwise circular polarization (magnetic field)

132   7 ISOTROPIC, ANISOTROPIC, GYROMAGNETIC CIRCULAR WAVEGUIDES

Whether $\bar{E}^+$ or $\bar{E}^-$ represents a clockwise or anticlockwise circular polarized wave may be readily established by constructing the real parts of these quantities and evaluating the same at $\omega t = 0$, $\beta z = 0$, $\pi/2$, π, $3\pi/2$, etc. or at $\beta z = 0$, $\omega t = 0$, $\pi/2$, $3\pi/2$, etc. This gives

$$\bar{E}^-(\omega t, \beta z) = \bar{a}_x E_0 \cos(\omega t - \beta z) + \bar{a}_y E_0 \sin(\omega t - \beta z) \qquad (7.74a)$$

$$\bar{E}^+(\omega t, \beta z) = \bar{a}_x E_0 \cos(\omega t - \beta z) - \bar{a}_y E_0 \sin(\omega t - \beta z) \qquad (7.74b)$$

Taking $\omega t = 0$, by way of an example gives

$$\bar{E}^-(0, \beta z) = \bar{a}_x E_0 \cos(\beta z) + \bar{a}_y E_0 \sin(\beta z) \qquad (7.75a)$$

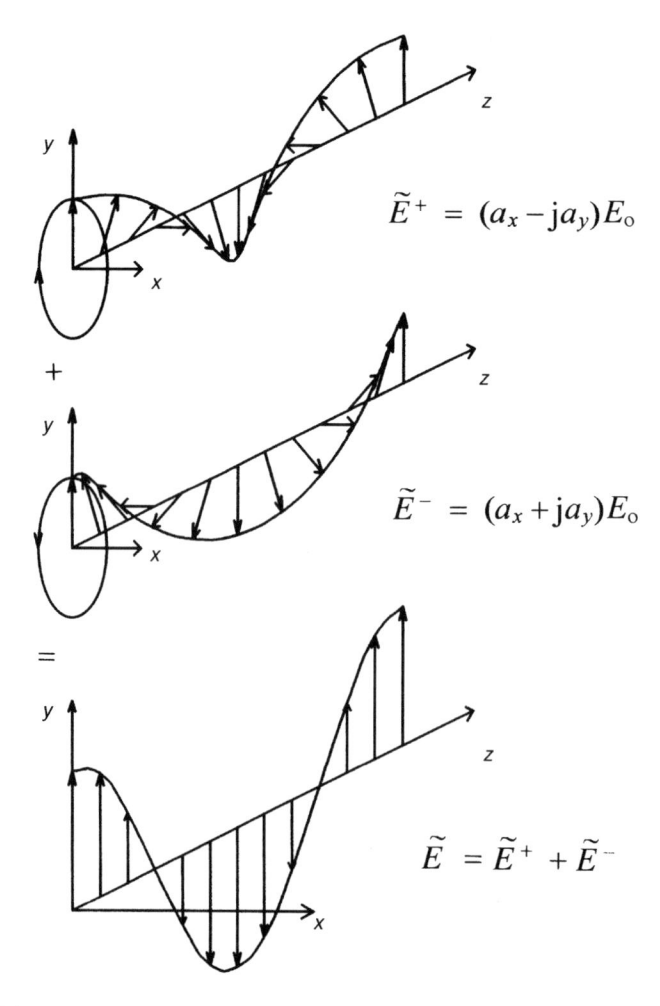

$$\tilde{E}^+ = (a_x - ja_y)E_o$$

$$+$$

$$\tilde{E}^- = (a_x + ja_y)E_o$$

$$=$$

$$\tilde{E} = \tilde{E}^+ + \tilde{E}^-$$

Figure 7.14 Linear polarization, clockwise, counter-clockwise circular polarization (electric field)

$$\bar{E}^+ (0, \beta z) = \bar{a}_x E_0 \cos (\beta z) + \bar{a}_y E_0 \sin (\beta z) \qquad (7.75b)$$

A scrutiny of the preceding equations also indicates that

$$\bar{E}^+ (\omega t, \beta z) + \bar{E}^- (\omega t, \beta z) = \bar{a}_x E_0 \cos (\omega t - \beta z) \qquad (7.76)$$

This result suggests that a linearly polarized wave can always be decomposed into a linear combination of counter-rotating circularly polarized waves. Figures 7.13(a) and (b) show pictorial displays of the three magnetic field patterns entering into the description of this problem region. Figure 7.14 indicates the corresponding solution for the electric field.

7.10 Circular Polarization in Round Waveguide

An important property of a round waveguide with a magnetic wall is the existence of counter-rotating circularly polarized waves on its axis. The demonstration propagating the dominant mode of this result starts by introducing the small argument approximations of $J'_1(x)$ and $J(x)$ into equation (7.42)

$$J_1(kr) \approx kr$$

$$J'_1(kr) \approx k$$

Introducing these identities into the description of the dominant TE_{11} mode in a circular waveguide with a magnetic wall indicates that the transverse fields on the axis of the waveguide are given by

$$E_r = \frac{-\omega \mu_0}{k_c} \qquad (7.77a)$$

$$E_\theta = \frac{j\omega \mu_0}{k_c} \qquad (7.77b)$$

and

$$H_r = \frac{-j\beta}{k_c} \qquad (7.77c)$$

$$H_\theta = \frac{-\beta}{k_c} \qquad (7.77d)$$

This result indicates that both the transverse electric and magnetic fields of this mode are circularly polarized on the axis of the waveguide. Since it applies equally well for n positive, a second solution is obtained with the hand of the circular polarization in the opposite direction. This solution therefore supports counter-rotating fields on the axis of the waveguide.

The preceeding equations satisfy

$$\frac{E_r}{H_\theta} = \frac{\omega\mu_0}{\beta} \tag{7.78a}$$

$$\frac{E_\theta}{H_r} = \frac{-\omega\mu_0}{\beta} \tag{7.78b}$$

7.11 Faraday Rotation in Circular Waveguides

One important phenomenon exhibited by a longitudinally magnetized circular gyromagnetic waveguide is that of Faraday rotation. This feature may be understood by recognizing that such a waveguide supports counter-rotating circularly

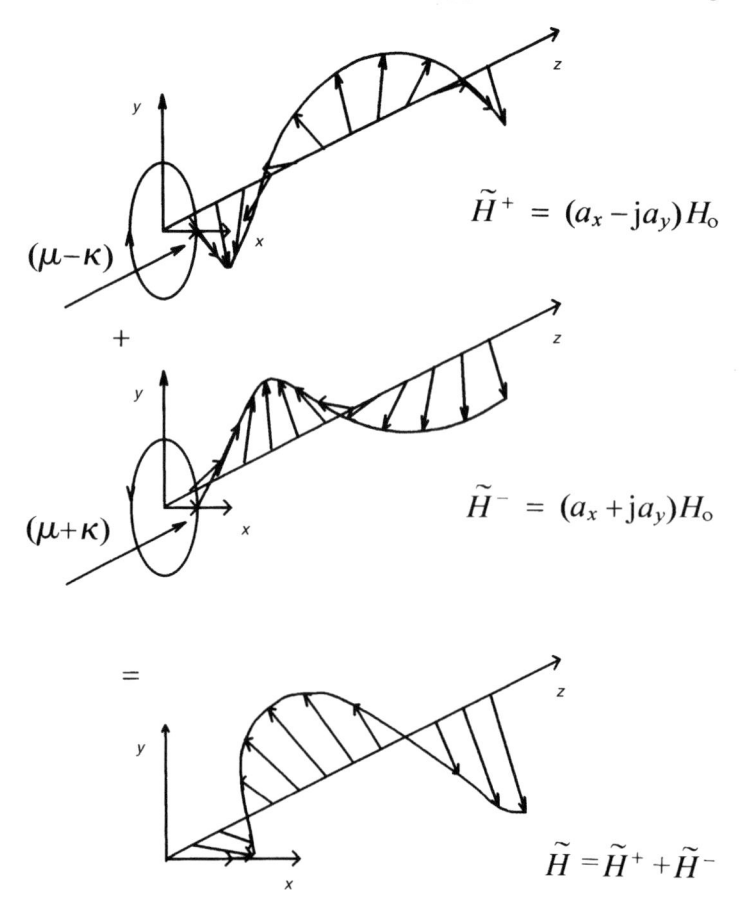

Figure 7.15 Faraday rotation (magnetic field)

7.11 FARADAY ROTATION IN CIRCULAR WAVEGUIDES

alternating magnetic fields on its axis which propagate with different propagation constants. Figure 7.15 illustrates the effect of removing the degeneracy between the periods of the counter-rotating magnetic fields. Figure 7.16 shows the corresponding results in the case of the electric field. The angle through which a linearly polarized wave is rotated in such a medium is:

$$\theta = \frac{\beta_+ - \beta_-}{2} z \quad (7.79)$$

and the average phase constant of the waveguide is:

$$\beta_0 = \frac{\beta_+ + \beta_-}{2} \quad (7.80)$$

$\beta_\pm$ may be exactly determined from the appropriate characteristic equation or from perturbation theory.

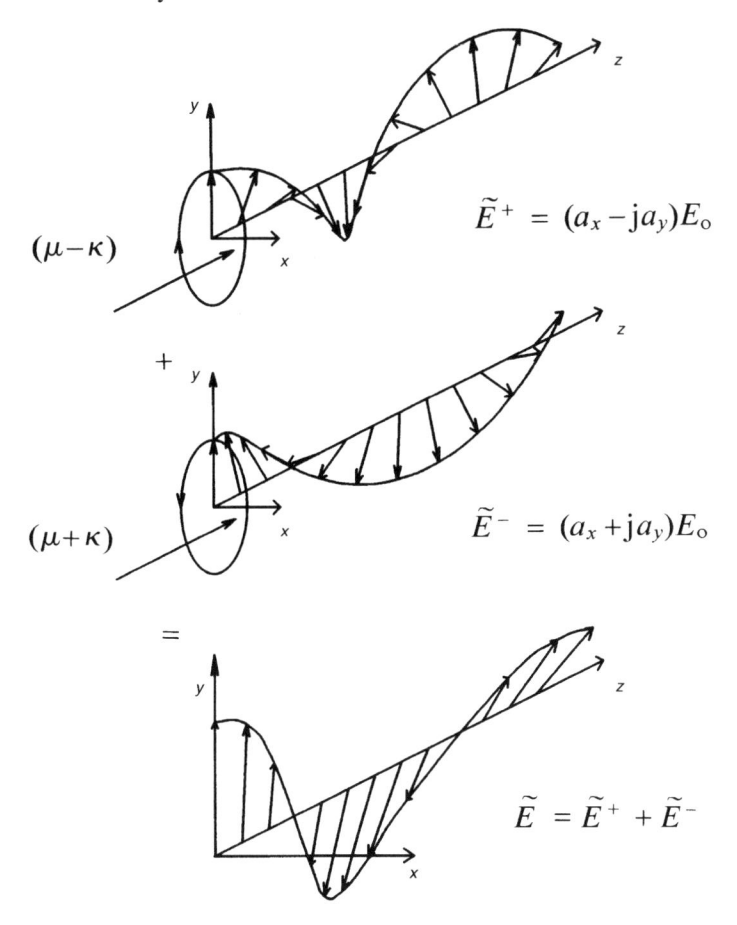

Figure 7.16 Faraday rotation (electric field)

7.12 Attenuation in Gyromagnetic Waveguides

If the waveguide has dielectric, magnetic or wall losses then its propagation constant γ is a complex quantity given by

$$\gamma = \alpha + j\beta$$

The presence of dielectric and magnetic losses is usually catered for by introducing complex constitutive parameters:

$$\mu = \mu' - j\mu'' \qquad \kappa = \kappa' - j\kappa''$$

$$\mu_z = \mu'_z - j\mu''_z \qquad \varepsilon_f = \varepsilon'_f - j\varepsilon''_f$$

If the losses are small then one description for the real part of the propagation constant which is particularly appropriate for computer-orientated practice is

$$\bar{\alpha} = \frac{\alpha}{k_0} = \varepsilon''_f \frac{\partial \bar{\beta}_\pm}{\partial \varepsilon_f} + \mu''_z \frac{\partial \bar{\beta}_\pm}{\partial \mu_z} + \mu'' \frac{\partial \bar{\beta}_\pm}{\partial \mu'} + \kappa'' \frac{\partial \bar{\beta}_\pm}{\partial \kappa'}$$

For the fully filled anisotropic waveguide with an electric wall

$$\bar{\beta}_0 = \frac{\beta_0}{k_0} = \sqrt{\varepsilon'_f \mu' - \left(\frac{1.84}{k_0 R}\right)^2 \frac{\mu'}{\mu'_z}}$$

$$\bar{\beta}_\pm = \frac{\beta_\pm}{k_0} = \sqrt{\left[\varepsilon'_f - \left(\frac{1.84}{k_0 R}\right)^2 \frac{1}{\mu'_z}\right](\mu' \pm \kappa')}$$

Evaluating each contribution gives

$$\varepsilon''_f \frac{\partial \bar{\beta}_\pm}{\partial \varepsilon''_f} = \frac{1}{2} \frac{\varepsilon''_f (\mu' \pm \kappa')}{\bar{\beta}_\pm}$$

$$\mu''_z \frac{\partial \bar{\beta}_\pm}{\partial \mu''_z} = \frac{1}{2} \frac{\mu''_z (\mu' \pm \kappa')}{(\mu'_z)^2 \bar{\beta}_\pm} \left(\frac{1.84}{k_0 R}\right)^2$$

$$\mu'' \frac{\partial \bar{\beta}_\pm}{\partial \mu'} = \frac{1}{2} \frac{\mu'' \bar{\beta}_0^2}{\mu' \bar{\beta}_\pm}$$

$$\kappa'' \frac{\partial \bar{\beta}_\pm}{\partial \kappa'} = \frac{1}{2} \frac{\kappa'' \bar{\beta}^2}{\kappa' \bar{\beta}_\pm}$$

References and Further Reading

Chambers, L. G. (1955) Propagation in a ferrite filled waveguide, *J. Mech. Appl. Math., USA*, 435–447.

Clarricoats, P. J. B. (1957) Some properties of circular waveguides containing ferrites, *Proc. IEE*, **104**, Part B, Suppl. 6, 286.

Duputz, A. M. and Priou, A. C. (1974) Computer analysis of microwave propagation in a ferrite loaded circular waveguide-optimization of phase-shifter longitudinal field sections, *IEEE Trans. on MTT*, **MTT-22**(6), 601–613.

Fox, A. G., Miller, S. E. and Weiss, M. T. (1955) Behaviour and applications of ferrites in the microwave region, *Bell System Tech. J.*, **34**, 5.

Gamo, H. (1953) The Faraday rotation of waves in a circular waveguide, *J. Phys. Soc., Japan*, **8**, 176–182.

Helszajn, J. and Gibson, A. P. (1987) Mode nomenclature of circular gyromagnetic and anisotropic waveguides with magnetic and open walls, *Proc. IEE*, **134**, Part H, No. 6.

Helszajn, J. and Gibson, A. P. (1989) Cutoff spaces of elliptical gyromagnetic planar circuits and waveguides using finite elements, *IEEE Trans. Microwave Theory Tech.*, **37**(1), 71–80.

Helszajn, J. and Sharp, J. (1997) Polynomial representations of split-phase constants of gyromagnetic waveguides with electric and magnetic walls, *IEEE Microwave and Guided Wave Letters*, **1**, 291.

Hogan, C. L. (1952) The ferromagnetic Faraday effect at microwave frequencies and its applications — the microwave gyrator, *Bell System Tech. J.*, **31**, 22–26.

Kales, M. L. (1953) Modes in waveguides containing ferrites, *J. Appl. Phys.*, **24**, 604–608.

Lee, C. S., Lee, S. W. and Chuang, S. L. (1985) Plot of modal field distribution in rectangular and circular waveguides, *IEEE Trans. Microwave Theory Tech.*, **MTT-33**, 271–274.

Luhrs, C. H. (1952) Correlation of the Faraday and Kerr magnetic-optical effects in transmission line terms, *Proc. IRE*, **40**, 76.

Luhrs, C. H. and Zull, W. J. (1953) Microwave polarization rotating device and coupling network, US Patent 2644930.

Melchor, J. L., Ayres, W. P. and Vartanian, R. H. (1956) Energy concentration effects in ferrite loaded waveguides, *J. Appl. Phys.*, **27**, 72–77.

Ohm, O. E. (1956) A broadband microwave circulator, *IRE Trans. on MTT*, **MTT-4**, 210.

Roberts, F. F. (1951) A note on the ferromagnetic Faraday effect of centimetre wavelengths, *J. Phys. Radium*, **12**, 305.

Suhl, H. and Walker, L. R. (1952) *Phys. Rev.*, **86**, 122.

Suhl, H. and Walker, L. R. (1954) Topics in guided wave propagation through gyromagnetic media, *Bell. System Tech. J.*, **33**, 579–659, 939–986, 1122–1194.

Van Trier, A. A. M. (1953) Guided electromagnetic waves in anisotropic media, *Appl. Sci. Res.*, **3**(B), 305–371.

Waldron, R. A. (1958) Electromagnetic wave propagation in cylindrical waveguides containing gyromagnetic media, *J. Br. IRE*, **18**, 597, 677, 733.

Waldron, R. A. (1960) Features of cylindrical waveguides containing gyromagnetic media, *J. Br. IRE*, **20**, 695–706.

Waldron, R. A. (1962) Properties of ferrite-loaded cylindrical waveguides in the neighbourhood of cut-off, *Proc. IEE*, **109**, Part B, Suppl. 21, 90–94.

Waldron, R. A. and Bowe, D. J. (1963) Loss properties of cylindrical waveguides containing gyromagnetic media, *J Br. IRE*, **25**, 321.

8

ISOTROPIC AND ANISOTROPIC OPEN CIRCULAR WAVEGUIDES

8.1 Introduction

While the open wall boundary condition met in the description of gyromagnetic resonators is often idealized by a magnetic wall recourse to an exact solution must always be preferable. The purpose of this chapter is to deal with some aspects of these sorts of problem regions. A property of the open demagnetized waveguide is that all modes except the symmetric ones are hybrid ones and that the dominant mode (designated HE_{11}) has no cutoff number. A feature of special interest in the design of gyromagnetic resonators is the fact that it supports degenerate counter-rotating circularly polarized magnetic fields along its axis. It is understood that this type of degeneracy may be split by perturbing the waveguide by a suitably magnetized ferrite medium. This feature may be employed in practice to rotate the standing wave pattern in gyromagnetic resonators. Such waveguides also exhibit so-called Faraday rotation as is readily appreciated.

8.2 Dielectric Waveguide

A open waveguide of some practical interest is the so-called dielectric waveguide (Figure 8.1). One feature of such a waveguide is that all modes except the symmetric ones are hybrid ones. Another is that the dominant one (designated HE_{11}) has no lower cut-off frequency. The modes in this type of waveguide are designated HE_{nm} or EH_{nm} depending on whether H_z or E_z, respectively, makes a larger contribution to the axial field.

Since the modes in this type of waveguide are hybrid ones separate wave equations must be solved for both E_z and H_z inside and outside the dielectric region. One suitable solution inside the dielectric region with ε_r replaced by ε_1 is

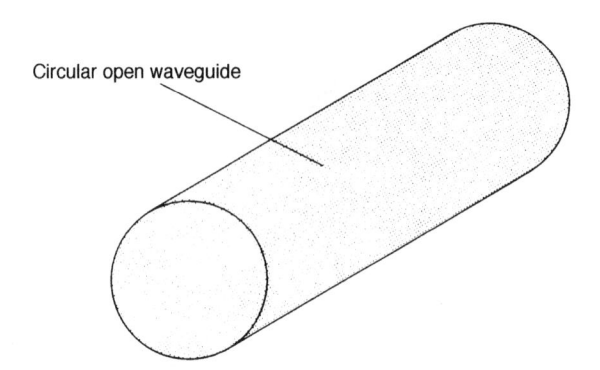

8 ISOTROPIC AND ANISOTROPIC OPEN CIRCULAR WAVEGUIDES

Figure 8.1 The open waveguide

$$E_{zd} = A_n J_n(k_1 r) \exp [j(\omega t - \beta z - n\theta)] \tag{8.1}$$

$$H_{zd} = B_n J_n(k_1 r) \exp [j(\omega t - \beta z - n\theta)] \tag{8.2}$$

where

$$k_1^2 = -\beta^2 + k_0^2 \varepsilon_1 \tag{8.3}$$

If the fields are to be confined to the dielectric region those in the other medium must be decaying exponentially from it. One solution that satisfies the wave equations for E_z and H_z with ε_r replaced by ε_2 is

$$E_{za} = C_n K_n(k_2 r) \exp [j(\omega t - \beta z - n\theta)] \tag{8.4}$$

$$H_{za} = D_n K_n(k_2 r) \exp [j(\omega t - \beta z - n\theta)] \tag{8.5}$$

where

$$k_2^2 = \beta^2 + k_0^2 \varepsilon_2 \tag{8.6}$$

$K_n(x)$ is the modified Hankel function of the second kind of order n which represents an outwardly travelling decaying wave. It can be written in terms of either Hankel functions $H_n^{(1)}$ or $H_n^{(2)}$

$$K_n(x) = \frac{\pi}{2} (j)^{(n+1)} H_n^{(1)}(jx) \tag{8.7}$$

or

$$K_n(x) = \frac{\pi}{2} (j)^{(n+1)} H_n^{(2)}(-jx) \tag{8.8}$$

8.2 DIELECTRIC WAVEGUIDE

The two Hankel functions $H_n^{(1)}$ and $H_n^{(2)}$ are related to the Bessel function of the first $(J_n(x))$ and second $(Y_n(x))$ kinds by

$$H_n^{(1)} j(x) = J_n(x) - jY_n(x) \tag{8.9}$$

$$H_n^{(2)} -j(x) = J_n(x) - jY_n(x) \tag{8.10}$$

in an analogous fashion to that existing between the common exponential $(\exp \pm x)$ and trigonometric functions $(\cos jx$ and $\sin jx)$.

$$\exp(x) = \cos(jx) - j\sin(jx) \tag{8.11}$$

$$\exp(-x) = \cos(jx) + j\sin(jx) \tag{8.12}$$

The second kind of Bessel functions are also sometimes referred to as Neuman functions.

For $x \gg n$

$$J_n(x) = \sqrt{\left(\frac{2}{\pi x}\right)} \cos\left(x - \frac{2n+1}{4}\pi\right) \tag{8.13}$$

$$J_n(x) = \sqrt{\left(\frac{2}{\pi x}\right)} \sin\left(x - \frac{2n+1}{4}\pi\right) \tag{8.14}$$

$K_n(x)$ is now obtained by having recourse to either equation (8.7) or (8.8).

$$K_n(x) = \sqrt{\left(\frac{\pi}{2x}\right)} \exp(-x) \tag{8.15}$$

Scrutiny of this latter equation indicates that the modified Hankel function represents a decaying outward travelling wave. The modified Hankel function is also sometimes referred to as a Bessel function of the third kind. It is also possible to define a modified Hankel function $(I_n(x))$ which represents a growing inward travelling wave but its definition is outside the result of this work.

The other field components are given in terms of E_z and H_z in the manner discussed in Chapter 7. The continuity of the tangential components of E and H at $r = R$ $(E_z, H_z, E_\theta,$ and $H_\theta)$ produces four equations from which the unknown coefficients A_n, B_n, C_n and D_n may be evaluated:

$$E_{\theta d} = E_{\theta a} \quad \text{at the dielectric-air boundary,}$$

$$H_{\theta d} = H_{\theta a} \quad \text{at the dielectric-air boundary,}$$

$$E_{zd} = E_{za} \quad \text{at the dielectric-air boundary,}$$

$$H_{zd} = H_{za} \quad \text{at the dielectric-air boundary.}$$

The suffix a in the descriptions of E and H indicates that the values are in the air region. In the dielectric region the suffix d is used to denote the fields.

The application of the above boundary conditions produces a homogeneous linear set in the variables A_n, B_n, C_n, D_n. The condition for consistency is that the determinant of the coefficients shall vanish, and this condition gives the characteristic equation:

$$\left[\frac{J'(u)}{uJ_n(u)} + \frac{K'_n(w)}{wK_n(w)}\right]\left[\left(\frac{\varepsilon_1}{\varepsilon_2}\right)\frac{J'_n(u)}{uJ_n(u)} + \frac{K'_1(w)}{wK_1(w)}\right]$$

$$= n^2\left[\frac{1}{u^2} + \frac{1}{w^2}\right]\left[\left(\frac{\varepsilon_1}{\varepsilon_2}\right)\frac{1}{u^2} + \frac{1}{w^2}\right] \tag{8.16}$$

where ε_1 is the dielectric constant of the ferrite and ε_2 is that of the surrounding material and

$$u^2 = (k_0^2 \varepsilon_1 - \beta^2)R^2 \tag{8.17}$$

$$w^2 = (\beta^2 - k_0^2 \varepsilon_2)R^2 \tag{8.18}$$

R is the radius of the ferrite, k_0 the wavenumber in free space given by $\omega\sqrt{\mu_0\varepsilon_0}$ and β the propagation constant of the structure. and is the unknown of the problem region. $J_n(u)$ is the Bessel function of the first kind and $K_n(w)$ is the modified Bessel function of the second kind representing outward decaying waves; n is the mode number.

When n = 0, the right-hand side of the characteristic equation vanishes, and each factor on the left-hand side must equal zero. These two factors give the characteristic equations for the axially symmetrical TM_{0m} and TE_{0m} modes as

$$\frac{J'_n(u)}{uJ_n(u)} + \frac{K'_n(w)}{wK_n(w)} = 0, \quad \text{TM mode} \tag{8.19}$$

$$\left(\frac{\varepsilon_1}{\varepsilon_2}\right)\frac{J'_n(u)}{uJ_n(u)} + \frac{K'_1(w)}{wK_n(w)} = 0, \quad \text{TE mode} \tag{8.20}$$

The symmetrical modes described by these equations do not exhibit Faraday rotation if perturbed by a gyromagnetic medium.

The dominant mode in this waveguide is the HE_{11} mode; the phase constant of this mode is depicted in Figure 8.2. This illustration suggests that such a solution is asymptotic to that of a closed circular waveguide with a magnetic sidewall propagating the TM_{11} mode. A feature of this mode is that the magnetic (and electric)

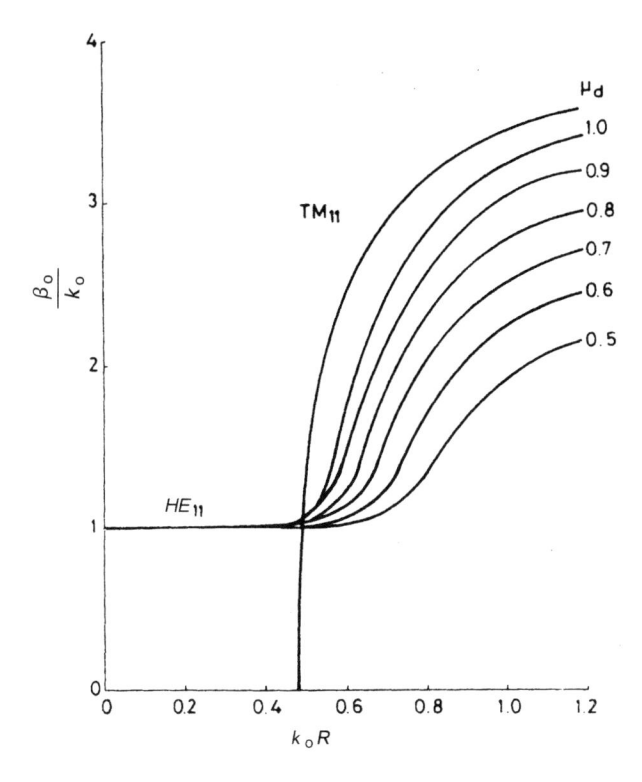

Figure 8.2 Phase constants of open and closed ferrite waveguides (reprinted with permission, Helszajn, J and Sharp, J. (1986) Dielectric and permeability effects in HE_{11} open demagnetized ferrite resonators, *IEE Proc.*, **133**, Part H, 271–276)

field is circularly polarized at the axis of the waveguide. It is therefore appropriate for use in Faraday rotation devices and for the construction of gyromagnetic resonators. Another property of this solution is that cut off occurs at $R = 0$. The field pattern for the dominant and some higher order modes in this type of waveguide are illustrated in Figure 8.3.

In the description of this type of waveguide n specifies the order of the Bessel function and m gives its roots.

8.3 Open Anisotropic Circular Waveguide

The characteristic equation of the open anisotropic waveguide with a diagonal permeability whose entries in the transverse plane corresponds to the eigenvalues of the tensor permeability is also of some interest. The form of the diagonal permeability is in this instance specified by

$$[\mu] = \begin{bmatrix} \mu \pm \kappa & 0 & 0 \\ 0 & \mu \pm \kappa & 0 \\ 0 & 0 & \mu_z \end{bmatrix} \quad (8.21)$$

8 ISOTROPIC AND ANISOTROPIC OPEN CIRCULAR WAVEGUIDES

The ensuing characteristic equation may be deduced without difficulty. The required result for $n = 1$ is

$$\left[\sqrt{\mu_z \mu_t} \frac{J'_1[\sqrt{(\mu_z/\mu_t)}\, u]}{\mu J_1[\sqrt{(\mu_z/\mu_t)}\, u]} + \frac{K'_1(w)}{w K_1(w)}\right]\left[\frac{\varepsilon_f J'_1(u)}{u J_1(u)} + \frac{K'_1(w)}{w K_1(w)}\right]$$

$$= \left[\frac{1}{u^2} + \frac{1}{w^2}\right]\left[\frac{\varepsilon_f \mu_t}{u^2} + \frac{1}{w^2}\right] \tag{8.22}$$

u and w have the meanings of equations (8.17) and (8.18) and

$$\mu_t = \mu \pm \kappa \tag{8.23}$$

If $\mu_t = \mu_z = 1$ then the characteristic equation reduces to that of the open isotropic waveguide.

Figure 8.4 illustrates the split phase constants of the dominant HE_{11} mode for this type of waveguide for some typical parameters. As κ increases to unity the lower branch is asymptotic to the condition for free space propagation, $\beta/k_0 = 1$, in keeping with the fact that the power in this instance resides outside the gyromagnetic rod. Figure 8.5 compares the split phase constants of open and closed gyromagnetic waveguides.

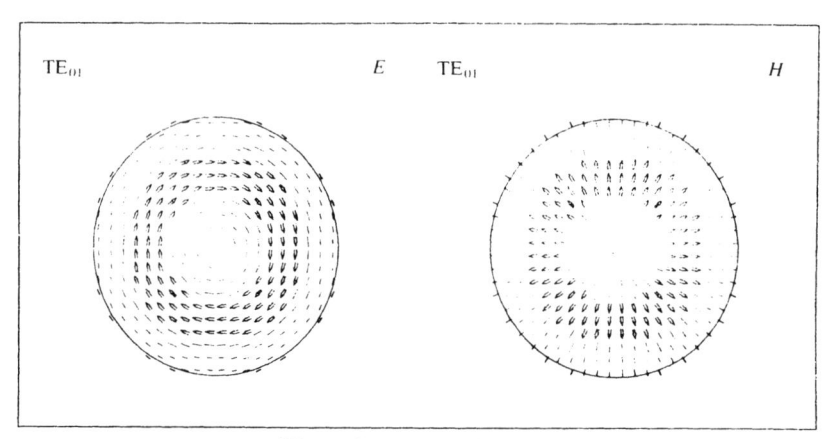

TE_{01} mode. $x = 3.2666$ $k_0 a = 1$

Figure 8.3 Transverse field distributions for some open dielectric waveguide modes (Source: Kajfez, D. (1983) Modal field patterns in dielectric rod waveguide, *Microwave J.*, 181)

8.3 OPEN ANISOTROPIC CIRCULAR WAVEGUIDE

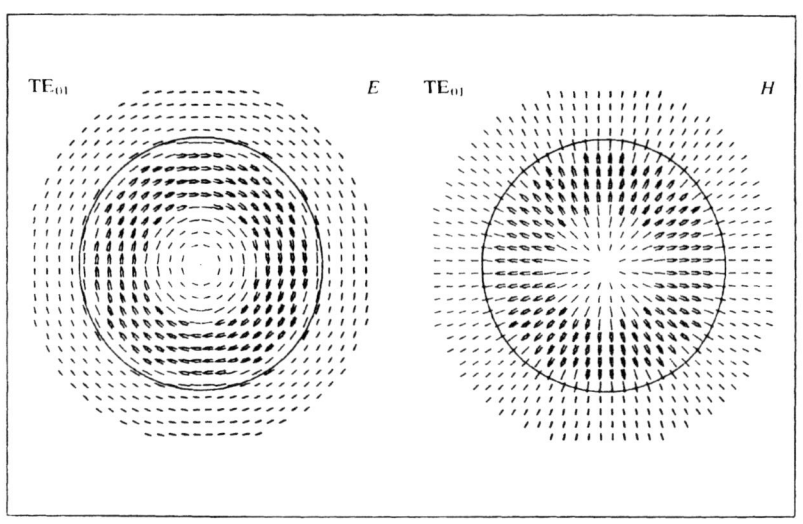

TE$_{01}$ mode, $x = 2.7345$, $k_0 a = 0.5$

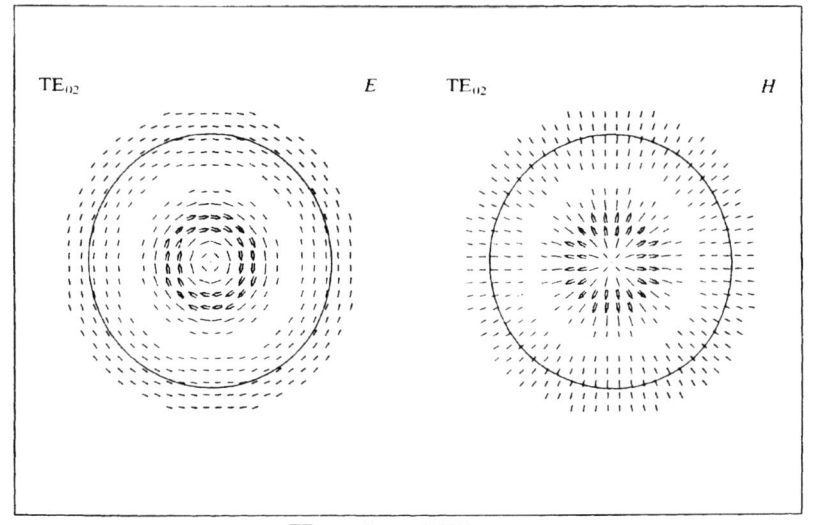

TE$_{02}$ mode, $x = 5.7753$, $k_0 a = 1$

Figure 8.3 *cont.*

TM$_{01}$ mode, $x = 3.8085$, $k_o a = 1$

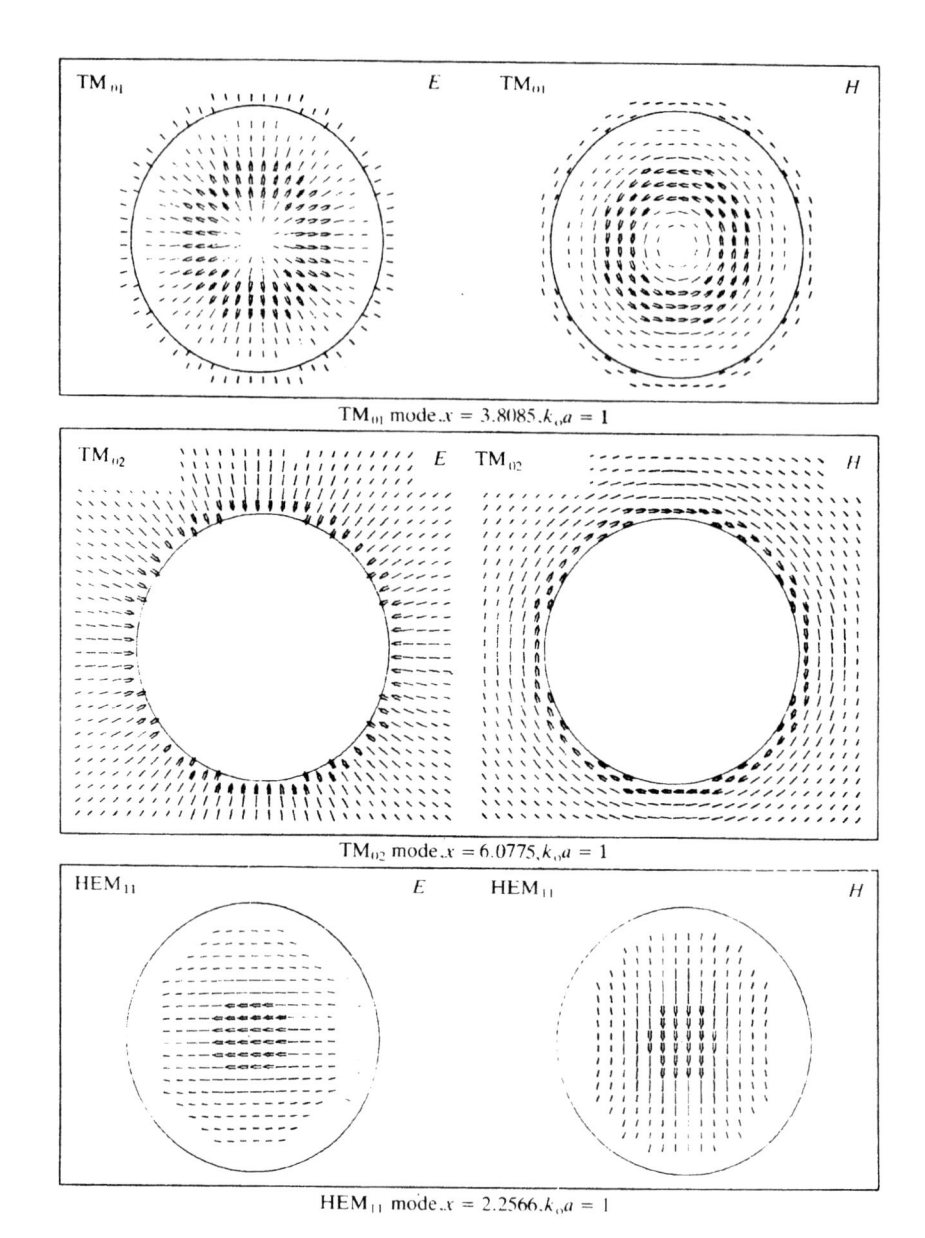

TM$_{02}$ mode, $x = 6.0775$, $k_o a = 1$

HEM$_{11}$ mode, $x = 2.2566$, $k_o a = 1$

Figure 8.3 *cont.*

8.3 OPEN ANISOTROPIC CIRCULAR WAVEGUIDE

HEM$_{11}$ mode, $x = 2.1860$, $k_o a = 0.5$

HEM$_{12}$ mode, $x = 4.4481$, $k_o a = 1.0$

HEM$_{13}$ mode, $x = 5.3483$, $k_o a = 1.0$

Figure 8.3 *cont.*

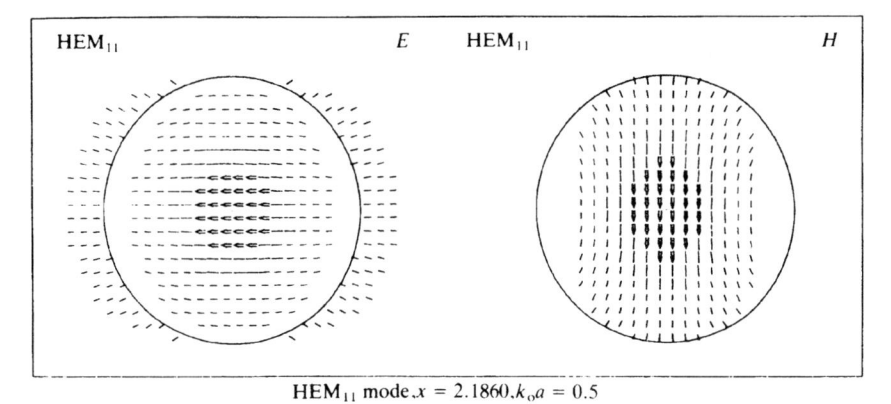

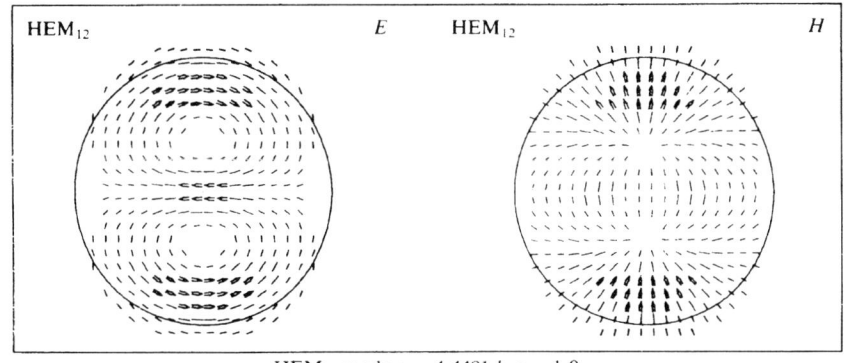

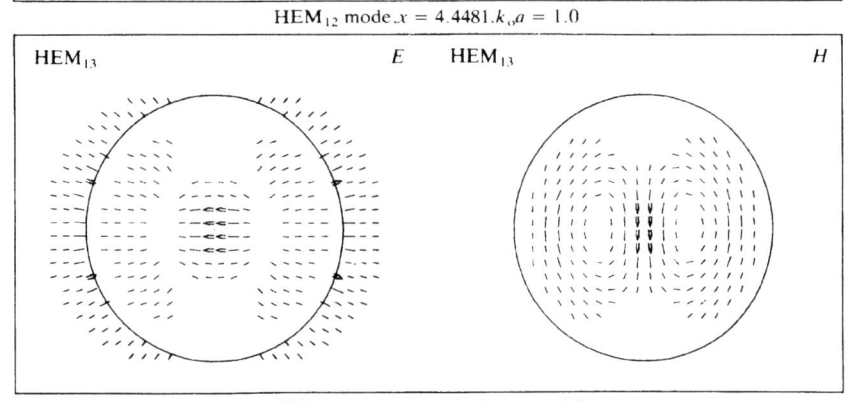

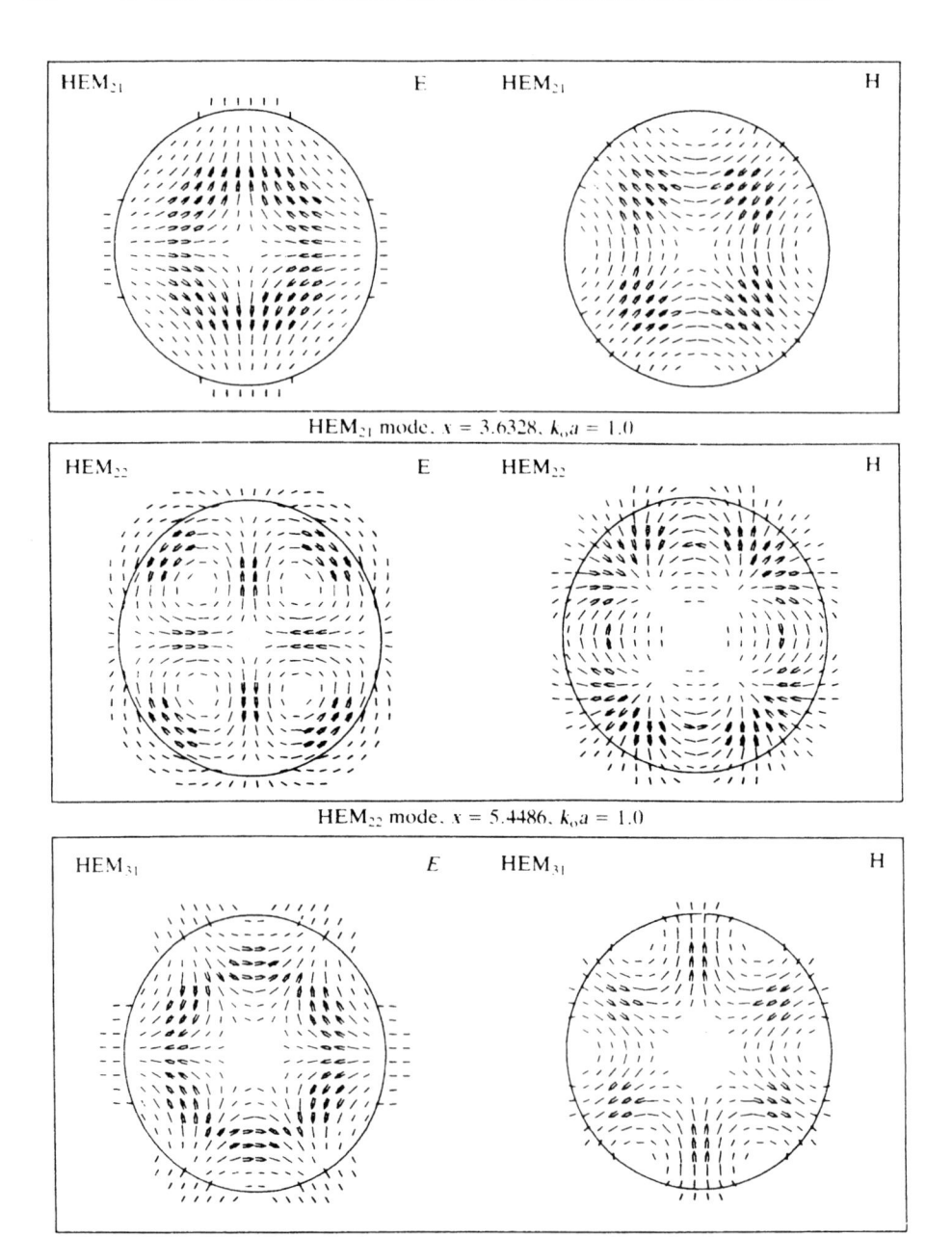

148   8 ISOTROPIC AND ANISOTROPIC OPEN CIRCULAR WAVEGUIDES

HEM$_{21}$ mode, $x = 3.6328$, $k_0 a = 1.0$

HEM$_{22}$ mode, $x = 5.4486$, $k_0 a = 1.0$

HEM$_{31}$ mode, $x = 4.9419$, $k_0 a = 1$

Figure 8.3 *cont.*

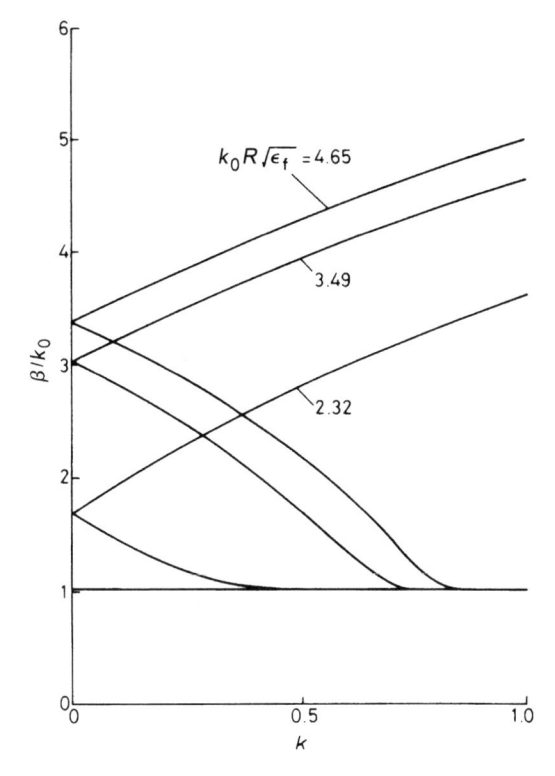

Figure 8.4 Split phase constants of the anisotropic open wall waveguide ($\mu_z = 1$, $\mu = 1$, $\varepsilon_f = 15$)

8.4 Effective dielectric constant of open dielectric waveguide

Equivalent waveguide models are often employed to represent inhomogenous waveguides or transmission lines; it is the purpose of this section to formulate one for the open dielectric waveguide. One possible equivalence between an open dielectric waveguide with a relative dielectric constant ε_r and radius R propagating the HE$_{11}$ mode and a circular waveguide with a magnetic wall boundary condition but with an effective dielectric constant ε_{eff} and effective radius R_{eff} propagating the TM$_{11}$ mode, is depicted in Figure 8.6. ε_{eff} and R_{eff} may be determined from a knowledge of the phase constant of the inhomogenous waveguide and the complex power flow. If only the phase constant is required, or if only matching between similar cross-sections is required, then an approximate waveguide model in terms of an effective dielectric constant is sufficient and will be adopted here in the first instance.

The effective dielectric constant (ε_{eff}) of the equivalent waveguide model with an idealized magnetic wall may now be evaluated from a knowledge of the phase constant (β) of the open waveguide by making use of the following relationship:

150 8 ISOTROPIC AND ANISOTROPIC OPEN CIRCULAR WAVEGUIDES

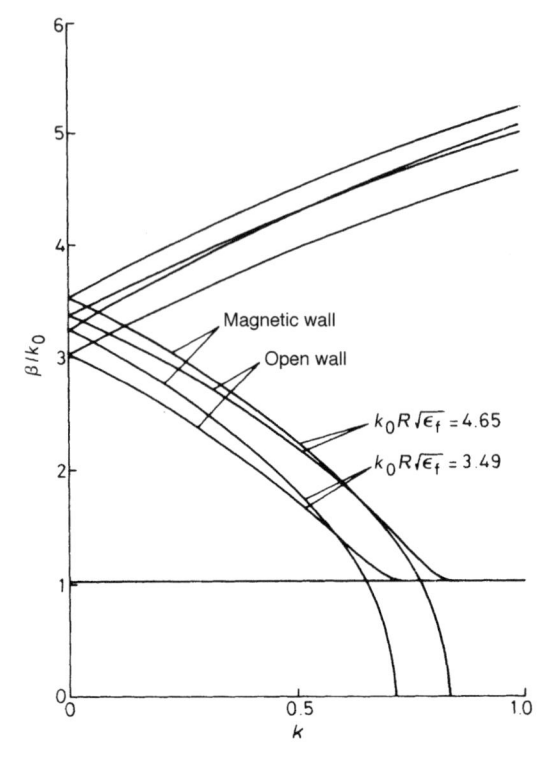

Figure 8.5 Comparison between phase constants in anisotropic open and closed (magnetic wall) waveguides ($\mu_z = 1$, $\mu = 1$, $\varepsilon_f = 15$)

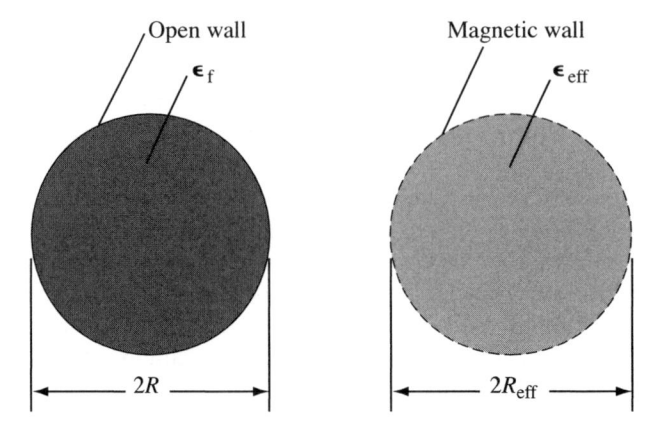

Figure 8.6 Equivalence between open dielectric waveguide with imperfect magnetic walls and equivalent waveguide model with idealized walls

8.4 EFFECTIVE DIELECTRIC CONSTANT

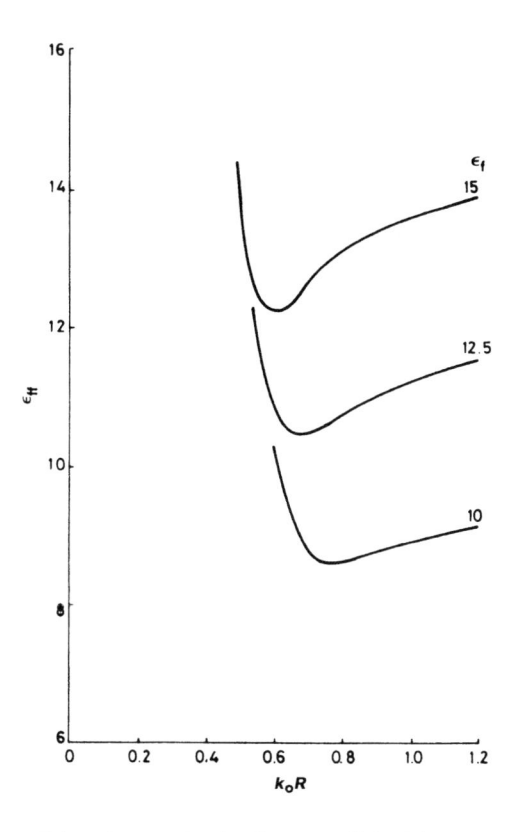

Figure 8.7 Effective dielectric constant of open dielectric waveguide (reprinted with permission, Helszajn, J. and Sharp, J. (1986) Dielectric and permeability effects in HE_{11} open demagnetized ferrite resonators, *IEE Proc.*, **133**, Part H, 271–276)

$$\beta^2 = k_0^2 \varepsilon_{\text{eff}} - k_c^2 \tag{8.24}$$

where

$$k_0 = \frac{2\pi}{\lambda_0} \tag{8.25}$$

$$k_c = \frac{1.84}{R} \tag{8.26}$$

Figure 8.7 illustrates the relationship between the effective dielectric constant ε_{eff} and radial wavenumber $k_0 R$ with the relative dielectric constant ε_f equal to 10, 12.5 and 15. This result indicates that for $\varepsilon_f = 15$, say, $\varepsilon_{\text{eff}} = 12.2$ for $k_0 R = 0.60$, 12.6 for $k_0 R = 0.70$, 13.1 for $k_0 R = 0.80$ and 13.4 for $k_0 R = 0.90$. The origin of this discrepancy may be separately understood by evaluating the power flow through the waveguide (P_i) and outside it (P_0). Such a calculation indicates that for $k_0 R = 0.80$, $\varepsilon_f = 16$ and $\mu = 1$, say, P_i/P_0 is of the order 25.

Since it is not realistic to form an equivalence between an open waveguide and one that is cutoff, the curves for ε_{eff} are not meaningful below the value of $k_0 R$ at which the slopes of the two solutions in Figure 8.7 diverge.

8.5 Effective Permeability of Open Demagnetized Ferrite Waveguide

It is also possible to derive an equivalence between an open demagnetized ferrite or garnet waveguide with a diagonal permeability described by

$$\mu_x = \mu_y = \mu_z = \mu_d \tag{8.27}$$

and a scalar dielectric constant ε_f and one with an idealised magnetic wall but with effective constitutive parameters. The characteristic equation of the open demagnetized waveguide is in this instance given by

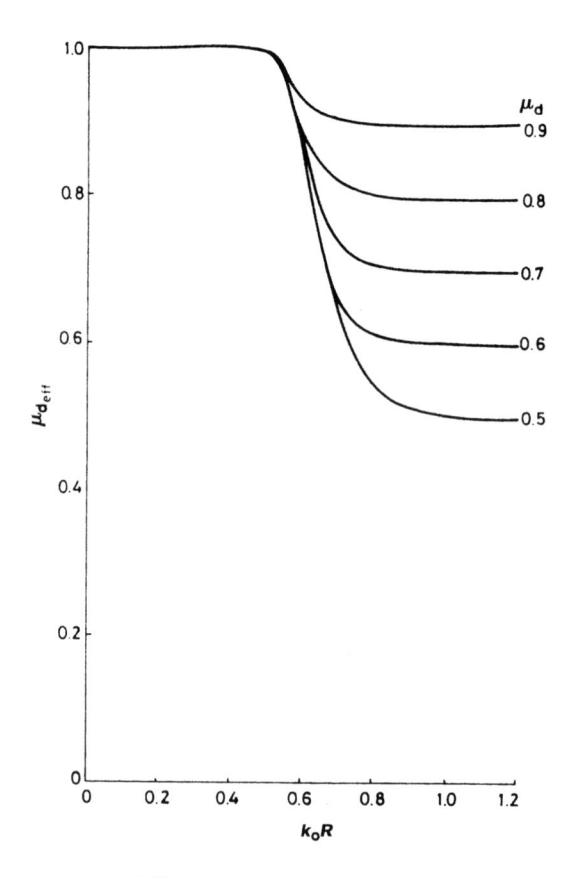

Figure 8.8 Effective permeability constant of open ferrite waveguide $\varepsilon_f = 15$ (reprinted with permission, Helszajn, J. and Sharp, J. (1986) Dielectric and permeability effects in HE_{11} open demagnetized ferrite resonators, *IEE Proc.*, **133**, Part H, 271–276)

$$\left[\mu_d \frac{J'_1(v)}{vJ_1(v)} + \frac{K'_1(w)}{wK_1(w)}\right]\left[\left(\frac{\varepsilon_1}{\varepsilon_2}\right)\frac{J'_1(v)}{vJ_1(v)} + \frac{K'_1(w)}{wK_1(w)}\right]$$

$$= \left[\frac{1}{v^2} + \frac{1}{w^2}\right]\left[\left(\frac{\varepsilon_1}{\varepsilon_2}\right)\frac{\mu_d}{u^2} + \frac{1}{w^2}\right] \qquad (8.28)$$

where

$$v^2 = (k_0^2 \varepsilon_1 \mu_d - \beta^2)R^2 \qquad (8.29)$$

and w is given by equation (8.18).

It is assumed here that the value of ε_{eff} derived from the solution of the open dielectric waveguide is not modified to first order by the perturbation in μ_d, especially since the magnetic field H_z is small or zero at the outer surface of the open waveguide. This approximation permits an equivalent waveguide model also to be adopted for the open demagnetized waveguide

$$\beta^2 = k_0^2 \varepsilon_{\text{eff}} \mu_{d,\text{eff}} - k_c^2 \qquad (8.30)$$

The effective permeability derived in this way is depicted in Figure 8.8. Scrutiny of this result indicates that it does not differ too much from its intrinsic value over the interval for which the equivalence between the open and equivalent waveguide models in valid.

References and Further Reading

Chatterjee, S. K. (1954) Propagation of microwave through a cylindrical metallic guide filled coaxially with two different dielectrics, *J. Indian Inst. Sci.*, **39**, 1.

Chatterjee, S. K. and Chatterjee, R. (1965) Dielectric loaded waveguides – a review of theoretical solutions, *Radio and Electronic Eng.*, **30**, 259–288.

Clarricoats, P. J. B. (1961) Propagation along bounded and unbounded dielectric rods, Pt. 2, *Proc. IEE*, **108**(C), 177.

Helszajn, J. and Sharp, J. (1986) Dielectric and permeability effects in HE_{11} open demagnetized ferrite resonators, *IEE Proc.*, **133**(H), 271–276.

Helszajn, J. and Gibson, A. P. (1989) Cutoff spaces of elliptical gyromagnetic planar circuits and waveguides using finite elements, *IEEE Trans. Microwave Theory Tech.* **37**(1), 71–80.

Kajfez, D. (1983) Modal field patterns in dielectric rod waveguide, *Microwave J.*, 181.

Nagelberg, E. R. and Hoffspiegel, J. M. (1967) Computer-graphic analysis of dielectric waveguides, *IEEE Trans. on Microwave Theory Tech.*, **MTT-15**, 187–189.

Snitzer, E. (1971) Cylindrical dielectric waveguide modes, *J. Optical Soc. Am.*, **51**, 491–498.

Zaki, K. A. and Chen, C. (1985) Intensity and distribution of hybrid-mode fields in dielectric-loaded waveguides, *IEEE Trans. on Microwave Theory Tech.*, **MTT-33**(12), 1442–1447.

9

THE DIELECTRIC CAVITY RESONATOR

9.1 Introduction

A classic resonator employed in the construction of many waveguide circulators consists of a pair of quarter wave long open gyromagnetic or ferrite waveguides propagating the HE_{11} hybrid mode separated by a suitable dielectric region. One flat face of each bit is closed by an electric wall and the symmetry plane may either support an electric or magnetic wall. Unlike the planar gyromagnetic resonator the cavity one supports a standing wave of the alternating radio frequency fields both along its axis and in its plane. This sort of resonator is, in the design of the waveguide circulator, quite often referred to as a turnstile one because it is usually embodied in a so called turnstile waveguide junction. It is, however, strictly speaking a gyromagnetic cavity with magnetic or quasi-open walls. The basic resonator in its elementary form is constructed from a gyromagnetic waveguide with a magnetic wall using a quarter long section open-circuited at one end and short-circuited at the other. A half-wave one short or open-circuited at each end is also a possibility. The usual geometry met in the design of waveguide circulators consists of a pair of coupled quarter-wave long resonators operating in either its even or odd mode. It fixes the midband frequency of this sort of circulator. An understanding of the dielectric resonator is therefore a prerequisite for design. The open problem region propagating the hybrid HE_{11} mode is often approximated for the purpose of engineering by that of a closed one with an ideal magnetic wall propagating a TM_{11} one with its relative dielectric constant replaced by an effective one. The purpose of this chapter is to consider some aspects of this problem.

9.2 Resonator Structures

An important class of resonators met in the construction of H-plane commercial waveguide circulators using E-plane turnstile resonators is one using either single or coupled quarter-wave long open circular gyromagnetic waveguides open-circuited at one end and short-circuited at the other, or half-wave long open waveguides open-circuited at both ends. Figure 9.1 depicts schematic diagrams of the three typical arrangements using disk resonators met in the design of this type of circulator. Introduction of an image plane or electric wall in the configurations in Figure 9.1(b) and (c) indicates that these are dual in that a single set of variables may be used to describe either geometry. The structure in Figure 9.1(a) is also equivalent to the other ones, except that its susceptance slope parameter is twice that of the other two. A quarter-wave long magnetized ferrite waveguide short-circuited at one end, and open-circuited or loaded by an image or electric wall at the other, is therefore a suitable prototype for the construction of this class of device. Such resonators are usually constructed from a section of an open gyromagnetic waveguide propagating a hybrid HE_{11} mode. The solution to each structure considered here coincides with the odd mode solution of two coupled

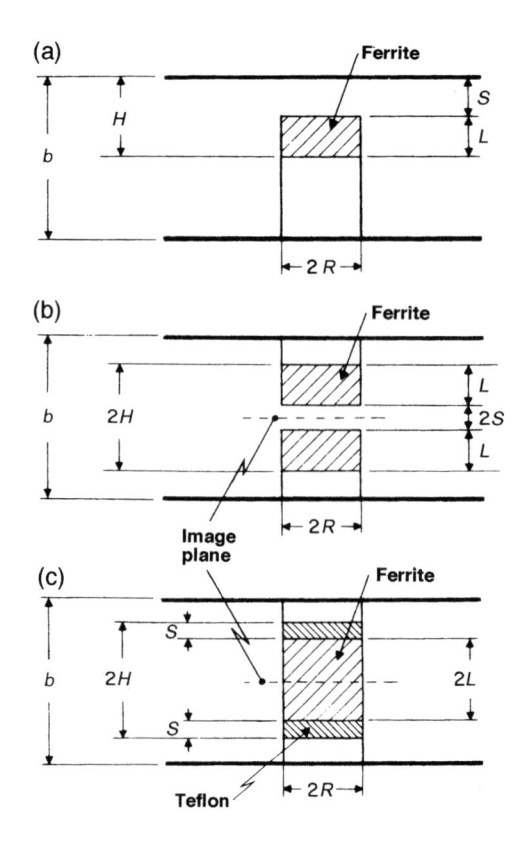

Figure 9.1 Schematic diagrams of (a) single disc, (b) coupled discs and (c) cylindrical resonator

quarter-wave long resonators supporting $HE_{1,1,1/2}$ resonances connected by a contiguous circular waveguide with a fictitious magnetic wall below cut-off. The resonator met in the dual E-plane circulator differs from that of the H-plane junction one in that its open flat faces are loaded by a magnetic instead of an electric wall. The characteristic equation for the frequency of the degenerate counter-rotating modes of the E-plane junction using H-plane turnstile resonators therefore

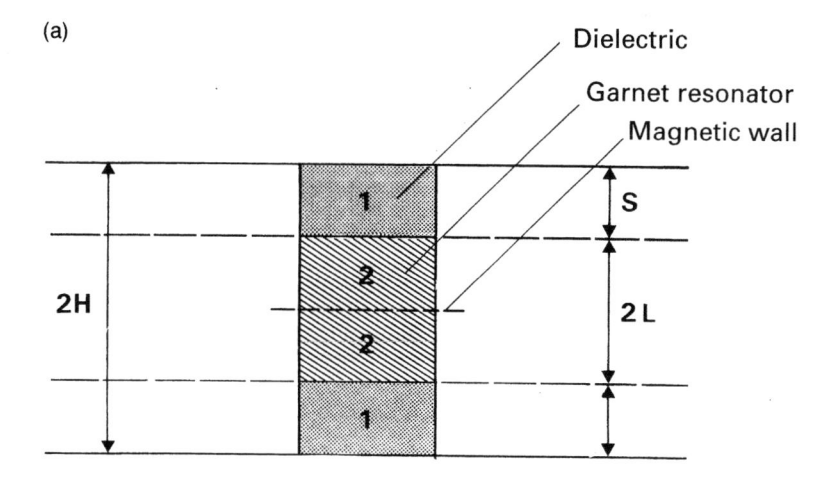

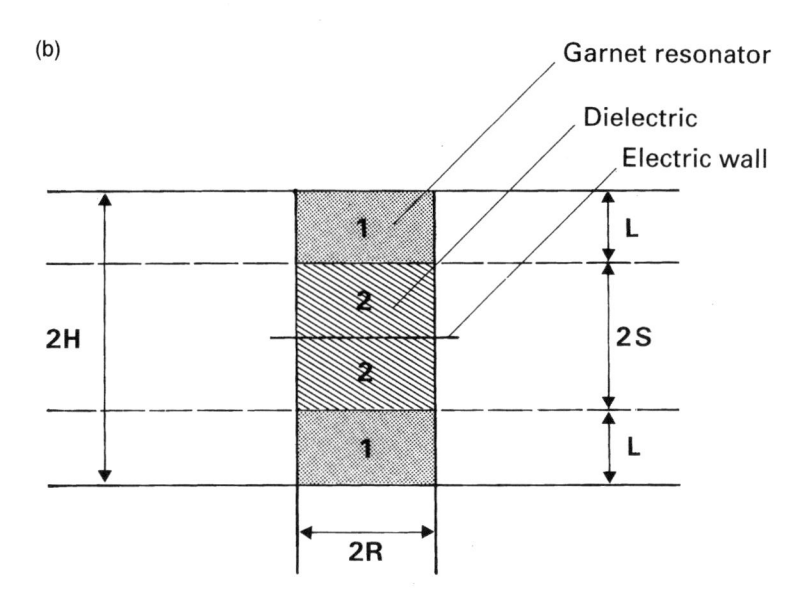

Figure 9.2 (a) Even-mode geometry of coupled turnstile resonators (magnetic wall arrangement), (b) Odd-mode geometry of coupled turnstile resonators (electric wall arrangement)

coincides with the even eigenvalue of two $HE_{1,1,1/2}$ resonators coupled by a section of round waveguide with a contiguous magnetic wall below cut-off. The duality between the H and E-plane arrangements is illustrated in Figures 9.2(a) and 9.2(b).

9.3 Closed Dielectric Resonator

An understanding of the open dielectric resonator may be obtained by first considering the more simple problem region consisting of a section of circular waveguide with an ideal magnetic wall with one flat face short-circuited and the other open-circuited. The characteristic equation of this arrangement from which its resonance frequencies may be deduced is

$$\cot(\beta_0 L_0) = 0 \tag{9.1}$$

The lowest resonance of this sort of resonator coincides with

$$\beta_0 L_0 = \pi/2 \tag{9.2}$$

β_0 may be calculated by either assuming a magnetic wall boundary condition, adopting an equivalent waveguide model or by solving the open dielectric waveguide problem region exactly. The solution adopted here for β_0 is based on the dominant TM_{11} mode of a waveguide with an idealized magnetic wall.

$$\left(\frac{\beta_0}{k_0}\right)^2 = \varepsilon_f - \left(\frac{k_{mn}}{k_0}\right)^2 \tag{9.3}$$

The separation constant of the dominant mode of a circular waveguide with an ideal magnetic wall is

$$k_{11} = 1.84/R \tag{9.4}$$

where

$$k_{11} = k_0\sqrt{\varepsilon_f} \tag{9.5}$$

and k_0 is the free space wave-number at its center frequency.

$$k_0 = \omega\sqrt{\mu_0 \varepsilon_0} \tag{9.6}$$

L_0 is the length of the closed resonator (m), R is the radius (m) of the waveguide, k_0 is the free-space wavenumber (rad/m), ε_f is the relative dielectric constant of the garnet or ferrite resonator, ω is the radian frequency (rad/s), μ_0 and ε_0 are the constitutive parameter of free space. Equation (9.3) is often written for the purpose of calculation as

$$(k_0R)^2 = \frac{\left(\frac{\pi}{2}\right)^2\left(\frac{R}{L}\right)^2 + (1.84)^2}{\varepsilon_f} \qquad (9.7)$$

If k_0R is taken as the independent variable then the aspect ratio of the resonator is given by

$$\left(\frac{R}{L}\right)^2 = \frac{(k_0R)^2\varepsilon_f - (1.84)^2}{\left(\frac{\pi}{2}\right)} \qquad (9.8)$$

It is, of course, separately necessary to ensure that the ensuing value of k_0R does not permit the intrusion of higher order modes in the passband of the required frequency response of the device or that it is not too close to the cut-off number of the equivalent waveguide. One way to deal with this problem is to define the quantity

$$\left(\frac{f_0}{f_c}\right) = \left(\frac{k_0R}{k_cR}\right) \qquad (9.9)$$

Values of f_0/f_c between

$$1.50 \le f_0/f_c \le 2.0 \qquad (9.10)$$

are appropriate for design.

Figure 9.3 depicts the standing wave solutions for the dominant modes in this sort of resonator.

9.4 The Approximation Problem

The most simple solution to the resonator problem is to approximate all the open walls by an idealized magnetic wall. Another approximation to the frequency of this sort of structure is to cater for the effect of the electric or magnetic wall at the symmetry plane of the problem region but to neglect that due to the open lateral wall of the cylindrical waveguide. While this latter approximation is adequate in some situations, it is not so for the design of high-quality junction circulators. A similar geometry is also encountered in the area of filters using dielectric resonators employing the first symmetric $TE_{0,1,1}$ mode in a dielectric waveguide. One approximation adopted in this latter problem, which goes a long way to reconcile theory and practice, is to satisfy the boundary condition between regions 1 and 4 and neglect those between 2 and 3 and 4 in Figure 9.4(a). A number of other solutions have also been described for this and similar resonator problem regions but are too numerous to be mentioned here. The method utilized in this chapter to determine the operating frequency of the $HE_{1,1,1/2}$ circuit in

160   9 THE DIELECTRIC CAVITY RESONATOR

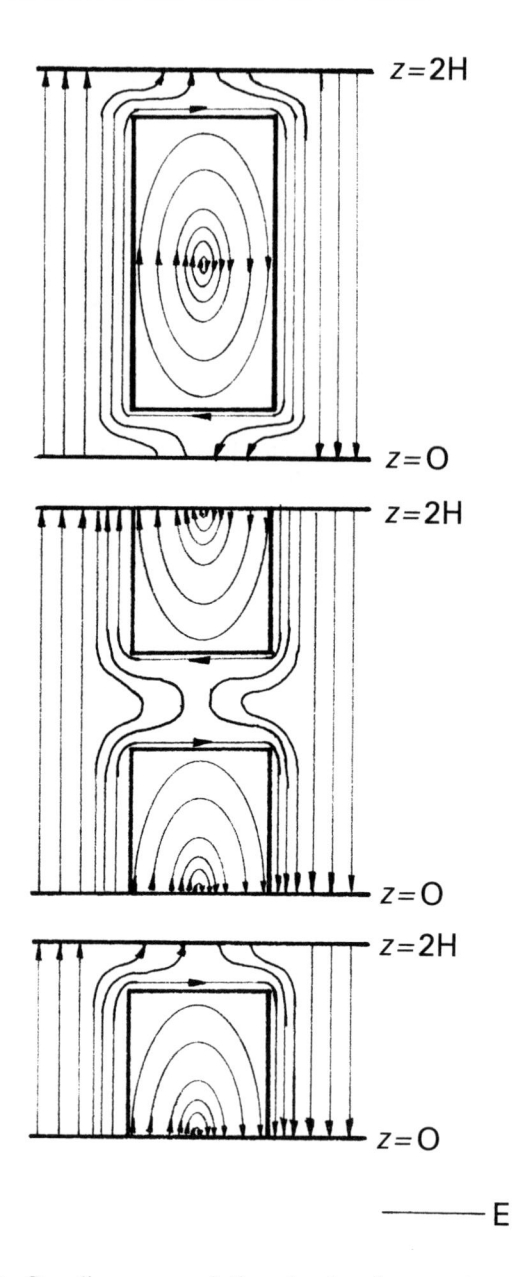

Figure 9.3 Standing wave solutions in closed resonator

Figure 9.4a rests on reconciling the boundary conditions between regions 1 and 4 by forming an equivalent waveguide model of the open dielectric waveguide having the same radius as the original waveguide but with an effective dielectric constant (ε_{eff}) instead of the constituent one (ε_f) and an ideal magnetic wall. A more complete equivalent waveguide model would, of course, be in terms of an effective dielectric constant and an effective radius, but this would require statements about both the phase constant and power flow of the open waveguide. In fact, the former approximation appears at first sight adequate for engineering purposes. The effect of the diagonal element μ of the tensor permeability on the frequency of the demagnetised ferrite waveguide may be separately catered for. The boundary condition between regions 1 and 2 is then satisfied by enclosing region 2 by a contiguous magnetic wall as a preamble to forming a transverse resonance relationship between regions 1 and 2. For TM-type solutions, only the effective dielectric constant of the equivalent waveguide appears explicitly in the transverse resonance condition; the permeability or effective permeability only enters in the description of the phase constant of the equivalent waveguide model of regions 1 and 4. Figure 9.4b illustrates the dual problem consisting of two coupled quarter-wave long resonators short-circuited at one end and open-circuited or loaded by an image wall. The related single quarter-wave long arrangement is also understood.

The influence of any coupling effects are omitted in the current discussion, so that the results outlined here, strictly speaking, apply to loosely coupled resonators only.

9.5 Odd Mode Solution of Coupled Dielectric Resonators

The operating frequency of the H-plane circulator coincides with the odd mode solution of two coupled isotropic resonators with an open sidewall. One solution to this problem is to replace the open waveguide wall by an ideal magnetic one and to replace the constitutive parameters by effective ones as a preamble to satisfying the boundary conditions at the flat faces of the resonator. The waveguide is characterized by a relative dielectric constant ε_f and a relative demagnetized permeability μ_{dem}. The effect of the dielectric spacer is separately catered for by extending the lateral walls of the resonator to form a cutoff circular waveguide section. The use of an effective dielectric constant for the end sections is, of course, not possible, since it is not realistic, to form an equivalence between propagating and cutoff sections of waveguides. Adopting the usual approximation to this problem, but replacing ε_f by ε_{eff} and noting that μ_{dem} or $\mu_{d,\text{eff}}$ does not explicitly appear in the immittance description of TM_{nm} type modes in round waveguide with an ideal magnetic wall, gives the standard result.

$$\frac{\varepsilon_{\text{eff}} k_0}{\beta_0} \cot(\beta_0 L) - \frac{\varepsilon_d k_0}{\alpha_0} \coth(\alpha_0 S) = 0 \tag{9.11}$$

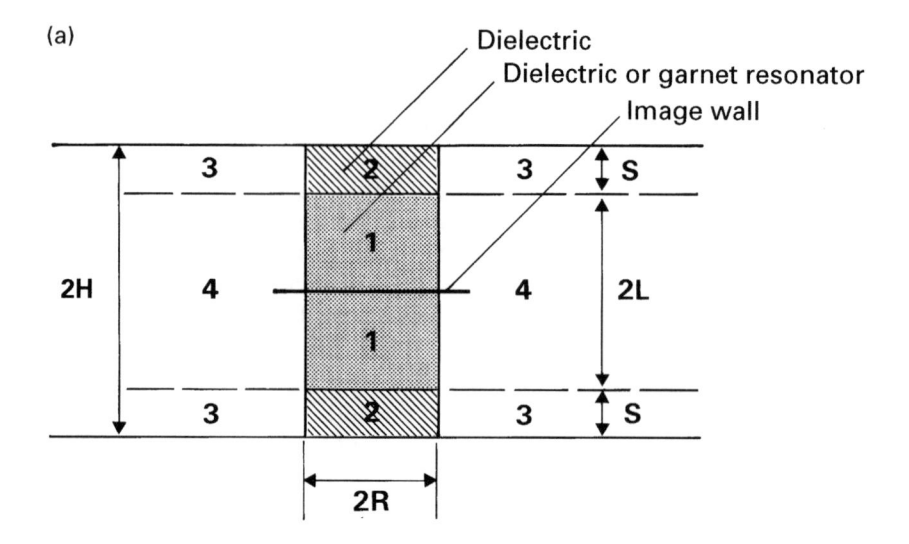

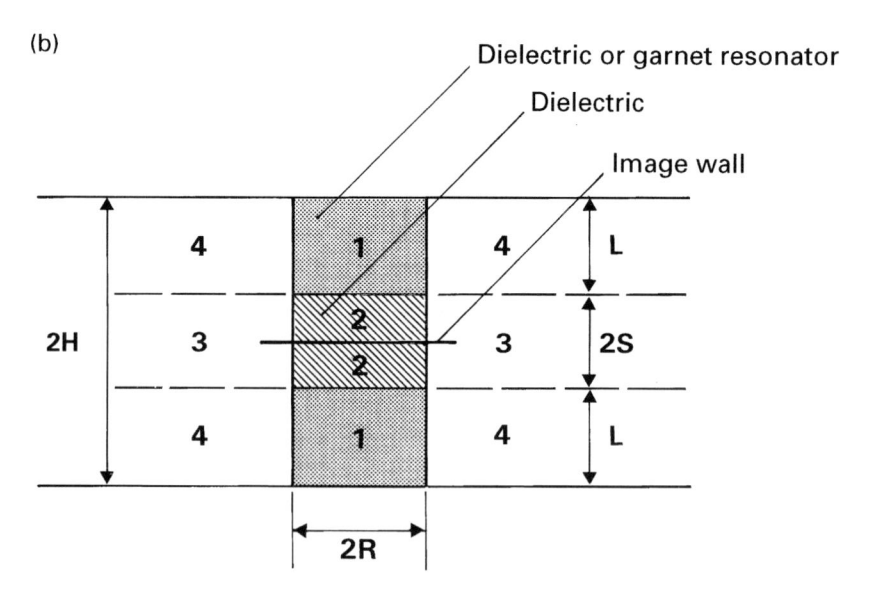

Figure 9.4 (a) Schematic diagram of coupled quarter-wave long ferrite or dielectric resonators loaded by image wall (b) Schematic diagram of half-wave long open ferrite or dielectric resonator loaded by image wall

9.5 ODD MODE SOLUTION

where

$$\left(\frac{\beta_0}{k_0}\right)^2 = \varepsilon_{\text{eff}} - \left(\frac{k_{mn}}{k_0}\right)^2 \tag{9.12}$$

$$\left(\frac{\alpha_0}{k_0}\right)^2 = \left(\frac{k_{mn}}{k_0}\right)^2 - \varepsilon_d \tag{9.13}$$

ε_{eff} is the effective dielectric constant of the gyromagnetic resonator; it is obtained from either measurement or calculation. ε_d is that of the relative dielectric constant of the spacers.

γ is the gyromagnetic ratio $(2.21 \times 10^5$ (rad/m)/(A/m)), M_0 is the saturation magnetization (T), μ_0 is the free space permeability $(4\pi \times 10^{-7}$ H/m) and ω is the radian frequency (rad/sec)

The separation constant of the dominant mode of a circular waveguide with an ideal magnetic wall propagating the TM_{11} mode is given by equation (9.5) and k_0 is the free space wave-number at its center frequency.

The condition in equation (9.11) may for computational purposes, be written as

$$\varepsilon_{\text{eff}}\left(\frac{k_0}{\beta_0}\right)\cot\left[\left(\frac{\beta_0}{k_0}\right)\left(\frac{L}{R}\right)k_0R\right] - \varepsilon_d\left(\frac{k_0}{\alpha_0}\right)\coth\left[\left(\frac{\alpha_0}{k_0}\right)\left(\frac{L}{R}\right)\left(\frac{1-k}{k}\right)k_0R\right] = 0 \tag{9.14}$$

k is the gap factor defined by

$$k = \frac{L}{L+S} \tag{9.15}$$

The above representation breaks down however, when the gap factor approaches zero since the open face of each resonator in Figure 9.4(a) is then loaded by a cutoff section of waveguide with a magnetic sidewall terminated an electric wall instead of the required magnetic wall as demanded by the physics of the circuit. The correct lower bound on the odd mode solution of the two resonators is that of the decoupled resonator. The actual radial wavenumber in any circulator design, as will be seen, is not unique but is fixed by the susceptance slope parameter and first circulation condition.

The characteristic equation in equation (9.11) is equally applicable to the dual geometry in Figure 9.1(b) and indeed to the single quarter-wave long resonator short circuited at one end and loaded by an image wall at the other. The length of the open resonator is $2L$, S is that of the end sections and ε_d is the relative constant of the end sections.

Some appreciation of the effect of the electric wall on the solution of this type of resonator may be deduced by replacing L by $L_0 + \Delta L$ in the characteristic equation. This gives

9 THE DIELECTRIC CAVITY RESONATOR

$$\frac{\varepsilon_{\text{eff}} k_0}{\beta_0} \cot(\beta_0 L_0) + \frac{\varepsilon_{\text{eff}} k_0}{\beta_0} \tan(\beta_0 \Delta L) - \frac{\varepsilon_d k_0}{\alpha_0} \coth(\alpha_0 S) = 0 \quad (9.16)$$

If this equation is now expanded in the vicinity of $\beta_0 L_0 = \pi/2$ then the characteristic equation reduces to

$$\frac{\varepsilon_{\text{eff}} k_0}{\beta_0} \tan(\beta_0 \Delta L) = \frac{\varepsilon_d k_0}{\alpha_0} \coth(\alpha_0 S) \quad (9.17)$$

$$\cot \beta_0 L_0 = 0 \quad (9.18)$$

One equivalent circuit of this type of resonator consists therefore of a distributed quarter-wave long resonator loaded by a lumped element shunt LC circuit. This arrangement is illustrated in Figure 9.5.

Theory and experiment for the odd mode solution of a pair of demagnetized garnet resonators with an aspect ratios (R/L) equal to 2.26 are separately compared in Figure 9.6. The experimental configuration consisted of a single quarter-wave long demagnetized garnet resonator short-circuited at one end and loaded by an electric wall at the other; the relative dielectric constant (ε_d) of the region between its open flat face and the image or waveguide wall is, in this arrangement, unity.

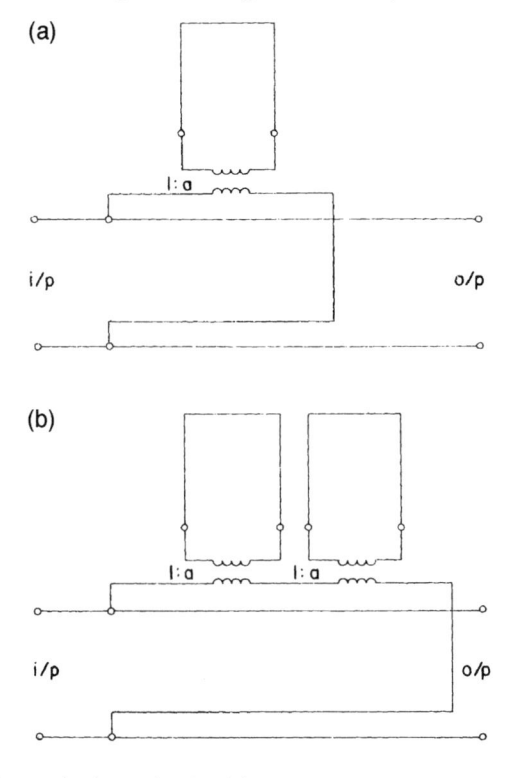

Figure 9.5 Odd mode equivalent circuit; (a) single disk resonator, (b) coupled disk or single cylinder resonator

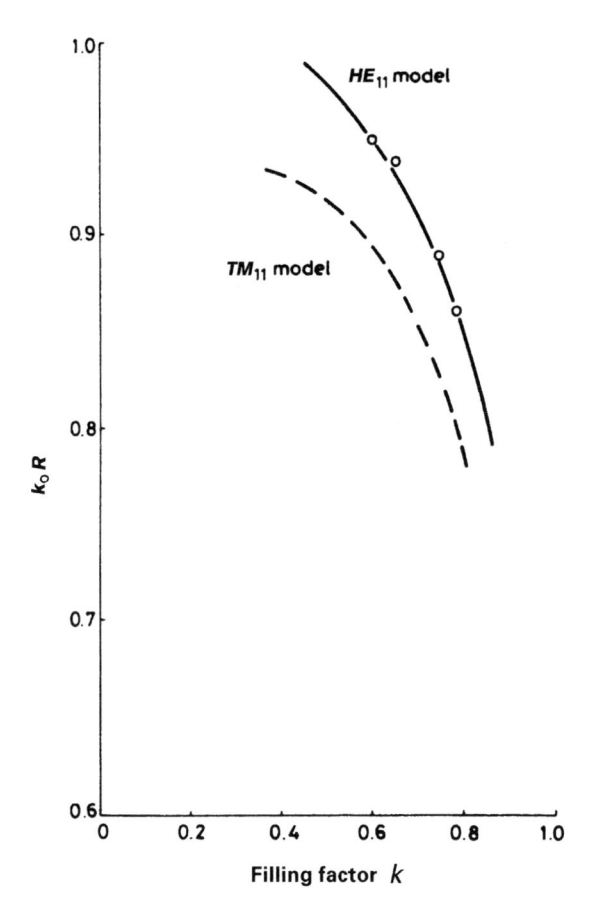

[Figure: plot of $k_0 R$ vs Filling factor k, showing HE₁₁ model (solid) and TM₁₁ model (dashed) curves with experimental data points]

Figure 9.6 Theoretical odd mode charts of coupled (closed and open magnetic walls) quarter-wave long garnet resonators and experimental results, $M_0 = 0.1600T$, $R/L = 2.26$, $\varepsilon_f = 15$, $\varepsilon_d = 1$ (reproduced with permission, Helszajn, J. and Sharp, J. (1986) Dielectric and permeability effects in $HE_{1,1,1/2}$ open demagnetized ferrite resonators, *IEE Proc.*, **133**, Part H, 271–276)

The magnetization of the garnet material (M_0) is 0.1600T and its nominal relative dielectric constant (ε_f) is 15.1. The experimental results superimposed on this illustration have been obtained by decoupling the in-phase and degenerate counter-rotating eigen-networks of a junction circulator. It suggests that, with the gap factor between 0.60 and 0.90, the formulation outlined here is more than adequate for everyday engineering work. The effect of ε_d is separately shown in Figure 9.7.

Some dedicated experimental results (again using single quarter-wave long resonators in junction circulators) for three different materials with nominal values of magnetizations of 0.0550T, 0.1200T and 0.1780T, respectively, are displayed in Figure 9.8; the values of the dielectric constant and magnetization of these materials have a maker's tolerance of ±5%.

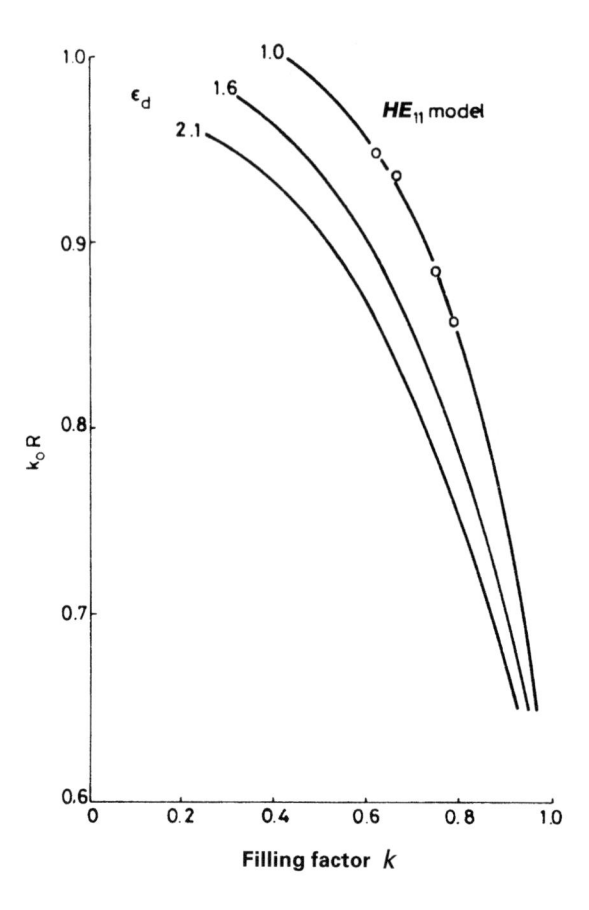

Figure 9.7 Theoretical odd mode charts of coupled quarter-wave long garnet resonators and experimental results showing effect of ε_d, $M_0 = 0.1600\text{T}$, $R/L = 2.26$, $\varepsilon_f = 151$, $\varepsilon_d = 1$ (reproduced with permission, Helszajn, J. and Sharp, J. (1986) Dielectric and permeability effects in HE_{11} open demagnetized ferrite resonators, *IEE Proc.*, **133**, Part H, 271–276)

9.6 Odd Mode Planar Solution

A quite different perspective of the odd mode solution of two-coupled dielectric resonators is, however, revealed by plotting R/L against k for parametric values of k_0R instead of k_0R versus k for different values of R/L. Figure 9.9 indicates one typical solution. The steep increase in R/L as k approaches unity indicates that the resonant frequency is independent of the thickness of the problem region there. This is of course the situation met in connection with a planar circuit. The odd mode transverse resonance condition of two coupled resonators therefore provides an appropriate representation of both the quarter-wave long resonator in the vicinity of k equal to zero and the planar solution in the vicinity of k equal to unity and the interval between the two regions. The steep increase in the susceptance slope parameter with k in the vicinity of the dispersion in R/L may not be

9.6 ODD MODE PLANAR SOLUTION

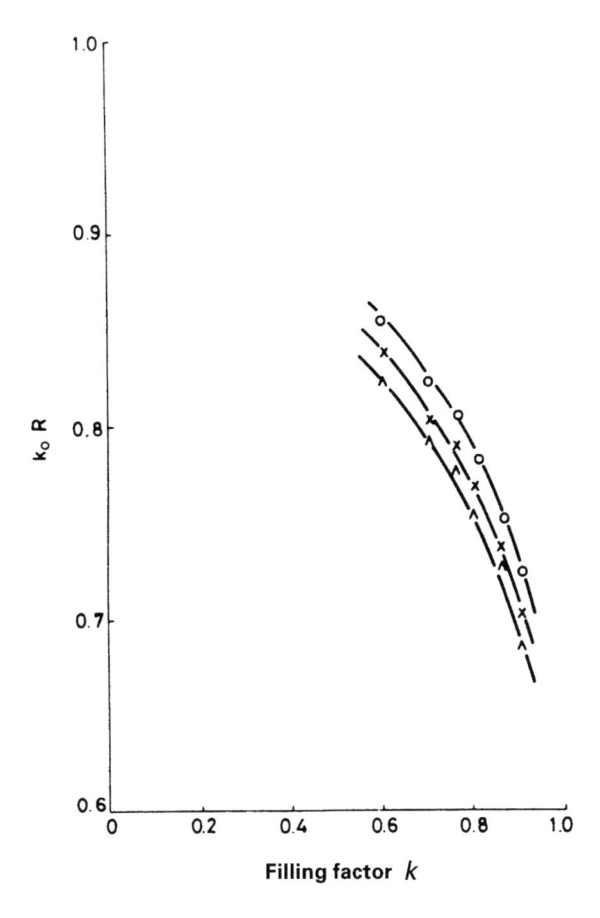

Figure 9.8 Experimental odd mode chart of quarter-wave long garnet resonators showing effect of magnetization (reproduced with permission, Helszajn, J. and Sharp, J. (1986) Dielectric and permeability effects in HE_{11} open demagnetized ferrite resonators, *IEE Proc.*, **133**, Part H, 271–276)

unconnected with this feature. The pronounced dispersion in the susceptance slope parameter in the description of the dual problem utilizing one or two cylindrical resonators has already been separately noted. The derivation of the planar solution begins by approximating the ordinary and hyperbolic trigonometric tangent quantities in the transverse resonance condition by the appropriate small angle descriptions. This gives

$$\left(\frac{\varepsilon_r}{\beta_0}\right)\left(\frac{\alpha_0(1-k)}{k}\right) = \left(\frac{\varepsilon_d}{\alpha_0}\right)(\beta_0) \tag{9.19}$$

This equation may be readily solved for k_0R without ado by expressing α_0 and β_0 in terms of the original variables. The required result is

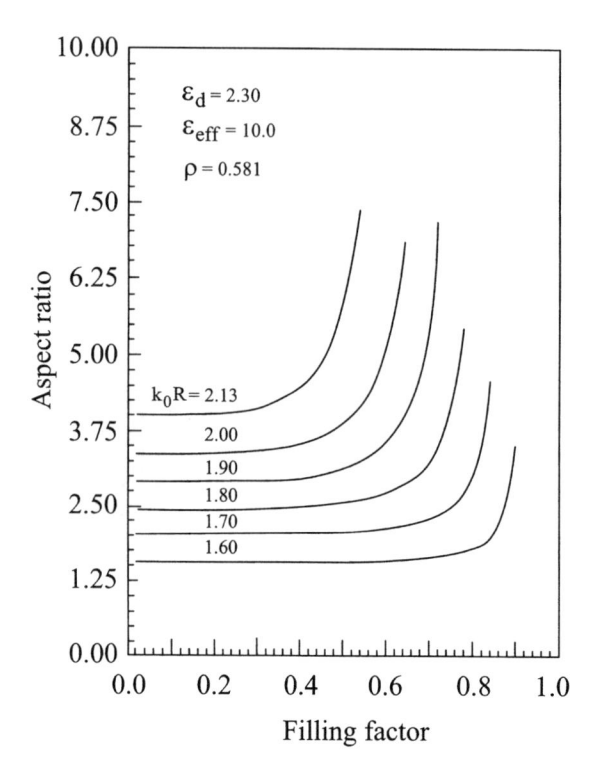

Figure 9.9 Aspect ratio R/L versus gap factor k for parametric values of k_0R

$$(k_0R)^2 \varepsilon_{\text{eff}} = (1.84)^2 \quad (9.20)$$

where

$$\varepsilon_{\text{eff}} = \frac{\varepsilon_r \varepsilon_d}{\varepsilon_e - k(\varepsilon_r - \varepsilon_d)} \quad (9.21)$$

The former equation is recognized as the cut-off number of the dominant mode of a triangular planar circuit and the latter one as the effective dielectric constant of a two dielectric planar capacitor. The latter quantity has of course already been directly constructed by forming the overall capacitance of a two-layer dielectric capacitor.

9.7 Even Mode Solution of Dielectric Resonator

The operating frequency of the E-plane circulator coincides with the even solution of a pair of quarter-wave long open gyromagnetic resonators resonating in the $HE_{1,1,1/2}$ mode spaced by a contiguous section of cut-off waveguide with a magnetic sidewall. The effect of the open wall of the resonator may again be

9.7 EVEN MODE SOLUTION

catered for by employing an effective dielectric constant (ε_{eff}) instead of the constitutive one (ε_f). The effect of the demagnetized permeability may also be catered for. There is, however, no need in practice to introduce an effective permeability in this instance. Since the even mode frequency of two coupled resonators is a flat function of the spacing between the two, its value may be estimated from that of the open resonator solution as a preamble to the exact calculation. The characteristic equation for the even-mode solution of the two coupled resonators in Figure 9.2(b) is

$$\frac{\varepsilon_{eff} k_0}{\beta_0} \cot(\beta_0 L) - \frac{\varepsilon_d k_0}{\alpha_0} \tanh(\alpha_0 S) = 0 \tag{9.22}$$

where α_0/k_0 and β_0/k_0 are defined by equations (9.12) and (9.13)

The propagation constant β/k_0 is again determined by solving the characteristic equation for the HE_{11} mode of the open demagnetized ferrite or dielectric waveguide from a knowledge of ε_{eff} and R. The parameter ε_{eff} is the effective dielectric constant of an equivalent round waveguide with an ideal magnetic wall; it is obtained from a knowledge of β_0/k_0 and $k_0 R$ in Chapter 8. The solution adopted here satisfies the boundary conditions between regions 1 and 4 and 1 and 2 in

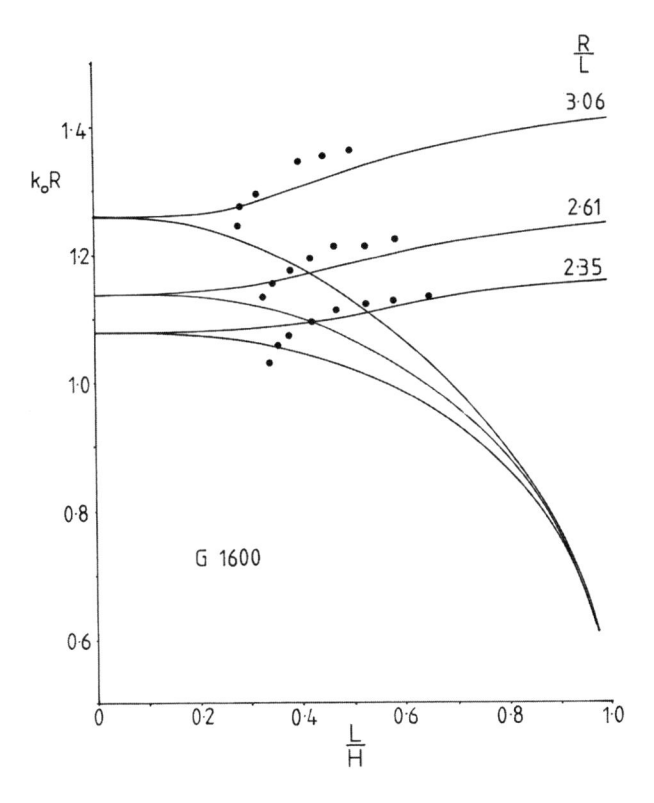

Figure 9.10 Even- and odd-mode charts of coupled $HE_{1,1,1/2}$ turnstile resonators ($\varepsilon_f = 12.7$)

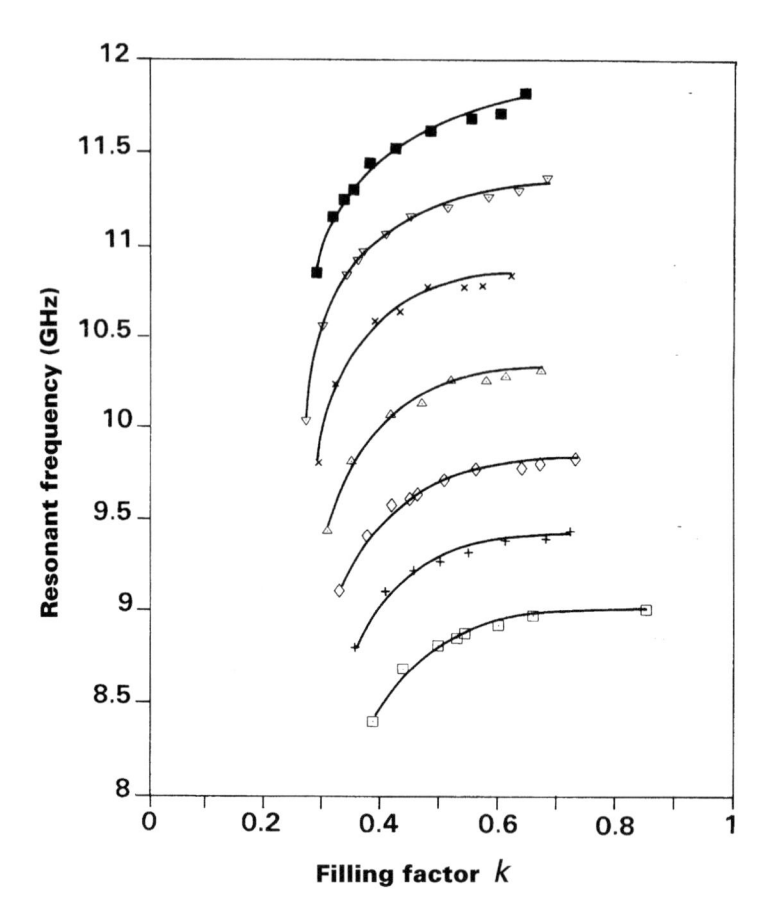

Figure 9.11 Resonant frequency of even mode resonator for different aspect ratios. $R = 4.77$mm, $\varepsilon_f = 15.1$; $\square$, $R/L = 1.60$; $+$, $R/L = 1.75$; $\diamond$, $R/L = 1.88$; $\triangle$, $R/L = 2.05$, $\times$, $R/L = 2.20$; $\triangledown$, $R/L = 2.35$; $\blacksquare$, $R/L = 2.50$. (reproduced with permission, Helszajn, J. and Cheng, S. (1990) Aspect ratio of open resonators in the design of evanescent mode E-plane circulators, *IEE Proc. Microwaves, Antennas and Propagation*, **137**, 55–60)

Figure 9.4(a) but neglects those between 3 and 4 and 2 and 3. It gives a relationship between the radial wave number ($k_0 R$) of each resonator and the spacing between the two (2S) for parametric values of the aspect ratio (R/L) of the decoupled resonator. The spacing between the resonators is also described by the so-called gap factor k in equation (9.19).

Figure 9.10 indicates the even mode solution of this sort of resonator for three different values of R/L. The odd mode solution is separately superimposed on this illustration for completeness sake.

When the gap factor tends to unity the solution degenerates to that of a planar resonator ($\beta = 0$). When it is equal to zero it reduces to that of a decoupled quarter wave long resonator.

Figure 9.11 indicates the relationship between the frequency and the gap factor of such a resonator with $R/R_0 = 1.0$ in a 3-port junction with a radius R_0 equal to

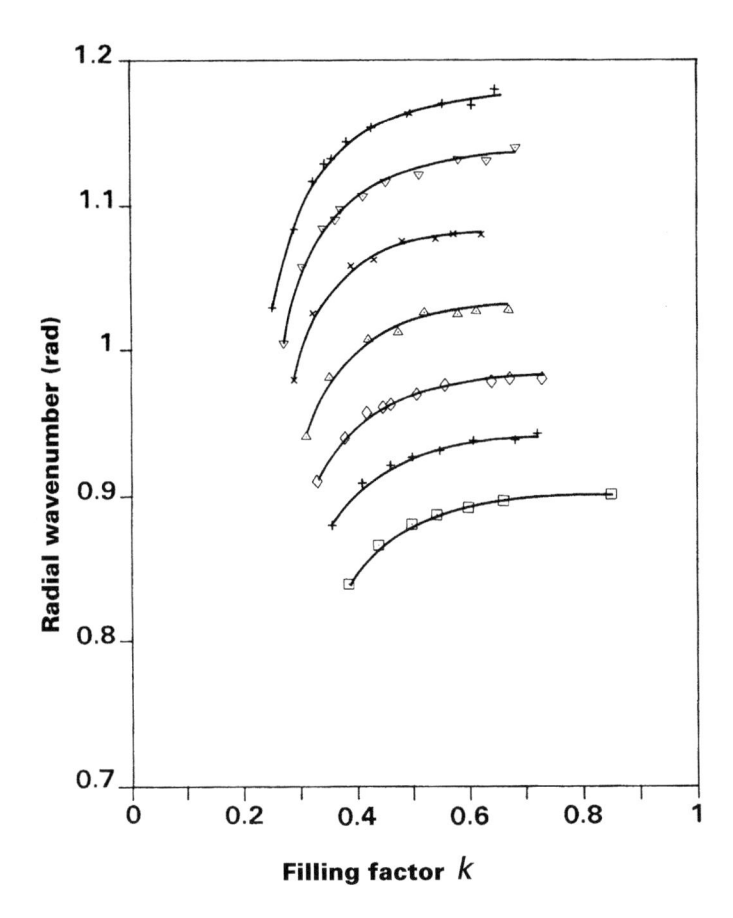

Figure 9.12 Normalized resonant frequency of even mode resonator for different aspect ratios. $R = 4.77$ mm, $\varepsilon_f = 15.1$; □ $R/L = 1.60$; + , $R/L = 1.75$; ◇, $R/L = 1.88$; ∆, $R = 2.05$; ×, $R/L = 2.20$; ▽, $R/L = 2.35$; +, $R/L = 2.50$ (reproduced with permission, Helszajn, J. and Cheng, S. (1990) Aspect ratio of open resonators in the design of evanescent mode E-plane circulators, *IEE Proc. Microwaves, Antennas and Propagation*, **137**, 55–60)

6.00 mm for different values of the aspect ratio R/L. Figure 9.12 depicts the same result in normalised form. One possible solution at a frequency of 9.375 GHz (say) is $k_0 R = 0.92$ rad and $R/L = 1.75$. Figure 9.13 depicts the normalised mode chart for the same junction but with $R = 4.37$ mm and $R/R_0 = 0.73$. One possible solution at 9.375 GHz in this instance given by $k_0 R = 0.81$ and $R/L = 1.49$. The garnet marital employed in this work has a magnetization M_0 of 0.1600T and a relative dielectric constant ε_f of 15.1. The effective dielectric constant (ε_{eff}) of the equivalent $TM_{1,1,1/2}$ waveguide mode is about 13.1 for the radial wavenumber employed here.

Scrutiny of the normalized mode charts obtained here suggests that $k_0 R$ is dependent to first order upon the aspect ratio of the resonator but only to second order on the gap factor of the junction. One possible engineering decision that appears adequate for practical purposes is to fix the radial waveguide of the device

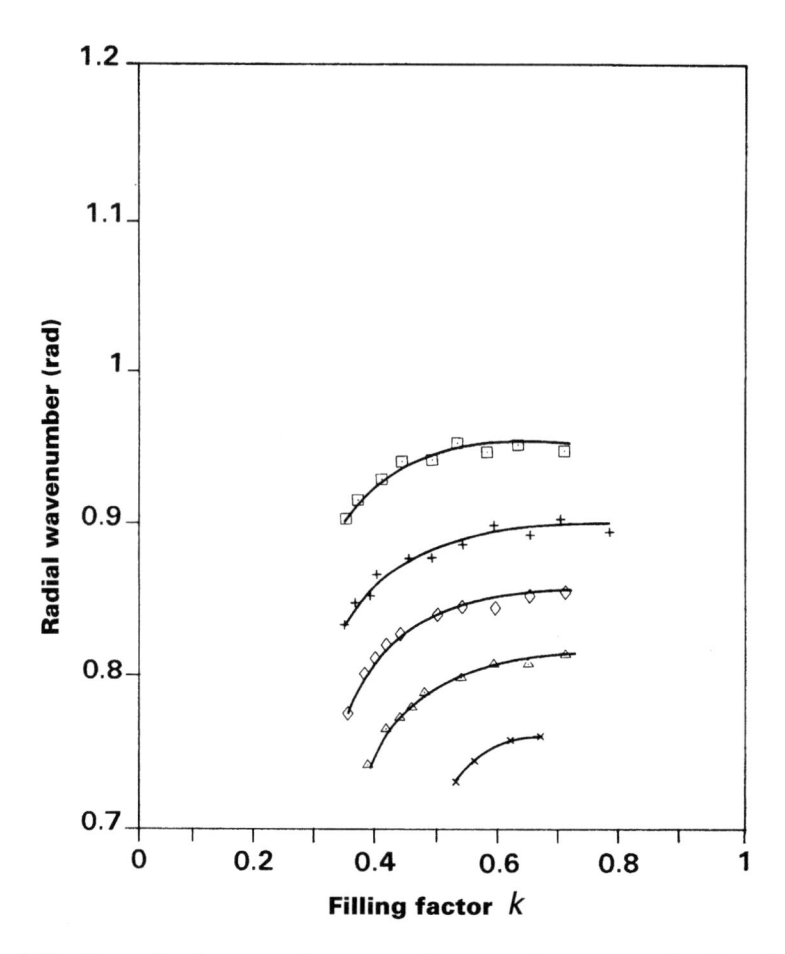

Figure 9.13 Normalized resonant frequency of even mode resonator circulator for different aspect ratios. $R = 4.37$ mm, $\varepsilon_f = 15.1$; ×, $R/L = 1.12$; △, $R/L = 1.36$; ◇, $R/L = 1.49$; +, $R/L = 1.63$; □, $R/L = 1.81$ (reproduced with permission, Helszajn, J. and Cheng, S. (1990) Aspect ratio of open resonators in the design of evanescent mode E-plane circulators, *IEE Proc. Microwaves, Antennas and Propagation*, **137**, 55–60)

by adjusting the aspect ratio of the resonator and to leave the choice of the gap factor to meet any other engineering requirements.

The discrepancy between the lower bound on the operating frequency of the resonator and experiment when the gap factor is small may in part be understood by recognizing that this frequency corresponds to that at which the in-phase and counter-rotating modes are in anti-phase rather than that at which these are commensurate. If the frequency variation of the in-phase eigen-network may be neglected compared with that of the counter-rotating ones then the former largely determines that of the junction. This condition is most likely to manifest itself in the vicinity of the cut off condition of the in-phase eigen-network.

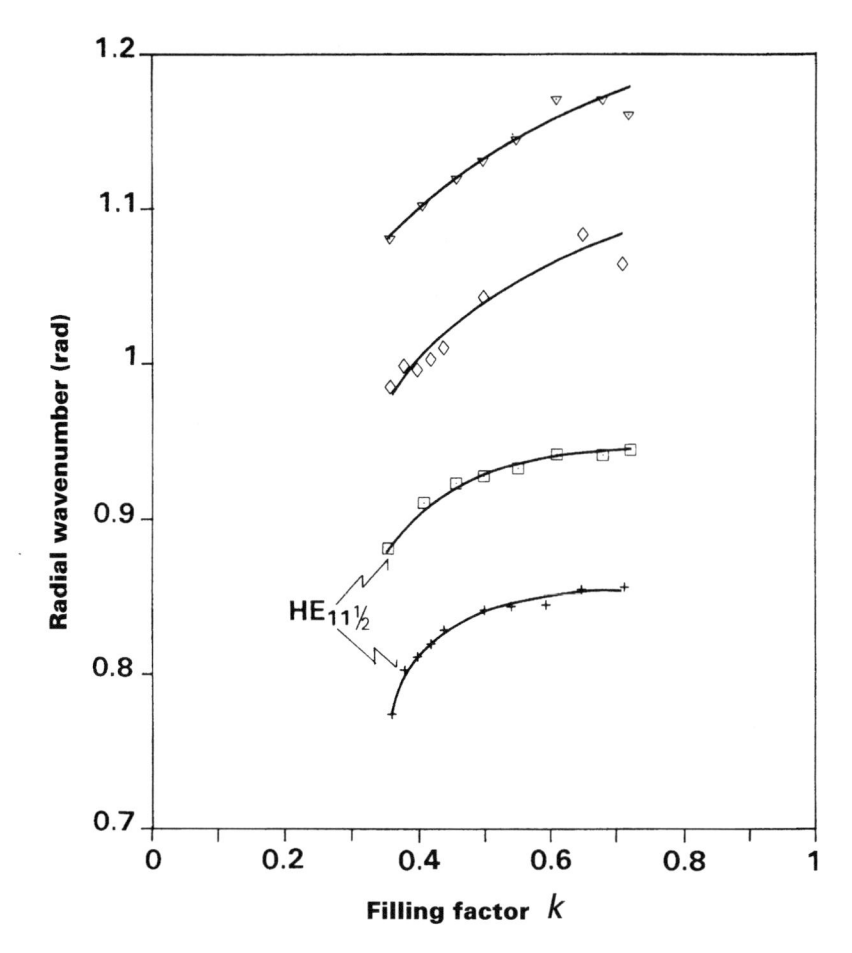

Figure 9.14 Normalized mode chart of even mode resonator showing dominant and first higher-order mode. +, $R = 4.37$ mm; $R/L = 1.49$; □, $R = 4.77$ mm, $R/L = 1.75$; ◇, $R = 4.37$; $R/L = 1.49$; ▽, $R = 4.77$ mm; $R/L = 1.75$ (reproduced with permission, Helszajn, J. and Cheng, S. (1990) Aspect ratio of open resonators in the design of evanescent mode E-plane circulators, *IEE Proc. Microwaves, Antennas and Propagation*, **137**, 55–60)

Some relationships between the radial wavenumbers of the dominant degenerate pair and symmetrical modes in this type of junction at an approximate frequency of 9.375 GHz are shown in Figure 9.14.

9.8 Dielectric Resonators in Metal Enclosures

A cylindrical half-wave long open dielectric or gyromagnetic resonator embedded in an oversized dielectric-filled metal enclosure with coupling apertures on its two lateral flat faces, provides one experimental means of evaluating the phase constants of open dielectric or gyromagnetic waveguides. The schematic diagram of this arrangement is illustrated in Figure 9.15. One possible model for its central region is a half-wave long dielectric section or gyromagnetic waveguide supporting a hybrid pair of degenerate or split HE$_{11}$ counter-rotating modes. The flat circular dielectric disks on either side of the main section ensure an open wall boundary condition at the two open flats faces of the resonator and may absorbed in the coupling mechanism at either end of the structure.

Propagation along the main section of this type of resonator is usually described by plotting β_0/k_0 versus $k_0 R_i$ with R_i/R_0 and $\varepsilon_f/\varepsilon_d$ as parameters. The required relationship may be deduced by making use of that between the propagation constant (β) and the guide wavelength (λ_g).

$$\beta_0 = \frac{2\pi}{\lambda_g} \tag{9.23}$$

If the reflection cavity resonator arrangement in Figure 9.16 is employed to deduce the relationship between β_0/k_0 and $k_0 R_i$ and if the effect of the semi-ideal lateral walls of the main section is neglected then

$$\lambda_g = 4L_i \tag{9.24}$$

so that

$$\beta_0 = \frac{\pi}{2L_i} \tag{9.25}$$

The wavenumber (k_0) is separately related to the free space wavelength (λ_0) of the resonator by equation (9.6)

The magnetic walls at the two flat faces for the cylindrical gyromagnetic region are enforced by dielectric spacers. The thickness of each end spacer is usually described by defining a gap factor k. This quantity is fixed as

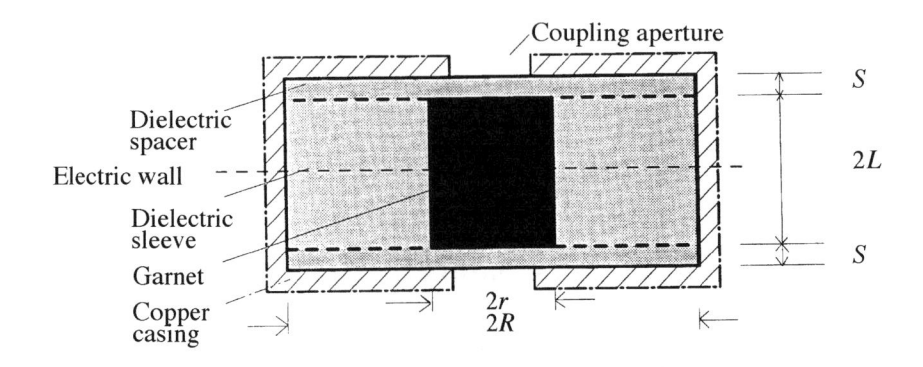

Figure 9.15 Transmission cavity resonator using half-wave long dielectric resonator

$$k = \frac{L_i}{L_0} \qquad (9.26)$$

ε_f and ε_d are the relative dielectric constants of the ferrite and that of the surrounding dielectric medium of the cavity, respectively. L_i and L_0 are related to the half spaces defined by introducing and electric wall at the symmetric plane of the problem region.

Although there is much practical and theoretical guidance in the literature in connection with the design of this sort of resonator, its details have been chosen semiempirically.

9.9 Experimental HE$_{1,1,1/2}$ Mode Chart in Metal Enclosure

An experimental mode chart of the sort of cavity under consideration can be constructed by having recourse either to the 2-port transmission geometry in Figure 9.15 or to the 1-port reflection arrangement in Figure 9.16. The 1-port arrangement is obtained by introducing an electric wall through the plane of symmetry of the cavity. A family of experimental mode charts in the 8–12 GHz band is summarized in Figure 9.17. While the mode chart of the dominant mode is not over-affected by the choice of the relative dielectric constant of the dielectric filler, those on either side of it are more influenced by it. This type of information is sometimes conveyed by plotting $k_0 R_i$ against R_i/L_i instead of $k_0 R_i$ versus β_0/k_0. A typical frequency response is indicated in Figure 9.18.

The metal enclosure employed in this work is described by its radius (R_0) and length (L_0) and by its aspect ratio: $R_0 = 7.00$ mm; $L_0 = 3.80$ mm; $R_0/L_0 = 1.84$.

The three resonator geometries employed here are described by

$R_i, L = 3.72$ mm, 3.30 mm; $R_i/R_0 = 0.53$

$R_i, L_i = 3.72$ mm, 2.89 mm; $R_i/R_0 = 0.53$

$R_i, L_i = 4.77$ mm, 3.00 mm; $R_i/R_0 = 0.68$

The ratio R_i/R_0 is 0.53 for two of the resonators and is 0.68 for the other. While it is not strictly appropriate to describe by a single curve assemblies with different values of R_i/R_0, it has nevertheless been done in this instance. The assumption employed in doing so may be deemed valid provided the fields in the resonator may be assumed to decay rapidly from the central region, so that the position of the outside waveguide wall may be neglected in calculating the phase constant of the main section.

The relative dielectric constants of the three dielectric fillers utilized here are $\varepsilon_d = 2.56$, $\varepsilon_d = 3$ and $\varepsilon_d = 3.47$, respectively. The magnetization (M_0) of the garnet material is 0.1600T and its relative dielectric constant (ε_f) is 15.1.

A typical solution may be summarized as

$$\beta/k_0 (k_0 R_i, R_i/R_0, \varepsilon_f/\varepsilon_d) = 2.40 \; (0.84, 0.53, 5.9)$$

9 THE DIELECTRIC CAVITY RESONATOR

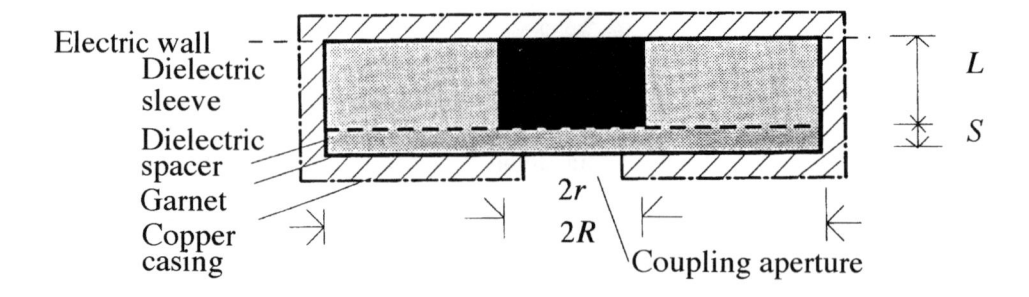

Figure 9.16 Reflection cavity resonator using quarter-wave long dielectric resonator

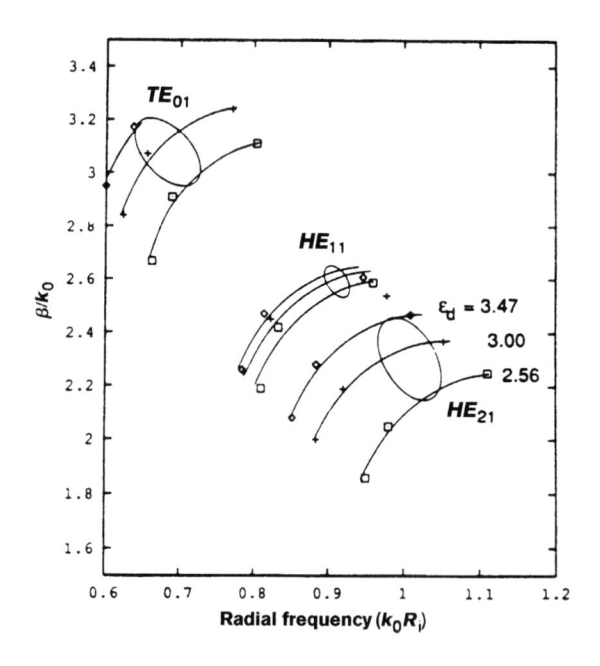

Figure 9.17 Relationship between $k_0 R_i$ and β/k_0 for demagnetized gyromagnetic resonator for parametric values of ε_d ($k \approx 0.80$) (reproduced with permission Helszajn, J., Cheng, C. S. and Wilcock, B. A. (1993) A nonreciprocal turnable waveguide directional filter using a turnstile open gyromagnetic resonator, *IEEE Trans. MTT*, **41**(11), 1950–1958) (© 1993 IEEE)

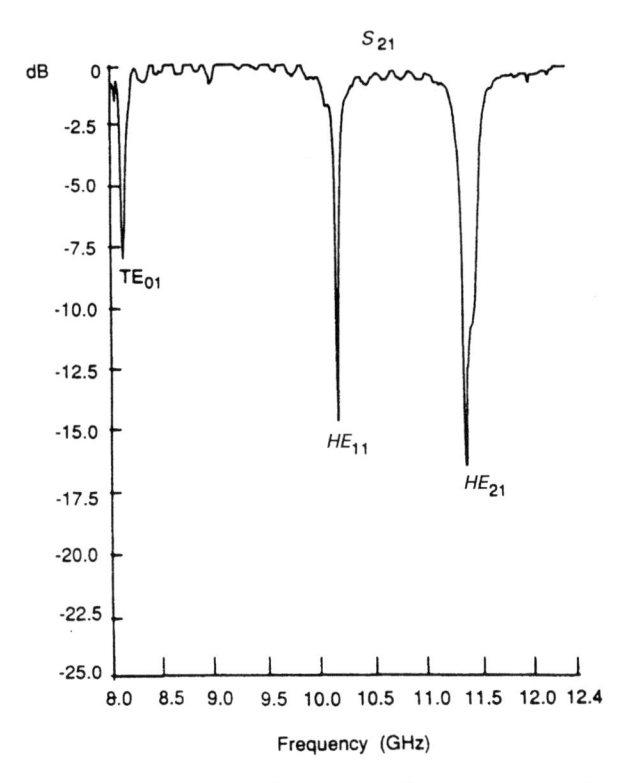

Figure 9.18 Frequency response of demagnetized quarter-wave long gyromagnetic resonator ($k \approx 0.80$, $\varepsilon_d = 3$, $\varepsilon_f = 15$, $R_i/L_i = 1.283$) (reproduced with permission Helszajn, J., Cheng, C. S. and Wilcock, B. A. (1990) A nonreciprocal turnable waveguide directional filter using a turnstile open gyromagnetic resonator, *IEEE Trans. MTT*, **41**(11), 1950–1958) (© 1993 IEEE)

Once $\varepsilon_f/\varepsilon_d$ is chosen, then R_i is fixed from a statement of k_0 and $k_0 R_i$, L_i is set from a knowledge of β/k_0. The cavity dimensions L_0 and R_0 are separately obtained from statements about k and R_i/R_0.

References and Further Reading

Akaiwa, Y. (1974) Operation modes of a waveguide Y-circulator, *IEEE Trans. Microwave Theory Tech.*, **MTT-22**, 954–959.

Fiedziusko, S. J. and Jelenski, A. (1971) The influence of conducting walls on resonant frequencies of the dielectric microwave resonator *IEEE Trans. Microwave Theory Tech.*, **MTT-19**, 778.

Guillon, P. and Garault, Y. (1977) Accurate resonant frequencies of dielectric resonators, *IEEE Trans. Microwave Theory Tech.*, **MTT-25**, 916–922.

Hakki, B. W. and Coleman, P. D. (1960) A dielectric resonator method of measuring inductive capacities in the millimetre range, *IRE Trans.*, **MTT-8**, 402–410.

Hauth, W. (1981) Analysis of circular waveguide cavities with partial-height ferrite insert, in *Proc. Eur. Microwave Conf.*, 383–388.

Helszajn, J. (1974) Common waveguide circulator configuration, *Electron. Eng.*, 66–68.

Helszajn, J. and Tan, F. C. C. (1975) Mode charts for partial-height ferrite waveguide junction circulators, *Proc. IEE*, **122**, 34–36.

Helszajn, J. and Sharp, J. (1986) Dielectric and permeability effects in HE_{11} open demagnetized ferrite resonators, *IEE Proc.*, **133**(H), 271–276.

Helszajn, J. and Cheng, S. (1990) Aspect ratio of open resonators in the design of evanescent mode E-plane circulators, *IEE Proc. Microwaves, Antennas and Propagation*, **137**, 55–60.

Helszajn, J., Leeson, W., Lynch, D. and O'Donnell, B. (1991) Normal mode nomenclature of quadruple gyromagnetic waveguide, *IEEE Trans. Microwave Theory Tech.*, **MTT-39**, 461–470.

Helszajn, J., Cheng, C. S. and Wilcock, B. A. (1993) A nonreciprocal turnable waveguide directional filter using a turnstile open gyromagnetic resonator, *IEEE Trans. Microwave Theory Tech.*, **41**(11), 1950–1958.

Itoh, T. and Rudokas, R. (1977) New method for computing the resonant frequencies of dielectric resonators *IEEE Trans. Microwave Theory Tech.*, **MTT-25**, 52–54.

Owen, B. (1972) The identification of modal resonances in ferrite loaded waveguide Y-junction and their adjustment for circulation, *Bell Syst. Tech. J.*, **51**(3), 595–627.

Pospieszalski, W. (1977) On theory of dielectric post resonator, *IEEE Trans. Microwave Theory Tech.*, **MTT-25**, 228–231.

Relsch, D. L., Webb, D. C., Moore, R. A. and Cowlishaw, J. D. (1965) A mode chart for accurate design of cylindrical dielectric resonators, *IEEE Trans. Microwave Theory Tech.*, **MTT-13**, 468–469.

10

THE GYROMAGNETIC CAVITY RESONATOR

10.1 Introduction

A classic feature of a junction circulator using a weakly magnetized gyromagnetic cavity resonator is that its gain-bandwidth product or quality factor is uniquely determined by the split frequencies of its resonator. An understanding of the split mode chart of gyromagnetic cavity resonators is therefore a prerequisite for the design of waveguide circulators using this class of resonators. The purpose of this chapter is to tackle this problem Such mode charts may either be obtained exactly or by having recourse to perturbation theory or by adopting an anisotropic model with an ideal magnetic wall boundary condition or by having recourse to measurements. An understanding of the exact or perturbation formulation of the closed and open anisotropic or gyromagnetic waveguide dealt with in Chapters 7 and 8 is therefore necessary for that of the gyromagnetic resonator. The gyromagnetic waveguide is of course the classic non-reciprocal Faraday rotation bit upon which many of the early ferrite devices rely. In practice some means of holding the resonator in place is necessary. The usual geometry met in the design of waveguide circulators consists of a pair of coupled quarter-wave long resonators operating in either its even or odd mode. While the discussion is restricted to the odd mode solution met in connection with the construction of H-plane junction circulator some remarks about the even one associated with the E-plane is also included. Another resonator configuration that has the symmetry of the problem region is the prism. Some remarks about this geometry is introduced on Chapter 18. While the chapter includes a polymonial representation of the problem region which is based on the exact solution the calculations presented in this chapter rely on its perturbation formulation.

10.2 Closed Gyromagnetic Resonator

The split frequencies of gyromagnetic resonators are important quantities in the description of weakly magnetized resonators. The simplest topology met in connection with this sort of problem region is illustrated in Figure 10.1. It consists of a quarter-wave long demagnetized or magnetized ferrite waveguide with an ideal or an open magnetic wall open-circuited at one end and short-circuited at the other. The characteristic equations associated with the two situations are

$$\cot(\beta_0 L_0) = 0 \quad (10.1)$$

$$\cot(\beta_\pm L_0) = 0 \quad (10.2)$$

respectively

The first of these two equations fixes the length of the resonator from a knowledge of $k_0 R$ and frequency and has been dealt with in Chapter 9. The first root of its characteristic equation is

$$\beta_0 L_0 = \frac{\pi}{2} \quad (10.3)$$

where

$$\beta_0^2 = k_0^2 \mu_{\text{eff}} \varepsilon_{\text{eff}} - \left(\frac{1.84}{R}\right)^2 \quad (10.4)$$

The second characteristic equation may be solved for the relationship between the split frequencies of the resonator and the magnetic variables in the neighbourhood of the demagnetized one once the characteristic equation for $\beta_\pm$ is at hand.

$$\beta_\pm L_0 = \frac{\pi}{2} \quad (10.5)$$

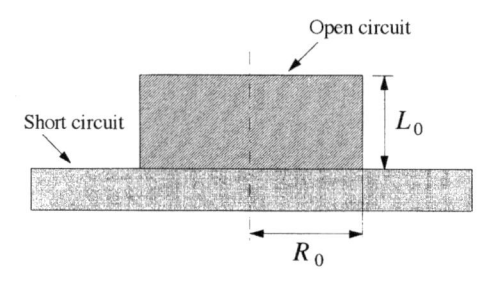

Figure 10.1 Closed gyromagnetic cavity resonator with ideal magnetic walls open-circuited (o/c) at one end and short-circuited (s/c) at the other

The simplest calculation met in this sort of circuit is the closed gyromagnetic resonator with an ideal magnetic wall and so it will be investigated here. The exact open problem may be derived from that of the related problem region of a partially filled circular waveguide with an electric side wall by allowing the outside radius to approach infinity, but its solution is outside the remit of this text.

10.3 Perturbation Theory of Closed Cylindrical Gyromagnetic Resonator

If the only engineering interest is a calculation of the split frequencies of the gyromagnetic resonator then is it sufficient to have recourse to a description of the split propagation constants based on perturbation theory. The required result for the split phase constants of a gyromagnetic waveguide with an ideal magnetic wall is

$$\beta_\pm^2 \approx k_0^2 \varepsilon_{\text{eff}}(\mu \mp C_{11}\kappa) - \left(\frac{1.84}{R}\right)^2 \tag{10.7}$$

where

$$C_{11} = \frac{2}{(1.84)^2 - 1} \tag{10.8}$$

ε_{eff} is an effective dielectric constant, μ and κ are the off-diagonal elements of the tensor permeability.

This solution correctly displays both the split phase-constants and split cut-off numbers of this type of waveguide. The cut-off condition for the lower split phase constant is given by

$$\varepsilon_{\text{eff}}(\mu - C_{11}\kappa) = \left(\frac{1.84}{k_+ R}\right)^2 \tag{10.8}$$

If a quarter-wave long cavity open-circuited at one flat face and short-circuited at the other is formed from such a waveguide then

$$(k_+ R)^2 = \frac{\left(\frac{\pi}{2}\right)^2 \left(\frac{R}{L}\right)^2 + 1.84^2}{(\mu - C_{11}\kappa)\varepsilon_f} \tag{10.9a}$$

$$(k_- R)^2 = \frac{\left(\frac{\pi}{2}\right)^2 \left(\frac{R}{L}\right)^2 + 1.84^2}{(\mu + C_{11}\kappa)\varepsilon_f} \tag{10.9b}$$

$$(k_0 R)^2 = \frac{\left(\frac{\pi}{2}\right)^2 \left(\frac{R}{L}\right)^2 + 1.84^2}{\mu_{\text{eff}} \varepsilon_f} \qquad (10.9c)$$

where

$$\mu_{\text{eff}} = \frac{\mu^2 - (C_{11}\kappa)^2}{\mu}$$

The split frequencies in such a gyromagnetic resonator are therefore described by

$$\frac{\omega_+ - \omega_-}{\omega_0} = C_{11}\kappa \qquad (10.10)$$

The upper bound on κ is of course fixed by the cut-off condition in equation (10.8). C_{11} is unity for an anisotropic waveguide.

10.4 Polynomial Representation of Split Phase Constants in Closed Gyromagnetic Resonator

The description of propagation in a circular waveguide containing a magnetized gyromagnetic rod and that in an open gyromagnetic resonators may be solved by having recourse to the exact formulation of the waveguide problem as is understood. The required characteristic equation from which the split propagation constants β_+ and β_- may be calculated is

$$\beta_\pm = \beta_0 \qquad (10.11)$$

The split frequencies $\omega_\pm$ of the gyromagnetic resonator may now be exactly evaluated once the description of $\beta_\pm$ is formulated. The polynomial representation of $\beta_\pm/k_0$ of the surface defined by $k_0 R$ and $\pm\kappa/\mu$ for $\varepsilon_r = 15.0$ based on the exact solution of the characteristic equation of the problem region with an ideal magnetic wall is reproduced below from Chapter 7.

$$\beta_\pm/k_0 = -11.668xy^2 + 24.657y^3 - 66.87y^2 + 61.548y + 21.809xy$$
$$+ 3.631yx^2 - 15.874 - 12.004x - 4.599x^2 - 1.181x^3,$$
$$0.6 \leq k_0 \leq 0.90 \qquad (10.12)$$

where $x = \kappa/\mu$; $y = k_0 R$

The cut-off space of the lower split branch of the magnetic wall problem region is as is understood dependent upon the gyrotropy. One polynomial representation for this feature which fixes the gyrotropy of the problem region has been given in Chapter 7.

Exact calculations of the split frequencies based on the perturbation formulation described by equations (10.9) is possible by equating $\beta_\pm/k_0$ in (10.7) and (10.12) and solving for C_{11}.

$$C_{11} = \frac{1}{\kappa} \left[\frac{\left(\frac{\beta_\pm}{k_0}\right)^2 + \left(\frac{1.84}{k_0 R}\right)^3}{\varepsilon_r} \right] \tag{10.13}$$

10.5 Partially Magnetized Gyrotropy

In practice, the gyromagnetic resonator is partially magnetized instead of saturated. The purpose of this section is to briefly revisit the entries of the tensor permeability under this condition. It is of separate note that the elements of the tensor permeability are frequency dependent so that the split frequency branches do not open up symmetrically.

In a partially magnetized material μ and κ are replaced by μ_p and κ_p

$$\mu_p = \mu_{\text{dem}} + (1 - \mu_{\text{dem}}) \left(\frac{M}{M_0}\right)^{3/2} \tag{10.14}$$

$$\kappa_p = \frac{\gamma M}{\mu_0 \omega} \tag{10.15}$$

where

$$\mu_{\text{dem}} = \frac{1}{3} + \frac{2}{3}\left[1 - \left(\frac{\gamma M_0}{\mu_0 \omega}\right)^2\right]^{1/2} \tag{10.16}$$

M is the actual magnetization (Tesla), M_0 is the saturation magnetization (Tesla), μ_0 is the free space permeability ($4\pi \times 10^{-7}$ H/m), ω is the radian frequency (rad/s), γ is the gyromagnetic ratio [2.21×10^5 (rad/s)/(A/m)].

When

$$M = M_0$$

$$\mu = 1 \tag{10.17a}$$

$$\kappa = \frac{\gamma M_0}{\mu_0 \omega} \tag{10.17b}$$

When $M = 0$

$$\mu = \mu_{\text{dem}} \qquad (10.18a)$$

$$\kappa = 0 \qquad (10.18b)$$

One common engineering solution is to take M equal to $M = 0.707 M_0$ and $(\gamma M_0/(\mu_0 \omega)) = 0.707$.

This then gives $\mu_p = 0.92$; $\kappa_p = 0.50$; and $\kappa_p/\mu_p = 0.54$.

10.6 Closed Gyromagnetic Resonator with Electric Wall

The development of the split frequencies of the closed gyromagnetic circular resonator with an electric wall proceeds as for the magnetic wall problem region. This gives

$$(k_+ R)^2 \varepsilon_f = \frac{\left(\frac{\pi}{2}\right)^2 \left(\frac{R}{L}\right)^2}{\mu - C_{11}\kappa} + 1.84^2 \qquad (10.19a)$$

$$(k_- R)^2 \varepsilon_f = \frac{\left(\frac{\pi}{2}\right)^2 \left(\frac{R}{L}\right)^2}{\mu + C_{11}\kappa} + 1.84^2 \qquad (10.19b)$$

$$(k_0 R)^2 \varepsilon_f = \frac{\left(\frac{\pi}{2}\right)^2 \left(\frac{R}{L}\right)^2}{\mu_{\text{eff}}} + 1.84^2 \qquad (10.19c)$$

Combining the proceeding equations gives the required result

$$\left(\frac{\omega_+ - \omega_-}{\omega_0}\right) = \left[\frac{\left(\frac{\pi}{2}\right)^2 \left(\frac{R}{L}\right)^2}{\left(\frac{\pi}{2}\right)^2 \left(\frac{R}{L}\right)^2 + 1.84^2}\right] \left(\frac{C_{11}\kappa}{\mu}\right) \qquad (10.20)$$

A scrutiny of this result indicates that the splitting in a circular gyromagnetic resonator with an electric wall is less than that in one with a magnetic wall one.

The constant C_{11} can again deduced by having recourse to perturbation theory or can be calculated by making use of the exact description of $\beta_\pm/k_0$. This gives

$$C_{11} = \frac{1}{\kappa} \left[\frac{\left(\frac{\beta_{\pm}}{k_0}\right)^2}{\varepsilon_r - \left(\frac{1.84}{k_0 R}\right)^2} \mp \mu \right] \qquad (10.21)$$

10.7 Higher Order Split Modes

The intersection between the upper and lower two split modes is also of some interest in the design of practical devices. For the TE_{11} and TE_{12} mode in the gyromagnetic waveguide with an electric wall condition

$$\beta_{11}^+ = \beta_{12}^- \qquad (10.22)$$

or

$$\frac{1 - C_{11}\kappa}{1 + C_{12}\kappa} \approx \frac{\varepsilon_r - \left(\frac{3.04}{k_0 R}\right)^2}{\varepsilon_f - \left(\frac{1.84}{k_0 R}\right)^2} \qquad (10.23)$$

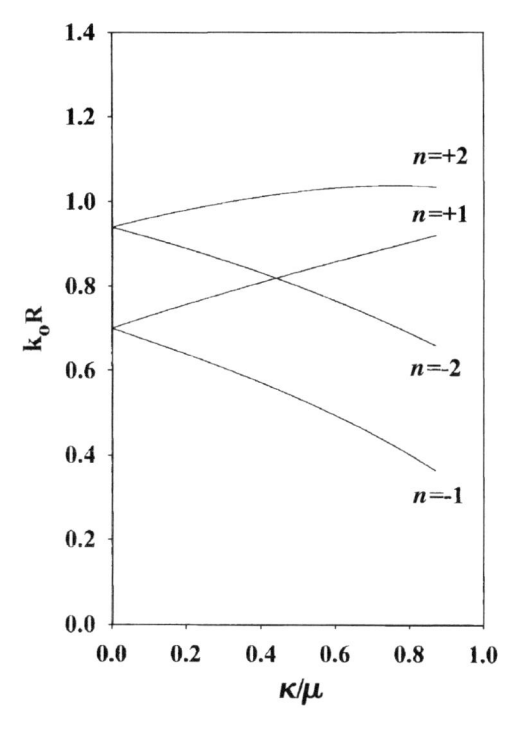

Figure 10.2 Intersection of first two split resonances in gyromagnetic resonator with a magnetic wall for $k_0 R = 0.82$ and 1.0 and $\varepsilon_r = 15$

For the TM$_{11}$ and TM$_{12}$ modes in the corresponding magnetic wall problem it is met by

$$\kappa \approx \frac{\left(\frac{3.04}{k_0 R}\right)^2 - \left(\frac{1.84}{k_0 R}\right)^2}{\varepsilon_f (C_{11} + C_{12})} \qquad (10.24)$$

C_{nm} is described by the cut-off frequencies of the related planar circuit with a magnetic wall:

$$C_{nm} = \frac{2n}{(kR)_{nm}^2 - n^2} \qquad (10.24)$$

The intersection between the branches of the TM$_{11}$ and TM$_{12}$ modes in a gyromagnetic waveguide with an ideal magnetic wall is indicated in Figure 10.2.

10.8 Quality Factor of Closed Gyromagnetic Resonator

One of the most important quantity in the theory of quarter-wave coupled circulators is its loaded Q-factor. This parameter is related to the split frequencies of the magnetized resonator in a particularly simple fashion by

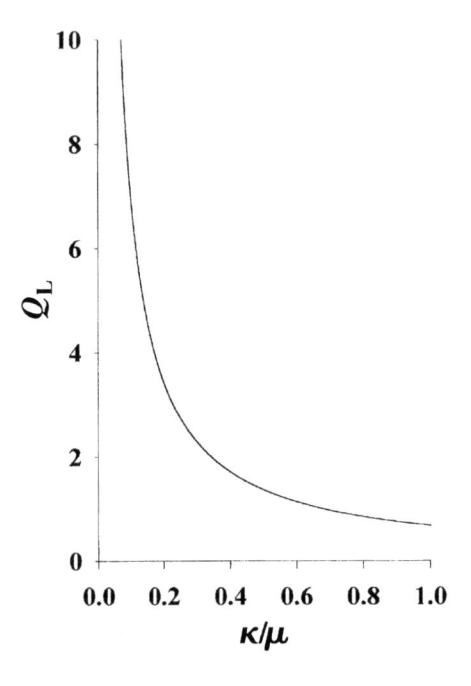

Figure 10.3 Quality factor of closed gyromagnetic resonator

10.9 ODD MODE GYROMAGNETIC RESONATOR

$$Q_L = \left[\sqrt{3}\left(\frac{\omega_+ - \omega_-}{\omega_0}\right)\right]^{-1} \tag{10.26}$$

Although the calculations of the split frequencies are exact, that of the quality factor neglects the influence of higher order modes (in a strongly magnetized resonator) on the description of the gyrator circuit. It separately assumes that the frequency variation of the in-phase eigen-network can be disregarded compared to those of the split counter-rotating ones and that only the first pair of counter-rotating modes need to be catered for in forming the complex gyrator circuit of the device. A knowledge of the onset of the first higher order split pair of modes is therefore desirable. Figure 10.3 depicts one result based on the perturbation formulation of the gyromagnetic resonator. Its description therefore applies to a weakly magnetized resonator only. For the purpose of this work, this condition is satisfied provided the quality factor has a lower bound equal to approximately two. Such a value of loaded Q-factor is compatible with the performance of many commercial devices.

10.9 Odd Mode Gyromagnetic Resonator

If the gyromagnetic cavity resonator is mounted in an H-plane junction then its solution corresponds to the odd mode solution of two coupled quarter long open gyromagnetic waveguides which are split by the gyrotropy. Figure 10.4 indicates the basic arrangement and Figure 10.5 three practical topologies. The coupling region in this sort of problem region is often taken as a cut-off section of circular waveguide with a fictitious magnetic wall. The solution of such a resonator is determined by

$$\frac{\varepsilon_f k_0}{\beta_0} \cot(\beta_0 L) - \frac{\varepsilon_d k_0}{\alpha} \coth(\alpha S) = 0 \tag{10.27}$$

$$\frac{\varepsilon_f k_0}{\beta_\pm} \cot(\beta_\pm L) - \frac{\varepsilon_d k_0}{\alpha} \coth(\alpha S) = 0 \tag{10.28}$$

where

$$\alpha^2 = \left(\frac{1.84}{R}\right)^2 - k_0 \varepsilon_d \tag{10.29}$$

and β_0 and $\beta_\pm$ are deduced by solving the open dielectric and gyromagnetic problem regions. The first of these equations again fixes $\beta_0 L$ and the second the quality factor.

At a fixed frequency, L_0 encountered in the decoupled resonator problem is replaced by

$$L = L_0 - \Delta L \tag{10.30}$$

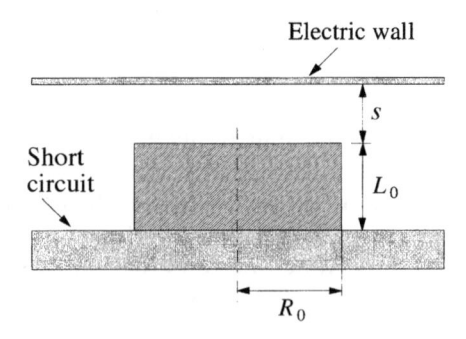

Figure 10.4 Schematic diagram of odd mode gyromagnetic resonator

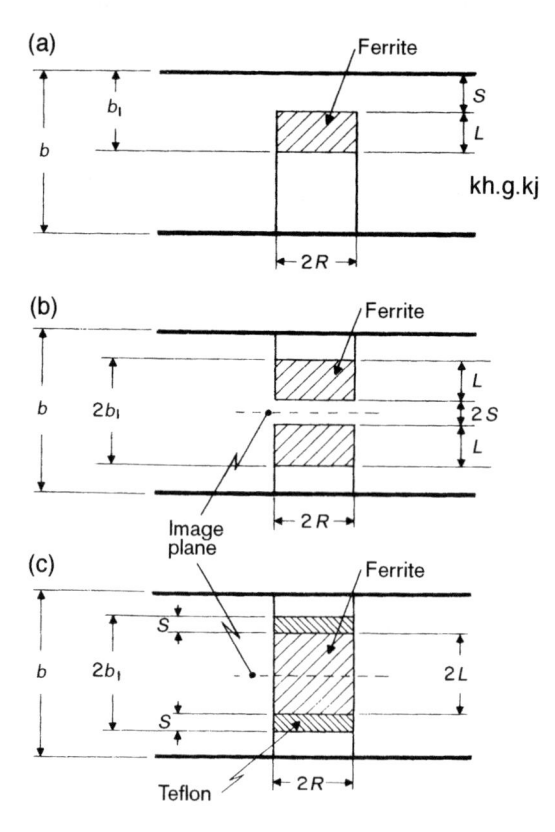

Figure 10.5 Schematic diagrams of practical gyromagnetic resonators (a) single disk resonator, (b) coupled disk resonators, (c) single cylinder resonator

10.9 ODD MODE GYROMAGNETIC RESONATOR

ΔL is a correction factor which accounts for the effect of the image or waveguide wall, S represents the spacing between the resonator and the image wall, ε_d is the relative dielectric constant of the region between the two. The second relationship may be evaluated for the two values of k_0 once β_0 and the dominant split phase constants $\beta_\pm$ of the gyromagnetic waveguide are calculated. The characteristic equation of the gyromagnetic resonator may for calculation purpose be written as

$$\varepsilon_{\text{eff}}\left(\frac{k_0}{\beta_\pm}\right)\cot\left[\left(\frac{\beta_\pm}{k_0}\right)\left(\frac{L}{R}\right)k_0 R\right] - \varepsilon_d\left(\frac{k_0}{\alpha_0}\right)\coth\left[\left(\frac{\alpha_0}{k_0}\right)\left(\frac{L}{R}\right)\left(\frac{1-k}{k}\right)k_0 R\right] = 0 \quad (10.31)$$

where

$$k = \frac{L}{L+S} \quad (10.32)$$

Scrutiny of this relationship suggests that the solution of this sort of problem may be presented in terms of R/L, $k_0 R$ and k.

One experimental plot of the split frequencies of an open gyromagnetic resonator using a garnet material with a dielectric constant of 15.3, a magnetiza-

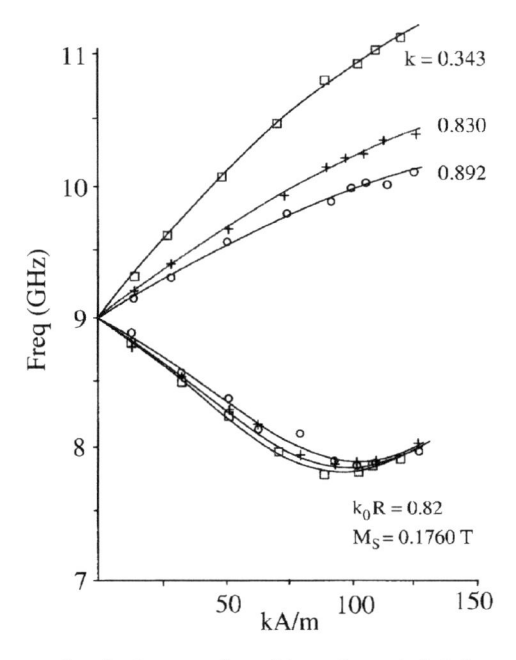

Figure 10.6 Experimental split frequencies of loosely and tightly coupled open triangular gyromagnetic resonator (reprinted with permission, Helszajn, J. and Sharp, J. (1983) Resonant frequencies, susceptance slope parameter and Q-factor of waveguide junctions using weakly magnetized open resonator *IEEE Trans. Microwave Theory Tech.*, **MTT-31**, 434–441) (© 1983 IEEE)

tion of 0.1760 T, for which $k_0R = 0.82$ at 9.0 GHz, is given in Figure 10.6. The lack of symmetry in the splitting at magnetic saturation is partly due to the form of κ in (10.16). The resonator is in direct contact with one waveguide wall to minimize the effect of the image wall on the result. The agreement between theory and experiment is adequate for engineering purposes.

10.10 Even Mode Gyromagnetic Resonator

The even mode solution met in connection with the design of E-plane circulators is readily determined by

$$\frac{\varepsilon_f k_0}{\beta} \cot(\beta L) - \frac{\varepsilon_d k_0}{\alpha} \tanh(\alpha S) = 0 \qquad (10.33)$$

$$\frac{\varepsilon_f k_0}{\beta_\pm} \cot(\beta_\pm L) - \frac{\varepsilon_d k_0}{\alpha} \tanh(\alpha S) = 0 \qquad (10.34)$$

The solution of the demagnetized resonator is again dealt with in Chapter 9.

10.11 Closed Composite Cavity Resonator

An important resonator geometry in the design of large mean power devices is a composite one consisting of a ferrite/dielectric assembly. Its topology is illustrated in Figure 10.7. The relationship between its split frequencies and its gyrotropy is therefore also of interest. It may be deduced by first determining the physical variables of the demagnetized resonator and, thereafter, experimentally or theoretically deducing the split frequencies of the magnetised one in terms of its physical and magnetic variables.

The resonant frequency of the demagnetized resonator may in the first instance be obtained by establishing a transverse resonance condition with electric and magnetic walls at the outside ferrite and dielectric flat faces of the composite

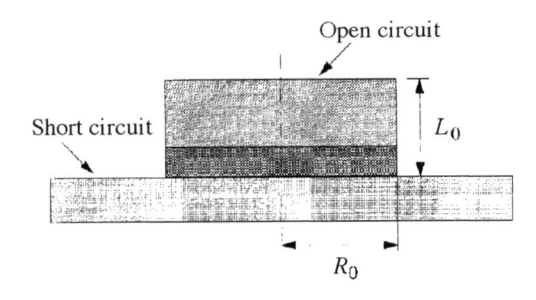

Figure 10.7 Schematic diagram of composite resonator

10.11 CLOSED COMPOSITE CAVITY RESONATOR

resonator. In this approximation, the influence of the image wall is omitted but the complete solution is outlined in the next section. Its characteristic equation is

$$\zeta_f \cot(\beta_f L_f) = \zeta_r \tan(\beta_r L_r) \tag{10.35}$$

This equation may be solved for $k_0 L_f$ in terms of $k_0 L_r$ and $k_0 R$

$$k_0 L_f = \frac{k_0}{\beta_f} \cot^{-1}\left[\frac{\zeta_r}{\zeta_f} \tan\left(\frac{\beta_r}{k_0}\right)(k_0 L_r)\right] \tag{10.36}$$

If the resonator is composed of waveguide sections with ideal magnetic walls, then the phase velocities of the demagnetized ferrite and dielectric waveguide sections are given by

$$\beta_r^2 = k_0^2 \varepsilon_r - \left(\frac{1.84}{R}\right)^2 \tag{10.37}$$

$$\beta_f^2 = k_0^2 \varepsilon_f \mu_f - \left(\frac{1.84}{R}\right)^2 \tag{10.38}$$

and the admittance ones by

$$\zeta_r = \zeta_0 \frac{\varepsilon_r k_0}{\beta_r} \tag{10.39}$$

$$\zeta_f = \zeta_0 \frac{\varepsilon_f k_0}{\beta_f} \tag{10.40}$$

where

$$\zeta_0 = \sqrt{(\varepsilon_0/\mu_0)} \quad \text{and} \quad k_0^2 = \omega_0^2 \mu_0 \varepsilon_0$$

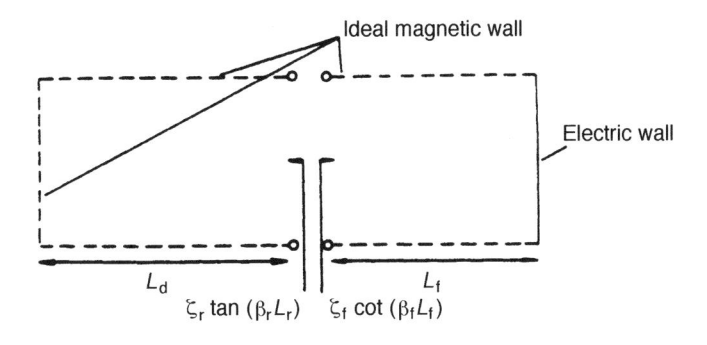

Figure 10.8 Circuit topology of closed composite gyromagnetic resonator

In the discussion of a composite resonator, it is also usual to define a filling factor to describe its details. It is given by

$$k_f = \frac{L_f}{L_f + L_r} \qquad (10.41)$$

The equivalent circuit of this arrangement is shown in Figure 10.8.

The relationship between $k_0 L_f$ and $k_0 L_r$ is separately illustrated in Figure 10.9 for three typical values of $k_0 R_0$. L_f is the length of the ferrite section, L_r is that of the dielectric one.

The split frequencies and quality factor of a composite resonator may be either established experimentally or theoretically by replacing β_f and $\beta_\pm$ and choosing some appropriate representation for $\beta_\pm$.

10.12 Odd Mode Composite Resonator

If the image wall is retained in the description of the overall resonator then the odd mode solution of two coupled resonators is again appropriate. The characteristic equation for the demagnetized and magnetized odd mode topology depicted in Figure (10.10) is then given by

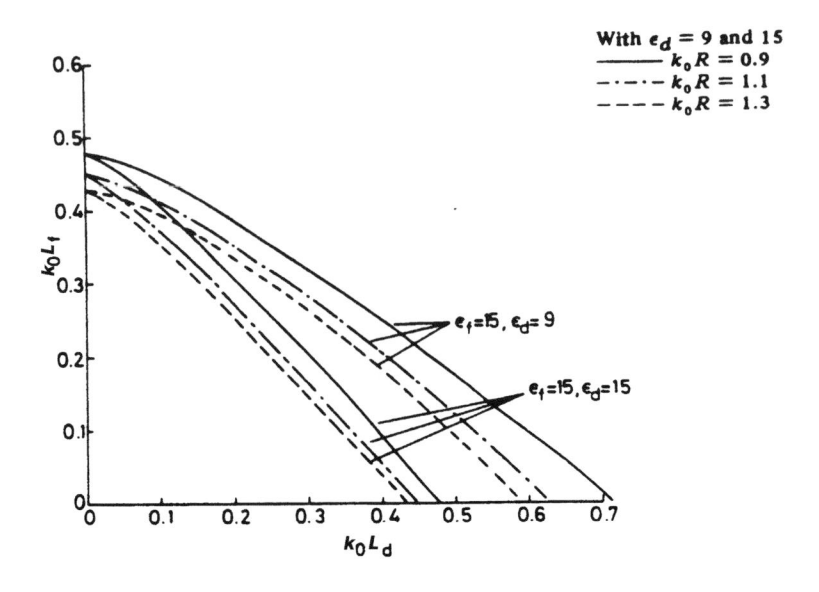

Figure 10.9 $k_0 L_f$ versus $k_0 L_r$ for closed composite resonator (reprinted with permission, Helszajn, J. (1981) High-power waveguide circulators using quarterwave long composite ferrite/dielectric resonators, *IEE Proc.*, **128**, Pt. H, No. 5, October)

10.12 ODD MODE COMPOSITE RESONATOR

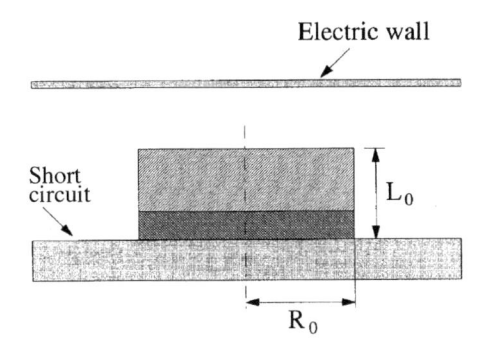

Figure 10.10 Odd mode topology of composite resonator

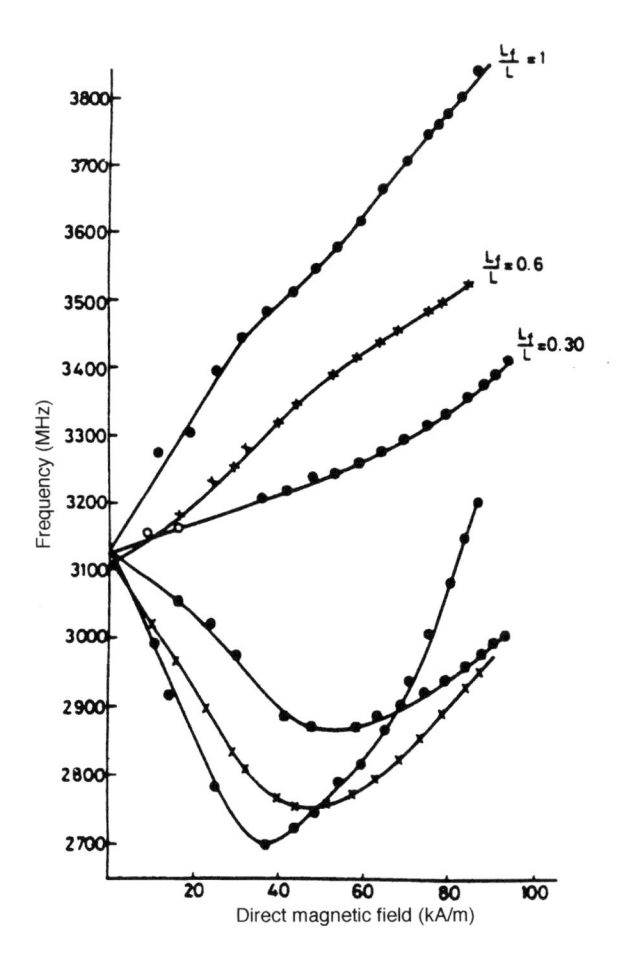

Figure 10.11 Split mode chart of composite resonator for $k = 0.34$, 0.60, 1.00 and $M_0 = 0.0680\text{T}$ (reprinted with permission, Helszajn, J. (1981) High-power waveguide circulators using quarterwave long composite ferrite/dielectric resonators, *IEE Proc.*, **128**, Pt. H, No. 5, October)

$$k_0 \left[\frac{1 - \dfrac{\beta_f \varepsilon_r}{\beta_r \varepsilon_f} \tan(\beta_f L_f) \tan(\beta_r L_r)}{\dfrac{\beta_r}{\varepsilon_r} \tan(\beta_r L_r) + \dfrac{\beta_f}{\varepsilon_f} \tan(\beta_f L_f)} \right] - \frac{\varepsilon_d k_0}{\alpha} \coth(\alpha S) = 0 \qquad (10.42)$$

where α has the meaning met in connection with the odd mode solution of the homogeneous resonator given by (10.27) and ε_r and β_r are the relative dielectric constant and phase constant of the dielectric region of the composite resonator. S is the distance of the gap between the open-circuited end of the resonator and the image wall.

For the demagnetized resonator β_f applies, whereas for the magnetized one $\beta_\pm$ must be used. The effect of the image wall may be discarded in the above equation by letting S approach infinity.

The odd mode solution of the composite resonator discussed here is usually described in terms of two filling factors. One filling factor is described by Equation (10.41) and the other which involves the gap region by

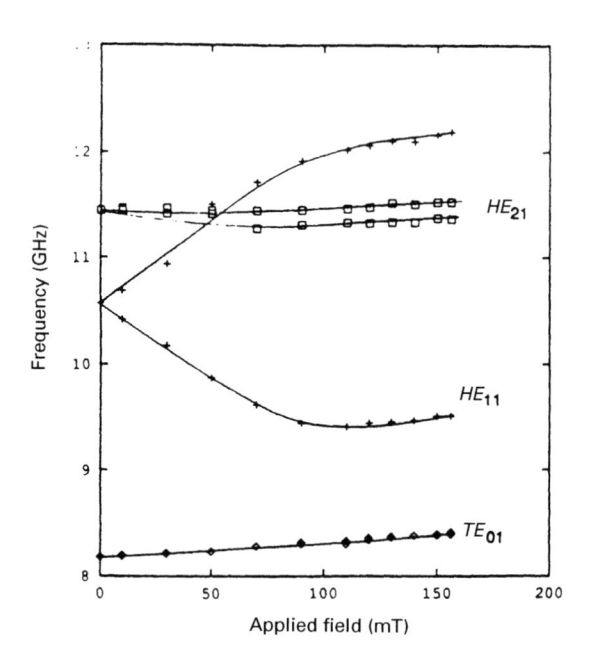

$$k = \frac{L_f + L_r}{L_f + L_r + S} \qquad (10.43)$$

Figure 10.12 Split modes in quarter-wave long gyromagnetic resonator in metal enclosure ($\varepsilon_d = 2.56$, $\varepsilon_f = 15.1$, $M_0 = 0.1600$ T, $R_i/L_i = 1.28$) (reprinted with permission, Helszajn, J., Cheng, C. S. and Wilcock, B. A. (1993) A nonreciprocal turnable waveguide directional filter using a turnstile open gyromagnetic resonator, *IEEE Trans. Microwave Theory Tech.*, **41**(11), 1950–1958) (© 1993 IEEE)

Some results obtained in a WR229 waveguide at about 3 GHz for three different values of filling factor for a garnet material with a saturation magnetisation of 0.0680 T and a dielectric constant of 14.3 are separately depicted in Figure 10.11. The relative dielectric constant of the material substituted for the garnet material was 15.0. The normalized split frequencies for these three values of filling factor at the direct magnetic field required to saturate the material are 0.267, 0.230 and 0.128 respectively. The result applies to a resonator with $k_0 R \approx 0.97$.

10.13 Gyromagnetic Cylindrical Resonator in Metal Enclosure

The odd mode split frequencies or phase constants of a gyromagnetic resonator may be experimentally determined by embodying it in an oversized metal enclosure in a 1-port cavity or 2-port arrangement. The required quantities may also

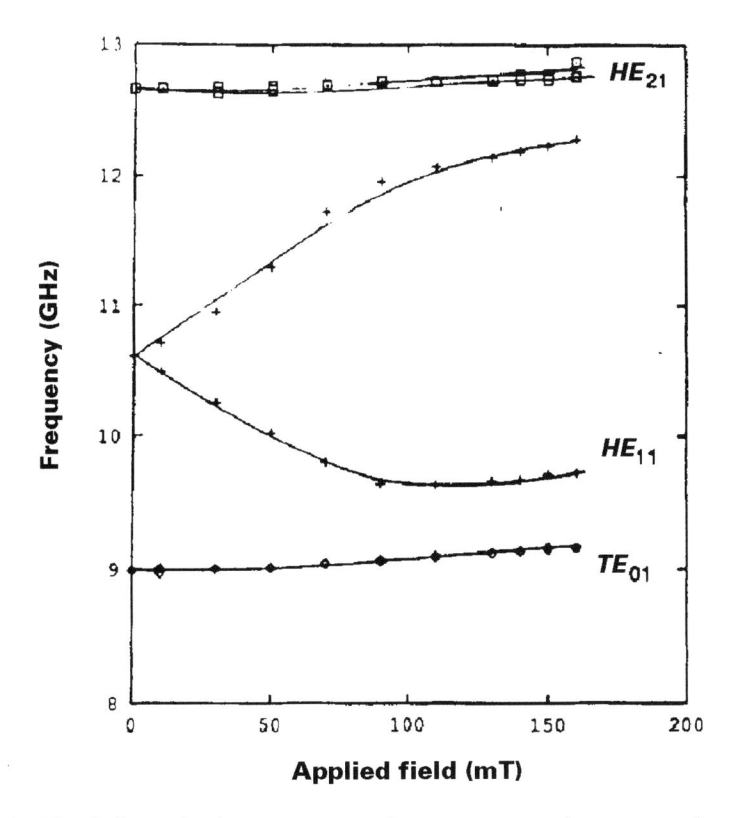

Figure 10.13 Split modes in quarter-wave long gyromagnetic resonator in metal enclosure ($\varepsilon_d = 2.56$, $\varepsilon_f = 15.1$, $M_0 = 0.1600$ T, $R_i/L_i = 1.28$) (reprinted with permission. Helszajn, J., Cheng, C. S. and Wilcock, B. A. (1993) A nonreciprocal turnable waveguide directional filter using a turnstile open gyromagnetic resonator, *IEEE Trans. Microwave Theory Tech.*, **41**(11), 1950–1958) (© 1993 IEEE)

be obtained by placing a 1-port geometry at one of the planes of circular polarization of the alternating magnetic field of a suitable rectangular waveguide. Figure 10.12 indicates the results obtained with the latter configuration for a gyromagnetic resonator for which the relative dielectric constant of the dielectric filler is $\varepsilon_d = 3.47$ and for which $R_i/L_i = 1.283$. Each branch on this illustration is obtained with a different sense of the direct magnetic field. Figure 10.13 depicts a similar result for same gyromagnetic resonator but embedded in a dielectric filler equal to $\varepsilon_d = 2.56$.

This sort of arrangement can also be utilized if so desired to realize tunable 1 or 2-port filter circuits. The overall frequency response of such a tunable filter can be extended if so desired to encompass both split branches. This may be done by sweeping the direct field between its positive and negative settings. This feature may be understood by recalling that a gyromagnetic waveguide displays one value of permeability for a circularly polarized wave with the direct magnetic field along that of propagation and a second value if the direct magnetic field is in the opposite sense from that of propagation. Figure 10.14 summarizes the relationships between frequency and direct magnetic field for the two arrangements considered here.

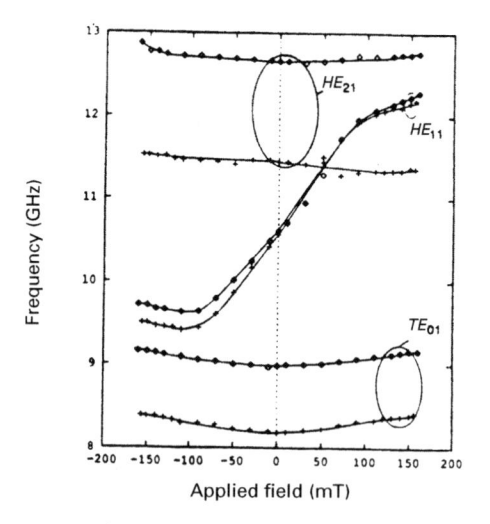

Figure 10.14 Tuning ranges of gyromagnetic resonator in metal enclosure ($\varepsilon_\delta = 2.56$ and 3.47, $\varepsilon_f = 1.51$, $k \approx 0.80$, $M_0 = 0.1600$ T, $R_i/L_i = 1.28$) (reprinted with permission, Helszajn, J., Cheng, C. S. and Wilcock, B. A. (1993) A nonreciprocal turnable waveguide directional filter using a turnstile open gyromagnetic resonator, *IEEE Trans. Microwave Theory Tech.*, **41**(11), 1950–1958) (© 1993 IEEE)

References and Further Reading

Akaiwa, Y. (1977) Mode classification of triangular ferrite post for Y-circulator, *IEEE Trans. Microwave Theory Tech.*, **MTT-25**, 59–61.

Bussey, H. E. and Steinert, L. A. (1957) An exact solution for cylindrical cavity containing a gyromagnetic material, *Proc. IRE*, **45**, 693.

Chin, G. C., Epp, L. W. and Wilkins, G. W. (1995) Determination of the eigenfrequencies of a ferrite-filled cylindrical cavity resonator using the finite element method, *IEEE Trans Microwave Theory Tech.*, **43**(5).

Godtmann, H. D. and Haas, W., (1967) Magnetodynamic modes in axially magnetized ferrite rods between two parallel conducting sheets, *IEEE Trans. Microwave Theory Tech.*, **MTT-15**, 476–481.

Heller, G. S. (1957) Ferrite loaded cavity resonators, *Onde Elec., Spec. Suppl.*, 588.

Heller, G. S. and Lax, B. (1957) Use of perturbation theory for cavities and waveguides containing ferrite, *IRE Trans. Antennas Propagat.*, **4**, 588.

Helszajn, J. (1981) High-power waveguide circulators using quarterwave long composite ferrite/dielectric resonators, *IEE Proc.*, **128**(H), (5), 268–273.

Helszajn, J. and Sharp, J. (1983) Resonant frequencies, susceptance slope parameter and Q-factor of waveguide junctions using weakly magnetized open resonator, *IEEE Trans. Microwave Theory Tech.*, **MTT-31**, 434–441.

Helszajn, J. and Sharp, J. (1986) Dielectric and permeability effects in HE_{11} open demagnetized ferrite resonators, *IEE Proc.*, **133**(H), 271–276.

Helszajn, J. and Cheng, S. (1990) Aspect ratio of open resonators in the design of evanescent mode E-plane circulators, *IEE Proc. Microwaves, Antennas Propagat.*, **137**, 55–60.

Helszajn, J., Leeson, W., Lynch, D. and O'Donnell, B. (1991) Normal mode nomenclature of quadrupole gyromagnetic waveguide, *IEEE Trans. Microwave Theory Tech.*, **MTT-39**, 461–470.

Helszajn, J., Cheng, C. S. and Wilcock, B. A. (1993) A nonreciprocal turnable waveguide directional filter using a turnstile open gyromagnetic resonator, *IEEE Trans. Microwave Theory Tech.*, **MTT-41**(11), 1950–1958.

Khilla, A. M. and Wolff, I. (1979) The point matching solution for magnetically turnable cylindrical cavities and ferrite planar resonators, *IEEE Trans. Microwave Theory Tech.*, **MTT-27**, 592–598.

Riblet, G., Helszajn, J. and O'Donnell, B. (1979) Loaded Q-factors of partial-height and full-height triangular resonator for use in waveguide circulators, *Proc. Eur. Microwave Conf.*, 420–424.

Schierlich, C. (1989) Mode charts for magnetized ferrite cylinders, *IEEE Trans. Microwave Theory Tech.*, **MTT-37**, 1555–1561, October.

11

IMPEDANCE IN RECTANGULAR, RIDGE AND RADIAL WAVEGUIDES

11.1 Introduction

The design of a degree-2 junction circulator relies on its realization upon a suitable unit element U.E. or impedance transformer. One practical structure consists of a radial transformer, another of a tapered section, still another upon a ridge one. These various structures as well as the use of a triangular plate are illustrated in Figure 11.1. All possibilities except the ridge one are so-called non-uniform lines. One property of such lines is that a short circuit is not mapped into an open circuit in the same length as an open circuit one is mapped into a short circuit. Another problem associated with this sort of line is the lack of uniqueness in the definition of impedance. One shortcoming of the use of wave impedance in the description of microwave structures adopted so far is that it does not embody the physical dimensions of the problem region. It is therefore unable to cope with steps in the heights of the different waveguide structures illustrated in Figure 11.2. In order to reconcile this difficulty it is usual to introduce voltage and current variables and the notion of characteristic impedance instead of wave impedance. The purpose of this chapter is to tackle some of the engineering aspects entering into the description of these types of lines. If a ridge structure is used for the U.E. in this sort of device then one possible way to fix the waveguide opening at its operating frequency is to adopt that of the corresponding ridge waveguide at the same frequency. A mitered arrangement met in some commercial arrangements is separately depicted in Figure 11.3.

11.2 Impedance in Waveguides

The characteristic impedance of a uniform transmission line supporting TEM propagation may be defined in one of three possible ways:

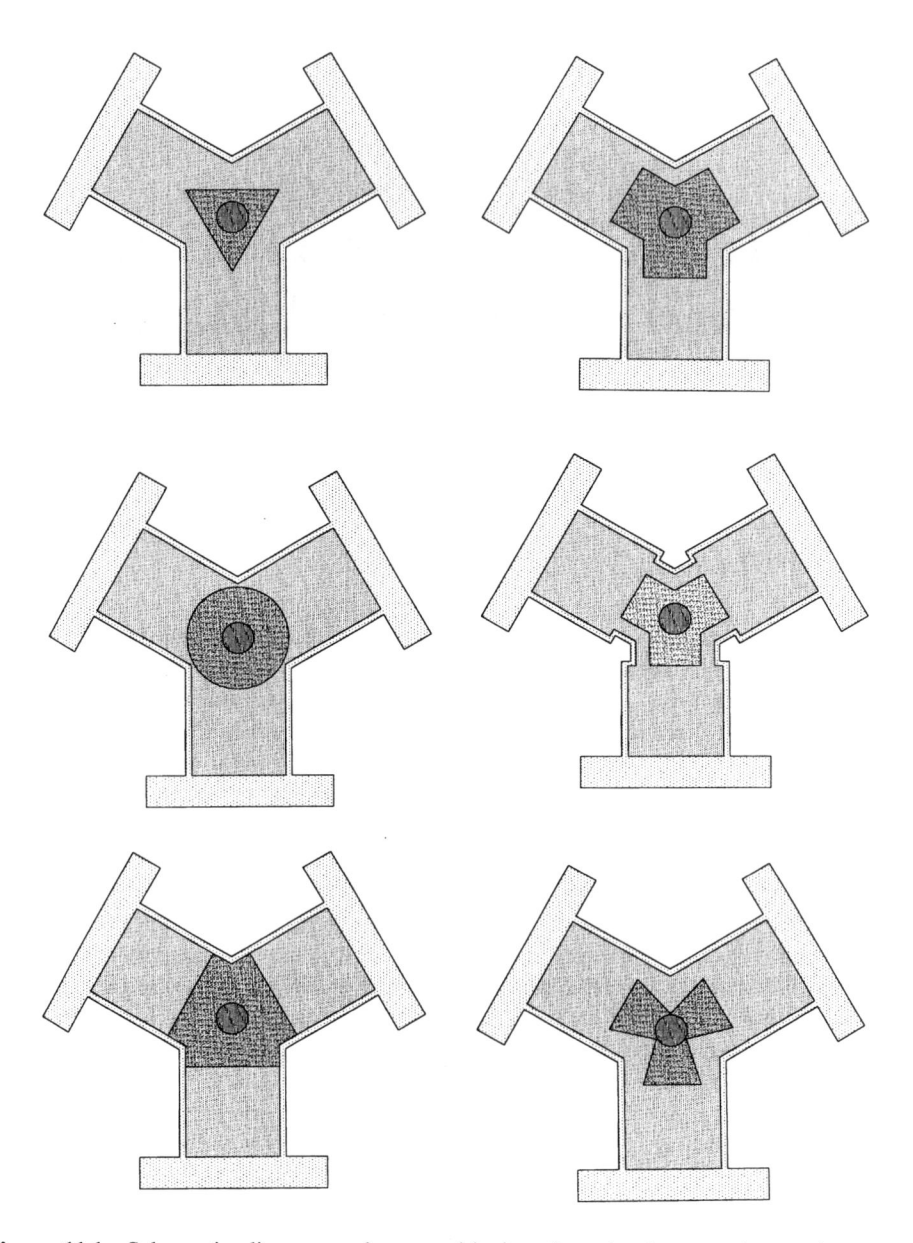

Figure 11.1 Schematic diagrams of waveguide junction circulators using various transformer structures

11.2 IMPEDANCE IN WAVEGUIDES

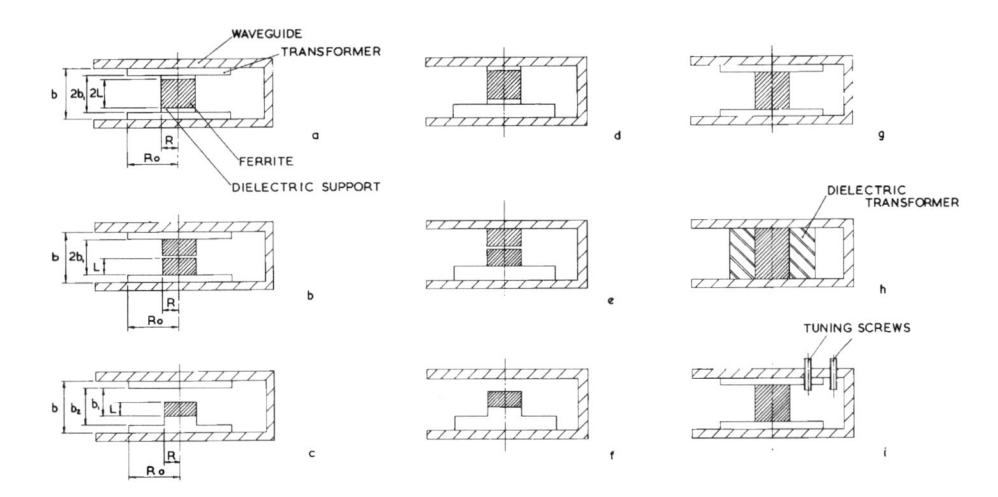

Figure 11.2 Schematic diagrams of waveguide junction circulators using various impedance structures; (a) Cross section of waveguide circulator using $\lambda/2$ resonator with symmetrical radial transformers, (b) Cross section of waveguide circulator using coupled $\lambda/4$ resonators with symmetrical radial transformers, (c) Cross section of waveguide circulator using single $\lambda/4$ resonator with symmetrical radial transformers, (d) Cross section of waveguide circulator using $\lambda/2$ resonator with asymmetrical radial transformers, (e) Cross section of waveguide circulator using coupled $\lambda/4$ resonators with asymmetrical radial transformers, (f) Cross section of waveguide circulator using single $\lambda/4$ resonator with asymmetrical transformers, (g) Cross section of waveguide circulator using ferrite post with symmetrical radial transformer, (h) Cross section of waveguide circulator using ferrite post with dielectric quarter wave transformer, (i) Cross section of waveguide circulator showing fine capacitive tuning

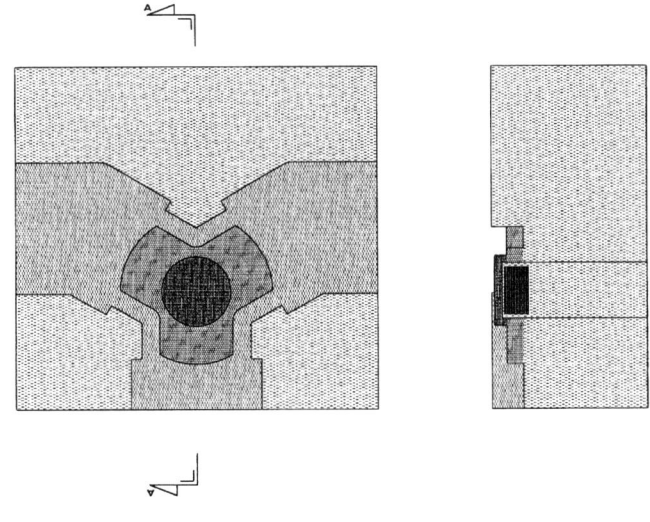

Figure 11.3 Schematic diagram of waveguide circulator using mitered bends

SECTION A-A

$$Z_{PV} = \frac{VV^*}{2P_t} \tag{11.1}$$

$$Z_{PI} = \frac{2P_t}{II^*} \tag{11.2}$$

$$Z_{VI} = \frac{V}{I} \tag{11.3}$$

and

$$Z_{PV} = Z_{PI} = Z_{VI} \tag{11.4}$$

However, in rectangular or circular waveguides, the definitions of voltage and current are not unique.

$$Z_{PV} \neq Z_{PI} \neq Z_{VI} \tag{11.5}$$

One relationship between the various definitions of impedance is

$$Z_{VI} = \sqrt{Z_{PV} Z_{PI}} \tag{11.6}$$

11.3 Power Transmission through Rectangular Waveguide

An important quantity in the description of the impedance of any waveguide is the average power transmitted (P_t) through it. It may be evaluated by making use of the complex Poynting theorem

$$P_t = \tfrac{1}{2} \int_s \operatorname{Re}(\overline{E} \times \overline{H}^*) \, d\overline{s} \tag{11.7}$$

where $\overline{s}$ is the total surface perpendicular to the direction of propagation.

For the dominant TE_{10} mode in a rectangular waveguide

$$\tfrac{1}{2}(\overline{E} \times \overline{H}^*) = \frac{1}{2} \begin{bmatrix} \overline{a}_x & \overline{a}_y & \overline{a}_z \\ 0 & E_y & 0 \\ H_x^* & 0 & H_z^* \end{bmatrix} \tag{11.8}$$

and

$$\tfrac{1}{2}(\overline{E} \times \overline{H}^*) = \tfrac{1}{2}[\overline{a}_x(E_y H_z^*) + \overline{a}_y(0) + \overline{a}_z(-E_y H_x^*)] \tag{11.9}$$

11.3 POWER TRANSMISSION

where

$$H_z = \cos\left(\frac{\pi x}{a}\right) \tag{11.10a}$$

$$H_y = 0 \tag{11.10b}$$

$$H_x = j\left(\frac{2a}{\lambda_g}\right)\sin\left(\frac{\pi x}{a}\right) \tag{11.10c}$$

$$E_z = 0 \tag{11.10d}$$

$$E_y = -j\left(\frac{2a}{\lambda_0}\right)\sqrt{\frac{\mu_0}{\varepsilon_0}}\sin\left(\frac{\pi x}{a}\right) \tag{11.10e}$$

$$E_x = 0 \tag{11.10f}$$

The cutoff wavelength for the TE_{10} mode with $m = 1$, $n = 0$ is

$$\lambda_c = 2a \tag{11.11}$$

The guide wavelength is separately described in the usual way by

$$\left(\frac{2\pi}{\lambda_g}\right)^2 = \left(\frac{2\pi}{\lambda_0}\right)^2 - \left(\frac{2\pi}{\lambda_c}\right)^2 \tag{11.12}$$

The transverse wave impedance of the waveguide is denoted by Z_{TE}

$$Z_{TE} = \frac{E_y}{H_x} = \left(\frac{\lambda_g}{\lambda_0}\right)\sqrt{\frac{\mu_0}{\varepsilon_0}} \tag{11.13}$$

Noting that $E_y H_z^*$ is a pure imaginary quantity gives

$$\tfrac{1}{2}\,\text{Re}\,(\overline{E} \times \overline{H}^*) = \frac{1}{2}\left(\frac{\lambda_c}{\lambda_g}\right)\left(\frac{\lambda_c}{\lambda_0}\right)\sqrt{\frac{\mu_0}{\varepsilon_0}}\sin^2\left(\frac{\pi x}{a}\right) \tag{11.14}$$

The derivation now proceeds by integrating this quantity over the waveguide cross-section.

$$P_t = \frac{1}{2}\left(\frac{\lambda_c}{\lambda_g}\right)\left(\frac{\lambda_c}{\lambda_0}\right)\sqrt{\frac{\mu_0}{\varepsilon_0}}\int_0^a\int_0^b \sin^2\left(\frac{\pi x}{a}\right) dx\,dy \tag{11.15}$$

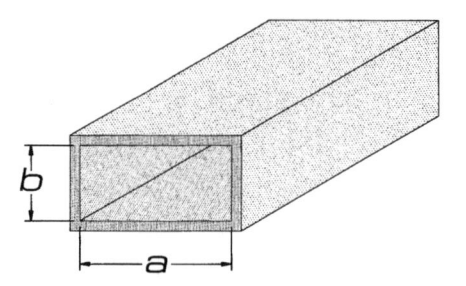

Figure 11.4 Schematic diagram of rectangular waveguide

The required result is:

$$P_t = \frac{ab}{4}\left(\frac{\lambda_c}{\lambda_g}\right)\left(\frac{\lambda_c}{\lambda_0}\right)\sqrt{\frac{\mu_0}{\varepsilon_0}} \qquad (11.16)$$

The rectangular waveguide considered here is indicated in Figure 11.4.

11.4 Impedance in Rectangular Waveguide

The derivations of each of the three definitions of impedance met in the case of a rectangular waveguide propagating the TE_{10} mode will now be illustrated by way of an example. The voltage V across the waveguide is defined as the line integral of the electric field at the midpoint of the waveguide and the current I as the total longitudinal current flowing in the wide surface of one of the waveguide walls. The power flow P_t is given by equation (11.16).

The derivation of the power–voltage definition of impedance Z_{PV} starts by defining V at the symmetry plane of the waveguide

$$V = \int_0^b E_y \, dy \qquad (11.17)$$

Evaluating this quantity at $x = a/2$ gives

$$V = -j\left(\frac{2ab}{\lambda_0}\right)\sqrt{\frac{\mu_0}{\varepsilon_0}} \qquad (11.18)$$

Combining this result with the description of the power flow in the waveguide produces the required result:

$$Z_{PV} = \left(\frac{2b}{a}\right) Z_{TE} \qquad (11.19)$$

The power–current definition for the impedance (Z_{PI}) is established by evaluating the total longitudinal current flowing in the wide dimension of the waveguide:

$$I = \int_0^a J_z \, dx \qquad (11.20)$$

This may be done by noting the relationship between the current density and the magnetic field:

$$J_z = H_x \qquad (11.21)$$

The total longitudinal current is therefore

$$I = j \left(\frac{\lambda_c^2}{\pi \lambda_g} \right) \qquad (11.22)$$

The corresponding power–current definition of impedance is

$$Z_{PI} = \left(\frac{\pi^2 b}{8a} \right) Z_{TE} \qquad (11.23)$$

The voltage–current definition of impedance in a rectangular waveguide is now deduced by making use of the relationship between it and the power–voltage and power–current definitions in equation (11.6). The result is

$$Z_{VI} = \left(\frac{\pi b}{2a} \right) Z_{TE} \qquad (11.24)$$

A scrutiny of the three descriptions of impedance in the waveguide suggests that Z_{PV} is of the order of Z_{TE} in standart rectangular waveguide.

11.5 Cut-Off Space of Ridge Waveguide

One important waveguide encountered in the realization of impedance transformers in the construction of waveguide circulators is the ridge one. Figure 11.5 illustrates the two possible arrangements. The physical variables entering into the description of these waveguides are separately shown on these diagrams. Figure 11.6 indicates the nature of the electric fields in these sorts of waveguides. Table 11.1 summarizes the details of some standard commercial double ridge waveguides.

The cut-off condition for either the single or the double ridge waveguide for the even TE_{no} modes based on a transverse resonance condition is given by

$$-\cot \theta_1 + \left(\frac{Y_{02}}{Y_{01}} \right) \tan \theta_2 + \frac{B}{Y_{01}} = 0 \qquad (11.25)$$

Table 11.1 Dimensions of a standard double ridge waveguide

WRD	Frequency range (GHz)	Attenuation for Copper at 1.73fc (dB/m)	A mm	B mm	C mm	D mm	E mm	F mm	G mm max	H mm ±10%
200	2.00 – 4.80	0.029	65.79 ± 0.10	30.61 ± 0.10	69.85 ± 0.10	34.67 ± 0.10	13.00 ± 0.05	16.46 ± 0.05	1.27	2.59
350	3.50 – 8.20	0.067	37.59 ± 0.08	17.48 ± 0.08	40.84 ± 0.10	20.73 ± 0.10	7.42 ± 0.05	9.40 ± 0.05	0.76	1.47
475	4.75 – 11.0	0.106	27.69 ± 0.08	12.85 ± 0.08	30.23 ± 0.08	15.39 ± 0.08	5.46 ± 0.05	6.91 ± 0.05	0.76	1.09
580	5.80 – 16.0	0.213	19.82 ± 0.08	9.40 ± 0.08	22.35 ± 0.08	11.94 ± 0.08	3.05 ± 0.05	5.08 ± 0.05	0.51	1.09
650	6.50 – 18.00	0.262	18.29 ± 0.08	8.15 ± 0.08	20.83 ± 0.08	10.69 ± 0.08	2.57 ± 0.05	4.39 ± 0.05	0.51	0.56
750	7.50 – 18.00	0.210	17.55 ± 0.08	8.15 ± 0.08	20.09 ± 0.08	10.69 ± 0.08	3.45 ± 0.05	4.39 ± 0.05	0.51	0.69
110	11.00 – 26.50	0.374	11.96 ± 0.08	5.56 ± 0.8	14.00 ± 0.08	7.59 ± 0.08	2.362 ± 0.05	2.997 ± 0.05	0.38	0.48
180	18.00 – 40.00	0.781	7.32 ± 0.08	3.40 ± 0.08	9.35 ± 0.08	7.59 ± 0.08	1.448 ± 0.05	1.829 ± 0.05	0.38	0.28

11.5 CUT-OFF SPACE OF RIDGE WAVEGUIDE

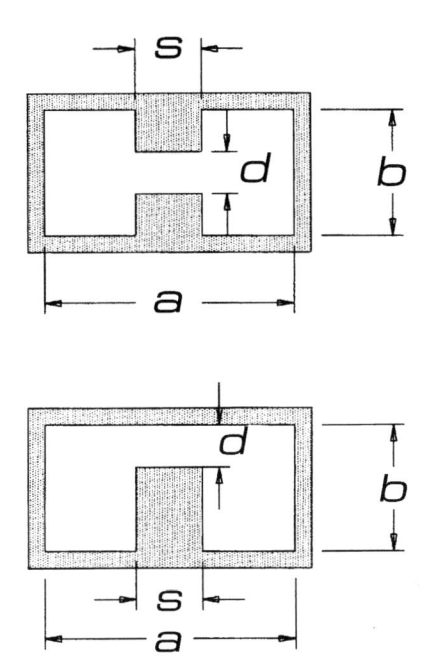

Figure 11.5 Schematic diagrams of single and double ridge waveguide

where

$$Y_{01} = \frac{k_c}{\omega\mu_0}\left(\frac{1}{b}\right) \tag{11.26a}$$

$$Y_{02} = \frac{k_c}{\omega\mu_0}\left(\frac{1}{d}\right) \tag{11.26b}$$

and

$$\theta_1 = \frac{\pi(a-s)}{\lambda_c} = \pi\left(1 - \frac{s}{a}\right)\left(\frac{a}{\lambda_c}\right) \tag{11.27a}$$

$$\theta_2 = \frac{\pi s}{\lambda_c} = \pi\left(\frac{s}{a}\right)\left(\frac{a}{\lambda_c}\right) \tag{11.27b}$$

k_c is equal to

$$k_c = \frac{2\pi}{\lambda_c} \tag{11.28}$$

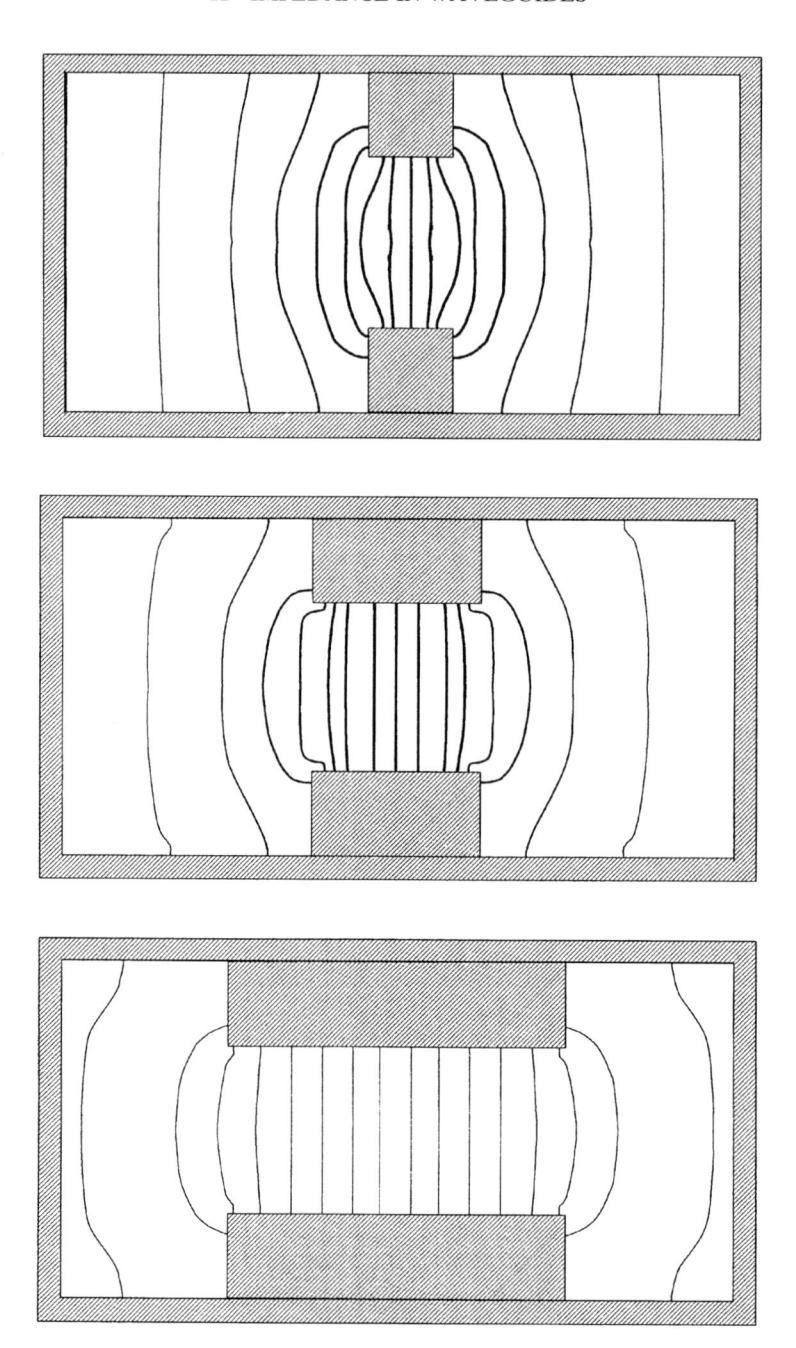

Figure 11.6 Electric fields in single and double ridge waveguide

11.5 CUT-OFF SPACE OF RIDGE WAVEGUIDE

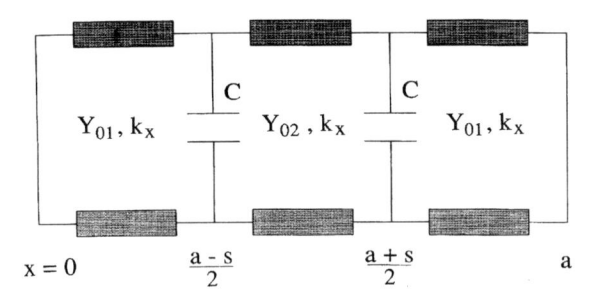

Figure 11.7 Equivalent circuit of single and double ridge waveguide

B/Y_{01} represents the susceptance of the step discontinuity on either side of the ridge. One approximation in the case of single ridge is (see Cohn 1947)

$$\frac{B}{Y_{01}} \approx \left(\frac{4b}{\lambda_c}\right) \ln \operatorname{cosec}\left(\frac{\pi d}{2b}\right) \tag{11.29}$$

The result for double ridge is

$$\frac{B}{Y_{01}} \approx \left(\frac{2b}{\lambda_c}\right) \ln \operatorname{cosec}\left(\frac{\pi d}{2b}\right) \tag{11.30}$$

The relationship between the fringing capacitances of the single ridge and double ridge waveguides may be understood by introducing an electric wall at the plane of symmetry of the latter arrangement. The introduction of the symmetry plane indicates that the two capacitors formed in this way are in series due to the fact that the top and bottom ridges are at different potentials in order to support the electric field. The normalized capacitance of the single ridge with the nomenclature used to label the details of the wavelength is therefore twice that of the double one as asserted. The equivalent circuit met in this problem region is indicated in Figure 11.7. The log term in these relationships is also sometimes written as

$$\ln \operatorname{cosec}\left(\frac{\pi a}{2}\right) = \frac{1}{2}\left[\left(\frac{1+\alpha^2}{\alpha}\right)\ln\left(\frac{1+\alpha}{1-\alpha}\right) - 2\ln\left(\frac{4\alpha}{1-\alpha^2}\right)\right] \tag{11.31}$$

where $\alpha = d/b$.

Figure 11.8 illustrates the cut-off space of the single ridge structure and Figure 11.9 that of the double ridge configuration. The cut-off condition at s/a equal to zero corresponds to that of an infinitely thin finline; at s/a equal to unity it reduces to that of standard rectangular waveguide.

One closed form approximation for the cut-off space of the dominant mode in the double ridge waveguide is (see Sharma and Hoefer, 1983)

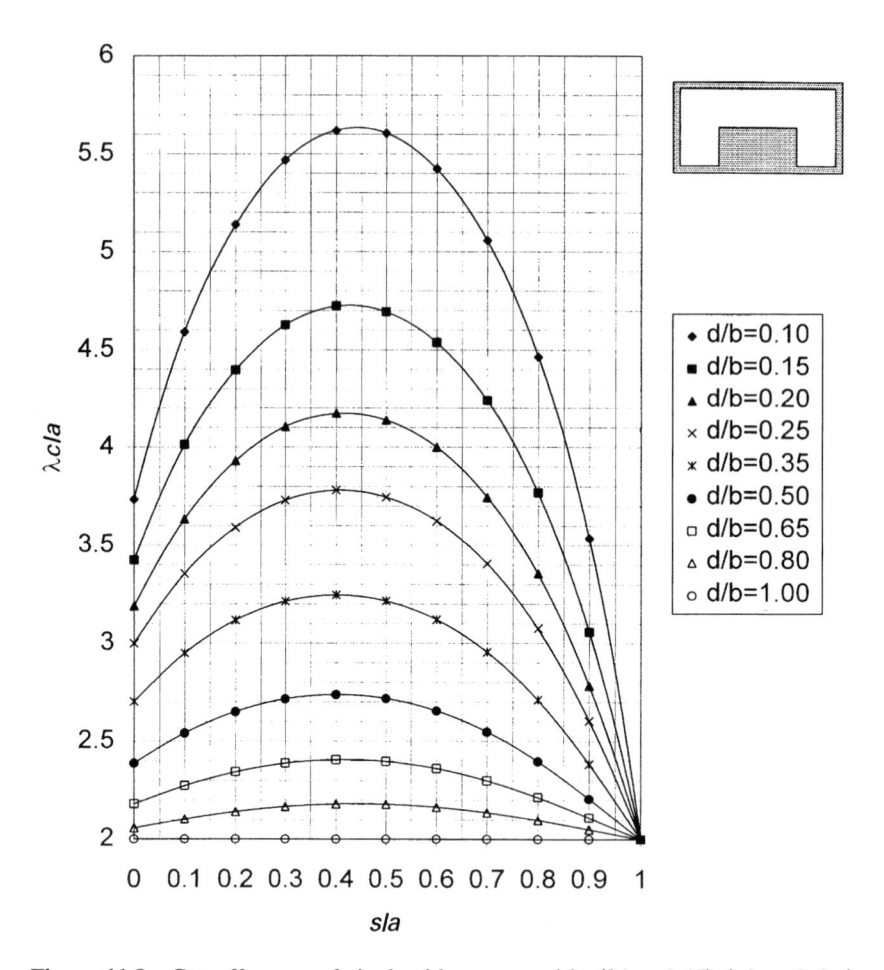

Figure 11.8 Cut-off space of single ridge waveguide ($b/a = 0.45$) (after Cohn)

$$\frac{a}{\lambda_c} = \frac{a}{2(a-s)}\left[1 + \frac{4}{\pi}\left(1 + 0.2\sqrt{\frac{b}{a-s}}\right)\left(\frac{b}{a-s}\right)\ln\csc\left(\frac{\pi}{2}\frac{d}{b}\right)\right.$$

$$\left. + \left(2.45 + 0.2\frac{s}{a}\right)\left(\frac{sb}{d(a-s)}\right)\right]^{-1/2} \qquad (11.32)$$

This solution, due to Hoefer, agrees with numerical methods to within 1% in the following ranges of parameters: $0.01 \le d/b \le 1$; $0 \le b/a \le 1$; $01 \le s/a \le 0.45$.

The approximate single ridge solution is obtained from that of the double one by replacing b by $2b$ in the coeffecnt multiplying the log term which may be taken to represent the susceptance one.

One possible computation procedure is to utilize the closed form solution for a/λ_c for either solution to initialize the root finding subroutine met in the exact transverse resonance condition of the problem region.

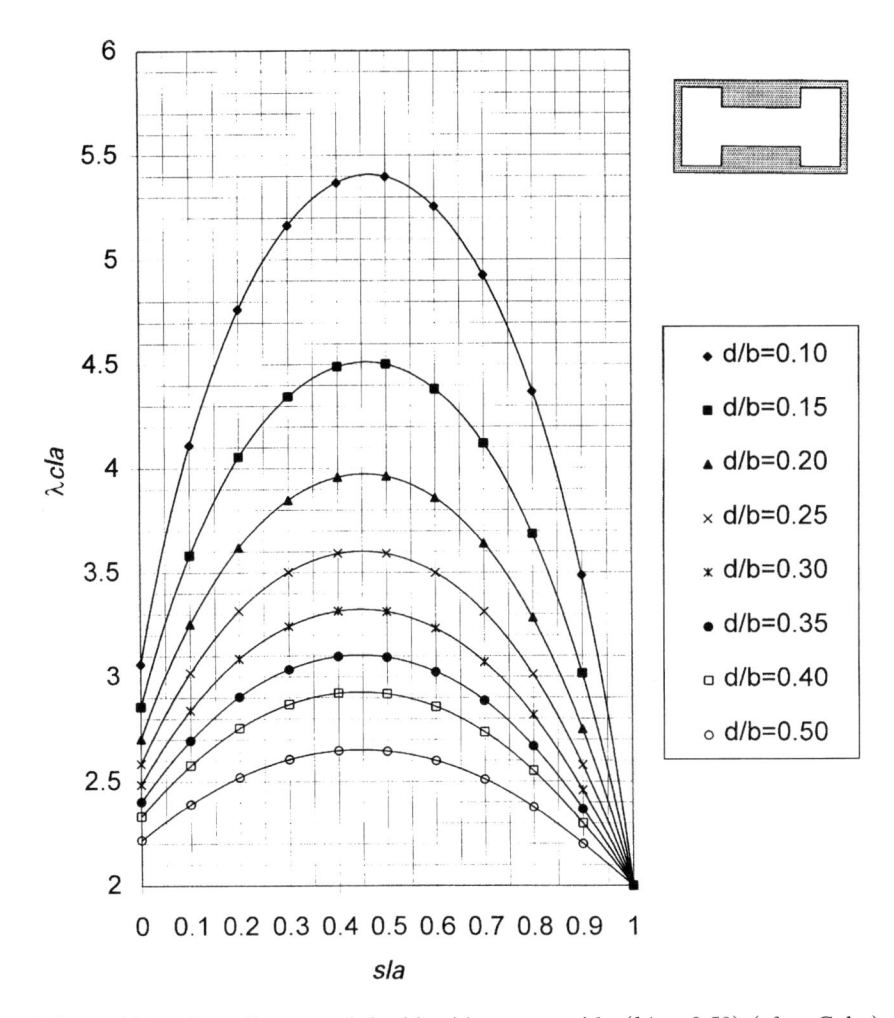

Figure 11.9 Cut-off space of double ridge waveguide ($b/a = 0.50$) (after Cohn)

11.6 Voltage–current Definition of Impedance in Ridge Waveguide

The impedance in ridge waveguide is again not unique. The approach used to calculate its impedance consists of initially forming this quantity at infinite frequency as a preamble to recovering it at finite frequency by introducing the dispersion factor λ_g/λ_0. Taking the voltage current definition of the problem region by way of example gives

$$Z_{VI}(\omega) = Z_{VI}(\infty)\left(\frac{\lambda_g}{\lambda_0}\right) \tag{11.33}$$

The origin of this technique relies on the fact that the fields at infinite frequency are purely transverse and that E_y and H_x are related there by the wave impedance of free space. This means that a knowledge of the distribution of E_y either at cut-off or at infinite frequency is sufficient for the solution of this sort of problem. One solution which omits the step discontinues on either sides of the ridge has been described by

$$Z_{VI}(\infty) = \frac{\pi \eta_0}{\sin \theta_2 + \left(\frac{d}{b}\right) \tan\left(\frac{\theta_1}{2}\right) \cos \theta_2} \left(\frac{b}{a}\right)\left(\frac{d}{b}\right)\left(\frac{a}{\lambda_c}\right) \quad (11.34)$$

In obtaining this result the voltage has been taken as the line integral over the electric field between the ridges and the current has been defined as the total longitudinal surface current in the three regions of the structure. θ_1 and θ_2 are defined in terms of λ_c in equations (11.27a) and (11.27b) and the physical variables are given in Figure 11.5.

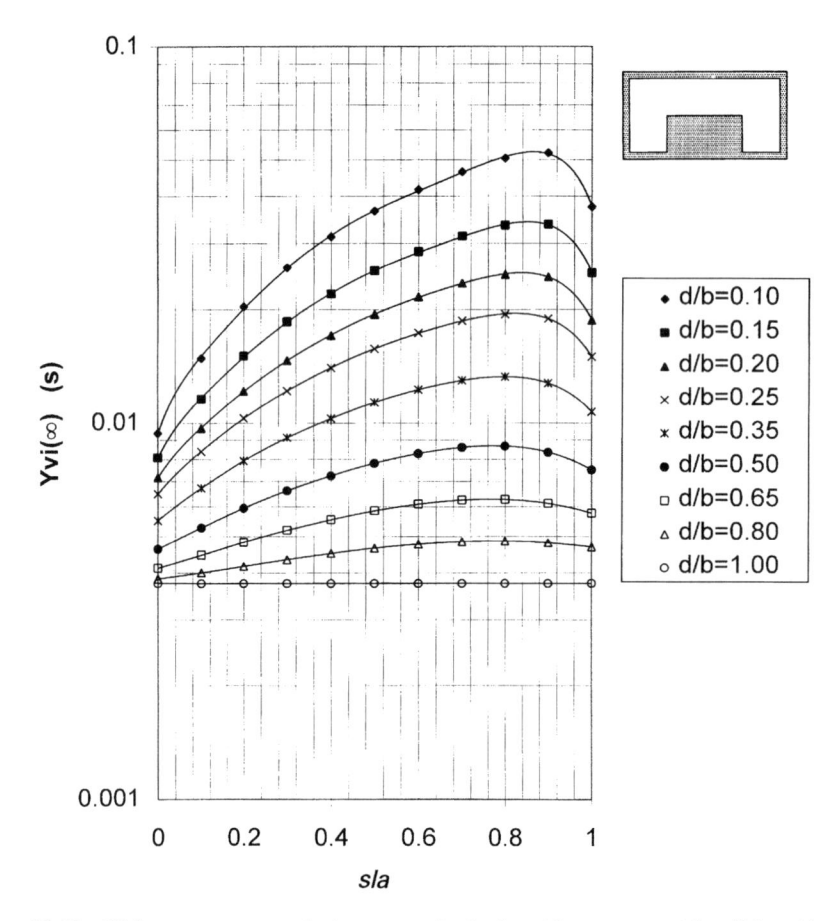

Figure 11.10 Voltage–current admittance of single ridge waveguide ($b/a = 0$) (after Hoefer)

11.6 VOLTAGE–CURRENT DEFINITION

One solution which applies to either the single or the double ridge arrangement and which caters for the fringing effect at the steps is

$$Z_{VI}(\infty) = \frac{\pi \eta_0}{\sin \theta_2 + \left(\dfrac{d}{b}\right)\left[\dfrac{B}{Y_{01}} + \tan\left(\dfrac{\theta_1}{2}\right)\right] \cos \theta_2} \left(\dfrac{b}{a}\right)\left(\dfrac{d}{b}\right)\left(\dfrac{a}{\lambda_c}\right) \quad (11.35)$$

The normalized susceptance given in equation (11.29) is used in evaluating the voltage current definition of impedance for single ridge and that given in equation (11.30) is utilized in calculating that of the double ridge. The normalized cut-off frequency a/λ_c is separately given by (11.32). The graphical result for single ridge is depicted in Figure 11.10 and that for double ridge waveguide in Figure 11.11.

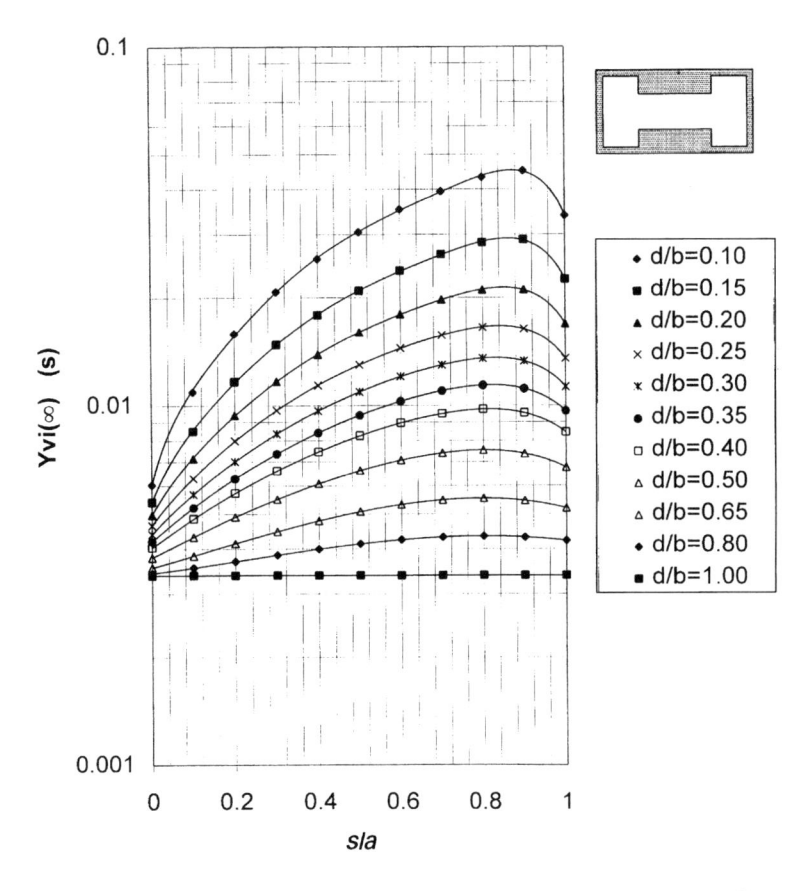

Figure 11.11 Voltage–current admittance of double ridge waveguide ($b/a = 0$) (after Hoefer)

11.7 Power–voltage Definition of Impedance in Ridge Waveguide

The power–voltage definition of impedance in ridge waveguide starts with a definition of power flow (P_t) in the waveguide at infinite frequency

$$P_t(\infty) = \left(\frac{E_0^2 d^2}{2\pi\eta_0}\right) \left\{ 2m \left(\frac{b}{a}\right)\left(\frac{d}{b}\right)\left(\frac{a}{\lambda_c}\right) \cos^2\theta_2 \ln \operatorname{cosec}\left(\frac{\pi d}{2b}\right) + \frac{\theta_2}{2} \right.$$

$$\left. + \frac{\sin 2\theta_2}{4} + \left(\frac{d}{b}\right)\left(\frac{\cos\theta_2}{\sin\theta_1}\right)^2 \left[\frac{\theta_1}{2} - \frac{\sin 2\theta_1}{4}\right] \right\} \left(\frac{1}{b}\right)\left(\frac{b}{d}\right)\left(\frac{\lambda_c}{a}\right) \quad (11.36)$$

The power flow at finite frequency is then given by

$$P_t(\omega) = P_t(\infty) \left(\frac{\lambda_0}{\lambda_g}\right) \quad (11.37)$$

$m = 1$ for double ridge and $m = 2$ for single ridge waveguide, E_0 is the peak electric field intensity (V/m) at the centre of the waveguide.

$$E_0 = \frac{V}{d} \quad (11.38)$$

While the absolute value of the electric field at the centre of the ridge (s) does not directly enter into the description of $Z_{PV}(\infty)$ a knowledge of its value is necessary in any calculation of $P(\infty)$. Its value at finite frequency in ordinary rectangular waveguide is

$$E_0 = \left(\frac{\lambda_c}{\lambda_0}\right) \sqrt{\frac{\mu_0}{\varepsilon_0}} \quad (11.39)$$

When $s = a$ and $d = b$ then $\theta_1 = \pi/2$ and $\theta_2 = 0$ and P_t reduces to the result for the ordinary waveguide as is readily verified.

The power–voltage definition of impedance at infinite frequency is now calculated by having recourse to its definition in equation (11.1)

$$Z_{PV}(\infty) = \frac{VV^*}{2P_t(\infty)} \quad (11.40)$$

The required result, in terms of the original variables, is

11.7 POWER–VOLTAGE DEFINITION

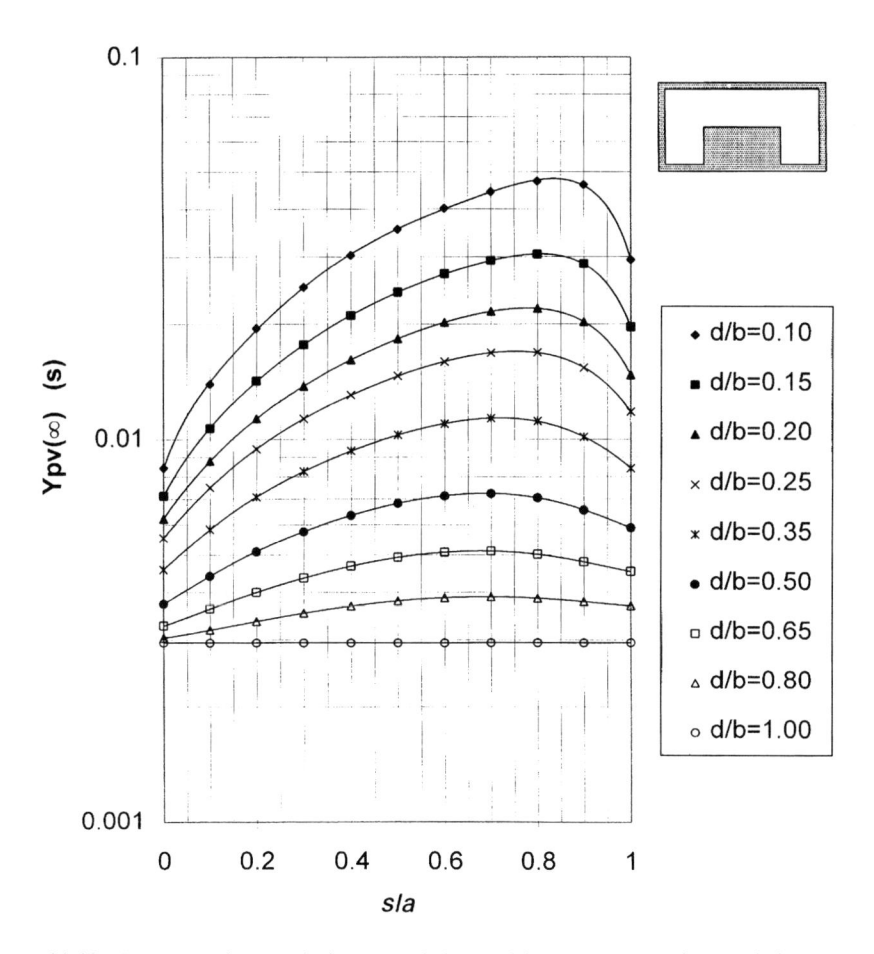

Figure 11.12 Power–voltage admittance of single ridge waveguide ($b/a = 0$) (after Hopfer)

$$Z_{\text{PV}}(\infty) = \left\{ \pi\eta_0 \left(\frac{b}{a}\right)\left(\frac{d}{b}\right)\left(\frac{a}{\lambda_c}\right) \right\} \Big/ \left\{ 2m\left(\frac{b}{a}\right)\left(\frac{d}{b}\right)\left(\frac{a}{\lambda_c}\right) \cos^2\theta_2 \ln \operatorname{cosec}\left(\frac{\pi d}{2b}\right) \right.$$

$$\left. + \frac{\theta_2}{2} + \frac{\sin 2\theta_2}{4} + \left(\frac{d}{b}\right)\left(\frac{\cos\theta_2}{\sin\theta_1}\right)^2 \left[\frac{\theta_1}{2} - \frac{\sin 2\theta_1}{4}\right] \right\} \quad (11.41)$$

Figure 11.12 gives the result for single ridge and Figure 11.13 that for double ridge.

$Z_{\text{PV}}(\omega)$ is related to $Z_{\text{PV}}(\infty)$ by a similar relationship to equation (11.33).

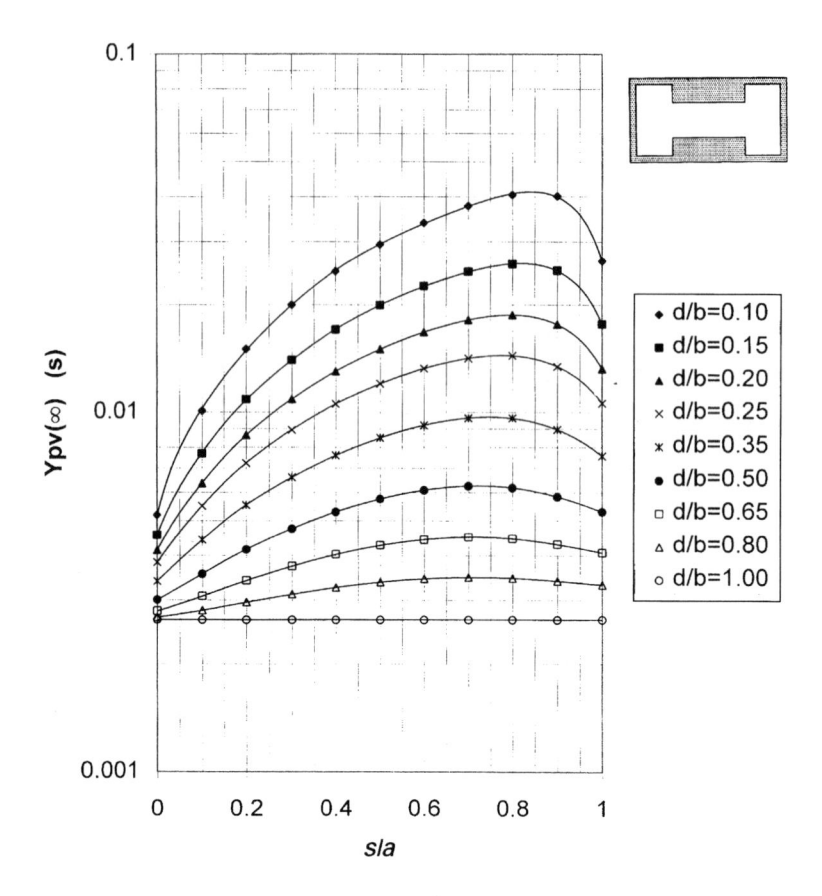

Figure 11.13 Power–voltage admittance of double ridge waveguide ($b/a = 0$) (after Hopfer)

11.8 Power–current Definition of Impedance in Ridge Waveguide

The power–current definition of impedance at infinite frequency in ridge waveguide may either be defined once those of the voltage–current and power–voltage are at hand or it may be developed from first principles. Adopting the first approach gives

$$Z_{\text{PI}}(\infty) = \frac{Z_{\text{VI}}^2(\infty)}{Z_{\text{PV}}(\infty)} \tag{11.42}$$

The result for the single ridge arrangement is indicated in Figure 11.14 and that for double ridge waveguide in Figure 11.15.

11.8 POWER–CURRENT DEFINITION

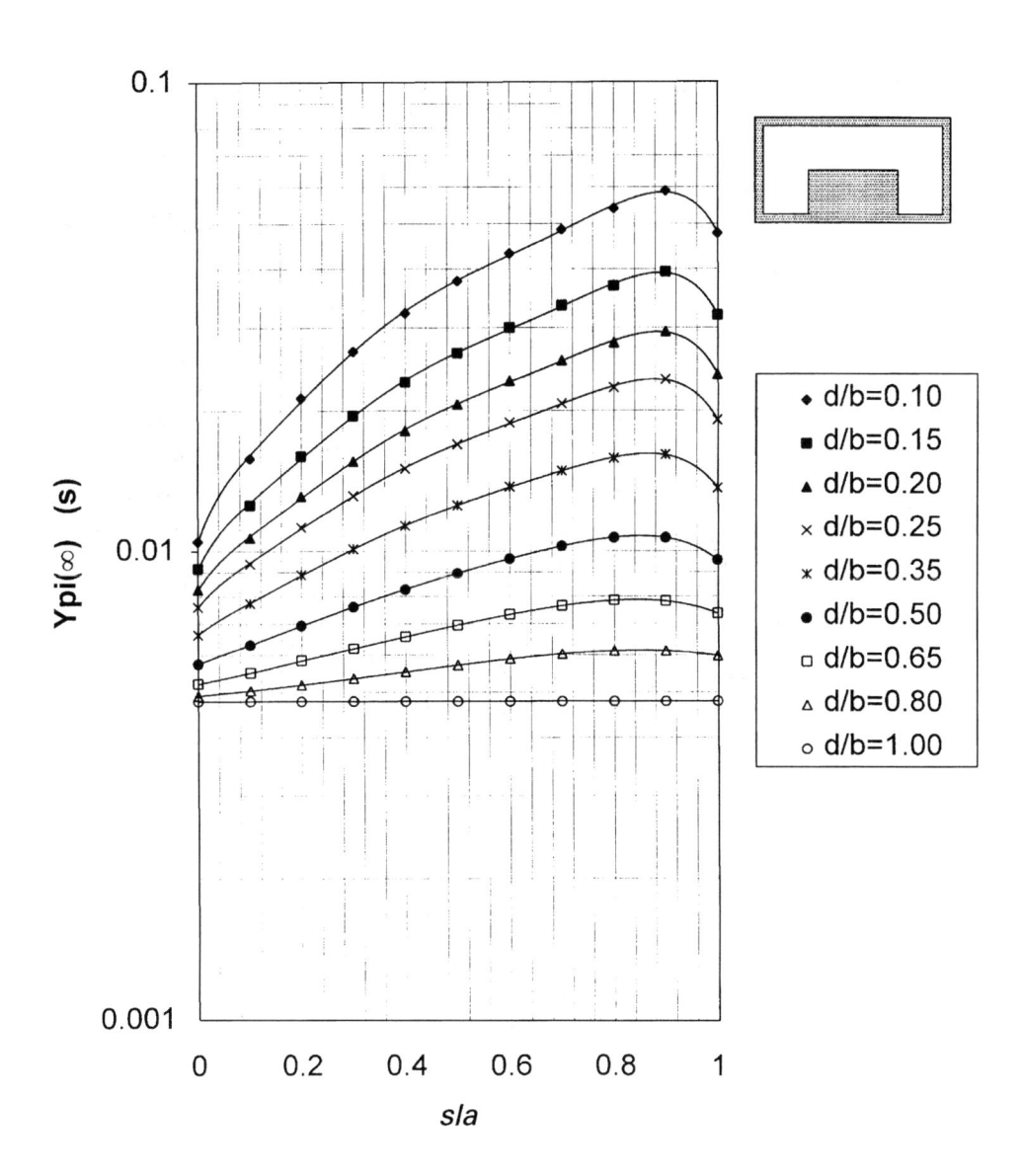

Figure 11.14 Power–current admittance of single ridge waveguide ($b/a = 0.45$)

218   11 IMPEDANCE IN WAVEGUIDES

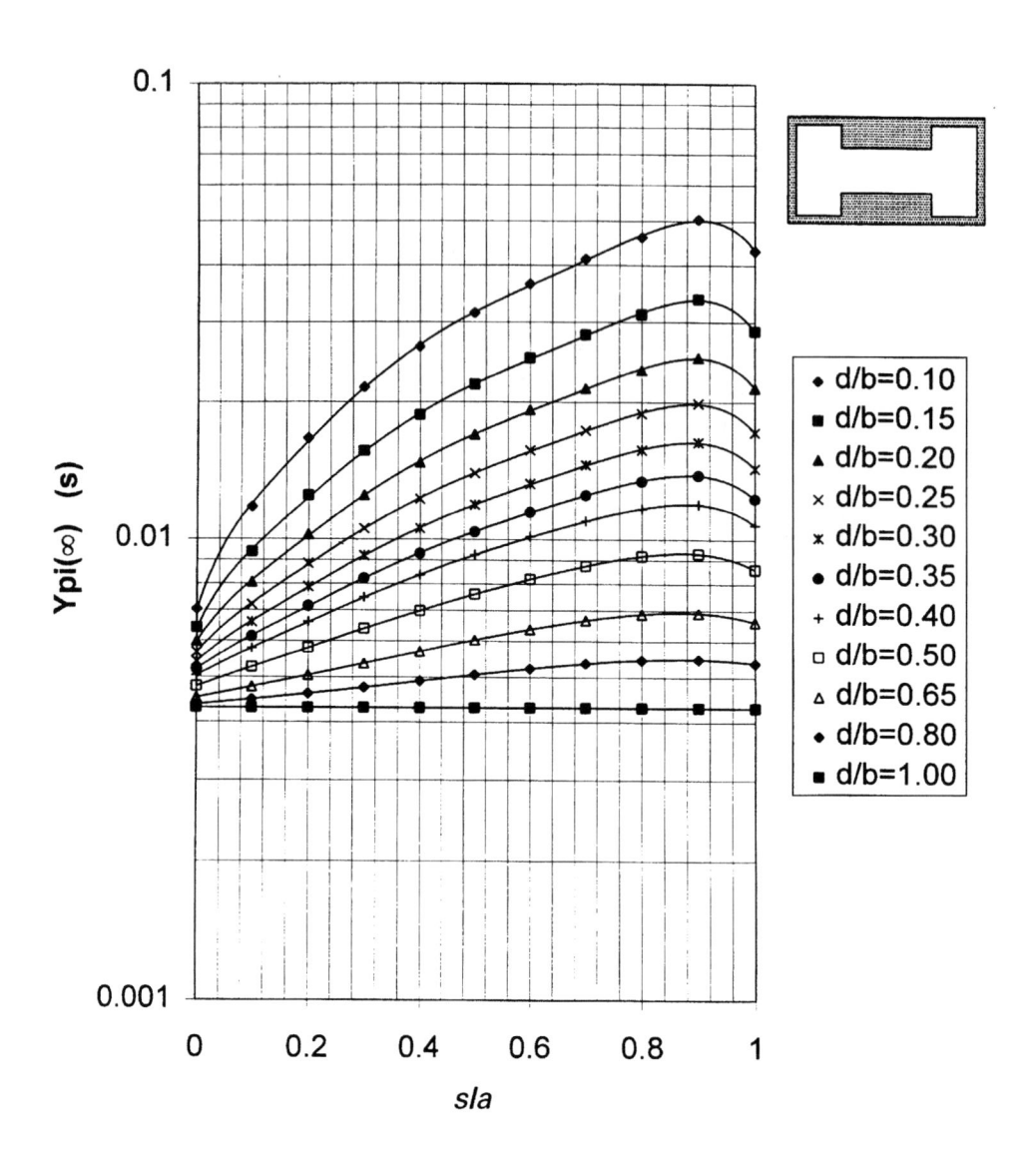

Figure 11.15 Power–current admittance of double ridge waveguide ($b/a = 0.50$)

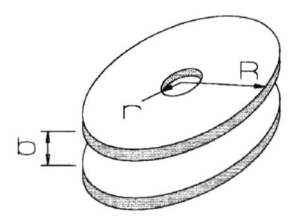

11.9 Radial Waveguide

Another important waveguide structure that enters into the description of waveguide junction circulators is the radial waveguide illustrated in Figure 11.16. One property of such lines is that a short circuit is not mapped into an open circuit in the same length as an open circuit one is mapped into a short circuit. Another problem associated with this sort of line is the lack of uniqueness in the definition of impedance. The $A\ B\ C\ D$ notation is often employed in the description of this waveguide and this is the approach utilized here. It is defined by

$$A = \frac{\pi(k_0 R)}{2} [J_n(k_0 R_0)\, Y'_n(k_0 R) - J'_n(k_0 R) Y_n(k_0 R_0)] \tag{11.43a}$$

$$B = \frac{\pi(k_0 R)}{2\zeta_0} [J_n(k_0 R_0)\, Y_n(k_0 R) - J_n(k_0 R) Y_n(k_0 R_0)] \tag{11.43b}$$

$$C = \zeta_0 \frac{\pi(k_0 R)}{2\zeta_0} [J'_n(k_0 R_0)\, Y'_n(k_0 R) - J'_n(k_0 R) Y'_n(k_0 R_0)] \tag{11.43c}$$

$$D = \frac{\pi(k_0 R)}{2} [J_n(k_0 R_0)\, Y'_n(k_0 R) - J'_n(k_0 R) Y_n(k_0 R_0)] \tag{11.43d}$$

with $n = 1$.

The relationship between the wave admittances at the input and output terminals of this kind of waveguide is given in terms of the $ABCD$ parameters by

$$Y_{\text{in}} = \frac{jC + DY_L}{A + jBY_L} \tag{11.44}$$

The load admittance is in general complex and is specified by

$$Y_L = G + jB \tag{11.45}$$

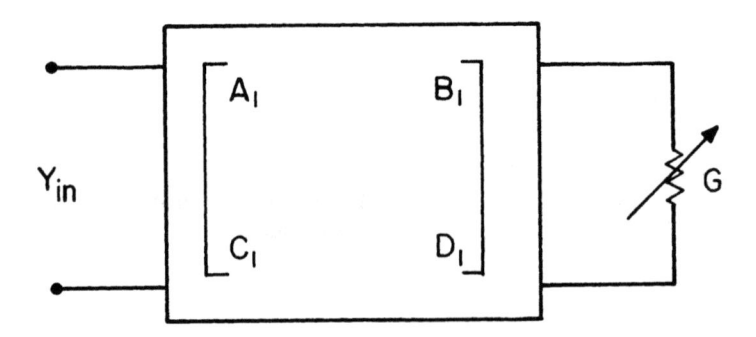

Figure 11.17 Schematic diagram of radial waveguide

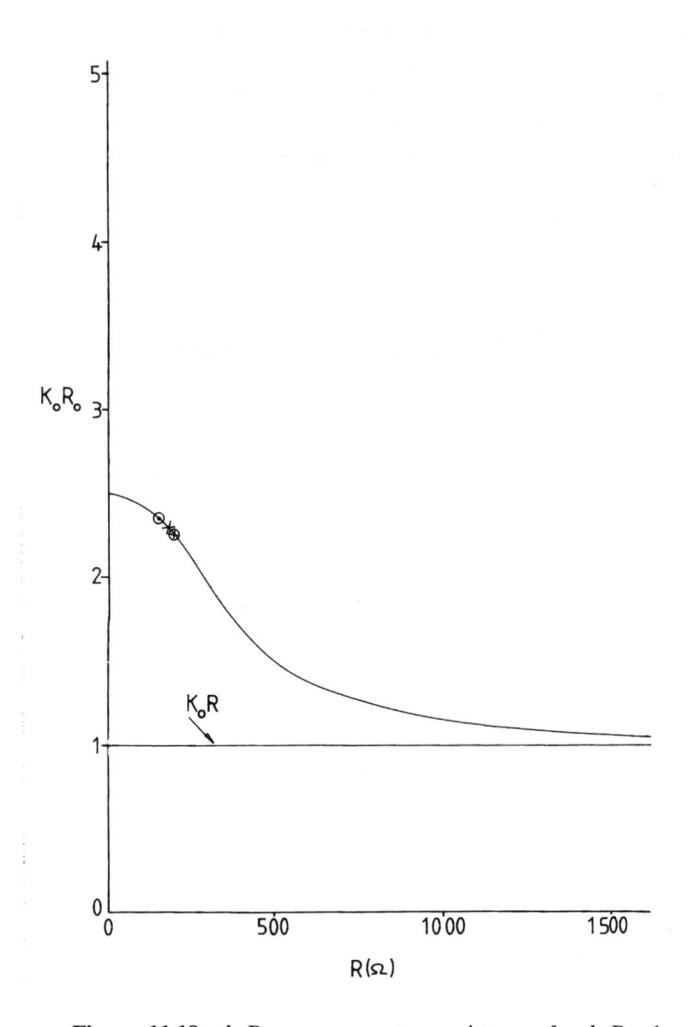

Figure 11.18 k_0R_0 versus gyrator resistance for $k_0R = 1$

11.10 Input Terminals of Radial Waveguide

One feature of a radial waveguide is that the input terminals at which its impedance is real is dependent upon the level of the real part at the output terminals. Forming the input admittance of the line in Figure 11.17 with Y_L real gives

$$Y_{in} = \frac{jC + DG}{A + jBG} \qquad (11.46)$$

The real and imaginary parts of this relationship are

$$Y_{in} = \frac{(AD + BC)G + j(AC - BDG^2)}{A^2 + B^2G^2} \qquad (11.47)$$

The input terminals of the device is now established by setting the imaginary part to zero.

$$AC - BDG^2 = 0 \qquad (11.48)$$

The corresponding real part is determined by

$$\frac{(AD + BC)G}{A^2 + B^2G^2} \qquad (11.49)$$

Figure 11.18 indicates one result.

11.11 Voltage–current Definition of Impedance in Radial Waveguide

In order to cater for steps in radial waveguides it is also necessary, to introduce the notion of characteristic impedance in its description. The voltage–current definition of this sort of waveguide starts by defining the voltage (V) between its top and bottom plates in terms of E_z.

$$V = bE_z \qquad (11.50)$$

It continues by forming the radial directed current (I) terms of H_ϕ

$$I = 2\pi r H_\phi \qquad (11.51)$$

everywhere in the radial region.

The voltage–current definition of a radial waveguide is therefore described by

$$Z_{VI}(r) = \left(\frac{b}{2\pi r}\right)\eta_0 \qquad (11.52)$$

Scrutiny of this relationship indicates that the definition of impedance in this type of waveguide is a function of the radial co-ordinate.

The corresponding voltage current definition of impedance in standard rectangular waveguide is given by equation (11.24)

$$Z_{VI} = \left(\frac{\pi b}{2a}\right)\eta_0 \qquad (11.53)$$

If b_1, b_2 and b_3 are the spacings between the top and bottom walls of the rectangular waveguide, the radial one and that of the post resonator respectively then the impedance of the different regions may for engineering purposes be assumed to be proportional to the corresponding spacings.

Another impedance formulation that has some merit is to take H_ϕ as

$$H_\phi = H_x \cos\phi - H_z \sin\phi \qquad (11.54)$$

and to form

$$I = \frac{1}{2\phi}\int_{-\phi}^{+\phi} H_\phi \, d\phi \qquad (11.55)$$

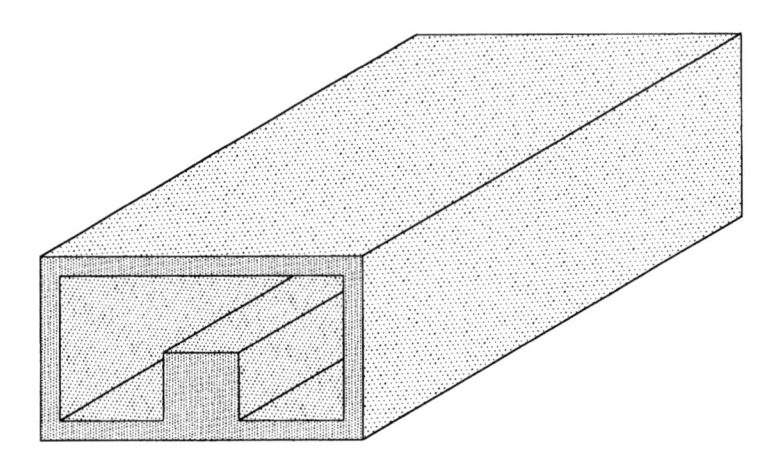

Figure 11.19 Half-wave ridge prototype

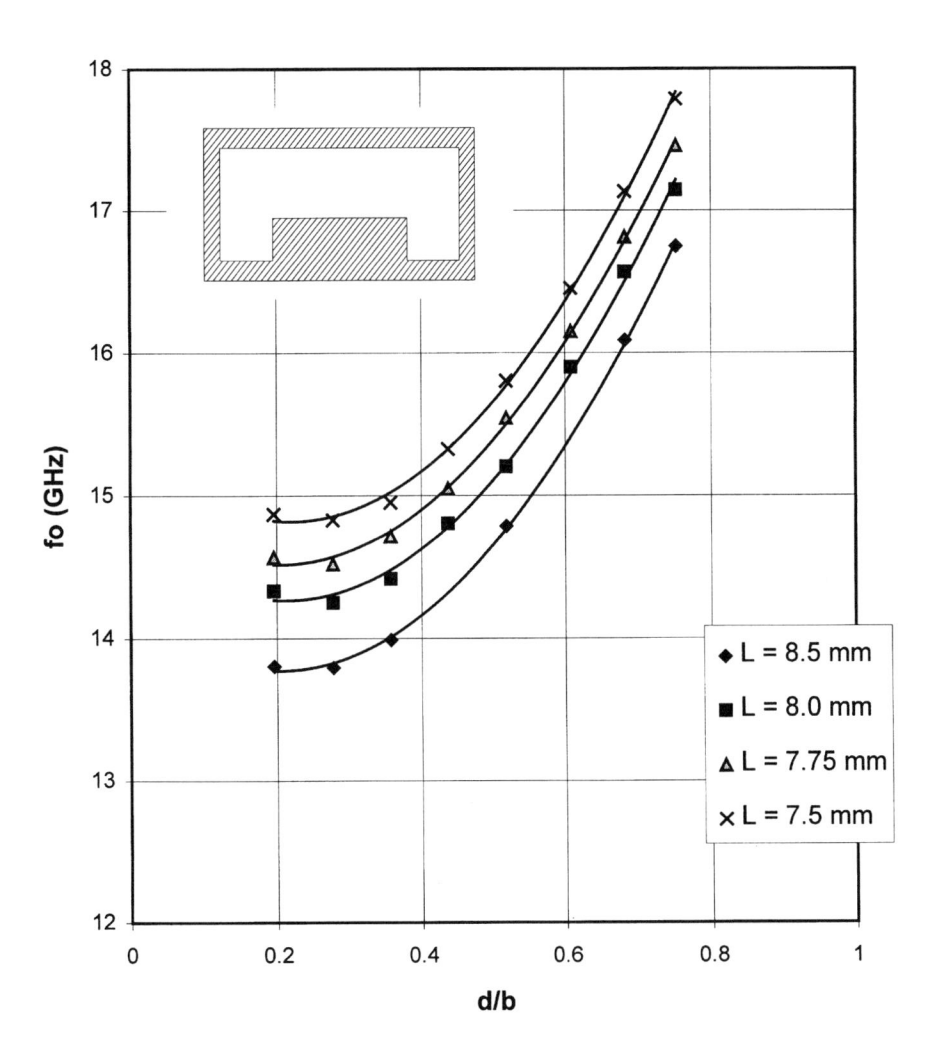

Figure 11.20 Relationship between frequency and aspect ratio of half-wave long ridge waveguide ($s/a = 0.50$)

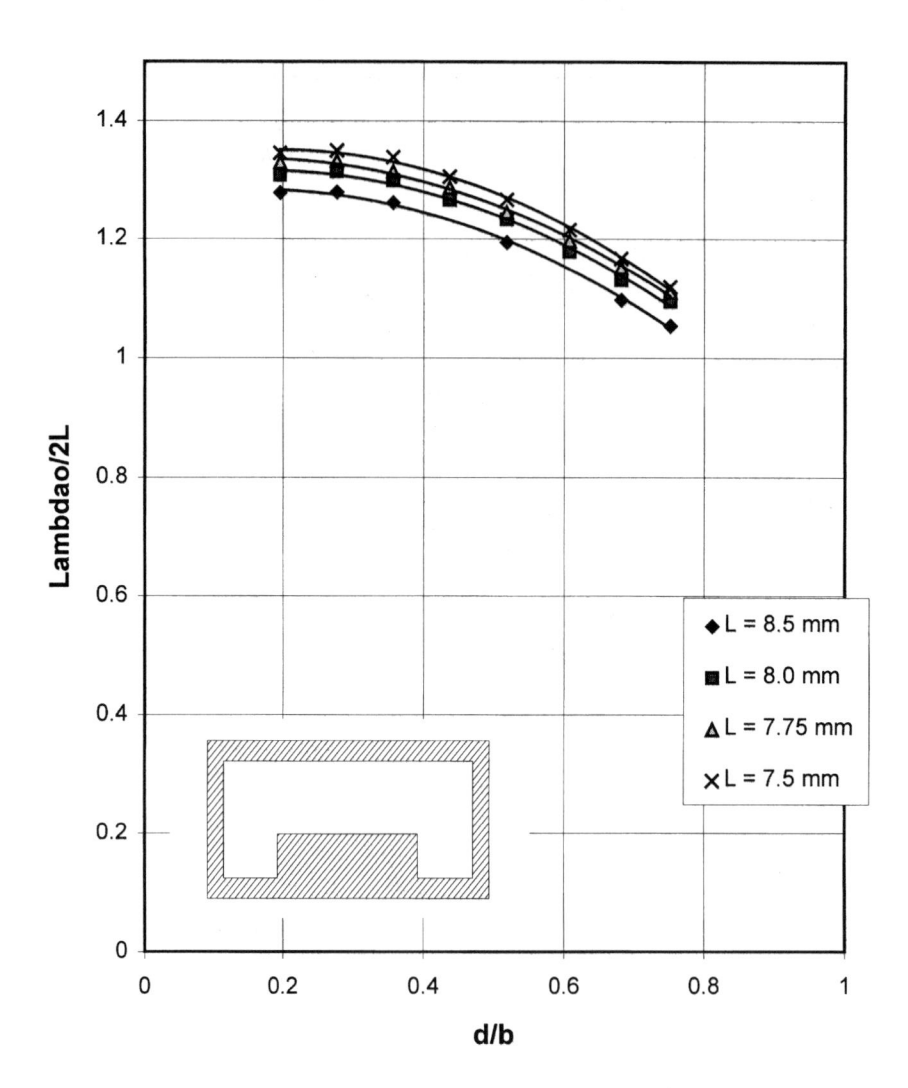

Figure 11.21 Relationship between wavelength and aspect ratio of half-wave long ridge waveguide ($s/a = 0.50$)

11.11 VOLTAGE–CURRENT DEFINITION

Quality Factor (Q) vs Aspect Ratio (d/b)

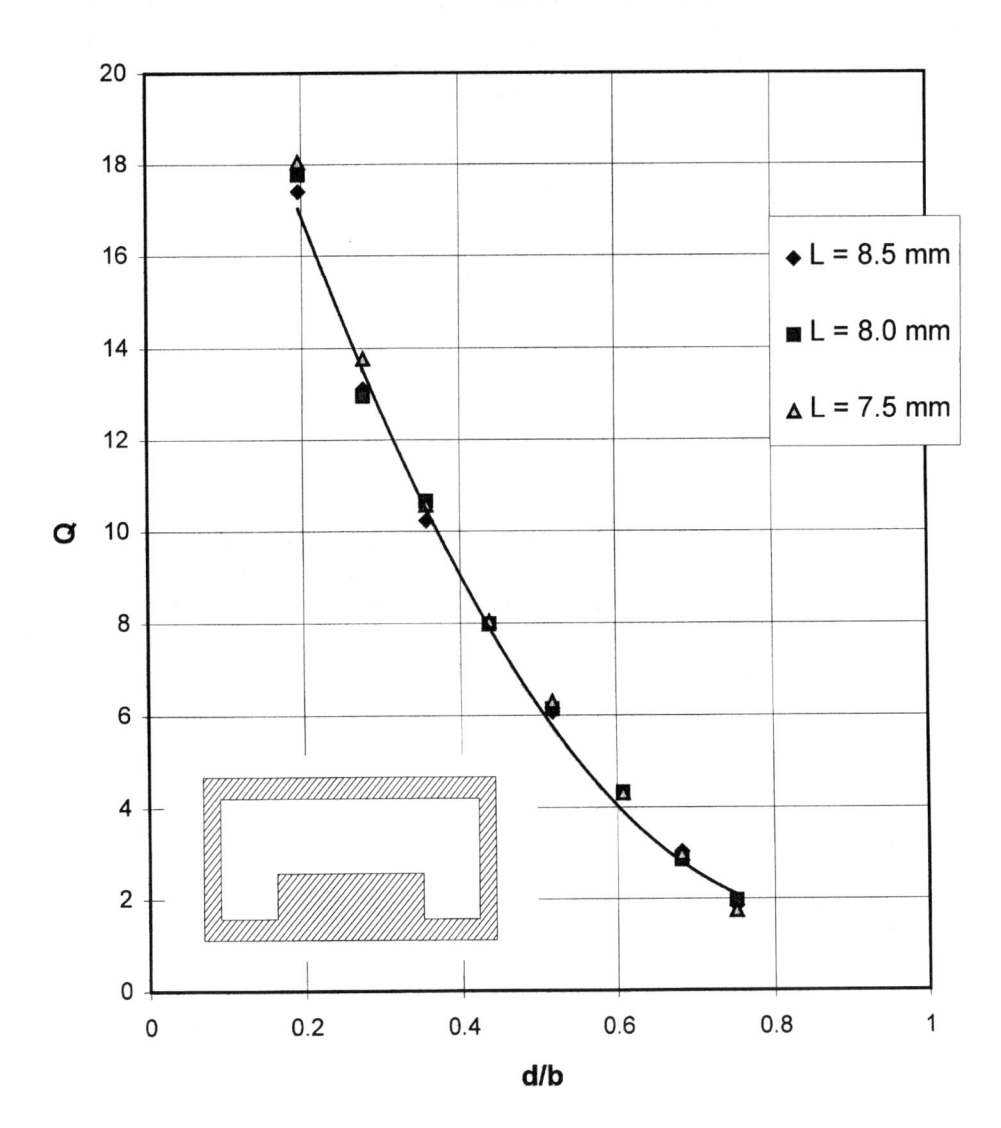

Figure 11.22 Relationship between loaded Q-factor and aspect ratio of half-wave long ridge waveguide

11.12 Half-wave Filter Prototype

One possible degree-2 filter topology met in the construction of waveguide circulators consists of a ladder network made up of half-wave long U.E.'s in cascade with a series STUB-R load. One element which exhibits a parallel resonator with a suitable value of loaded Q-factor is a half-wave long section of ridge waveguide illustrated in Figure 11.19.

The quality factor of this sort of resonator is obtained from a knowledge of the 20 dB return loss points by treating the network as a 1-port LCR circuit.

$$Q_L = \frac{(VSWR - 1)}{2\delta_0 \sqrt{VSWR}} \tag{11.56}$$

where

$$2\delta_0 = \frac{\omega_2 - \omega_1}{\omega_0} \tag{11.57}$$

Figure 11.20 indicates one relationship between the frequency of this sort of arrangement and the physical variable entering into the descriptions of its geometry. Figure 11.21 gives the normalized frequency of the same circuit. The experimental relationship between the quality factor of this type of element and its aspect ratio is separately indicated in Figure 11.22. This database was obtained in WR62 waveguide.

The normalized frequency is represented by defining the following quantity

$$\frac{\beta_g}{k_0} = \frac{\lambda_0}{2L} \tag{11.58}$$

The quantity $2L$ differs from λ_g due to the fringing contribution at each end of the network.

Scrutiny of these latter two illustrations suggests that the quality factor of this element is essentially fixed by the aspect ratio of the ridge waveguide and that its frequency is thereafter set by its length. When d/b equals unity the frequency is equal to that of a standard WR62 waveguide. While these measurements may be used to extract end effect parameters for this type of discontinuity this problem is outside the remit of this work.

References and Further Reading

Chen, T. S. (1957) Calculation of the parameters of ridge waveguide, *IRE Trans. Microwave Theory Tech.*, **MTT-5**, 12–17.

Cohn, S. B. (1947) Properties of ridge waveguide, *Proc. IRE*, **35**, 783–788.

Getsinger, W. J. (1962) Ridge waveguide field description and application to directional couplers, *IRE Trans. Microwave Theory Tech.*, **MTT-10**, 41–50.

Hoefer, W. J. R. and Burton, M. N. (1982) Closed form expressions for parameters of finned and ridge waveguide, *IEEE Trans. Microwave Theory Tech.*, **MTT-30**, 2190–2194.

Hopfer, S. (1955) The design of ridged waveguides, *IRE Trans. Microwave Theory Tech.*, **MTT-3**, 20–29.

Marcuvitz, N. (1964) Waveguide Handbook, MIT Radiation Lab. Series, No. 10, Boston Technical Publishers Inc., New York.

Mihran, T. G. (1949) Closed and open-ridged waveguide, *Proc. IRE*, **37**, 640–644.

Pile, J. R. (1966) The cut-off wavelength of the TE_{10} mode in ridged rectangular waveguide of any aspect ratio, *IEEE Trans. Microwave Theory Tech.*, **MTT-14**, 175–183.

12

JUNCTION CIRCULATOR USING POST RESONATORS

12.1 Introduction

The topology of one early classic waveguide circulator consists of a ferrite or garnet gyromagnetic cylindrical resonator with no variations of the electromagnetic fields along its axis in an H-plane 3-port junction. While such a junction may either be directly or quarter-wave coupled, commercial devices are more often than not quarter-wave coupled. Typical impedance transformers consist of radial or ridge structures and dielectric rings. Figure 12.1 depicts the basic radial arrangement. A phenomenological description of this sort of circulator is usually based on the rotation of a suitable standing wave in the gyromagnetic resonator under the application of a direct magnetic field. Figure 12.2(a) indicates the standing wave pattern of the dominant mode in such a resonator and Figure 12.2(b) depicts the same field pattern rotated by 30°. An understanding of the field patterns of planar resonators is therefore a prerequisite to the understanding of this class of device. Such resonators have been investigated in Chapter 5. Gyromagnetic prism resonators which also have the symmetry of the problem region have also been employed in the construction of this class of circulator. While the electromagnetic description of a ferrite post in a radial cavity formed by the junction of three waveguides represents one model of this sort of problem its circuit topology has also much to comment it. The purpose of this chapter is to outline some circuit and field features of this class of circulators. A typical pole of each eigen-network of the device is realised in terms of an impedance one at the terminals of the ferrite or gyromagnetic post, a $ABCD$ matrix which describes the region between the ferrite post and the rectangular waveguides and an ideal transformer which represents the boundary condition between the radial and rectangular waveguides. This notation is particularly helpful in the realization of equivalent circuits, the construction of immittance matrices and the notion of radial and ridge waveguide impedance transformers. The synthesis problem separately requires a description of the immittances of the circuit.

12 JUNCTION CIRCULATOR USING POST RESONATORS

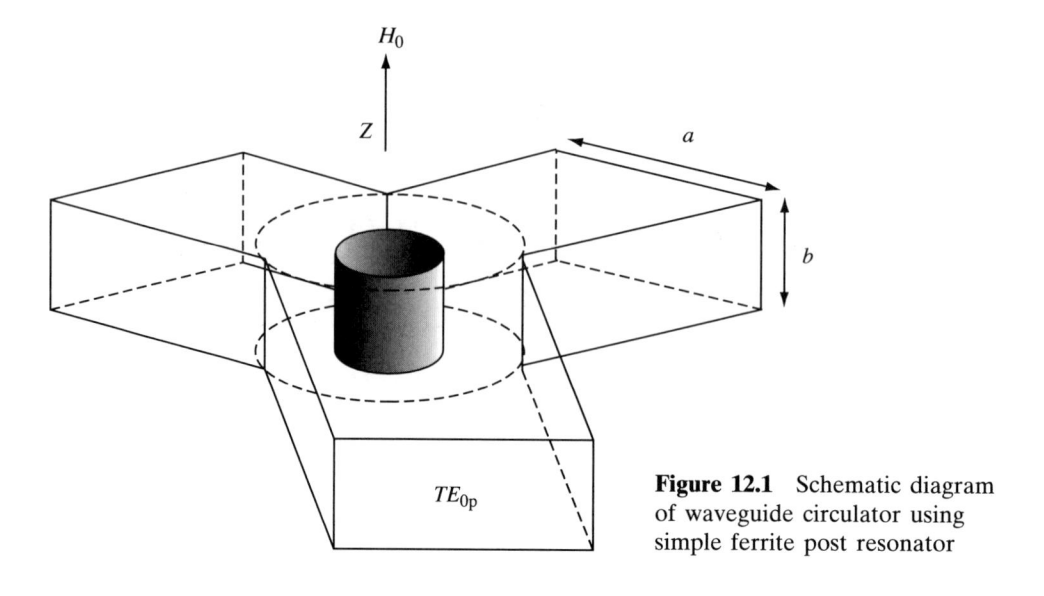

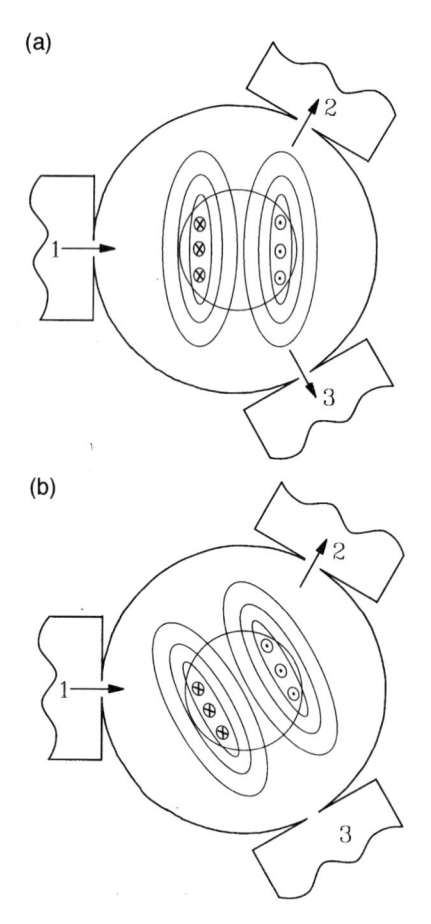

Figure 12.1 Schematic diagram of waveguide circulator using simple ferrite post resonator

Figure 12.2 Phenomenological operation of waveguide circulator (Butterweck)

One difficulty encountered with the adjustment of this sort of circulator is the fact that a radial waveguide is a so called nonuniform line so that the eigen-networks of the problem region are non-commensurate. This feature of the problem region makes it difficult to uniquely fix the terminals of the overall junction. Its description is facilitated by adopting the characteristic planes of the junction as the terminals of the problem region.

12.2 Admittance Matrix

There are in general six different ways in which circulator boundary conditions can be introduced. This may be done in terms of the scattering, impedance and admittance matrices of an ideal circulator or in terms of the corresponding triplets of eigenvalues. The actual choice is usually determined by the physical problem. It is observed that although it is always possible to construct the scattering matrix of a junction it is not always possible to construct an impedance or admittance one. The use of an impedance formulation is appropriate at the terminals of the resonator and an admittance one at those of the rectangular waveguides.

The description of the conventional waveguide circulator starts by deriving its admittance matrix at the terminals of the rectangular waveguides.

$$\bar{\zeta} = \begin{bmatrix} \zeta_{11} & \zeta_{12} & \zeta_{13} \\ \zeta_{13} & \zeta_{11} & \zeta_{12} \\ \zeta_{12} & \zeta_{13} & \zeta_{11} \end{bmatrix} \quad (12.1)$$

The short-circuit parameters of the junction are completely described once the admittance eigenvalues of the in-phase (ζ^0) and counter-rotating (ζ^+ and ζ^-) are available. The required relationships are described in Chapter 3 by

$$\zeta_{11} = \frac{\zeta^0 + \zeta^+ + \zeta^-}{3} \quad (12.2a)$$

$$\zeta_{12} = \frac{\zeta^0 + \zeta^+ \exp(j2\pi/3) + \zeta^- \exp(-j2\pi/3)}{3} \quad (12.2b)$$

$$\zeta_{13} = \frac{\zeta^0 + \zeta^+ \exp(-j2\pi/3) + \zeta^- \exp(j2\pi/3)}{3} \quad (12.2c)$$

Each of the three one-port susceptances ζ^0, ζ^+ and ζ^- appearing in the descriptions of the short-circuit parameters of a junction circulator may be expanded in terms of its poles in a second Foster or partial fractions form in the manner indicated in Figure 12.3. The symmetric poles and those with the three-fold symmetry of the junction are identified with the in-phase eigen-network and the other split poles with the two counter-rotating eigen-networks. Adopting this nomenclature permits the desired 1-port susceptances to be defined by

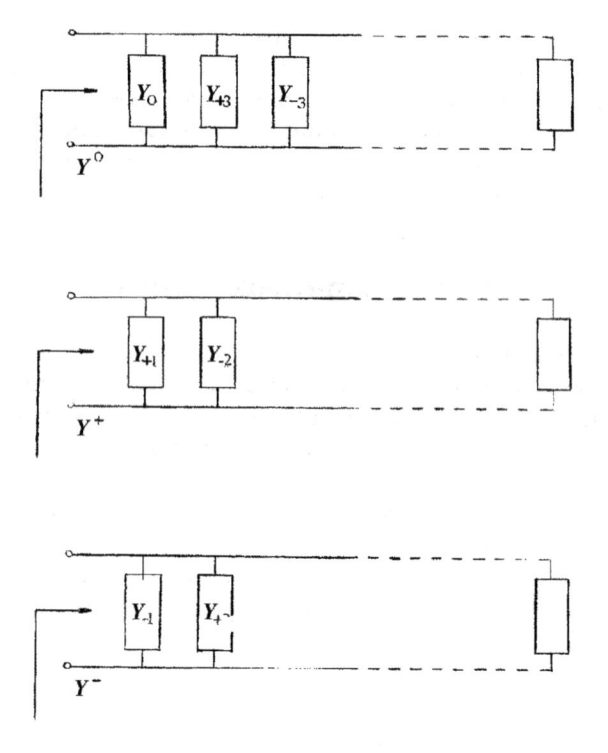

Figure 12.3 Second Foster realizations of in-phase and counter-rotating eigen-networks at junction terminals

$$\zeta^0 = \sum \zeta_n \quad n = 0, \pm 3, \pm 6, \pm 9, \text{etc} \tag{12.3}$$

$$\zeta^+ = \sum \zeta_n \quad n = 0, +1, -2, +4, \text{etc} \tag{12.4}$$

$$\zeta^- = \sum \zeta_n \quad n = 0, -1, +2, +5, \text{etc.} \tag{12.5}$$

The exact nature of a typical pole in the definitions of the eigenvalues of the junction requires a knowledge of the electromagnetic problem. This is the main purpose of this chapter.

A scrutiny of the eigenvectors associated with each pole readily reveals that those of the in-phase eigenvalue establishes a magnetic wall on the axis of the problem region whereas those of the split counter-rotating ones establishes electric walls there. The first circulation adjustment is established when the demagnetized eigen-networks at the terminals of the junction exhibit dual walls; the second adjustment is obtained by removing the degeneracy between the counter-rotating ones.

While the immittance eigenvalues can always be expanded in a first or second Foster form in terms of the poles of the problem region no such rules exist for its scattering variables. These must be deduced by having recourse to the usual bilinear transformations between the various quantities.

12.2 ADMITTANCE MATRIX

$$s^0 = \frac{\zeta_0(\lambda_0/\lambda_g) - \zeta^0}{\zeta_0(\lambda_0/\lambda_g) + \zeta^0} \tag{12.6a}$$

$$s^+ = \frac{\zeta_0(\lambda_0/\lambda_g) - \zeta^+}{\zeta_0(\lambda_0/\lambda_g) + \zeta^+} \tag{12.6b}$$

$$s^- = \frac{\zeta_0(\lambda_0/\lambda_g) - \zeta^-}{\zeta_0(\lambda_0/\lambda_g) + \zeta^-} \tag{12.6c}$$

The scattering matrix of the overall circuit may be assembled from the reflection eigenvalues. The first Foster realization of the eigen-networks at the resonator terminals are indicated for completeness in Figure 12.4.

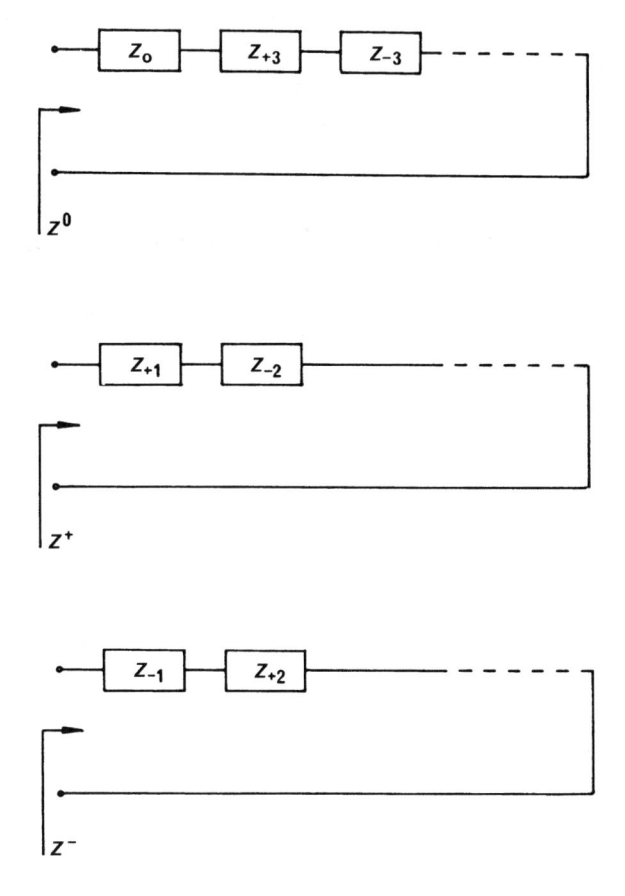

Figure 12.4 First Foster realizations of in-phase and counter-rotating eigen-networks at resonator terminals

12.3 Characteristic Planes of 3-Port Junctions

Some simplification of the waveguide circulator boundary problem may be obtained by introducing the notion of the characteristic planes of the 3-port junction. These planes coincide with those at which a short circuit placed in one port will cause a wave at the input port to be completely reflected, none entering the third one. The reflection coefficient at port 1 is then the degenerate eigenvalue s_1. One property of such planes is that these need not coincide with the physical terminals of the junction. Another property of such planes is that when a wave incident on the junction at one port is totally reflected by the location of a short circuit at a characteristic plane of a second port, the electric field vanishes at all characteristic planes in the other ones. Still another property of the characteristic planes is that the values of the admittance in each port extrapolated to the characteristic planes in each port are related in a standard way. These properties of a symmetrical 3-port junction provides one method of measuring s_1.

Since the characteristic planes display electric walls then the degenerate counter-rotating eigen-networks may be realised by halfwave-long short-circuited transmission lines there. These eigen-networks therefore coincide with the counter-rotating ones of demagnetized ideal quarterwave coupled circulators and may be used to establish the input terminals of such circulators. At these terminals the admittance matrix exists, while the impedance one does not.

12.4 Cavity Fields of Waveguide Circulator

The schematic diagram of the waveguide circulator considered in this text is illustrated in Figure 12.5. Solutions are sought with the electric field purely in the z-direction and the magnetic fields purely in the x–y plane. The fields are independent of z and are taken to have time dependence $\exp(j\omega t)$.

The electric field inside the ferrite or garnet region is described in terms of circular variables by

$$E_{zfn} = \sum_{n=-\infty}^{n=\infty} A_{fn} J_n(k_e r) \exp(jn\phi) \qquad (12.7)$$

The azimuthal (ϕ) component of the magnetic field is separately obtained from Maxwell's equation as

$$H_{\phi fn} = -j\zeta_0 \zeta_{\text{eff}} \sum_{n=-\infty}^{n=\infty} A_{fn} \left[J_n'(k_e r) - \frac{\kappa n}{\mu} \frac{J_n(k_e r)}{k_e r} \right] \exp(jn\phi) \qquad (12.8)$$

H_{rfn} which also exists does not enter into the final result and so is omitted for brevity. The other field components (E_r, E_ϕ, H_z) are equal to zero.

The corresponding alternating fields in the radial waveguide are given by

12.4 CAVITY FIELDS OF WAVEGUIDE CIRCULATOR

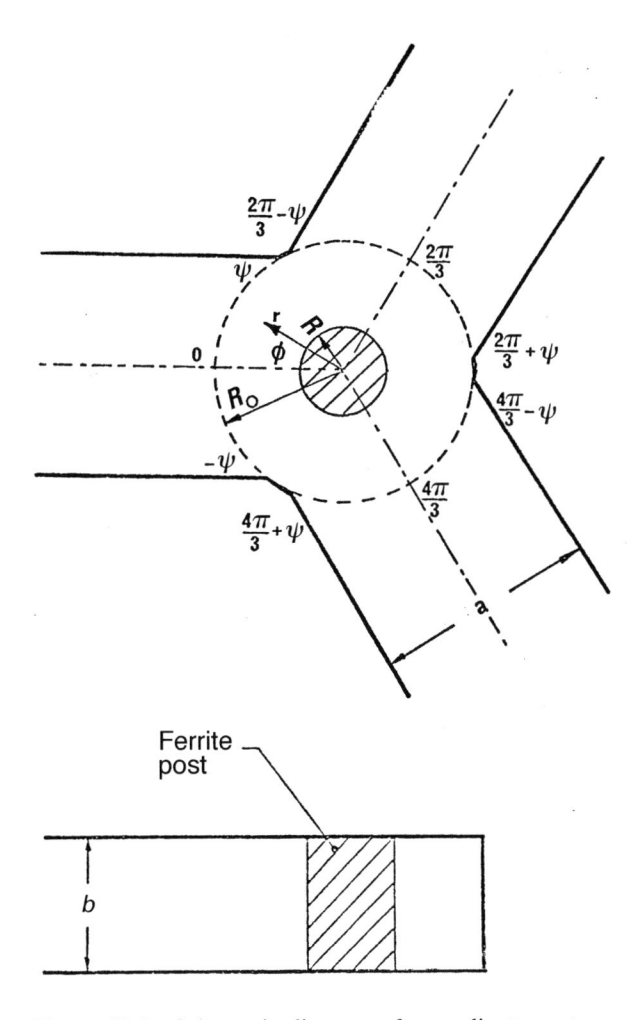

Figure 12.5 Schematic diagram of co-ordinate system

$$E_{zdn} = \sum_{n=-\infty}^{n=\infty} \left[A_n E_{zfn} + jB_n H_{\phi fn} \right] \exp(jn\phi) \tag{12.9}$$

$$H_{\phi dn} = \sum_{n=-\infty}^{n=\infty} \left[jC_n E_{zfn} + D_n H_{\phi fn} \right] \exp(jn\phi) \tag{12.10}$$

A, B, C and D are the $ABCD$ parameters of the radial region

$$A_n = \frac{\pi(k_0 R)}{2} [J_n(k_0 r) Y'_n(k_0 R) - J'_n(k_0 R) Y_n(k_0 r)] \tag{12.11a}$$

$$B_n = \eta_0 \frac{\pi(k_0 R)}{2} [J_n(k_0 R) Y_n(k_0 r) - J_n(k_0 r) Y_n(k_0 R)] \qquad (12.11b)$$

$$C_n = \frac{1}{\eta_0} \frac{\pi(k_0 R)}{2} [J'_n(k_0 R) Y'_n(k_0 r) - J'_n(k_0 r) Y'_n(k_0 R)] \qquad (12.11c)$$

$$D_n = \frac{\pi(k_0 R)}{2} [J_n(k_0 R) Y'_n(k_0 r) - J'_n(k_0 r) Y_n(k_0 R)] \qquad (12.11d)$$

ζ_0 and ζ_{eff} are the absolute wave admittance of free space and the relative wave admittance of the ferrite region respectively

$$\zeta_0 = \sqrt{\frac{\varepsilon_0}{\mu_0}} \qquad (12.12a)$$

$$\zeta_{\text{eff}} = \sqrt{\frac{\varepsilon_f}{\mu_{\text{eff}}}} \qquad (12.12b)$$

k_0 and k are the free space and ferrite region wave-numbers.

$$k_0 = \omega \sqrt{\mu_0 \varepsilon_0} \qquad (12.13a)$$

$$k = \omega \sqrt{\mu_0 \mu_{\text{eff}} \varepsilon_0 \varepsilon_f} \qquad (12.13b)$$

respectively. $J_n(x)$, $Y_n(x)$, $J'_n(x)$ and $Y'_n(x)$ are Bessel functions of the first and second kind of order n and the corresponding derivatives, μ_0, μ_{eff}, ε_0 and ε_f are constitutive parameters and have the usual meanings, μ and κ are the diagonal and off-diagonal entries of the tensor permeability.

12.5 Port Relationships

The field variables at the ports may be deduced once the unknown constant A_{fn} appearing in the description of the electric field in the cavity region is available. In order to obtain this quantity it is necessary to establish the boundary conditions at the terminals between the radial and rectangular waveguides. This requirement may be either directly realized in terms of those of an ideal circulator or in terms of the eigenvalue problem. The choice between the two approaches lies between the solution of a single 3-port network and that of three 1-port circuits.

The exact boundary condition is obtained by matching the electric and magnetic fields at the ports along the arc of radius R_0. The approximate method used here

12.5 PORT RELATIONSHIPS

assumes that the boundary conditions are satisfied on the chords of the rectangular waveguides and that these support single dominant mode propagation only

$$-\psi \leq \phi \leq \psi \qquad E_{z1} \approx E_1 \cos(3\phi/2) \qquad (12.14a)$$

$$-120 - \psi \leq \phi \leq -120 + \psi \qquad E_{z1} \approx E_2 \cos[(3\phi/2) - \pi] \qquad (12.14b)$$

$$+120 - \psi < \phi < +120 + \psi \qquad E_{z1} \approx E_3 \cos[(3\phi/2) + \pi] \qquad (12.14c)$$

with $E_z = 0$ elsewhere

If the triplet of electric field settings E_1, E_2 and E_3 at the three ports corresponds to one of the possible eigenvectors associated with the symmetry of the problem region then the task reduces to that of satisfying the boundary conditions at a typical port. The solution is then the eigenvalue associated with the particular eigenvector under consideration.

The electric field on the boundary between the radial and rectangular waveguides is now expanded in a Fourier series.

$$E_z = \sum_{n=-\infty}^{n=\infty} c_n \exp(jn\phi) \qquad (12.15)$$

where

$$c_n = \frac{1}{2\pi} \int_{-\psi}^{+\psi} E_1 \cos\left(\frac{3\phi}{2}\right) \exp(-jn\phi) \, d\phi$$

$$+ \frac{1}{2\pi} \int_{-120-\psi}^{-120+\psi} E_2 \cos\left(\frac{3\phi}{2} - \pi\right) \exp(-jn\phi) \, d\phi$$

$$+ \frac{2}{2\pi} \int_{120-\psi}^{120+\psi} E_3 \cos\left(\frac{3\phi}{2} + \pi\right) \exp(-jn\phi) \, d\phi \qquad (12.16)$$

In order to cater for the fact that the contour around the periphery of the problem region is defined by a number of disjointed arcs the range of integration in the Euler formula has been broken up to correspond to the various segments of the function

Integrating and applying the integration limits gives

$$c_n = b_n^2 \left[E_1 + E_2 \exp{-j\left(\frac{2\pi n}{3}\right)} + E_3 \exp{j\left(\frac{2\pi n}{3}\right)} \right] \qquad (12.17)$$

where

12 JUNCTION CIRCULATOR USING POST RESONATORS

$$b_n^2 = \frac{1}{2\pi}\left[\frac{\sin(3/2 - n)\psi}{(3/2 - n)} + \frac{\sin(3/2 + n)\psi}{(3/2 + n)}\right] \quad (12.18)$$

ψ is defined in terms of the waveguide opening (a) and outside radius of the cavity (R_0) formed by the junction of the three waveguides illustrated in Figure 12.5 by

$$\sin\psi = \frac{a}{2R_0} \quad (12.19)$$

Taking ψ as $\pi/3$ by way of example gives $b_0 = 0.46$ and $b_{+1} = b_{-1} = 0.437$.

Comparing the two solutions for E_z at $r = R_0$ in (12.9) and (12.15) yields the constant A_{fn} as

$$A_{fn} = \frac{b_n^2\left[E_1 + E_2 \exp-j\left(\frac{2\pi n}{3}\right) + E_3 \exp j\left(\frac{2\pi n}{3}\right)\right]}{[A_n E_{zfn} + jB_n H_{\phi fn}]} \quad (12.20)$$

The magnetic field on the contour of the radial waveguide is therefore specified by

$$H_\phi = jb_n^2\zeta_0\left\{\sum_{n=-\infty}^{\infty}\frac{jC_n E_{zfn} + D_m H_{\phi fn}}{A_n E_{zfn} + jB_n H_{\phi fn}}\right.$$

$$\left.\left[E_1 + E_2 \exp-j\left(\frac{2\pi n}{3}\right) + E_3 \exp j\left(\frac{2\pi n}{3}\right)\right]\exp(jn\phi)\right\} \quad (12.21)$$

The required solution is now deduced by constructing the magnetic field in each rectangular waveguide. One solution is to take the value of the magnetic field at the centre of the waveguide and to take the value of the electric field there also. Another, which is adopted here is to evaluate the average magnetic field (longitudinal current) at each port ($\phi = 0, 2\pi/3, -2\pi/3$) at the boundary between the radial and rectangular waveguides

$$H_1 = \frac{1}{2\psi}\int_{-\psi}^{\psi} H_\phi\, d\phi = \zeta_{11}E_1 + \zeta_{12}E_2 + \zeta_{13}E_3 \quad (12.22a)$$

$$H_2 = \frac{1}{2\psi}\int_{2\pi/3-\psi}^{2\pi/3+\psi} H_\phi\, d\phi = \zeta_{13}E_1 + \zeta_{11}E_2 + \zeta_{12}E_3 \quad (12.22b)$$

$$H_3 = \frac{1}{2\psi} \int_{-2\pi/3-\psi}^{-2\pi/3+\psi} H_\phi \, d\phi = \zeta_{12} E_1 + \zeta_{13} E_2 + \zeta_{11} E_3 \quad (12.22c)$$

Scrutinizing these port relationships for the in-phase eigenvector with $n = 0, \pm 3, \pm 6$ etc., indicates that each port displays the same eigenvalue ζ_0.

$$\zeta^0 = \Sigma \, \zeta_n \quad n = 0, \pm 3, \pm 6, \text{ etc} \quad (12.23)$$

where ζ_n is a typical pole at any port of the junction

$$\zeta_n = -j3a_n^2 \zeta_0 \left[\frac{jC_n + D_n y_n}{A_n + jB_n y_n} \right] \quad (12.24)$$

y_n are the split admittance poles at the terminals of the gyromagnetic resonator.

$$y_n = j\zeta_0 \zeta_{\text{eff}} \left[\frac{J_n'(kR)}{J_n(kR)} \mp \frac{\kappa}{\mu} \left(\frac{n}{kR} \right) \right] \quad (12.25)$$

and

$$a_n^2 = \left(\frac{\sin n\psi}{n\psi} \right) b_n^2 \quad (12.26)$$

In a like manner

$$\zeta^+ = \Sigma \, \zeta_n \quad n = +1, -2, +4, -5, +7 \text{ etc} \quad (12.27)$$

$$\zeta^- = \Sigma \, \zeta_n \quad n = -1, +2, -4, +5, -7 \text{ etc} \quad (12.28)$$

ζ^0 and $\zeta^\pm$ are known as in-phase and counter-rotating eigenvalues in that these correspond to in-phase and counter-rotating eigen-vectors (fields) at the ports of the problem region, ζ_n is a typical admittance pole of the problem region.

The entries of the short-circuit parameters at the terminals between the rectangular and radial waveguides may now be written in terms of the eigenvalues by having recourse to the appropriate relationship between the two in equation (12.2).

12.6 The Ideal Transformer

A scrutiny of a typical pole entering into the description of any eigen-network suggests that it may be realized in terms of an immittance pole of the ferrite resonator, a radial region which is described by a *ABCD* matrix and an ideal transformer which describes the boundary conditions at any port of the junction in the manner indicated in Figure 12.6. In order to synthesise the required topology

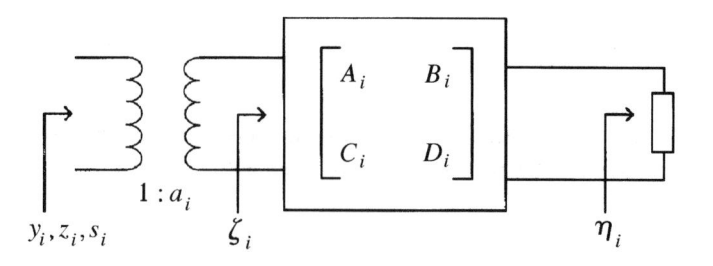

Figure 12.6 Realization of typical pole of demagnetized junction waveguide

it is necessary to extract an ideal transformer with a turns ratio $1 : a$ from ζ_n. This may be done by recognizing that ζ_n may be related to y_n by an overall $ABCD$ matrix given by

$$\zeta_n = \frac{jC + Dy_n}{A + jBy_n} \qquad (12.29)$$

where

$$\begin{bmatrix} A & jB \\ jC & D \end{bmatrix} = \begin{bmatrix} 1/a_n & 0 \\ 0 & a_n \end{bmatrix} \times \begin{bmatrix} A_n & jB_n \\ jC_n & D_n \end{bmatrix} \qquad (12.30)$$

or

$$A = A_n/a_n \qquad (12.31a)$$

$$B = B_n/a_n \qquad (12.31a)$$

$$C = a_n C_n \qquad (12.31a)$$

$$D = a_n D_n \qquad (12.31a)$$

In the preceding equation, A_n, B_n, C_n and D_n are the entries of the $ABCD$ matrix of the radial waveguide section. b_n represents the turns ratio of an ideal transformer with a turns ratio $1 : a_n$.

12.7 Eigen-networks

The eigen-networks of the junction containing a simple gyromagnetic post resonator can be formed once the topology of a simple pole is available. A First Foster form is appropriate at the terminals of the resonator and a Second Foster one is suitable at the terminals of a typical waveguide. Figure 12.7 indicates the solution at the terminals of a typical waveguide. Figure 12.8 the topology at the same terminals but with the in-phase poles idealized by electric walls at the resonator terminals.

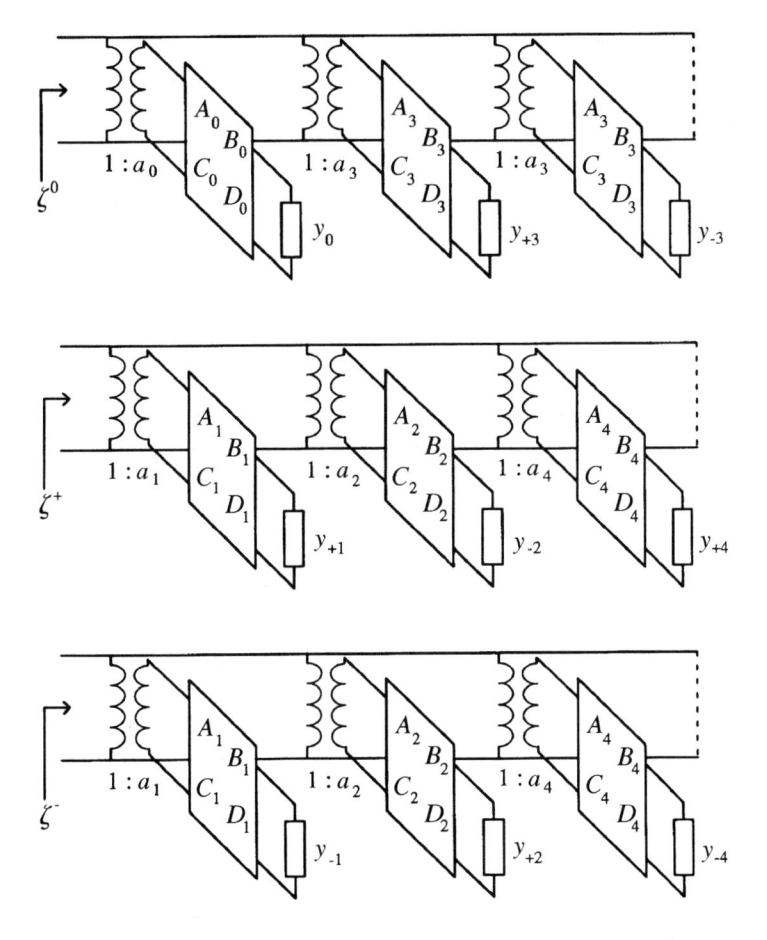

Figure 12.7 Second Foster realization of in-phase and counter-rotating eigen-networks of waveguide circulators

Since radial waveguides are so-called nonuniform lines the poles appearing in the descriptions of the eigen-networks are not commensurate. One feature of such lines is that the poles and zeros of the problem region do not display the periodicity associated with uniform lines. One serious consequence of this property is that if a pair of degenerate poles of the counter-rotating eigen-networks coincide with a zero of the in-phase one at the terminals of the gyromagnetic resonator the dual situation will not prevail at some other terminals. One possible way to reconcile this difficulty is to assume that the plane at which the degenerate counter rotating eigen-networks maps a magnetic wall at the terminals of the resonator into an electric one (characteristic plane) represents an upper bound on the reference terminals of the network and that the plane at which the in-phase eigen-network places a magnetic wall represents a lower bound on the problem region. If the frequency response of the in-phase eigen-network can be neglected compared to that of the generate counter-rotating ones then the characteristic

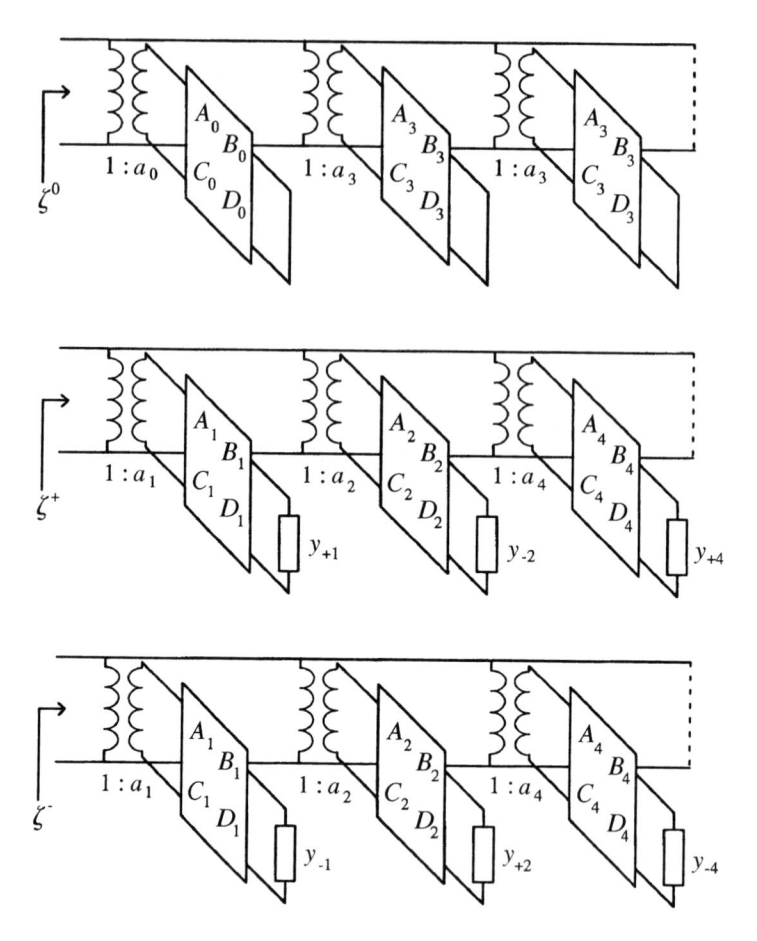

Figure 12.8 Approximate Second Foster realization of in-phase and counter-rotating eigen-networks of waveguide circulators

plane can be taken to correspond to the actual terminals of the junction. The exact plane corresponds, of course, to that for which the imaginary part of the gyrator impedance is zero. The possibility of loading one or the other eigen-networks with inductive or capacitive elements in order to overcoming this shortcoming is readily appreciated.

12.8 Gyrator Circuit

The two classic circulation conditions are determined by setting the imaginary part of the complex gyrator impedance to zero and thereafter adjusting its real part to coincide with the terminal impedance of the rectangular waveguide.

$$\text{Im}(\eta_{\text{in}}) = 0 \qquad (12.32)$$

$$\operatorname{Re}(\eta_{in}) = R \tag{12.33}$$

The complex gyrator admittance ζ_{in} at a typical waveguide port is determined in the usual way by writing $V_3 = I_3 = 0$.

$$\zeta_{in} = \zeta_{11} - \frac{\zeta_{12}^2}{\zeta_{13}} \tag{12.34}$$

η_{in} is then related to ζ_{in} by

$$\eta_{in} = \frac{1}{\zeta_{in}} \tag{12.35}$$

The first circulation condition determines the ferrite and radial waveguide radii, while the second defines the gyrator conductance or gyrotropy at the characteristic plane of the device.

$$R = Z_0 \tag{12.36}$$

While the imaginary part condition satisfies the so-called first circulator condition it does not necessarily ensure that the in-phase and degenerate counter-rotating eigen-network are commensurate and that an impedance zero (poles) of the former corresponds to a pair of degenerate poles (zero) of the latter ones.

12.9 Approximate Circulation Conditions for $n = \pm 1$

For small magnetic splitting it is possible to introduce a selection rule whereby circulation is taken to occur in the vicinity of a pair of degenerate admittance poles provided the frequency variation of the in-phase poles may be neglected compared to those of the counter-rotating ones. In the vicinity of any such pairs of degenerates poles it is furthermore assumed that the amplitude of the others are small and can be neglected. The eigen-networks in this approximation are depicted in Figure 12.9 and those of the magnetized ones in Figure 12.10.

For small magnetic splitting the approximate circulation conditions at the input terminals of the rectangular waveguides are therefore

$$\frac{\eta_{+1} + \eta_{-1}}{2} = 0 \tag{12.37}$$

$$\frac{-j\sqrt{3}\,(\eta_+ - \eta_-)}{2} = Z_0 \tag{12.38}$$

provided the in-phase eigenvalue ζ_0 is approximately equal to zero at the terminals of the rectangular waveguides.

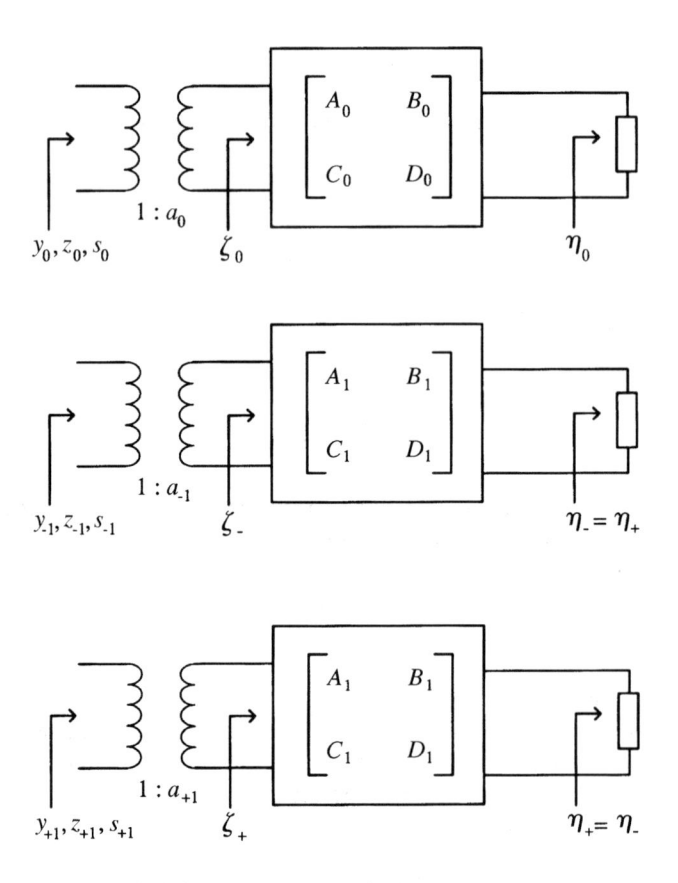

Figure 12.9 Realization of a typical in-phase and counter-rotating poles of demagnetized waveguide junction

The counter-rotating impedance eigenvalues $\eta_\pm$ are given in terms of the original variables

$$\eta_{\pm 1} = a_1^2 \cdot \frac{jC_1 + D_1 y_{\pm 1}}{A_1 + jB_1 y_{\pm 1}} \tag{12.39}$$

where $y_\pm$ are the split admittance eigenvalues at the terminals of the gyromagnetic resonator.

$$y_\pm = j\zeta_0 \zeta_{\text{eff}} \left[\frac{J_1'(kR)}{J_1(kR)} \pm \frac{\kappa}{\mu}\left(\frac{1}{kR}\right) \right] \tag{12.40}$$

A_1, B_1, C_1 and D_1 are given in (12.11a) to (12.11d) with $n = 1$, and a_1 is given in (12.26).

12.9 APPROXIMATE CIRCULATION CONDITIONS

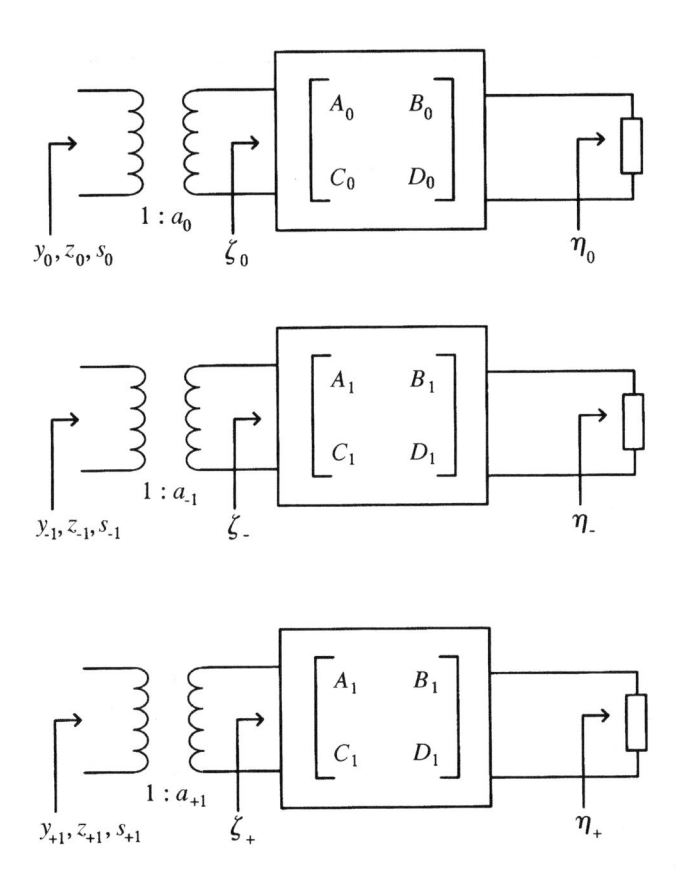

Figure 12.10 Realization of a typical in-phase and counter-rotating poles of magnetized waveguide junction

The first circulation condition is met with κ/μ equal to zero provided

$$J_1'(kR) = 0 \tag{12.41}$$

and

$$A_1 = 0$$

or

$$\frac{J_1(k_0 R_0)}{Y_1(k_0 R_0)} = \frac{J_1'(k_0 R)}{Y_1'(k_0 R)} \tag{12.42}$$

These last two conditions place a magnetic wall at the ferrite terminals and an electric wall at the circulator ones.

The real part condition at the terminals of the ferrite region and radial waveguide is now given by

$$Z_0 = \left(\frac{-j\sqrt{3}\,a^2}{2}\right)\left(\frac{C_1}{B_1}\right)\left(\frac{y_- - y_+}{y_+ y_-}\right) \quad (12.43\text{a})$$

where

$$y_\pm = \mp \left(\frac{j\zeta_0 \zeta_{\text{eff}}}{kR}\right)\left(\frac{\kappa}{\mu}\right) \quad (12.43\text{b})$$

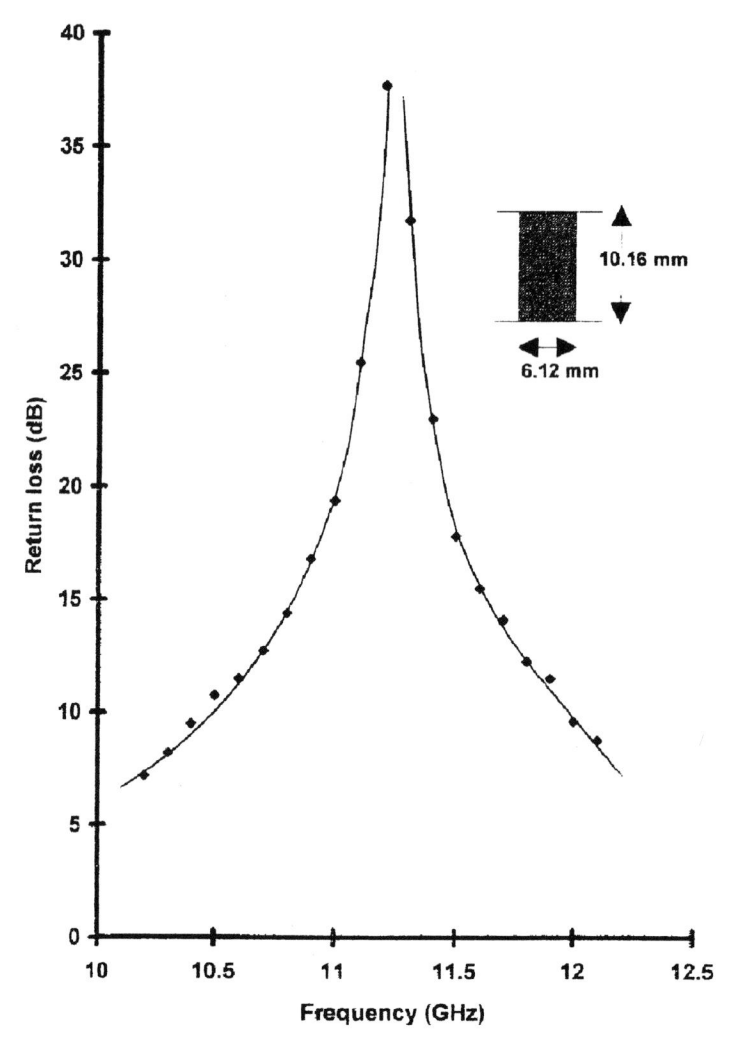

Figure 12.11 Frequency response of directly coupled circulator in WR90 waveguide

This last equation fixes κ/μ since all the other quantities are specified by the first circulation condition. Here, $kR = 1.84$ and $k_0 R_0$ is defined in Figure 12.5.

The split frequencies of the gyromagnetic resonator from which the quality factor of the resonator may be calculated, an important quantity in the description of any circulator, corresponds to the condition for which

$$y_{\pm 1} = 0 \tag{12.44}$$

This gives

$$\left[\frac{J_1'(kR)}{J_1(kR)}\right] \mp \frac{\kappa}{\mu}\left(\frac{1}{kR}\right) = 0 \tag{12.45}$$

and

$$(\Delta kR)_{11} = \left[\frac{2(kR)_{11}}{(kR)_{11}^2 - 1}\right]\left(\frac{\kappa}{\mu}\right) \tag{12.46}$$

where

$$(kR)_{11} = 1.84 \tag{12.47}$$

Figure 12.11 depicts the frequency response of a typical WR90 junction using a simple ferrite post.

12.10 Degree-2 Solution

While the frequency response of the sort of arrangement considered so far usually displays a degree-1 response it may be also adjusted to have a degree-2 one. Such a response is realized by enforcing a degree-2 filter topology on the circulator structure with a frequency response akin to that met in connection with the more familiar quarter-wave coupled G-STUB gyrator circuit. While the physical arrangement resembles that of a quarter-wave coupled one there is in fact no need to make any such connection. Figure 12.12 illustrates some possible configurations. The appropriate equivalent circuit is reproduced in Figure 12.13. In realizing this 1-port complex gyrator circuit the usual assumption that the frequency variation of the in-phase eigen-network may be neglected compared to those of the counter-rotating ones has been put aside. The former eigen-network is now a short-circuited radial waveguide section as is readily understood, whose immittance slope parameter may be adjusted by varying the impedance of the waveguide.

One shortcoming of the use of wave impedance in the description of microwave structures adopted so far is that it does not embody the physical dimensions of the problem region. It is therefore unable to cope with steps in the heights of the

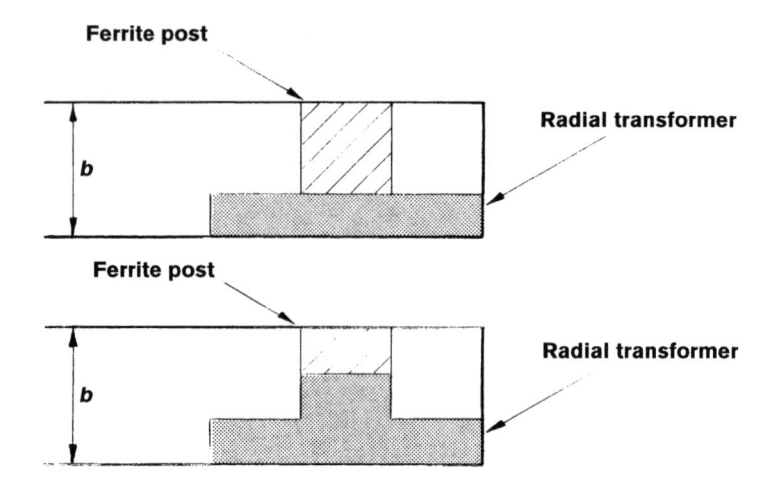

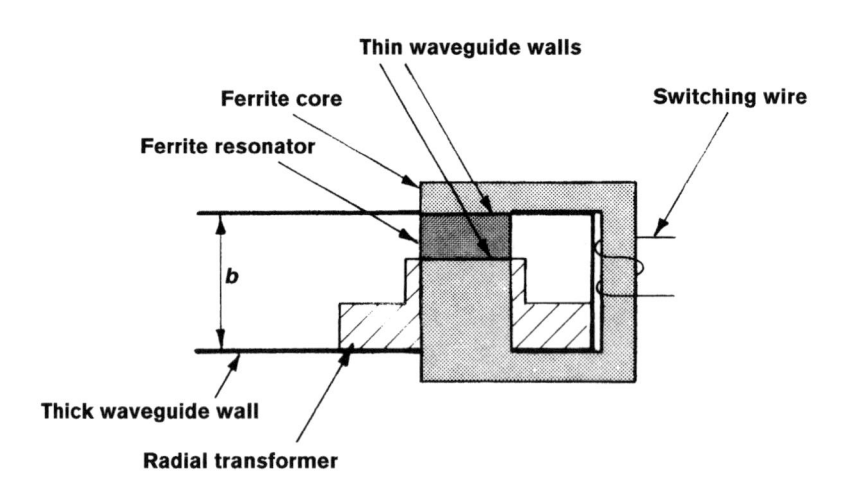

Figure 12.12 Schematic diagram of waveguide circulator using radial transformers

different waveguide structures illustrated in Figure 12.12. In order to reconcile this difficulty it is usual to introduce voltage and current variables and the notion of characteristic impedance instead of wave impedance. This problem is separately tackled in Chapter 11.

12.11 Characteristic Plane

The characteristic plane of a radial waveguide at the junction of three waveguides is obtained by locating the position of a short circuit plane at one port with another one terminated in a short circuit piston. This is repeated for a number of different radii until the position of the $VSWR$ minimum from the centre of the device

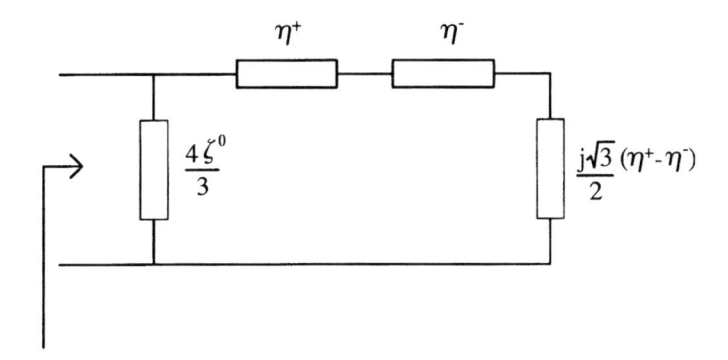

Figure 12.13 Degree-2 gyrator circuit

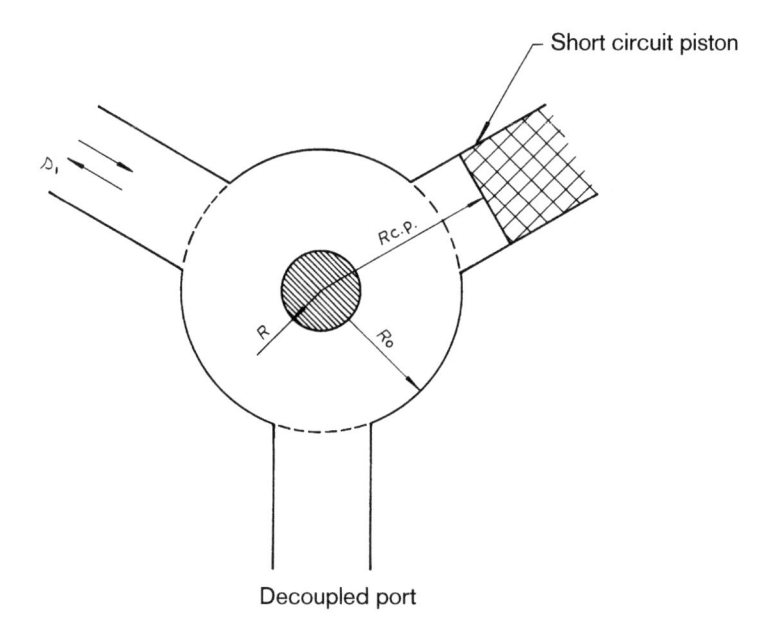

Figure 12.14 Definition of characteristic plane

coincides with the radius of the junction. The experimental arrangement is indicated in Figure 12.14. Some experimental work on the characteristic plane of a waveguide junction containing a dielectric post is summarized in this section.

The experimental work was carried out in WR90 waveguide at 9.50 GHz for $\varepsilon_r = 1, 3.2,$ and 10. The radius of the dielectric post was each time chosen to coincide with $kR = 1.84$. It was also repeated for a demagnetized ferrite with $\varepsilon_r = 12.7$ and $\mu_e = 0.78$. The experimental results are superimposed on the theoretical ones in Figure 12.15. A comparison between theory and practice suggest that the theoretical result for $k_0 R_0$ provides an upper bound on the experimental one.

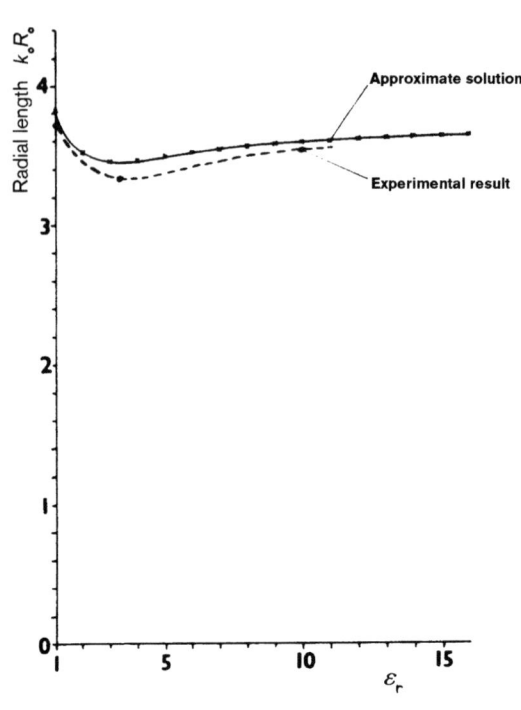

Figure 12.15 Characteristic plane of waveguide junction (reproduced with permission, Helszajn, J. and Tan F. C. F. (1975) Design data for radial waveguide circulators using partial-height ferrite resonators, *IEEE Trans Microwave Theory Tech.*, **MTT-23**, 288–298, March) (© 1975 IEEE)

This discrepancy is probably due to the fact that the higher order modes necessary to fully match the junction to the waveguide have been omitted in the theoretical calculation made here.

The characteristic plane of the junction of the three rectangular waveguides is separately defined in Figure 12.16. Here, however, the characteristic plane lies in the waveguide outside the actual junction terminals. A similar result was obtained using a dielectric resonator with a relative dielectric constant $\varepsilon_r = 10$ in which a variable short-circuit piston in one of the rectangular waveguides was used in conjunction with a standard junction instead an oversized radial one.

12.12 The Constituent Resonator

The analytical results may be verified in a number of different ways, some of which are separately enunciated in the chapter on measurements. The most simple arrangement which may be employed to evaluate the degenerate counter-rotating variables at the rectangular waveguide terminals is to plot the frequency response of the constituent resonator at the characteristic plane of the circuit and to compare the same with the theoretical description. Such an arrangement may also

12.12 THE CONSTITUENT RESONATOR

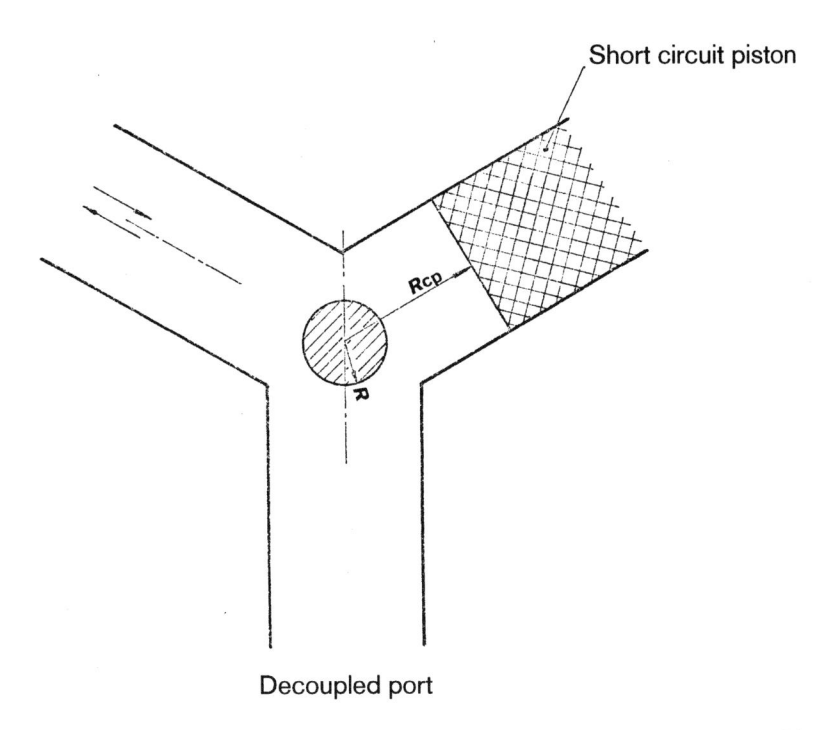

Figure 12.16 Definition of characteristic plane of junction of three rectangular waveguides

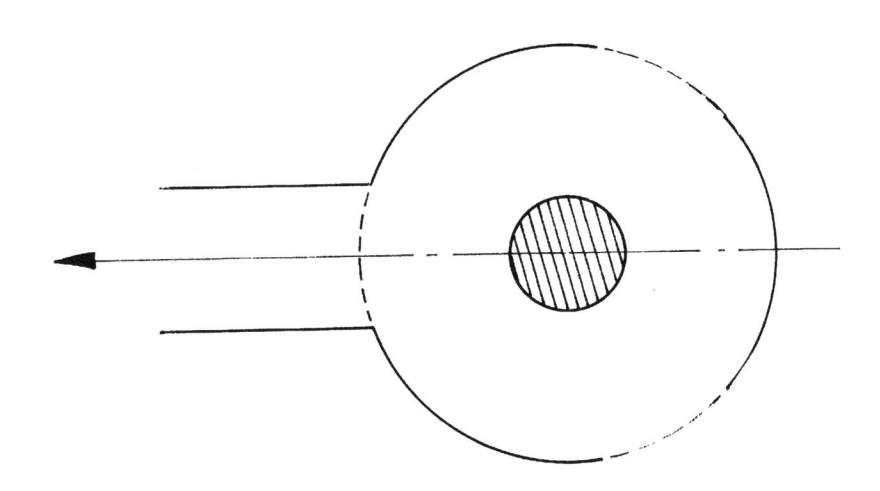

Figure 12.17 Constituent resonator

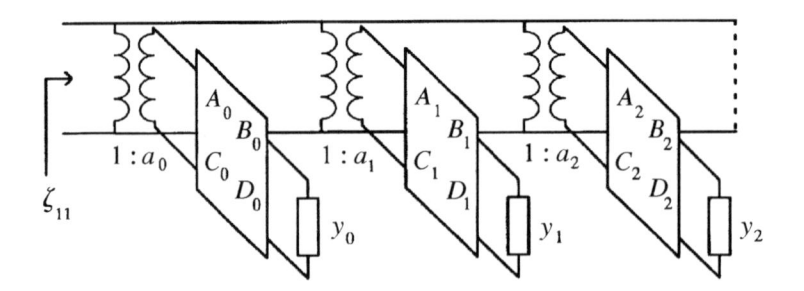

Figure 12.18 Second Foster realization of Y_{11}

be separately used to form the reactance slope parameter at the same terminals and the split frequencies of the junction. Figure 12.17 indicates a typical constituent resonator. These same quantities may of course be deduced by making appropriate measurements of a 3-port structure.

One feature of the constituent resonator is that it displays all the degenerate or split poles of the problem region. Figure 12.18 illustrates the Second Foster expansion of the 1-port demagnetized problem obtained in this way.

References and Further Reading

Butterweck, H. J. (1963) The Y-circulator *Arch. Elek. Ubertragung*, **17**(4), 163–176.
Castillo, J. B. and Davis, L. E (1970) Computer-aided design of 3-port waveguide junction circulators, *IEEE Trans. Microwave Theory Tech.*, **MTT-18**, 25–34.
Castillo, J. B. and Davis, L. E (1972) Identification of spurious modes in circulators, *IEEE Trans. Microwave Theory Tech.*, **MTT-19**, 112–113.
Castillo, J. B. and Davis, L.E. (1973) A higher order approximation for waveguide circulators, *IEEE Trans. Microwave Theory Tech.*, **MTT-20**, 410–412.
Chait, H. N. and Curry, T. R. (1959) Y-circulator, *J. Appl. Phys.*, **30**, 152.
Collins, J. H. and Watters, A. R. (1963) A field map for the waveguide Y-junction circulator, *Electron. Engr. London*, **35**, 540–543.
Davies, J. B. (1962) An analysis of the m-port symmetrical *H*-plane waveguide junction with central ferrite post, *IRE Trans. Microwave Theory and Techniques*, **MTT-10**, 596–604.
Davies, J. B. (1965) Theoretical design of wideband waveguide circulators, *Electronics Lett.*, **1**, 60–61.
Davis, L. E. (1966) Theoretical design of static and latching ferrite 3-port and 4-port symmetrical waveguide circulators, *Digest of 1966 International Microwave Symp.*, 281–285.
Davis, L. E. (1968) Central-pin tolerances in broadband 3-port waveguide circulators, *Electronics Lett.*, **4**(15).
Dou, W. B. and Rong, Y. (1991) Analysis of unsymmetrically waveguide Y-junction circulators, *Microwave and Opt. Technol. Lett.*, **4**(3), 134–138.
El-Shandwily, M. E., Kamal, A. A. and Abdullah, E. A. F. (1973) General field theory treatment of *H*-plane waveguide junction circulators, *IEEE Trans.*, **MTT-21**(6), 392–403.

Khilla, A. and Wolff, I. (1978) Field theory treatment of H-plane waveguide junction with triangular ferrite post, *IEEE Trans.*, **MTT-26**(4), 279–287.

Koshiba, M. and Suzuki, M. (1986) Finite-element anlaysis of H-plane waveguide junction with arbitrarilly shaped ferrite post, *IEEE Trans.*, **MTT-34**(1), 103–109.

Okamoto, N. (1979) Computer-aided design of H-plane waveguide junctions with full-height ferrites of arbitrary shape, *IEEE Trans.*, **MTT-27**(4), 315–321.

Parsonson, C. G., Longley, S. R. and Davies, J. B. (1968) The theoretical design of broadband 3 port waveguide circulators, *IEEE Trans. Microwave Theory Tech.*, **MTT-16**, 256–258.

Stolyarov, A. K. and Tyukov, I. P. (1967) Electrical parameters of waveguide Y-circulators, *Radio Eng. Electron. Phys. (USSR)*, **12**, 1954–1962.

Stolyarov, A. K. and Tyukov, I. P. (1967) On the analysis of waveguide Y-circulators in terms of their extrinsic characteristics, *Radio Eng. Electron. Phys.* **12**, 346–348.

Tyukov, I. P. (1967) Analysis of symmetrical Y-junctions filled with ferrite dielectrics, *Radio Eng. Electron. Phys. (USSR)*, **12**, 341–345.

Tyukov, I. P. (1967) Theory of symmetrical waveguide circulators, *Radio Eng. Electron. Phys. (USSR)*, **12**, 244–253.

13

COMPLEX GYRATOR CIRCUIT OF A WAVEGUIDE JUNCTION CIRCULATOR USING AN OKADA RESONATOR

13.1 Introduction

By and large most low-power high-quality waveguide circulators are based on turnstile junctions which embody one or two quarter-wave long resonators open at one end and short-circuited at the other. High mean power circulators using composite resonators have been separately described. Such circulators may, however, also be constructed by mounting a planar high mean power resonator originally developed in the design of a stripline circulator into a waveguide arrangement. The schematic diagram of one circulator is illustrated in Figure 13.1 (a). The quasi-planar resonator consists of top and bottom circular plates upon which a pair of ferrite disks separated by a dielectric one are mounted. An arrangement which only utilizes a single ferrite region is separately depicted in Figure 13.1(b). Circulators using multi-chamber Okada resonators are dealt with in Chapters 6, 14 and 15.

The circulator model adopted in this paper is based on the classic 1-port gyrator description at the resonator terminals of the junction in terms of its conductance, susceptance slope parameter and loaded Q-factor. Any two of these quantities being sufficient to describe the third. In this model, the loaded Q-factor of the junction (or equivalently, the split frequencies of the magnetized resonator) is completely fixed by the magnetic variables, the susceptance slope parameter by the shape of the resonator and its aspect ratio, and the correct value of the gyrator conductance is assumed to be met by the proper design of the matching network without further ado. Each of these variables is evaluated, in this chapter, in terms of effective constitutive parameters using the weakly magnetized model of the junction. Specifically, the loaded Q-factor of the junction is fixed by the filling factor of the resonator and gyrotropy of its garnet or ferrite material, the radial dimension of the resonator by the filling factor and its centre frequency, and the

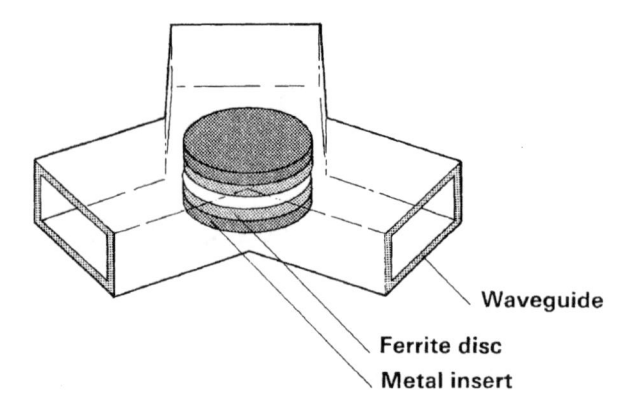

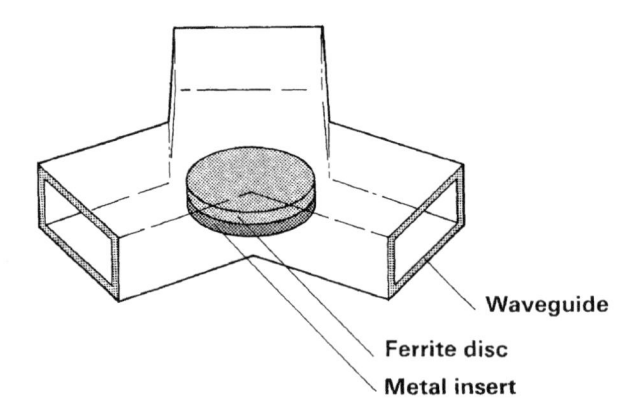

Figure 13.1 (a) Schematic diagram of waveguide junction circulator using double planar resonator. (b) Schematic diagram of single planar resonator

susceptance slope parameter by the filling factor, the configuration of the resonator and its aspect ratio. A filling factor of about 0.62 appears to be adequate for design of below resonance devices.

13.2 Complex Gyrator Circuit of Junction Circulator

The nomenclature adopted in the description of the planar resonator investigated in this work is based on the variables entering in its 1-port complex gyrator circuit. If the in-phase mode is idealized by a frequency independent short circuit boundary condition, the normalized immittance of this 1-port equivalent circuit is described by

$$y_{in} = g + jb \tag{13.1}$$

13.3 RESONANT FREQUENCY

where

$$g = \sqrt{3}\, b' \left(\frac{\omega_+ - \omega_-}{\omega_0} \right) \qquad (13.2)$$

$$b = 2b' \left(\frac{\omega - \omega_0}{\omega_0} \right) \qquad (13.3)$$

g and b are the real and imaginary parts of the normalized complex gyrator circuit, $\omega_\pm$ are the split frequencies of the counter-rotating modes of the gyromagnetic resonator, ω_0 is the centre frequency of the junction, b' is the normalized susceptance slope parameter of the network.

$$b' = \left(\frac{\omega_0}{2} \right) \cdot \left. \frac{db}{d\omega} \right|_{\omega = \omega_0} \qquad (13.4)$$

The independent variables in this relationship are the split frequencies and the susceptance slope parameter; the dependent one is the gyrator conductance. The split frequencies of the network are fixed by the details of the gyromagnetic resonator, the susceptance slope parameter by the configuration of the demagnetized resonator.

Equation (13.2) is consistent with the classic relationship between the loaded Q-factor, Q_L, of the junction, and the split frequencies of the magnetized resonator as already noted more than once in the text.

$$\frac{1}{Q_L} = \sqrt{3} \left(\frac{\omega_+ - \omega_-}{\omega_0} \right) \qquad (13.5)$$

Since the network problem is usually expressed in terms of these same variables, this notation is particularly apt provided these can be readily experimentally evaluated. This is fortunately indeed the case. The network problem is separately summarized in Chapter 23.

Whereas an exact realization of the loaded Q-factor of the junction specified by the network problem is essential in meeting the required maximum ripple level and bandwidth of the specification, the absolute values of the real and imaginary parts of the complex gyrator circuit are less critical than once supposed, provided the minimum ripple level in the pass-band is not restricted to zero.

13.3 Resonant Frequency of Quasi-planar Resonators

The planar resonator utilized in this work consists of top and bottom circular plates on which are mounted thin gyromagnetic disks separated by a free space or a suitable dielectric region in the manner indicated in Figure 13.2. A feature of this resonator which makes it particularly attractive for use in the design of

258 13 COMPLEX GYRATOR CIRCUIT

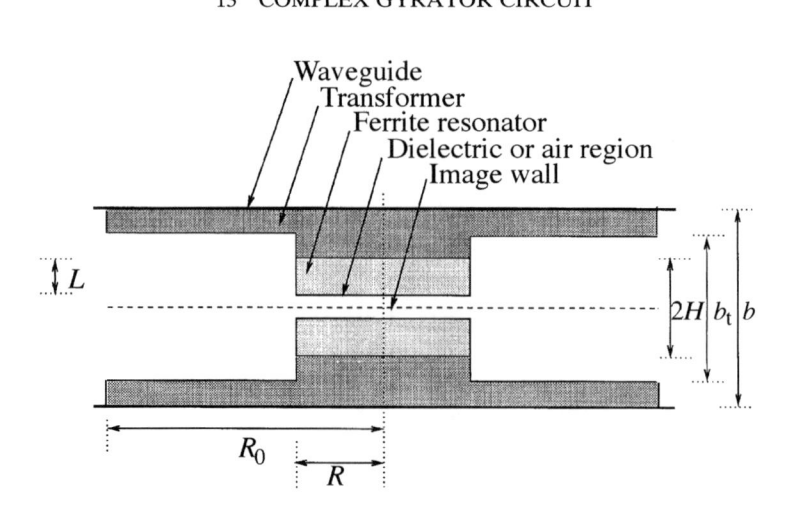

Figure 13.2 Definition of filling factor

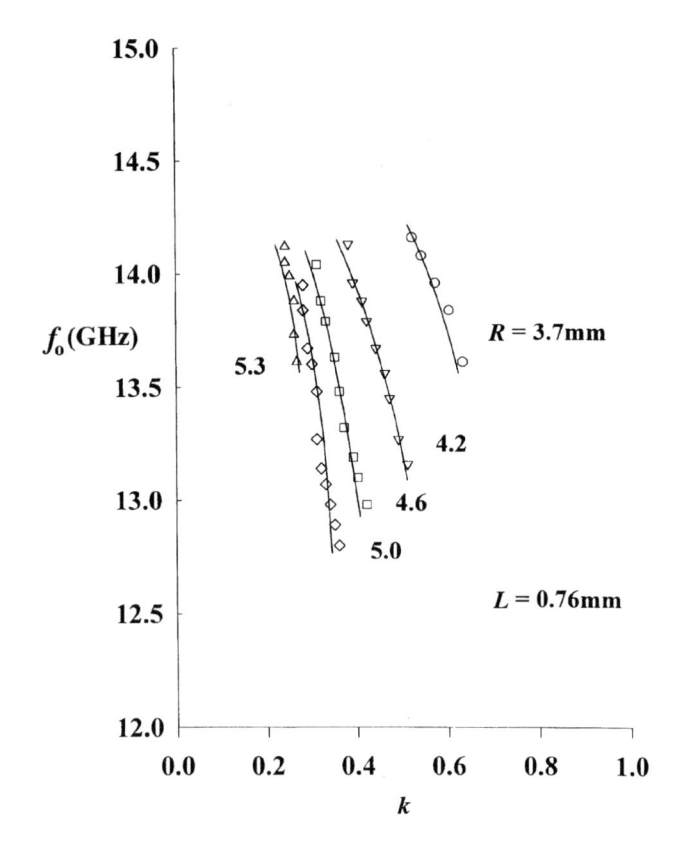

Figure 13.3 Frequency versus filling factor of planar resonator for parametric values of radius (L = 0.76 mm, ε_f = 12.7)

13.3 RESONANT FREQUENCY

high mean power devices is the fact that the each ferrite disk has a relatively large surface area and a relatively thin thickness. Its resonant frequency may be calculated in terms of the effective constitutive parameters introduced in Chapter 6. The radial wavenumber of the resonator is then described by

$$k_{eff} R_{eff} = 1.84 \qquad (13.6)$$

where

$$k_{eff} = \frac{2\pi}{\lambda_0} \sqrt{\varepsilon_f(eff)\, \mu_e(eff)} \qquad (13.7)$$

Figure 13.3 indicates some experimental data obtained on a three region structure mounted in a 2-port WR62 waveguide section. The thickness of each gyromagnetic layer was fixed at 0.76 mm and the filling factor was varied between zero and unity. The radius of the resonator was separately varied between 3.7 mm and 5.3 mm. The ferrite material used in this work was a manganese magnesium

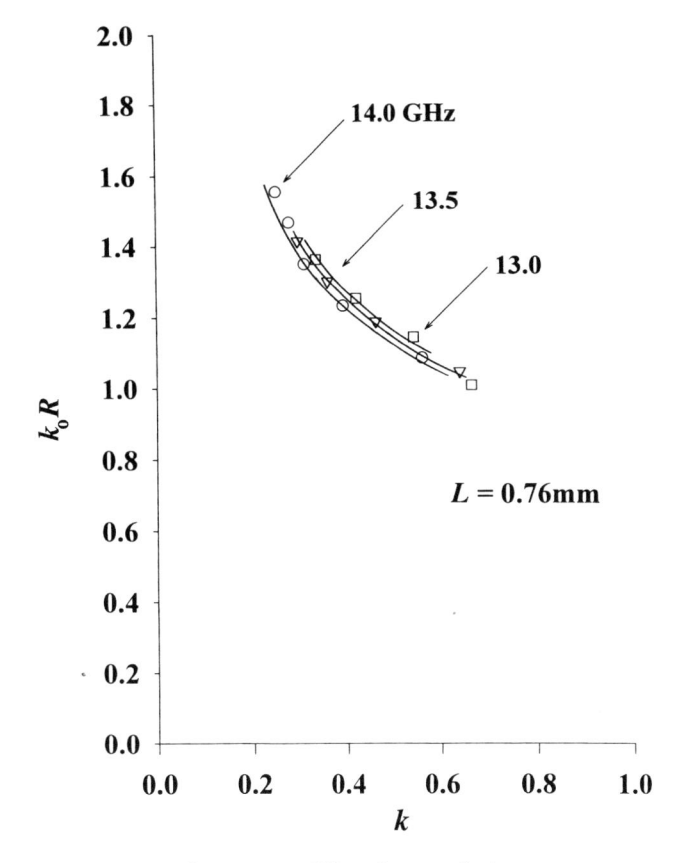

Figure 13.4 Radial wavenumber versus filling factor of planar resonator at 13, 13.5 and 14 GHz ($L = 0.76$ mm, $\varepsilon_f = 12.7$)

one with a magnetization M_0 of 0.2150 T and a relative dielectric constant ε_f of 12.7. Some additional experimental work using a planar resonator with the thickness of each gyromagnetic layer equal to 0.50 mm is omitted for brevity.

Figure 13.4 gives the experimental relationship between the filling factor and $k_0 R$ at frequencies of 13, 13.5 and 14 GHz for the resonator under consideration. This relationship is compatible with the effective radius of this type of resonator and the constitutive parameters introduced in Chapter 6.

13.4 Quality Factor

In order to proceed with the synthesis of any junction circulator it is essential to have a knowledge of its quality factor. This quantity is at hand once the split frequencies of the gyromagnetic resonator are established. One experimental procedure relies on the fact that the resonant frequencies of the split counter-rotating modes of a magnetized resonator may be shown to coincide with those at which the VSWR at one port with the other two ports terminated is given by

$$VSWR = 2 \qquad (13.8)$$

The split frequencies of one such a resonator for two different values of filling factor are depicted in Figure 13.5. A similar result for another resonator with a

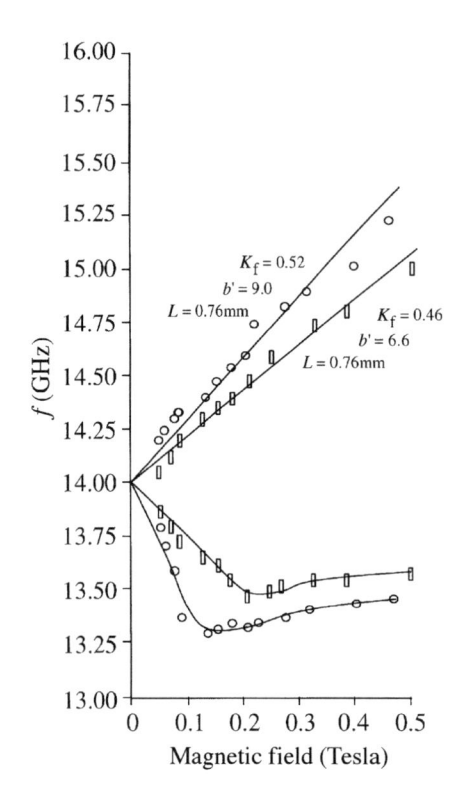

Figure 13.5 Split frequencies versus direct magnetic field of waveguide junction circulators using planar resonators with $k_f = 0.52$, $L = 0.76$ mm and $k_f = 0.46$, $L = 0.76$ mm, $M_0 = 0.2150$ T

13.4 QUALITY FACTOR

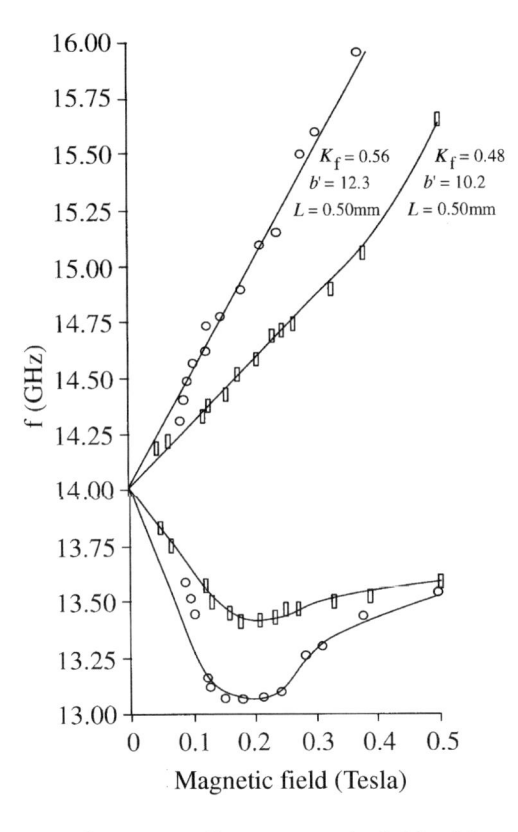

Figure 13.6 Split frequencies versus direct magnetic field of junction circulators using planar resonators with $k_f = 0.56$, $L = 0.50$ mm and $k_f = 0.48$, $L = 0.50$ mm, $M_0 = 0.2150$ T

different aspect ratio is indicated in Figure 13.6. Scrutiny of these two illustrations indicates that the splitting between the degenerate frequencies in this type of resonator is mainly dependent upon the filling factor of the resonator. Figure 13.7 indicates the quality factors associated with the two arrangements under discussion. The magnetization of the material is again 0.2150 T and its relative dielectric constant is 12.7.

One approximate description of the quality factor of this sort of arrangement which represents a lower bound on experiment has been separately given in Chapter 6. It is reproduced here for completeness.

$$\frac{1}{Q_L} = \frac{2\sqrt{3}}{(k_{\text{eff}} R_{\text{eff}})^2 - 1} \left(\frac{\kappa}{\mu}\right)_{\text{eff}} \quad (13.9)$$

where

$$\left(\frac{\kappa}{\mu}\right)_{\text{eff}} = \frac{\mu k}{1 + k(\mu - 1)} \left(\frac{\kappa}{\mu}\right) \quad (13.10)$$

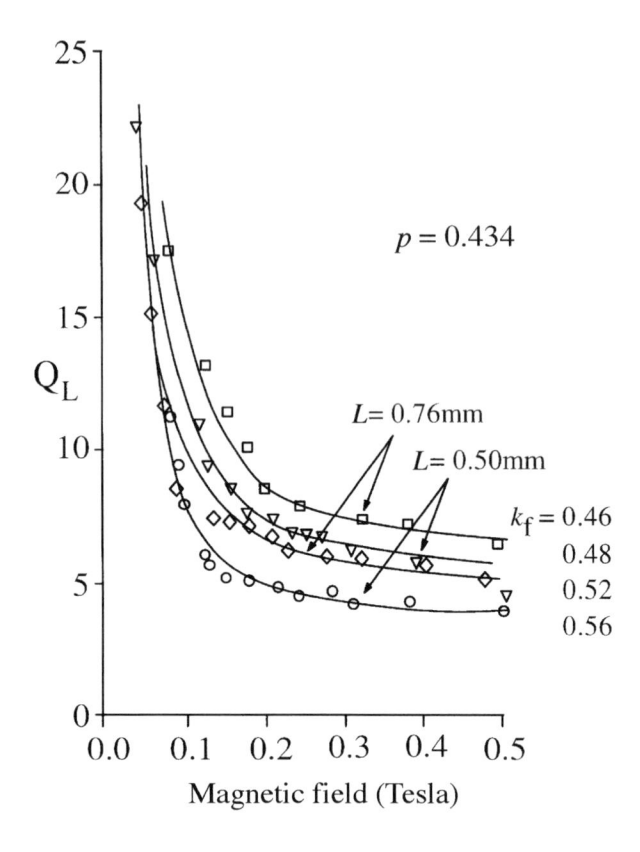

Figure 13.7 Quality factor of planar resonator ($M_0 = 0.2150$T)

μ and κ are the actual diagonal and off diagonal elements of the tensor permeability, k is the axial filling factor. In practice the theoretical description of Q_L represents a lower bound on experiment.

13.5 Susceptance Slope Parameter

One independent adjustment in the design of any junction circulator is its quality factor. The other is its susceptance slope parameter. Once these two adjustments are met then the desired gyrator conductance is automatically satisfied. This quantity fixes the overall thickness of the resonator and that of each ferrite disk once its filling factor is specified. It is also therefore related to the peak power rating of the device. One possible approximate form for this quantity which is surprisingly good is obtained by assuming that b' has the same dependence upon the effective constitutive parameters as that met in the case of a microstrip geometry.

13.5 SUSCEPTANCE SLOPE PARAMETER

$$b' = \frac{0.74\, \zeta_0 \zeta_{eff}(k_{eff}R_{eff})^2}{Y_{PV}(2k_{eff}H)} \tag{13.11}$$

where

$$\zeta_0 = \sqrt{\frac{\varepsilon_0}{\mu_0}} \tag{13.12}$$

$$\zeta_{eff} = \sqrt{\frac{\varepsilon_f(eff)}{\mu_e(eff)}} \tag{13.13}$$

$k_{eff}\, R_{eff}$ is defined in equation (13.6) and k_{eff} in equation (13.7).

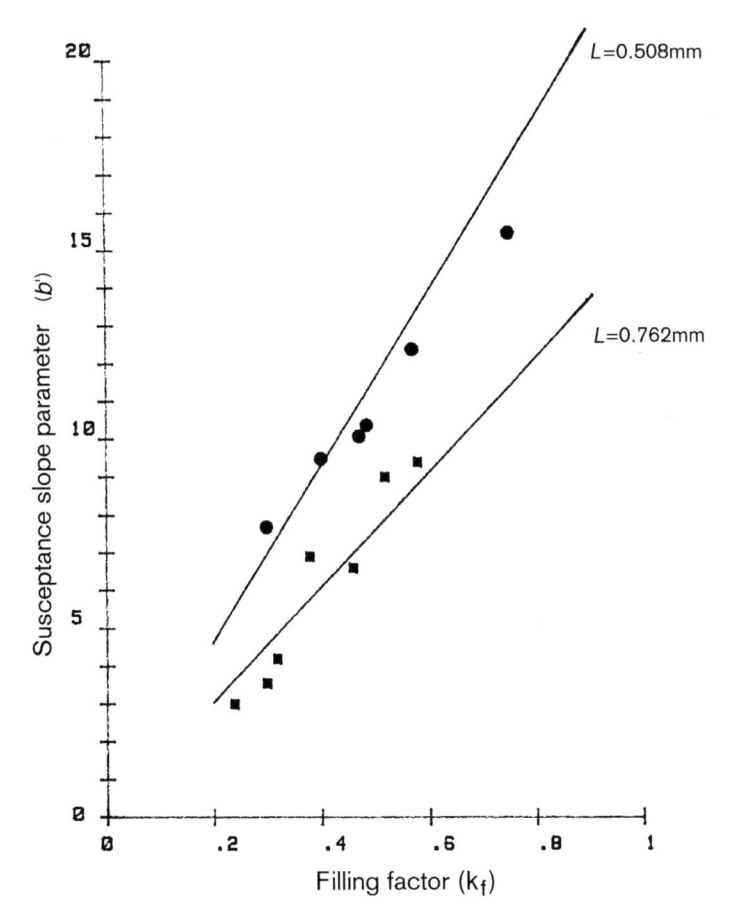

Figure 13.8 Susceptance slope parameter versus filling factor (k_f) of waveguide junction circulator using planar resonators at 14 GHz for $L = 0.50$ mm and $L = 0.76$ mm

Y_{PV} is the power voltage definition of the admittance of standard full height waveguide

$$Y_{PV} = \left(\frac{2b}{a}\right)\left(\frac{\lambda_0}{\lambda_g}\right)\zeta_0 \tag{13.14}$$

a is the width and b is the height of the waveguide, R_{eff} is the effective radius defined earlier. λ_0 and λ_g are the free space and waveguide wavelengths respectively.

The susceptance slope parameter may also be expressed, for calculation purposes, in terms of the thickness of the individual ferrite regions instead of that of the overall resonator. The required relationship is deduced by having recourse to the definition of the filling factor. The required result is

$$b' = \frac{0.74k\zeta_0\zeta_{eff}(k_{eff}R_{eff})^2}{Y_{PV}(2k_{eff}L)} \tag{13.15}$$

Adopting this form for the susceptance slope parameter allows L to be fixed from a specification of b' and H to be fixed thereafter from a knowledge of k and L. This completes the physical adjustment of the junction.

A scrutiny of the variables entering into the description of the susceptance slope parameter indicates that it has a higher value below resonance than above it. This property may be readily revealed by evaluating the quantity ζ_{eff}/k_{eff} in the formulation of b' in either equation (13.11) or (13.15). Taking the former relationship by way of an example gives

$$b' = \frac{0.74k\zeta_0(k_{eff}R_{eff})^2}{\mu_e(\text{eff})Y_{PV}(2k_0H)} \tag{13.16}$$

The normalized susceptance slope parameter of the junction may be experimentally deduced from either the frequency response of a demagnetized resonator in a 2-port junction or from that of a 3-port junction circulator

$$b' = \frac{VSWR - 1}{3\delta_0\sqrt{VSWR}}, \quad \text{2-port} \tag{13.18}$$

$$b' = \frac{VSWR - 1}{2\delta_0\sqrt{VSWR}}, \quad \text{3-port} \tag{13.19}$$

where $2\delta_0$ is usually taken as the normalized bandwidth at which the $VSWR$ is 1.22 (20 dB return loss).

$$2\delta_0 = \frac{\omega_2 - \omega_1}{\omega_0} \tag{13.20}$$

The agreement between the theoretical relationship and the experimental result for the two resonators employed here is surprisingly good. This is indicated in Figure 13.8.

13.6 Complex Gyrator Conductance of Planar Junction

One experimental arrangement whereby the complex gyrator conductance of a junction circulator may be deduced is obtained by decoupling port-3 from port-1 by placing a variable mismatch at port-2 and measuring the corresponding admittance at port-1. A more simple measurement based on some idealization of the junction with both ports 2 and 3 terminated by matched loads has also been described in Chapter 5 and is employed here.

The normalized gyrator conductance is determined from a knowledge of the VSWR at port-1 at the centre frequency of the junction with ports 2 and 3 terminated by matched loads by.

$$g = \sqrt{2VSWR - 1}, \quad g > 1 \tag{13.21a}$$

$$g = \sqrt{\frac{2 - VSWR}{VSWR}}, \quad g < 1 \tag{13.21b}$$

The relationships between the gyrator conductance and the direct magnetic field obtained in this way are shown in Figure 13.9 for the two different planar resonators employed here at a frequency of 14 GHz. The corresponding solutions obtained by having resource to equation (13.2) are separately superimposed on this experimental work. The agreement is seen to be excellent.

13.7 Degree-2 Junction Circulator Using Planar Resonators

Before proceeding with the construction of a quarter-wave coupled device it is necessary to investigate the matching problem. The optimum response is one for which both the minimum and maximum values of the voltage standing wave ratio $S(\min)$, $S(\max)$ are different from unity. A feature that is particularly attractive is the design of the class of circulators investigated here is that choice of $S(\min)$ permits a significant control of the absolute values of the complex gyrator circuit.

An appropriate specification in WR75 waveguide is: $f = 12.0$ GHz; $2\delta_0 = 0.25$; $S(\max) = 1.15$; $S(\min) = 1.08$.

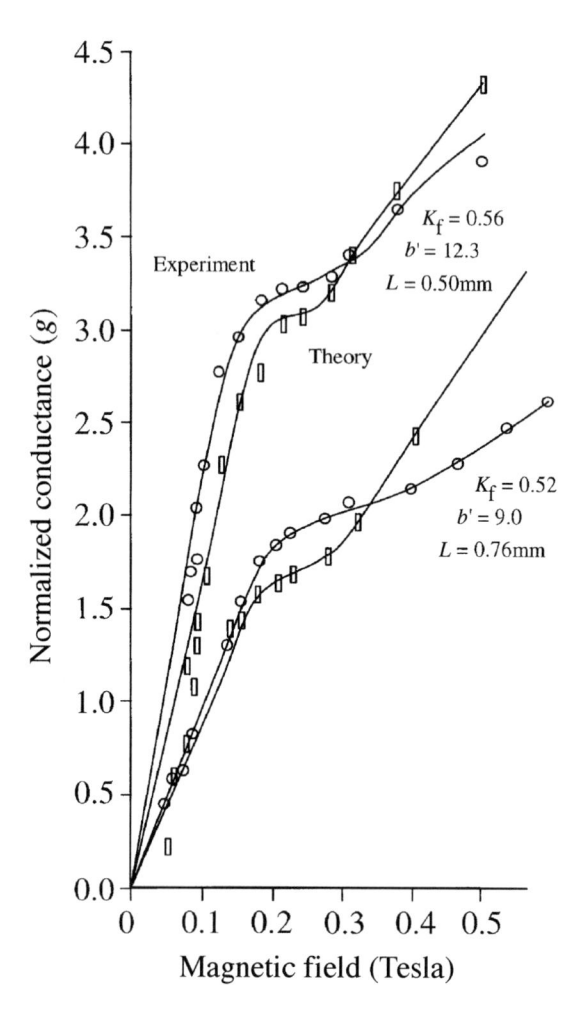

Figure 13.9 Comparison between theoretical and experimental conductances of waveguide junction circulators using planar resonators with $k_f = 0.52$ and $L = 0.76$ mm and with $k_f = 0.56$ and $L = 0.50$ mm

The normalized network variables for this specification are: $g = 5.33$; $b' = 10.81$; $Q_L = 2.03$; $Y_t = 2.48$.

A perusal of the experimental data assembled in the previous sections indicates that this requirement cannot be met with the physical variables employed so far. The first task is therefore to adjust the value of the quality factor (Q_L) of the junction to coincide with that dictated by the network problem. One possible solution is described by: $M_0 = 0.3000$T; $B/M_0 = 0.70$; $\kappa_p = 0.50$; $\mu_p = 0.917$; $(\kappa/\mu)_{\text{eff}} = 0.325$. The required filling factor is $k_f = 0.62$.

The second need is to trim the susceptance slope parameter of the junction to satisfy the network problem. This may be done by adjusting the thickness of each

13.7 DEGREE-2 JUNCTION CIRCULATOR

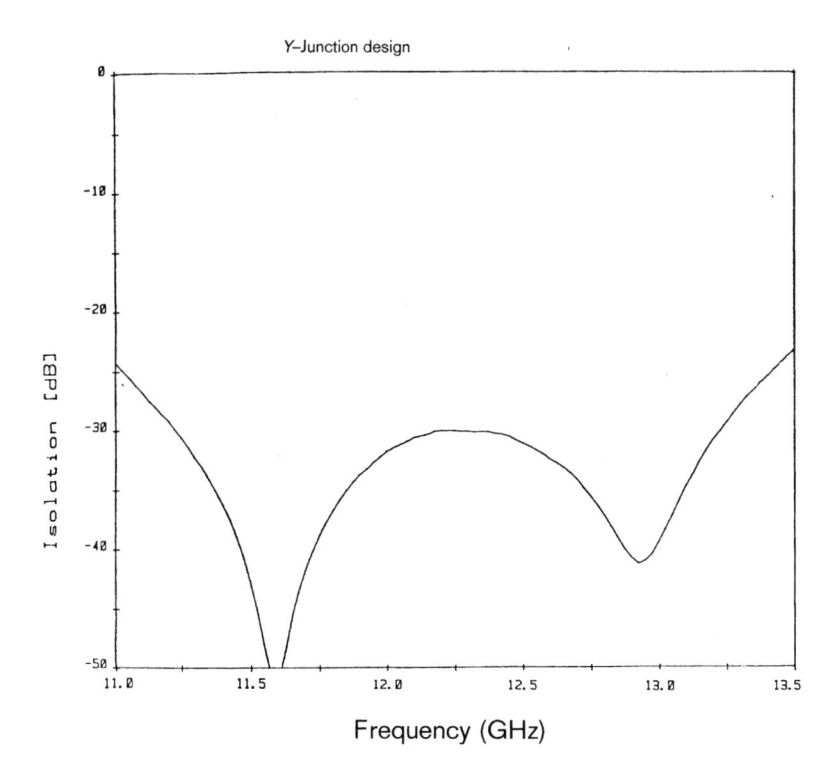

Figure 13.10 Experimental frequency response of 12 GHz quarter-wave coupled junction using quasi-planar resonator

ferrite disk (L). At 12 GHz this gives: $\varepsilon_f = 12.9$; $\varepsilon_e(\text{eff}) = 2.336$; $\mu_e(\text{eff}) = 0.842$; $\zeta_{\text{eff}} = 1.66$; $k_{\text{eff}}L = 0.158$ rad.

At k_0 equal to 0.251 rad/mm; $k_{\text{eff}} = 0.353$/nm and $L = 0.49$ mm.

Once Q_L and b' are established the required value of g is met by definition.

Half the overall thickness of the resonator (H) is now calculated from a knowledge of L and k_f as $H = 0.72$ mm.

The radius corresponding to this value of k_f is obtained by first evaluating R_{eff} in equation (13.6). The result is: $R = 4.74$ mm

Figures 13.10 and 13.11 indicate the experimental return loss and insertion loss at 12 GHz of such a quarter-wave coupled device.

The power rating of this and some other devices are summarized in Figure 13.12. A photograph of one commercial circulator is reproduced in Figure 13.13.

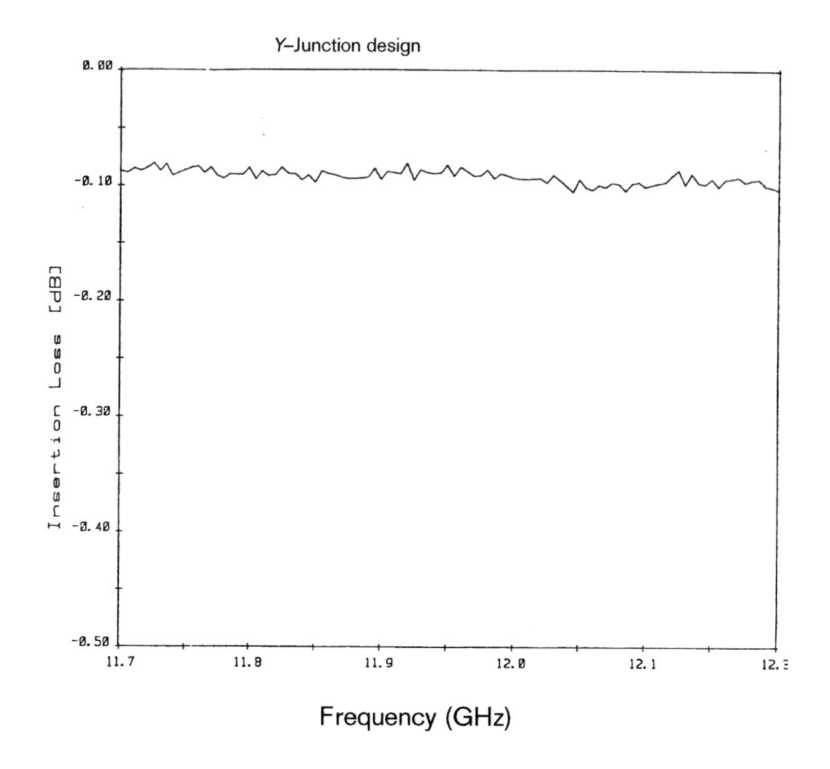

Figure 13.11 Insertion loss of 12 GHz quarter-wave coupled junction using quasi-planar resonator

13.8 15°, 30° and 60° Planar Resonators

A scrutiny of some commercial practice suggests that the electrical length of the individual ferrites (θ_f) used in planar resonators can be broken down to three families. One of these corresponds with (θ_f) equal to 15° and the other two to 30° and 60° respectively. The geometry for which θ_f equals to 15° is appropriate for the design of high-mean power devices with modest bandwidth specifications whereas that for which it is 60° is suitable for the construction of devices with more demanding high-peak power and bandwidth specifications. The solution for which θ is equal to 30° is applicable for everyday commercial practice.

13.9 Temperature Stability of Quality-Factor Curve

The temperature stability of any junction circulator is an important quantity in fixing its average power rating. While the saturation magnetization of some ferrite

13.9 TEMPERATURE STABILITY

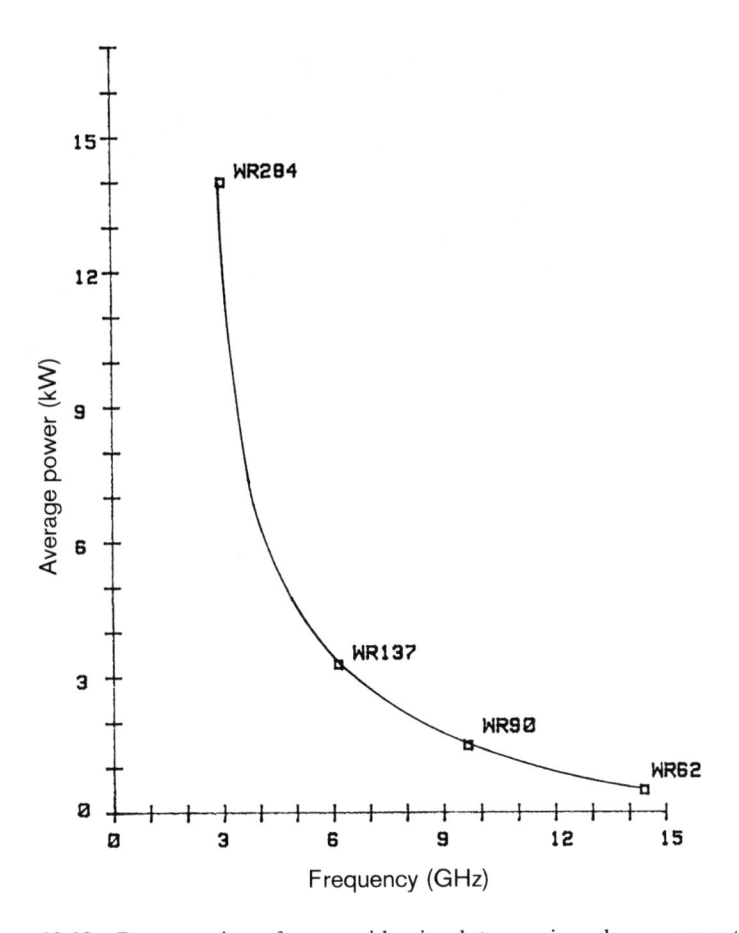

Figure 13.12 Power rating of waveguide circulators using planar resonators

and garnet materials may be stabilized by suitable doping this approach may not always be possible. Another solution is to employ series or shunt compensation of the direct magnetic circuit. Still another one is to enforce the condition between the direct magnetic field and the direct magnetization of the material at which the magnetization curves at different temperatures intersect. This possibility may be understood by noting that a typical hysterisis loop of a magnetic material at room temperature both shrinks and collapses as the temperature approaches its Curie value.

A unique relationship between the direct magnetic field (H_{appl}), and the direct saturation magnetization (M_0) which ensures a degree of temperature stability in a ferrite material has been proposed on the basis of some measurements on the magnetization curves of some typical materials. The required condition may be readily expressed in terms of the ratio of the applied direct magnetic field and the saturation magnetization by starting with the definition of the internal field (H_{in}) in a magnetic insulator.

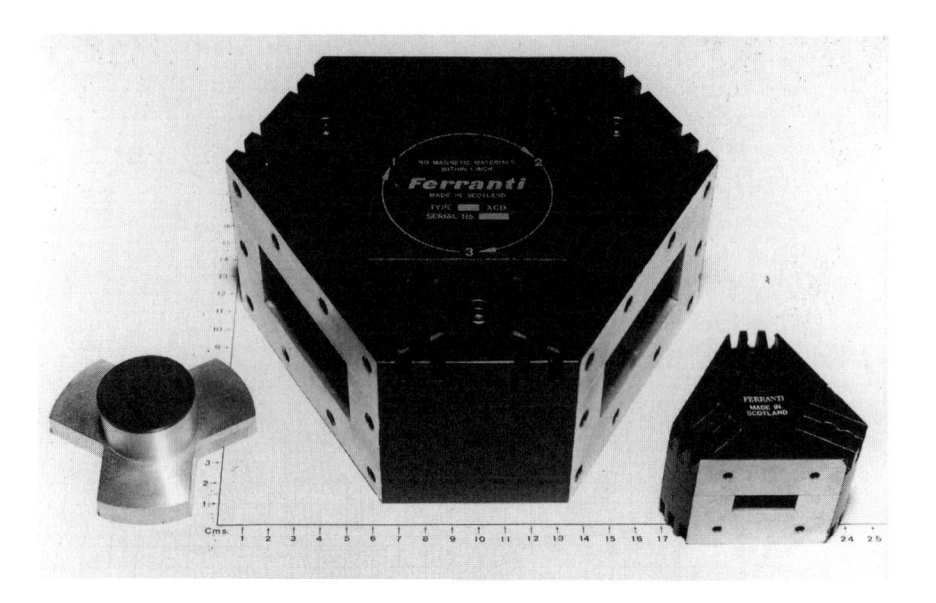

Figure 13.13 A commercial circulator

$$H_{\text{in}} = H_{\text{appl}} - \left(\frac{N_z M}{\mu_0}\right) \qquad (13.22)$$

where N_z is the direct demagnetizing factor along the axis of the resonator and M is the actual direct magnetization.

For a ferrite disk, of the sort used in this work, N_z is approximately given by

$$N_z \approx 1 - \left(\frac{L}{2R}\right)\left[1 + \left(\frac{L}{2R}\right)^2\right]^{-1/2} \qquad (13.23)$$

L is the length of the disk and R is its radius.

If the direct internal field is small compared to the other quantities then the relationship in equation (13.21) may be written as

$$\frac{\mu_0 H_{\text{appl}}}{M_0} \approx N_z \left(\frac{M}{M_0}\right) \qquad (13.24)$$

Some experimental work on a number of different ferrite materials in Helszajn (1981) indicates that the magnetization curves at different temperatures approximately intersect at the direct magnetic field for which

$$\frac{M}{M_0} = 0.70 \qquad (13.25)$$

13.9 TEMPERATURE STABILITY

A temperature stable condition for the purpose of design is therefore given by

$$\frac{\mu_0 H_{appl}}{N_z M_0} \approx 0.70 \qquad (13.26)$$

Since the single most important quantity in the description of the gain-bandwidth of a junction circulator is its quality factor (Q_L) its temperature stability is of some import. The temperature stability of this quantity has therefore been examined for one typical waveguide circulator using a CVG resonator. Scrutiny of this data suggests that the rule proposed in connection with that of the magnetization curves at different temperatures also coincides with that of the Q-curves of the junction circulator.

Some dedicated measurements on the quality factor of a C-band waveguide junction circulator using a CVG garnet resonator at a frequency of 5.75 GHz over

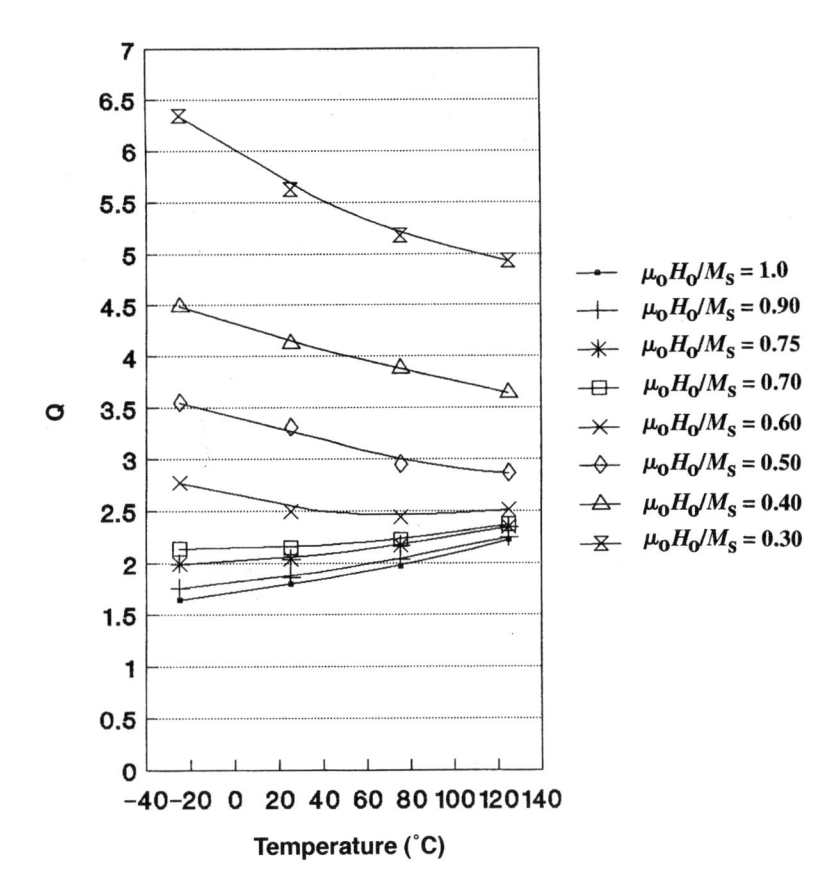

Figure 13.14 Quality factor of complex gyrator circuit versus temperature for parametric values of $\mu_0 H_0/M_0$. ($k = 0.60$, $M_0 = 0.1600$T) (reproduced with permission, Helszajn, J. and Tsounis, B. (1995) Temperature stability of quality factor of junction circulation, *IEE Proc., Microwave, Antennas and Propagation*, **142**, 67–70)

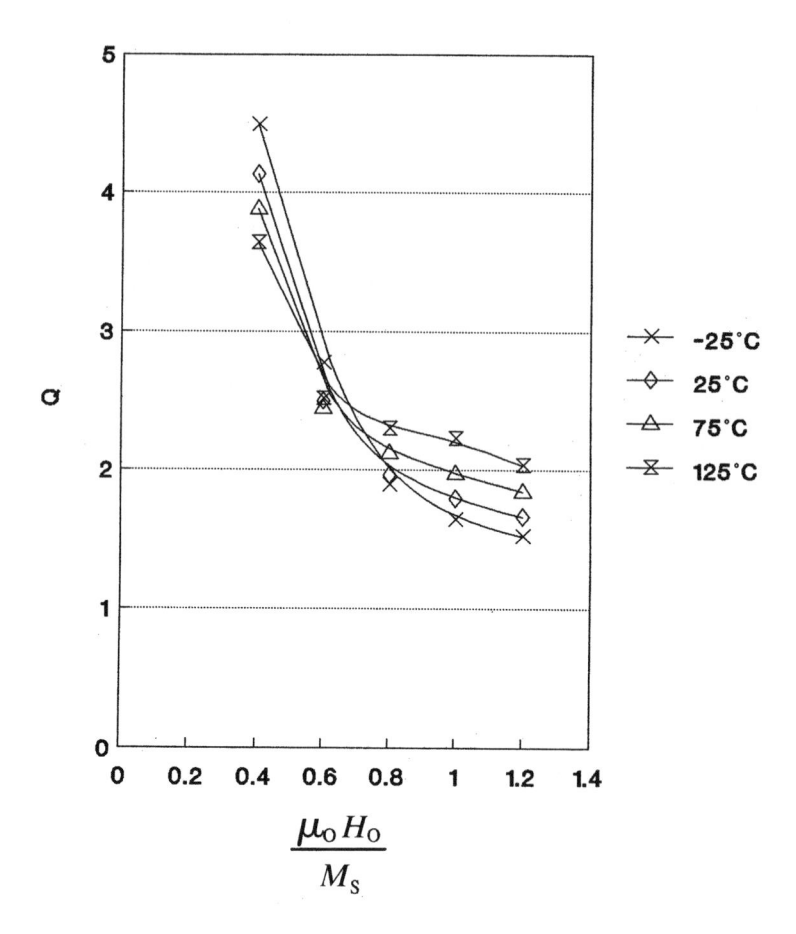

Figure 13.15 Quality factor of complex gyrator circuit versus $\mu_0 H_0/M_0$ at −25°C, 25°C, 75°C and 125°C. ($k = 0.60$, $M_0 = 0.1600$T) (reproduced with permission, Helszajn, J. and Tsounis, B. (1995) Temperature stability of quality factor of junction circulation, *IEE Proc., Microwave, Antennas and Propagation*, **142**, 67–70)

a typical range of temperature is depicted in Figure 13.14. It is obtained by making use of the relationship between the split counter-rotating frequencies of a loosely coupled resonator and the quality factor (Q_L) in equation (13.26).

Figure 13.15 depicts the same result plotted against the quantity $\mu_0 H_{appl}/M_0$ instead of temperature. Scrutiny of the intersection of the different Q-curves indicates that the conditions in equation (13.26) may be taken as an upper bound on the purpose of design. A more detailed scrutiny of this result indicates, however, that the actual temperature stable solution is in fact dependent upon the temperature interval in question. It is of separate note that while the resonator is essentially temperature stable up to the direct magnetic field at which the Q-curves intersect the synthesis problem indicates that it should be based at the intersection.

13.9 TEMPERATURE STABILITY

Another quantity that enters into the description of a ferrite or garnet material is the squareness of its hysterisis loop. It is defined in terms of the saturation and remanent magnetizations, a typical value is

$$\frac{M_r}{M_0} = 0.62 \tag{13.27}$$

Writing M_0 in terms of M_r suggests that an equally good or possibly better criteria for the temperature stability of a junction circulator is in this instance given by

$$H_{appl} = N_z M_r \tag{13.28}$$

Figure 13.16 illustrates a typical plot of the split modes over the same temperature interval. A feature of this result is that the angle between the split frequencies is independent of the magnetization of the material up to the direct magnetic field required to establish the knee of the magnetization curves.

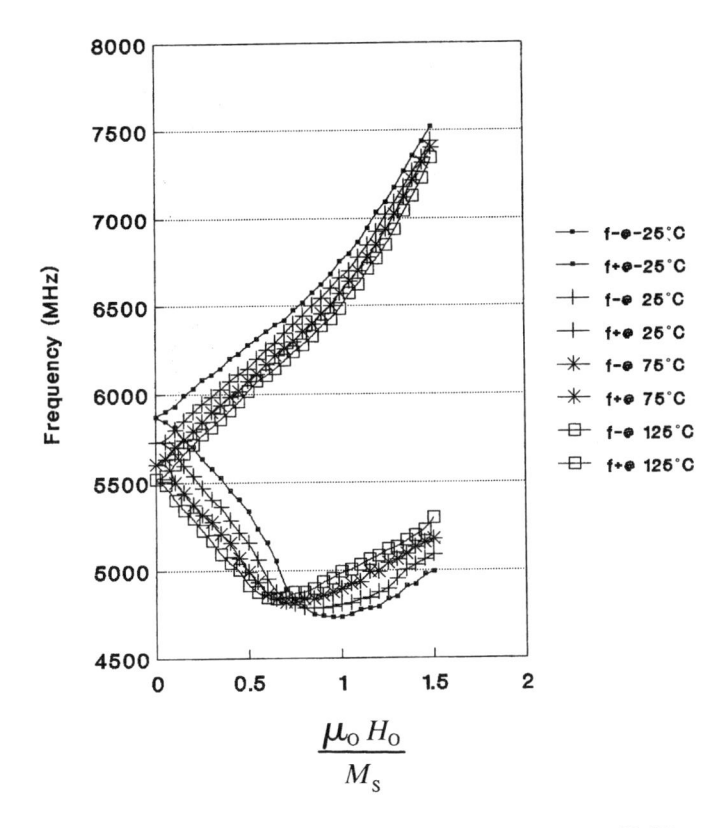

Figure 13.16 Split frequencies of gyromagnetic resonator versus $\mu_0 H_0/M_0$ at −25°C, 25°C, 75° C and 125°C ($k = 0.60$, $M_0 = 0.1600$T) (reproduced with permission, Helszajn, J. and Tsounis, B. (1995) Temperature stability of quality factor of junction circulation, *IEE Proc., Microwave, Antennas and Propagation*, **142**, 67–70)

Scrutiny of Figure 13.15 suggests that the intersection of the Q-curves divides the space into a weakly magnetized interval over which series compensation of the direct magnetic field is possible and a strongly magnetized one over which shunt compensation may be feasible. Since the gain-bandwidth product over the former interval is restricted the main interest is limited to the latter interval. However, due to the fact that the Q-curves have flattened out in this interval a relative large change in direct magnetic field is necessary in order to achieve effective temperature compensation. This observation indicates that the use of a temperature compensated material when available still remains of significant interest in design.

If the internal field may still be assumed negligible then $\mu_0 \Delta(H_{appl})$ is of the order of $N_z \Delta(M)$

$$\mu_0 \Delta(H_{appl}) = N_z \Delta(M) \tag{13.29}$$

If the largest change in ΔQ occurs at the direct magnetic field at which the material is saturated then

$$\frac{\Delta Q_L}{Q_L} \approx \frac{\Delta(M)}{M_0} \tag{13.30}$$

The resonator employed in obtaining the data in Figures 13.14 to 13.16 is a CVG material with a magnetization M_0 (T) equal to 0.1600 T and a Curie temperature T_c (°C) equal to 230°C. The radius (R) of the resonator is equal to 9.42 mm, it length (L) is equal to 0.76 mm, its filling factor (k) is 0.60 and its demagnetizing factor (N_z) is equal to 0.985.

13.10 Gain-Bandwidth of Quarter-Wave Coupled Circulator

The effect of a change in the quality factor of a circulator on the frequency response of a quarter-wave coupled device may be examined by way of an example. Its equivalent circuit is illustrated in Figure 13.17. One typical network specification is: $(VSWR)_{max} = 1.08$; $(VSWR)_{min} = 1.0$; $Q_L = 1.847$; $2\delta_0 = 0.20$; $b' = 12.895$; $g = 6.980$; $y_t = 2.746$, where g is the normalized gyrator conductance, b' is its normalized susceptance slope parameter and the other quantities have the usual meanings.

The change $\Delta(VSWR)$ in the midband or maximum value of the $VSWR$ brought about by a one in the quality factor from Q to $Q + \Delta Q$ is

$$\frac{\Delta(VSWR)_{max}}{(VSWR)_{max}} \approx \frac{\Delta Q_L}{Q_L} \tag{13.31}$$

This condition is readily deduced by having recourse to the two standard relationships below

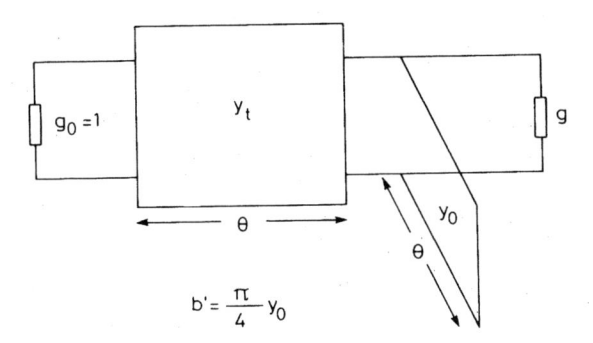

Figure 13.17 Equivalent circuit of quarter-wave coupled gyrator circuit

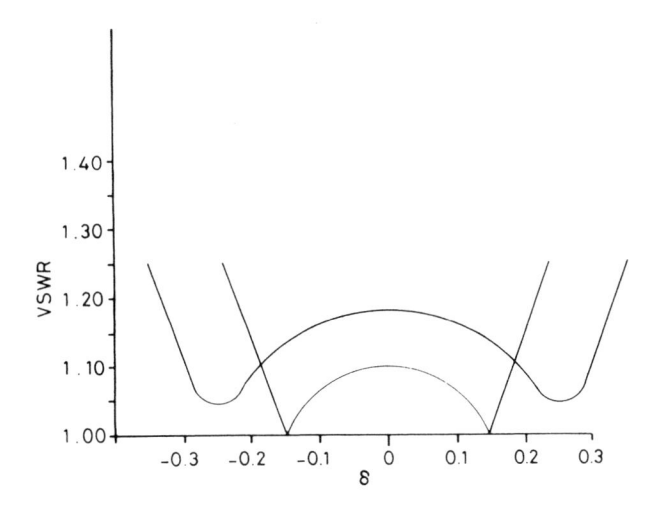

Figure 13.18 Frequency responses of network solutions

$$(VSWR)_{max} g = y_t^2 \qquad (13.32)$$

$$Q_L = \frac{b'}{g} \qquad (13.33)$$

In obtaining this result it has also been assumed that y_t and b' do not vary with temperature.

If $\Delta Q_L/Q_L$ is taken as 0.10 then the new network problem is defined by: $(VSWR)_{max} = 1.188$; $(VSWR)_{min} = 1.039$; $Q_L = 2.021$; $2\delta_0 = 0.287$; $b' = 12.895$; $g = 6.347$; $y_t = 2.746$.

The frequency responses of these two solutions are summarized in Figure 13.18. This example suggests that a change in the order of 10% in the quality factor of the junction produces a significant modification in its frequency response. This feature, is of course, well understood by workers in the field.

References and Further Reading

Akaiwa, Y. (1974) Operation modes of a waveguide Y-circulator, *IEEE Trans. Microwave Theory Tech.*, **MTT-22**, 954–959.

Bufler, C. R. and Helszajn, J. (1968) The use of composite junctions in the design of high power circulators, *IEEE Intl. Microwave Symp. on MTT*, May, 239–249.

Denlinger, E. J. (1975) Design of partial-height ferrite waveguide circulators, *IEEE Trans. Microwave Theory Tech.*, **MTT-22**, 810–813.

Fay, C. E. and Cornstock, R. L. (1965) Operation of the ferrite junction circulator, *IEEE Trans. Microwave Theory Tech.*, **MTT-13**, 15–27.

Hauth, W. (1981) Analysis of circular waveguide cavities with partial-height ferrite insert, *Proc. European Microwave Conference*, 383–388.

Hauth, W., Lenz, S. and Pivit, E. (1986) Design and realization of very-high-power waveguide junction circulators, *Frequency*, **40**, 2–11.

Helszajn, J. (1967) An H-plane high-power TEM ferrite junction circulator, *Radio and Electron. Eng.*, **33**, 257–261.

Helszajn, J. (1970) Frequency and bandwidth and H-plane TEM junction circulator, *Proc. IEE*, **117**(7), 1235–1238.

Helszajn, J. (1973) Microwave measurement techniques for below-resonance junction circulators, *IEEE Trans. Microwave Theory Tech.*, **MTT-21**, 347–351.

Helszajn, J. (1974) Common waveguide junction circulators, *Electronic Eng*, September.

Helszajn, J. (1975) Scattering matrices of junction circulators with Chebychev characteristics, *IEEE Trans. Microwave Theory Tech.*, **MTT-23**, July.

Helszajn, J. (1981) High-power waveguide circulators using quarter-wave long composite ferrite/dielectric resonators, *IEEE Proc. H, Microwaves, Opt. Antennas*, **128**(5), 268–273.

Helszajn, J. (1982) Quarter-wave coupled junction circulators using weakly magnetized disk resonators, *IEEE Trans. Microwave Theory Tech.*, **MTT-30**, 800–806.

Helszajn, J. (1995) Synthesis of quarter-wave coupled semi-tracking octave band stripline junction circulators, *IEEE Trans. Microwave Theory Tech.*, **MTT-43**, 310–318.

Helszajn, J. and Sharp, J. (1983) Resonant frequencies, Q-factor, and susceptance slope parameter of waveguide circulators using weakly magnetized open resonators, *IEEE Trans. Microwave Theory Tech.*, **MTT-31**(6), 434–441.

Helszajn, J. and Tan, F. C. F. (1975) Design data for radial waveguide circulators using partial-height ferrite resonators, *IEEE Trans. Microwave Theory Tech.*, **MTT-23**, 288–298.

Helszajn, J. and Tan, F. C. F. (1975) Susceptance slope parameters of waveguide partial-height ferrite circulators, *Proc. IEE*, **122**(72), 1329–1332.

Helszajn, J. and Tsounis (1995) Temperature stability of quality factor of junction circulation, *IEE Proc., Microwave, Antennas and Propagation*, **142**, 67–70.

Helszajn, J., Walker, P. N. and Davidson, E. (1980) Design of low loss waveguide circulators at large peak and mean power levels, *Military Microwave*, October.

Hlawiczka, P. and Mortis, A. R. (1963) Gyromagnetic resonance graphical design data, *Proc. IEE*, **110**(4), 665–670.

Konishi, Y. (1969) A high power UHF circulator, *IEEE Trans. Microwave Theory Tech.*, **MTT-15**, 700–708.

Kotoh, H., Takamizawa, H. and Itano, T. (1974) Temperature stabilization of 1.7 GHz-band lumped-element circulator, *Paper of Prof. Group on Microwave, Inst. Electron. Commun. Eng., (Japan)* **MW 74-7**, April.

Levy, R. and Helszajn, J. (1982) Specific equations for one and two section quarter-wave matching networks for stub-resistor loads, *IEEE Trans. Microwave Theory Tech.*, **MTT-30**, 55–62.

Okada, F. and Ohwi, K. (1978) Design of a high power CW Y-junction waveguide circulator, *IEEE Trans. Microwave Theory Tech.*, **MTT-26**, 364–369.

Okada, F. and Ohwi, K. (1981) High power circulators for industrial processing systems, *IEEE MAG-17* **6**, S. 2957–2960.

Okada, F. and Chwit, K. (1975) The development of a high power microwave circulator for use in breaking of concrete and rock, *J. Microwave Power*, **10**(2).

Okada, F., Chwit, K. and Yasude, M. (1978) Design procedure of high power CW Y-junction waveguide circulators, *IMPI*, 118–120.

Okada, F., Chwit, K., Mori, M. and Yasude, M. (1977) A 100 kW waveguide Y-junction circulator for microwave power systems at 915 MHz, *J. Microwave Power*, **12**(3).

Tokumitsu, Y., Kasahara, T. and Konizo, H. (1976) A new temperature stabilized waveguide circulator, *IEEE-MTT-S, Int. Symp.*, Cherry Hill, N.J.

14

DEGREE-1 AND 2 OKADA CIRCULATORS

14.1 Introduction

An important resonator in the design of high mean power junction circulators is the Okada resonator. It consists of a number of stacked circular plates supported by triplets of struts between which single or pairs of ferrite disks separated by an air or a dielectric region are located. The mean power rating of this sort of resonator is fixed by the number of chambers and by its filling factor; the gain-bandwidth by its complex gyrator circuit.. The purpose of this chapter is to deal with some aspects of a 2-chamber arrangement supported by three struts. Figure 14.1 depicts its geometry This resonator has its origin in the design of high mean power stripline devices. One consideration in the design of this sort of junction is the effect of the mounting struts of the resonator upon its susceptance slope parameter and frequency. The fact that both quantities are essentially independent of its details is of note. Degree-2 arrangements based on quarter-wave and half-wave prototypes can readily be realized using such resonators. The half-wave topology consists of half-wave U.E.'s (unit elements) at the characteristic planes of the junction where its complex gyrator displays a series STUB-R network. Although this topology is not particularly compact it is appropriate for the design of high-frequencies devices with high peak and mean power ratings. It is also suitable for matching junction circulators with relatively high Q complex gyrator circuits. The topology of the quarter-wave arrangement consists of a ladder network of quarter-wave long U.E.'s in cascade with a shunt STUB-G network. Its geometry is more appropriate at lower frequencies and for the design of devices with very high mean power ratings but more modest peak power ones. Since the quarter-wave coupled arrangement has the nature of an impedance transducer, the real part of its gyrator circuit is different from unity. The half-wave topology has, however, the nature of a filter circuit. The real part of its gyrator circuit is therefore of the order of unity. The gain-bandwidth of this class of circuits is

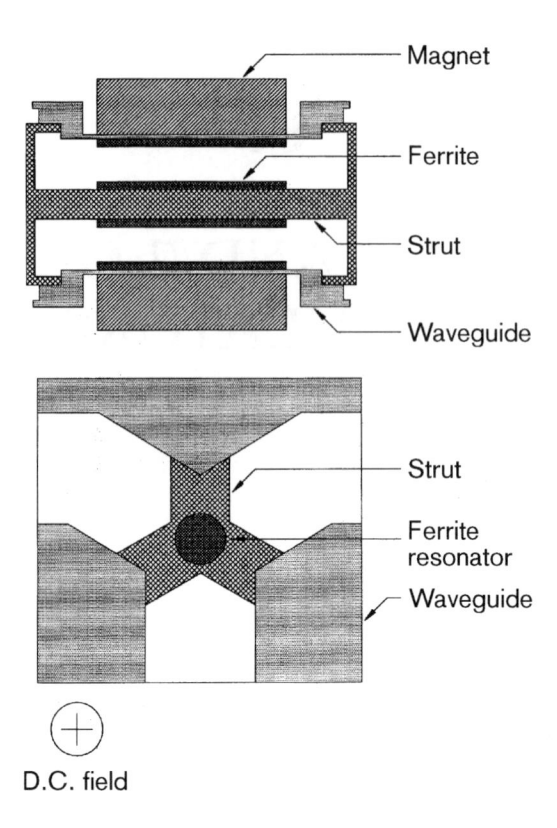

Figure 14.1 Schematic diagram of junction circulator using 2-chamber Okada resonator

realised by reconciling the gyrotropy of the resonator with the susceptance slope parameter of the junction. If the gyrotropies of the two arrangements are the same then the synthesis problems indicate that the susceptance slope parameter required for the adjustment of the former geometry is larger than that necessary for the latter one. The aspect ratios of the two resonators are therefore quite different. A degree-1 arrangement has been evaluated at 14 GHz in WR75 waveguide using convection cooling at a mean power of 1800 W c.w. with an insertion loss of 0.10 dB.

14.2 Okada Resonator

One way to increase the power rating of a planar resonator consisting of ferrite circular tiles on its top and bottom walls is to have recourse to the Okada resonator. It consists of stacked circular metal plates on which are mounted ferrite disks separated by a dielectric or free space region. The equivalence between a single resonator chamber and a dual one using two chambers which leaves every electrical parameters unchanged is reproduced in Figure 14.2. It is met provided the effective dielectric constants of the two structures are equal. This condition is

14.2 OKADA RESONATOR

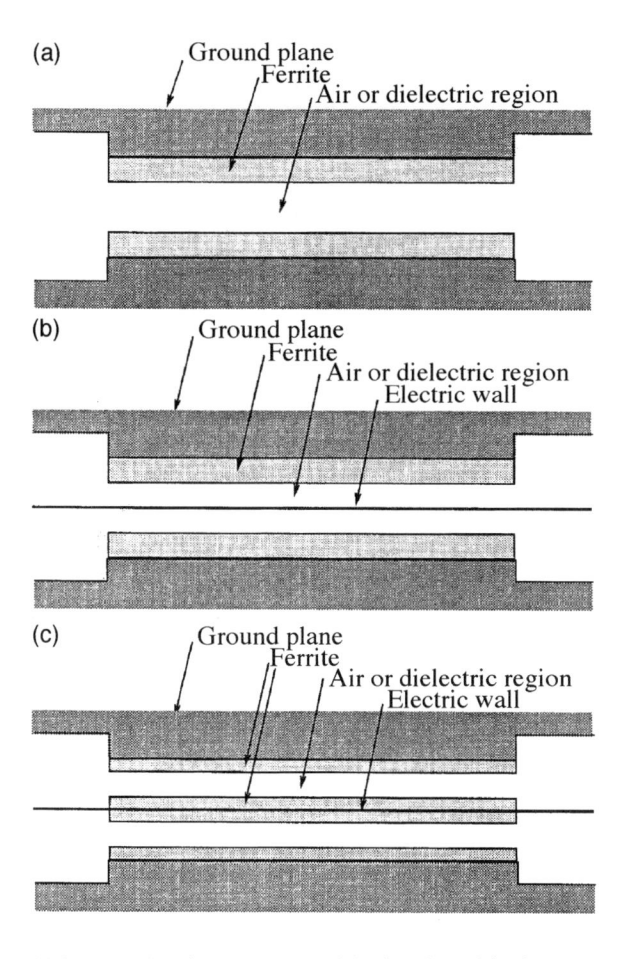

Figure 14.2 Mapping between 1 and 2-chamber Okada resonators

achieved if the thickness of each ferrite layer in the reorganized arrangement ($L/2$) is half that of the original structure (L) and if the filling factor is unaltered.

The step-by-step derivation of the equivalence between the two arrangements may be understood by considering a circuit with a filling factor of 0.50 by way of an example. The mapping between the two starts by introducing an electric wall at the symmetry plane of the original circuit. The required demonstration is immediately completed by noting that the effective dielectric constant (and consequently the capacitance of the circuit) is unaltered if the dielectric disks on the top and bottom of the original circuit are equally divided between the original walls and the surfaces revealed by the electric wall. The required steps are illustrated in Figures 14.2(a)–(c). Reorganizing the parallel plate capacitance in this manner doubles the surface area of the resonator which is in contact with the heat sink and halves its thickness. The power rating of the device is therefore increased by a factor of four. In a practical arrangement, the infinitely thin electric wall may be replaced by a conducting plate which is water cooled.

Since the possibility exists of introducing an electric wall along the symmetry plane of the prototype resonator the details of an n-chamber resonator may be described by the half space bounded by its top and bottom plates. A convenient description of the structure is therefore one based on the original geometry. It is given by the radial wavenumber ($k_{eff}R_{eff}$) of the circular plates, the thickness of each original ferrite disk (L) or aspect ratio (R/L) and the aspect ratio of the half space (H) between the two plates (R/H). If the details of the n-chamber prototype are expressed in terms of those of the original network then L is replaced by L/n and H by H/n in the actual structure. The details of each strut is separately described by its normalized thickness (s/b), Figure 14.3 gives one schematic diagram. The effects of the struts on the details of the complex gyrator circuit represent the main effort of this work.

Instead of specifying the spacing between the ferrite plates it is usually more convenient to introduce the notion of a filling factor (k). This latter quantity is described in terms of the physical parameters entering into the description of the half space defined by introducing a symmetry wall midway between the top and bottom walls of the original resonator

$$k = L/H \tag{14.1}$$

It specifies the ratio of the thickness of each ferrite and half the spacing between the top and bottom walls of both the original and derived resonators.

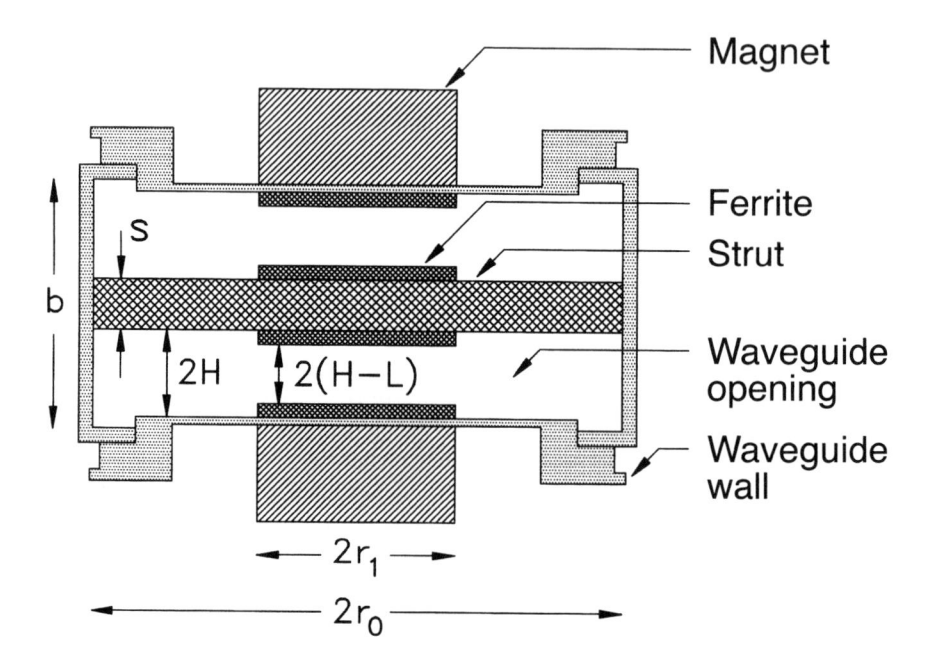

Figure 14.3 Schematic diagram of struts

14.3 Complex Gyrator Circuit

A description of any junction circulator requires in addition to a description of its frequency a statement of its 1-port complex gyrator circuit in the vicinity of its midband frequency as is now understood. If the frequency variation of the in-phase eigen-network of its junction may be neglected compared to those of its degenerate counter-rotating ones then it consists of a quarter-wave long short-circuited stub in shunt with the gyrator conductance. If the admittance of the stub is taken as Y_s than it is possible to define the gyrator circuit in terms of a normalized susceptance slope parameter b'

$$b' = \left(\frac{\pi}{4}\right)\frac{Y_s}{Y_0} \qquad (14.2)$$

a normalized conductance (g)

$$g = \frac{G}{Y_0} \qquad (14.3)$$

and a loaded Q-factor (Q_L)

$$Q_L = \frac{b'}{g} \qquad (14.4)$$

The quality factor is separately related to the stored energy in the resonator (U_0) and the power dissipated at the output port (P_0) by

$$Q_L = \frac{\omega U_0}{P_0} \qquad (14.5)$$

The susceptance slope parameter is fixed by the number of individual chambers and the overall thickness of the junction and the half space between the top and bottom plates of a typical chamber of the junction or equivalently, by the half-space of the original resonator prototype. In a directly coupled device g is unity and b' and Q_L have the same values.

$$Q_L = b' \qquad (14.6)$$

The bandwidth of the circulator is in this instance set by its susceptance slope parameter.

Another relationship for Q_L which may be derived in this type of circuit is in terms of the split frequencies of the gyromagnetic resonator or counter-rotating eigen-networks. It is given in the usual way by

$$\frac{1}{Q_L} = \sqrt{3}\left(\frac{\omega_+ - \omega_-}{\omega_0}\right) \qquad (14.7)$$

The first circulation condition in the adjustment of the circuit is met by adjusting the electrical details of the resonator and the second one is satisfied provided the two definitions of Q_L are reconciled. This condition fixes the relationship between the physical and magnetic variables of the resonator.

14.4 Struts, Susceptance Slope Parameter and Frequency

The Okada resonator is held in place by three struts in the manner illustrated in Figure 14.3. It's geometry and that of the struts must be such that it must either be capable of removing the heat from the ferrites in contact with either of its surfaces or permit cooling pipes to be inserted in its body. The perturbation of the susceptance slope parameter and the frequency of this sort of resonator due to the introduction of the struts are the main objects of this section.

The parameters entering into the description of the resonator are the filling factor (L/H) of the original or derived circuit, the aspect ratio of each original ferrite disk (R/L), the aspect ratio of the original resonator (R/H) defined by the half space obtained by introducing an electric wall between the top and bottom walls of the resonator, the number of chambers (n) and its radial wavenumber ($k_{eff}R_{eff}$). The detail of each strut is separately fixed by its normalized thickness (s/b).

The filling factor (k) of this type of resonator may be either varied by altering H while keeping L constant or by altering L and maintaining H constant. If the former possibility is adopted then the susceptance slope parameter is unaffected; if the latter one is employed then it is altered by any such an adjustment. In the work dealt with here L has been kept constant while H has been adjusted. This means that each piece of data must separately identify the filling factor and the aspect ratio of the half space defined by the symmetry plane between the top and bottom walls of the original resonator.

The relationship between the susceptance slope parameter of this sort of circulator and the geometry of the struts in WR75 waveguide is summarized in Figure 14.4 for three different values of the filling factor.

Since the gyrotropy required to establish a gyrator conductance of unity is dependent upon susceptance slope parameter of the circuit, each such curve is associated with a different direct magnetic field (H_0). This parameter is best represented by defining the quantity

$$\frac{\mu_0 H_0}{M_0}$$

The radius and thickness of the original resonator employed in obtaining this result was 3.90 mm and 1.0 mm respectively. The actual thickness of each ferrite in the 2-chamber arrangement is therefore 0.50 mm. The saturation magnetization

14.4 STRUTS, SUSCEPTANCE SLOPE PARAMETER AND FREQUENCY 285

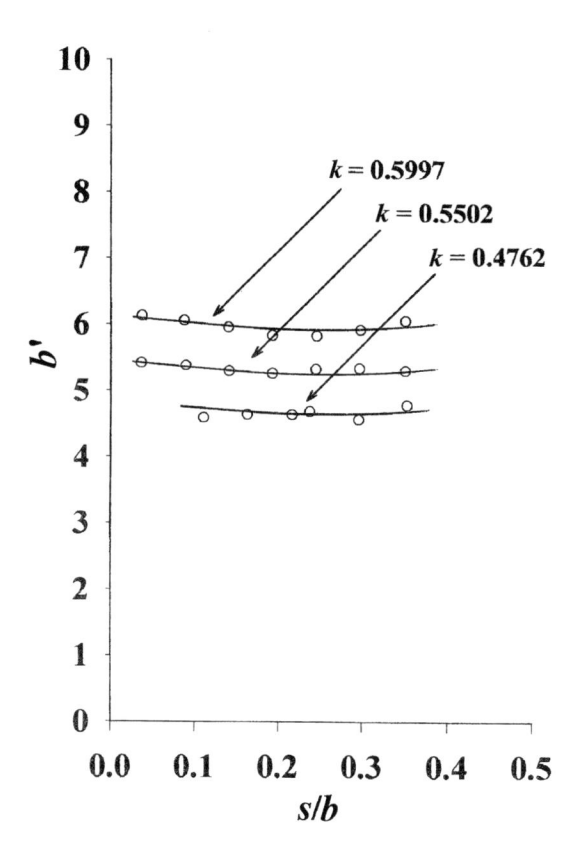

Figure 14.4 Susceptance slope parameter versus geometry of strut WR75 waveguide for parametric values of filling factor

of the material was 0.3000 T. Its dielectric constant was 15.3. The initial spacing between the top and bottom plates of each chamber was chosen on the basis of experience and the filling factor was modified by altering the gap of each chamber.

Figure 14.5 indicates the dependence of the frequency on the same physical variables. The fact that both the frequency and the susceptance slope parameter of this sort of device are independent of the details of the struts is of special note.

The discrepancy in the radian frequency between theory and experiment may be represented by defining an effective radius in addition to effective constitutive parameters. This gives

$$k_{\text{eff}} R_{\text{eff}} = 1.84 \tag{14.8}$$

where

$$k_{\text{eff}} = k_0 \sqrt{\varepsilon_{\text{f}}(\text{eff}) \, \mu_{\text{e}}(\text{eff})} \tag{14.9}$$

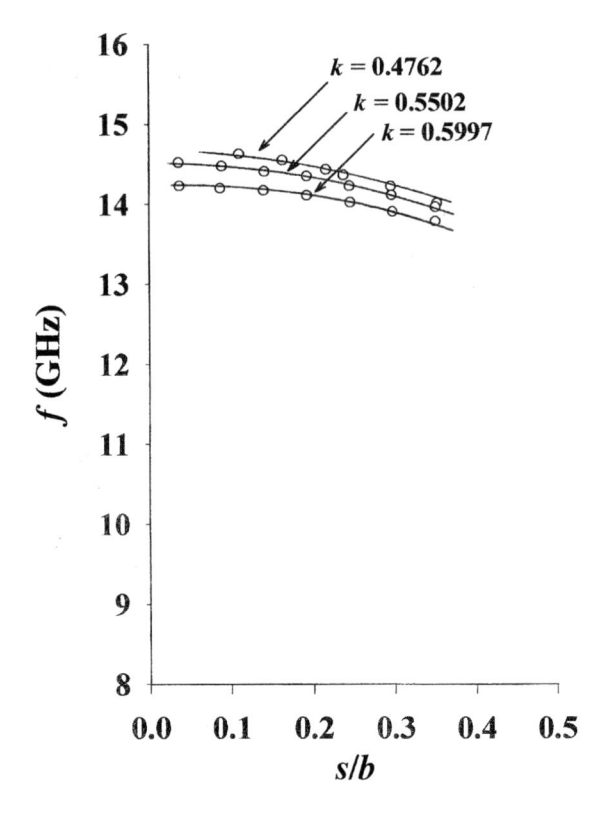

Figure 14.5 Frequency versus geometry of strut in WR75 waveguide for parametric values of filling factor

ε(eff) and μ(eff) have the meanings introduced in connection with the Okada resonator in Chapter 6

Figure 14.6 depicts the relationship between R_{eff}/R and the filling factor k of a 2-chamber Okada resonator in WR75 waveguide. The effective radius obtained from this data, with μ_e(eff) corresponding to $k = 0.55$ is in this instance given by

$$\frac{R_{eff}}{R} = 1.103$$

The performance of one experimental directly coupled circulator based on this data is illustrated in Figure 14.6.

The nominal physical parameters employed in this work are defined below in terms of the parameters of the original prototype.

$k_{eff} R_{eff}, k, R_{eff}/R, R/H, s/b$: 1.84, 0.476, 1,145, 1.76, 0.236

$k_{eff} R_{eff}, k, R_{eff}/R, R/H, s/b$: 1.84, 0.550, 1.103, 2.03, 0.243

$k_{eff} R_{eff}, k, R_{eff}/R, R/H, s/b$: 1.84, 0.600, 1.053, 2.22, 0.245

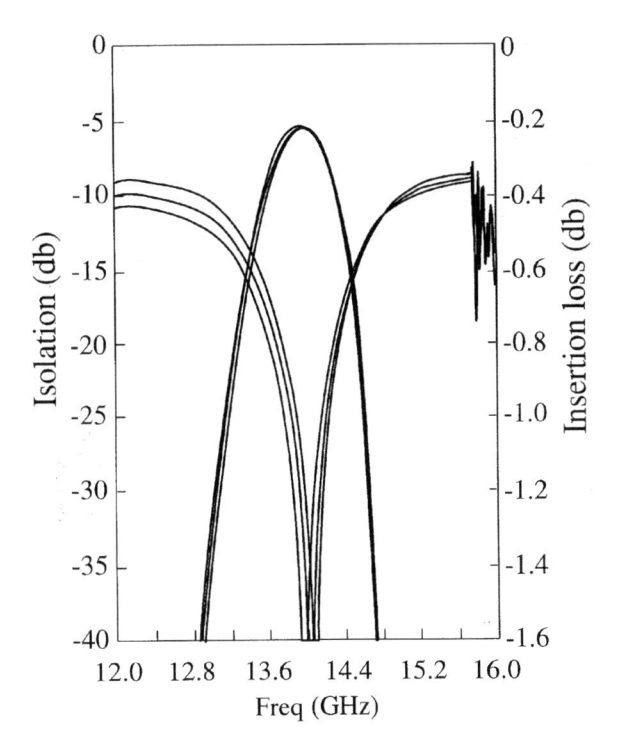

14.4 STRUTS, SUSCEPTANCE SLOPE PARAMETER AND FREQUENCY 287

Figure 14.6 Frequency response at ports 1, 2 and 3 of degree = 1 junction circulator using a 2-chamber Okada resonator

The corresponding material parameters are

$$\varepsilon_f, \varepsilon_{eff}, p, \mu_0 H_0/M_0 : 15.3, 1.880, 0.600, 0.400$$

$$\varepsilon_f, \varepsilon_{eff}, p, \mu_0 H_0/M_0 : 15.3, 2.059, 0.600$$

$$\varepsilon_f, \varepsilon_{eff}, p, \mu_0 H_0/M_0 : 15.3, 2.273, 0.600$$

The complex gyrator is defined by

$$b', Q_L, g; 4.86, 4.86, 1$$

$$b', Q_L, g; -, -, 1$$

$$b', Q_L, g; -, -, 1$$

p is the normalized magnetization.

$$p = \frac{\gamma M_0}{\mu_0 \omega_0}$$

M_0 is the direct saturation magnetization (T) and H_0 the direct magnetic field. (A/m), $\gamma = 2.21 \times 10^5$ (rad/s)/(A/m), $\mu_0 = 4\pi \times 10^{-7}$ (H/m.)

14.5 Quality Q-Factor

A complete experimental description of any junction circulator, as is now well understood, also requires a statement of its loaded Q-factor. This quantity is defined in the usual way by equation (14.7). It may, in practice, be experimentally deduced by either directly measuring the split frequencies of a loosely coupled constituent resonator or by determining the two frequencies at which the return loss of a directly coupled junction displays values of 9.50 dB on either side of the midband frequency of the resonator. It may also be obtained by recognizing that the parameters b' and Q_L have the same values at $g = 1$. The value obtained from a measurement of the 9.50 dB frequencies is $Q_L = 4.86$. This value is in good agreement with a separate measurement of Q_L. Figure 14.7 indicates the relationship between the quality factor and the normalized direct magnetic field.

If temperature stability is paramount then the minimum realisable bound on the quality factor may for engineering purposes be taken at the value at which the direct magnetic field is related to the magnetization by

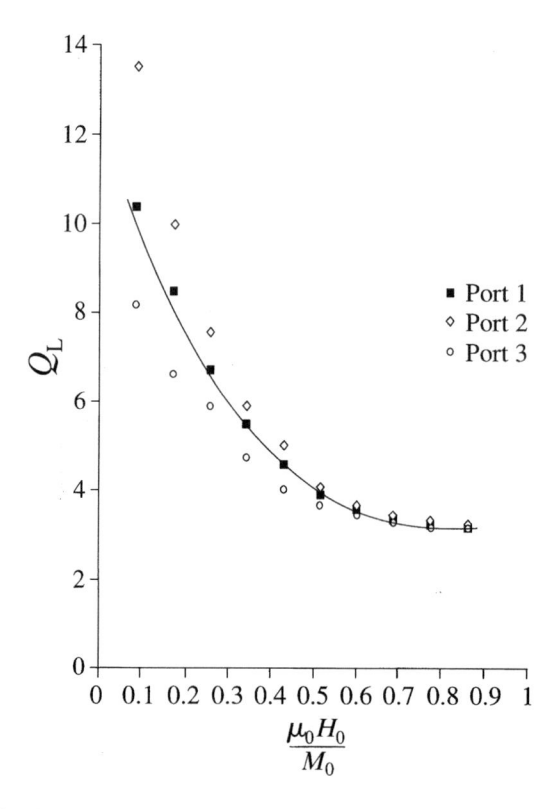

Figure 14.7 Quality factor of 2-chamber Okada resonator in WR75 waveguide ($k = 0.476$)

$$\frac{\mu_0 H_0}{M_0} = 0.70$$

While the lower value employed here has not been specifically established at this time, it may be estimated by assuming that the relationship between $1/Q_L$ and $\mu_0 H_0/M_0$ is linear over the interval below

$$0 \le \frac{\mu_0 H_0}{M_0} \le 0.70$$

If this assumption is adopted, then the value of the quality factor for a resonator with a gap factor (k) equal to 0.476 (say) is $Q_L = 4.86$ and the realizable lower bound for the same resonator is $Q_L \approx 2.77$.

The actual value in Figure 14.7 at a direct magnetic field $\mu_0 H_0/M_0$ equal to 0.70 is 3.25.

14.6 High Power Rating

The main attraction of the Okada resonator is, of course, its average power rating. This section summarizes some measurements on one 14.05 GHz device in WR75 waveguide using a 1800 W c.w. TWT source. Figure 14.8 depicts the variation of the return loss of one directly coupled device with convection cooling only as the

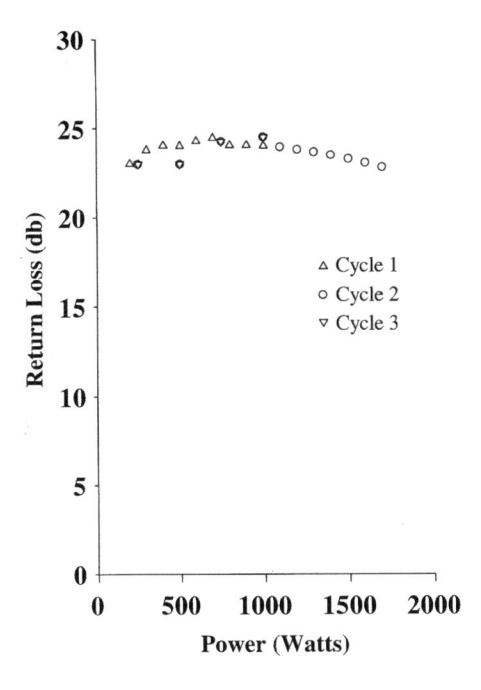

Figure 14.8 R.L. versus mean power of 14 GHz circulator

output power of the TWT tube was varied between 100 and 1800 W c.w. It suggests that the mean power rating of the device is adequate at 1800 W c.w. The milliwatt insertion loss of the device is typically 0.10 dB.

14.7 Half-Wave Filter Prototype

One way to broadband a junction circulator using an Okada resonator is to embody its complex gyrator circuit into a degree-2 filter network. One possible topology is a ladder network made up of half-wave long U.E.'s in cascade with a series STUB-R load. This circuit is indicated in Figure 14.9. Its synthesis is dealt with in Chapter 23. The required nature of the gyrator circuit is established by adopting a typical characteristic plane of the junction as the terminal of the problem region. Such a plane coincides with the position of a short-circuit piston at one port for which the other two ports are decoupled. In an H-plane junction symmetrically loaded by a planar resonator, one characteristic plane is located approximately a quarter-wave from the terminals of the gyromagnetic resonator. Additional planes are displayed from it by half-wave long waveguide sections. The result obtained here is $l = 0.22 \lambda g$. It may be separately demonstrated that this sort of transformation leaves the gain-bandwidth of the load unchanged.

One possible element which exhibits a parallel resonator with a suitable value of loaded Q-factor is a half-wave long section of ridge waveguide. The electrical details of this arrangement are summarized in Chapter 11. Figure 14.10 gives a schematic diagram of one structure.

An important practical difference between the quarter-wave and half-wave structures is that the gain-bandwidth of the gyrator circuit of the latter topology is fixed by the lowest realizable value of the susceptance slope parameter of the junction rather than by the gyrotropy of the resonator. The relationship between the susceptance slope parameter and the gain-bandwidth is in this instance given by $(VSWR)Q_L = b'$.

The optimum value is obtained by either fully extending the resonator between the top and bottom walls of the waveguide or by recessing the junction into the

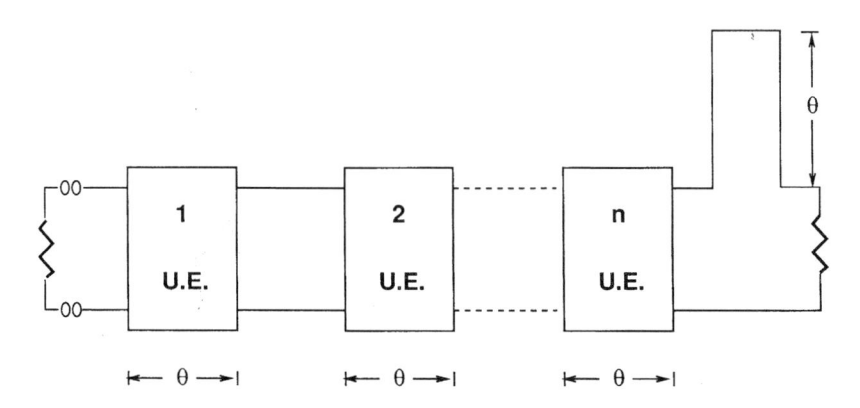

Figure 14.9 Topology of n half-wave long U.E.'s terminated in shunt STUB-G load

14.7 HALF-WAVE FILTER PROTOTYPE

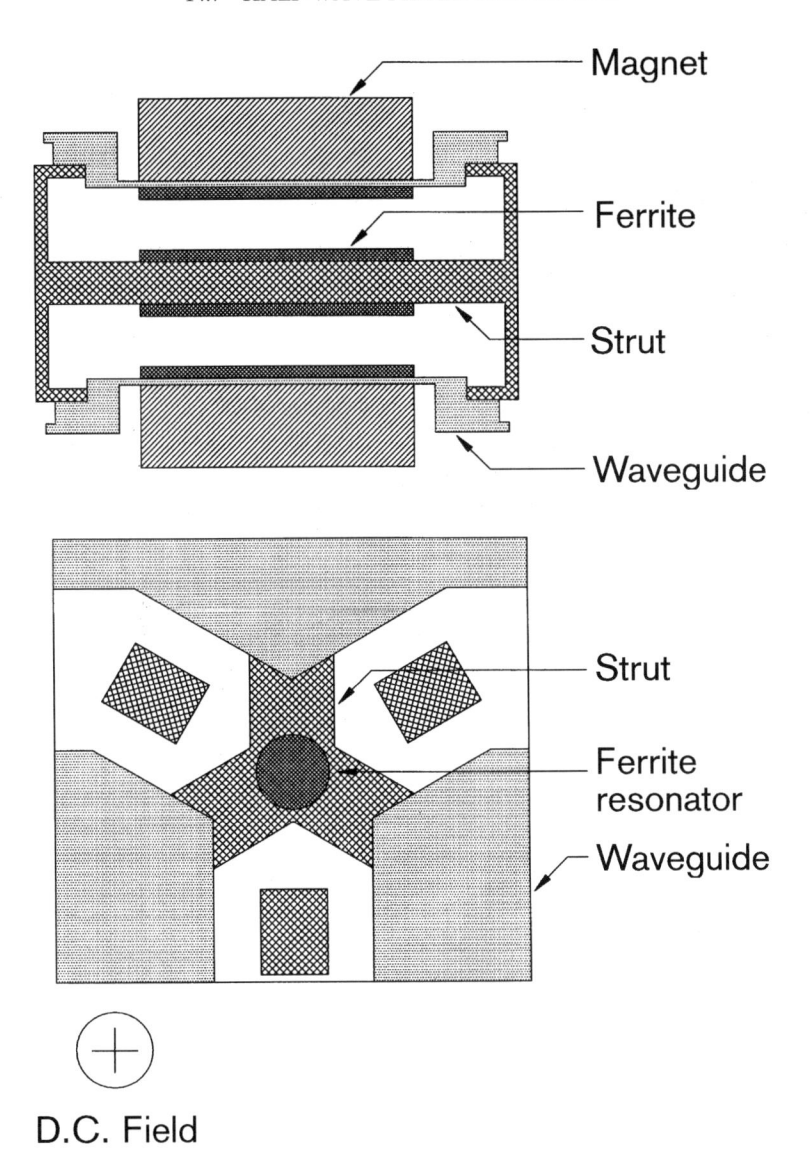

Figure 14.10 Schematic diagram of half-wave coupled Okada resonator

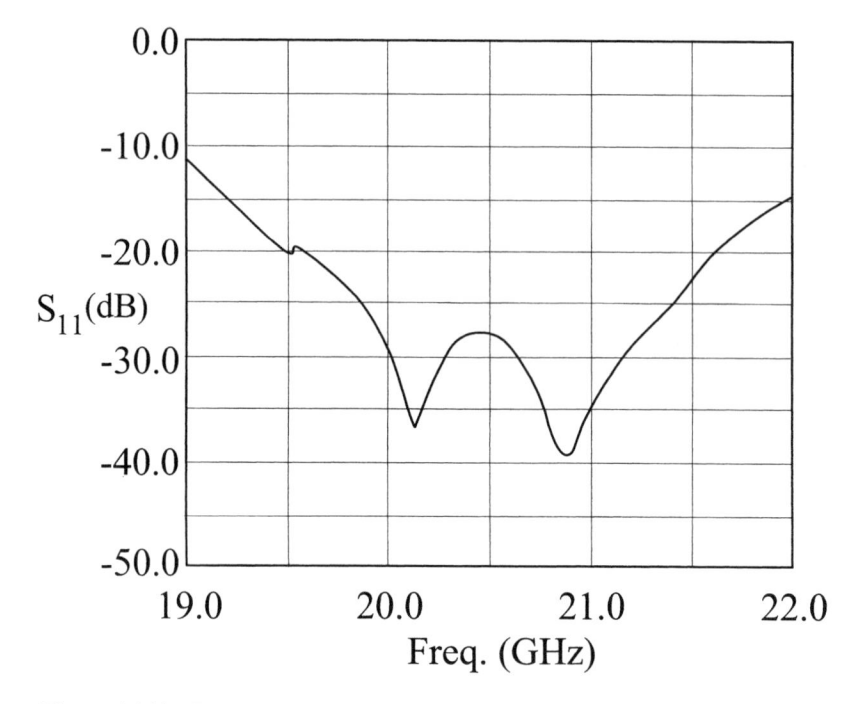

Figure 14.11 Frequency response of half-wave coupled Okada resonator

top and bottom walls of the waveguide. The value of the susceptance slope parameter realized here is $b' = 3.5$. This value contrasts with that of 12 used in the design of the quarter-wave coupled device to be described next.

Figure 14.11 indicates the frequency response of one circulator assembly at 20 GHz in WR42 waveguide. Its specification is defined by: R.L. = 26 dB; I.L. = 0.15 dB; B.W. = 7.0%.

This junction has been evaluated at a mean power of 300 W.

14.8 Quarter-Wave Coupled Device

While the half-wave filter topology has much to recommend it for the design of high frequency devices with high peak and average power ratings the quarter-wave one is still perhaps the preferred geometry at lower frequencies and at large mean power levels. The network topology of this arrangement is illustrated in Figure 14.12. The purpose of this section is to describe one practical version at 6.0 GHz which was largely designed on the basis of experience. Figure 14.13 shows its physical structure. The main consideration in the design of this type of device is the need to achieve some compromise between thermal capacity (robustness of struts), transformer impedance (separation between gap of ridges) and susceptance slope parameter (details of resonator geometry). Figure 14.14 indicates the

14.8 QUARTER-WAVE COUPLED DEVICE

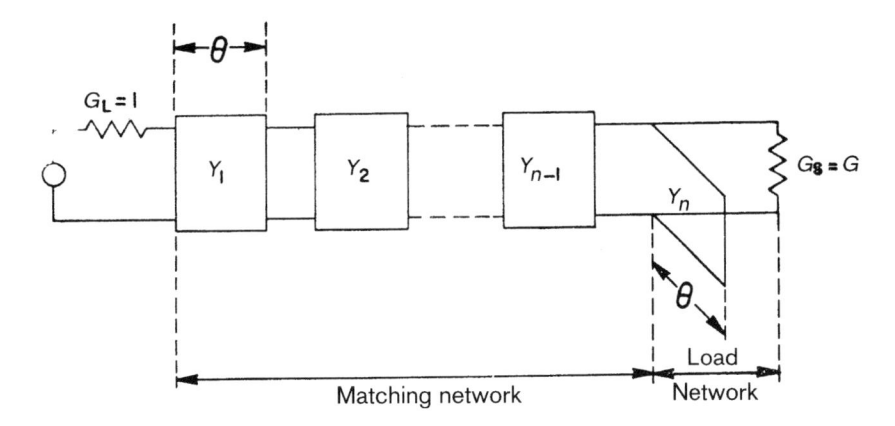

Figure 14.12 Topology of n quarter-wave long U.E.'s terminated in series STUB-R load

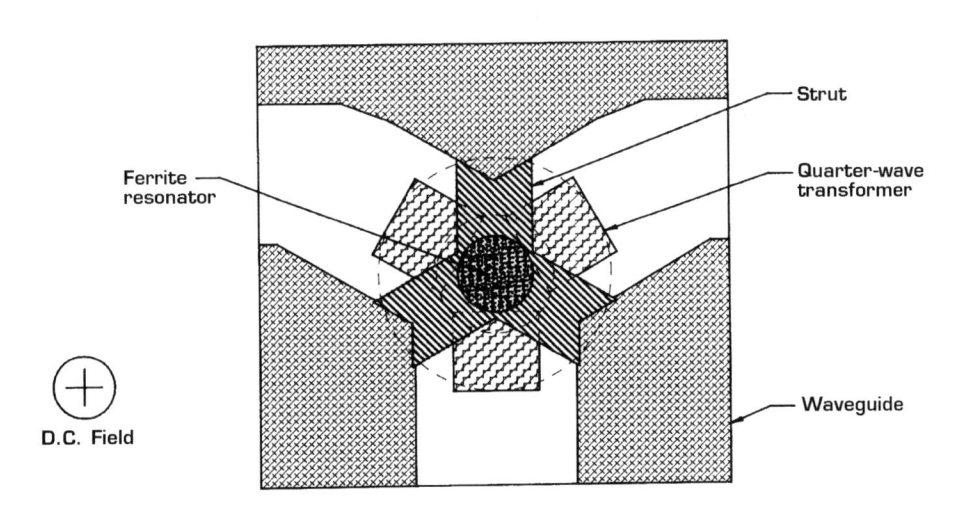

Figure 14.13 Schematic diagram of quarter-wave coupled Okada resonator

frequency response of one quarter-wave coupled device. Its insertion loss is separately superimposed on the illustrations. Its specification may be summarised by: R.L. = 26 dB; I.L. = 0.15 dB; B.W. = 20%.

The normalized susceptance slope parameter b' of the directly coupled 2-chamber Okada resonator used to obtain this data is $b' \approx 11.90$.

The detailed description of the adjustment of its resonator geometry is outside the remit of this week.

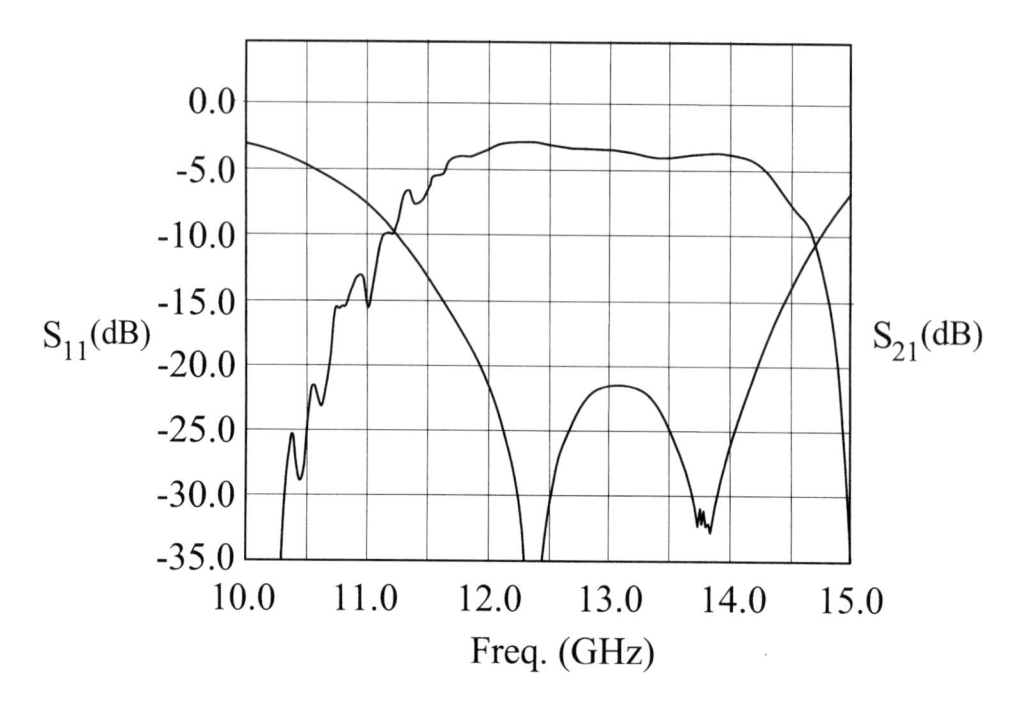

Figure 14.14 Frequency response of junction circulator using quarter-wave coupled 2-chamber Okada resonator

References and Further Reading

Buffler, C. and Helszajn, J. (1968) The use of composite junctions in the design of high-power ciculators *IEEE MT. Microwave Symp., Microwave Theory Tech.*, **20**, May.

Damon, R. W., (1953) Relaxation effects in ferromagnetic resonance, *Rev. Mod. Phys.*, **25**, 239–45.

Hauth, W., Lenz, S. and Pivit, E. (1986) Design and realization of very-high-power waveguide junction circulators, *Frequency*, 2–11.

Helszajn, J. (1967) An H-plane high-power TEM ferrite junction circulator, *Radio and Electron. Eng.*, **33**, 257–261.

Helszajn, J. (1969) Frequency and bandwidth and H-plane TEM junction circulator, *Proc. IEE*, **117**(7).

Helszajn, J. (1973) Microwave measurement techniques for below-resonance junction circulators, *IEEE Trans. Microwave Theory Tech.*, **MTT-19**, 347–351.

Helszajn, J. (1981) High-power waveguide circulators using quarter-wave long composite ferrite/dielectric resonators, *IEE Proc. H, Microwaves, Opt & Antennas*, **128**(5), 268–273.

Kajfez, D. and Wheless, J. R. (1986) Invariant definition of unloaded Q-factor, *IEEE Trans. Microwave Theory Tech.*, **MTT-37**, 840–841.

Konishi, Y. (1969) A high power UHF circulator, *IEEE Trans. Microwave Theory Tech.*, **MTT-15**, 700–708.

Levy, R. and Helszajn, J. (1982) Specific equations for one and two section quarter-wave matching networks for stub-resistor loads, *IEEE Trans. Microwave Theory Tech.*, **MTT-30**, 55–62.

Levy, R. and Helszajn, J. (1993) Synthesis of series STUB-R complex gyrator circuits using halfwave long impedance transformers, *IEE Proc. Part H*, **140**, 426–432.

Okada, F. and Ohwit, K. (1975) The development of a high power microwave circulator for use in breaking of concrete and rock, *J. Microwave Power*, **10**(2).

Okada, F., Ohwit, K., Mori, M. and Yasude, M. (1977) A 100 kW waveguide Y-junction circulator for microwave power systems at 915 MHz, *J. Microwave Power*, **12**(3).

Okada, F., Ohwit, K., and Yasude, M. (1978) Design procedure of high power Y-junction waveguide circulators, *IMPI*, 118–120.

Okada, F. and Ohwi, K. (1978) Design of a high power CW Y-junction waveguide circulator, *IEEE Trans. Microwave Theory Tech.*, **MTT-26**, 364–369.

Suhl, H., (1956) The non-linear behaviour of ferrites at high signal level, *Proc. Inst. Radio Engineers*, **44**, 1270.

15

AN EVANESCENT MODE OKADA JUNCTION CIRCULATOR

15.1 Introduction

One means of reducing the size and weight of microwave components is to use evanescent mode practice. The development of filters based on this type of topology is now well established. Two evanescent mode circulators employing cut-off waveguide ports in conjunction with either propagating or evanescent mode resonators have been separately described. The geometry considered here consists of a 2-chamber Okada resonator on the axis of an evanescent mode junction comprising three waveguides symmetrically loaded by thick metal vanes or spokes. It may be described by the ratio of the radius of the gyromagnetic resonator (R_i) and that of the inscribed radius of the cavity formed by the metal vanes (R_o). The physical variables of the Okada resonator such as its filling factor, its aspect ratio and the thickness of the struts are defined in more detail elsewhere in the text. A property of note of this type of resonator is of course its C.W. power rating. The purpose of this chapter is to investigate the complex gyrator circuit of one evanescent arrangement. A detailed investigation of its geometry indicates that its quality factor deteriorates somewhat when the ratio of the radii used to define the cut-off region approaches unity. The spurious modes which can on occasion intrude into the pass-band of the device is also given some attention. The range of R_i/R_o investigated in this work is bracketed between about 0.50 and 0.90.

15.2 The Evanescent Mode Okada Junction

The resonator employed in this work consists of the junction of three waveguides symmetrically loaded by thick metal vanes. Its arrangement is illustrated in Figure 15.1. While the resonator employed in this work is an Okada one, the development outlined here is not restricted to it. Figure 15.2 depicts the side view

15 AN EVANESCENT MODE OKADA JUNCTION CIRCULATOR

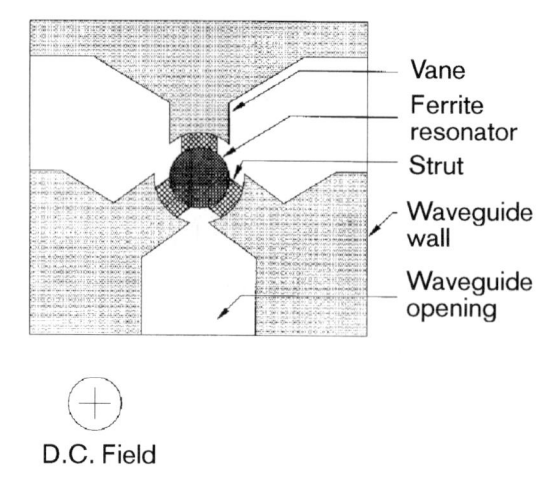

Figure 15.1 Schematic diagram of evanescent waveguide junction

of the structure under consideration. The openings of the input waveguides of this configuration may be readily reduced to form cut-off lines which may be incorporated into evanescent-mode filter sections if so desired. Its physical variables may be described by the radius of the resonator (R_i), that of cavity formed by the three spokes (R_o) and by the width of a typical spoke (W).

The details of the actual Okada resonator is described in the usual way by its filling factor $k = L/H$; the aspect ratio of each of its ferrite or garnet elements R/L and the ratio of the thickness of its struts to that of the height of the waveguide S/b. The ferrite or garnet material is separately specified by its relative dielectric constant (ε_f) and its saturation magnetization (M_0).

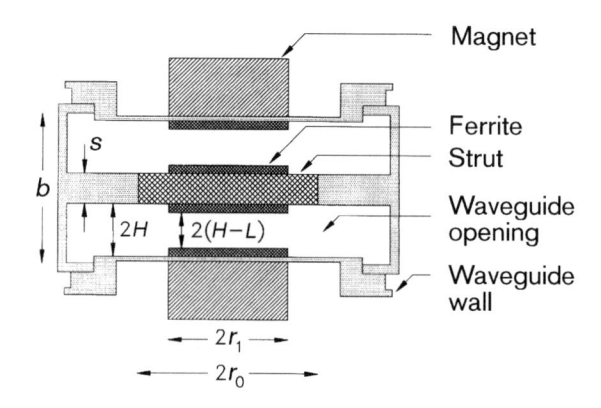

Figure 15.2 Schematic diagram of 2-chamber Okada resonator

15.3 Mode Spectrum of Evanescent Okada Junction

An understanding of the mode chart of any gyromagnetic resonator for use in the construction of a junction circulator is essential for design. The purpose of this section it to investigate that of an evanescent mode Okada junction. One means of doing so is to vary the details of the cavity formed by the three symmetric vanes of the junction and to observe the effect upon the magnetized mode spectrum. Figure 15.3 illustrates the mode chart of a conventional 2-chamber Okada resonator at the junction of three standart WR137 waveguides versus the normalised direct magnetic field. Figures 15.4 and 15.5 depict the split frequencies of two evanescent assemblies in the same waveguide. A scrutiny of these results suggest that both the opening between the split frequencies and the midband frequency are dependent upon the quantity R_i/R_o.

The origin of the two split pairs of degenerate modes has already been discussed in Chapter 6. It may be put down to small differences in the mechanical symmetry of the two chambers of the resonator. This feature restricts somewhat the band-

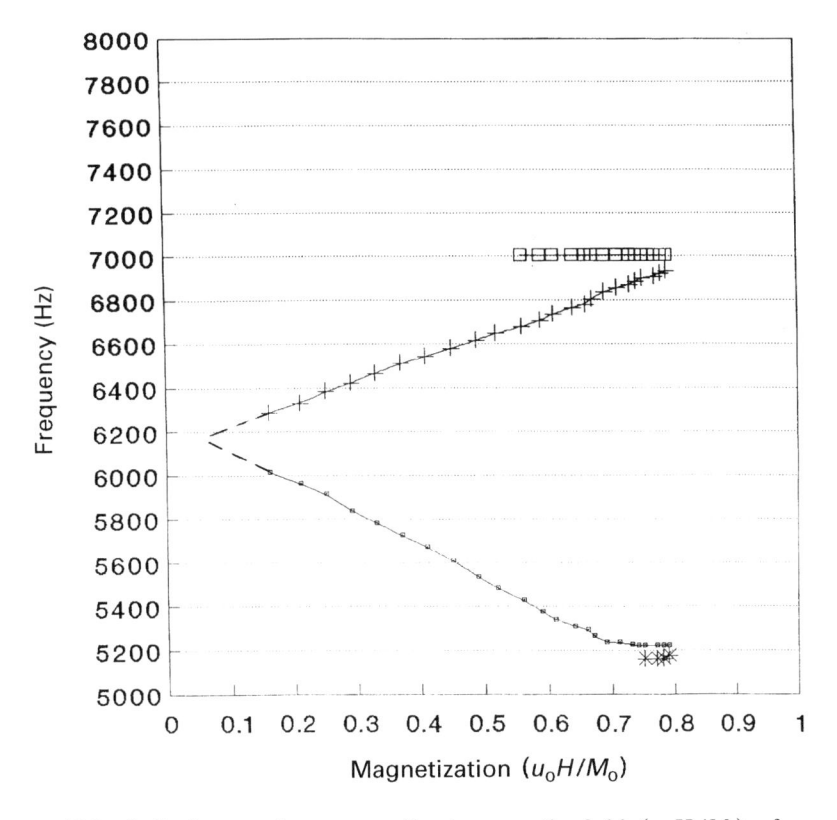

Figure 15.3 Split frequencies versus direct magnetic field ($\mu_0 H_0/M_0$) of conventional 2-chamber Okada resonator at junction of three H-plane waveguides (R_i = 0.389″, R_i/R_0 = 0.48) (reprinted with permission, see Helszajn and Tsounis (1998)

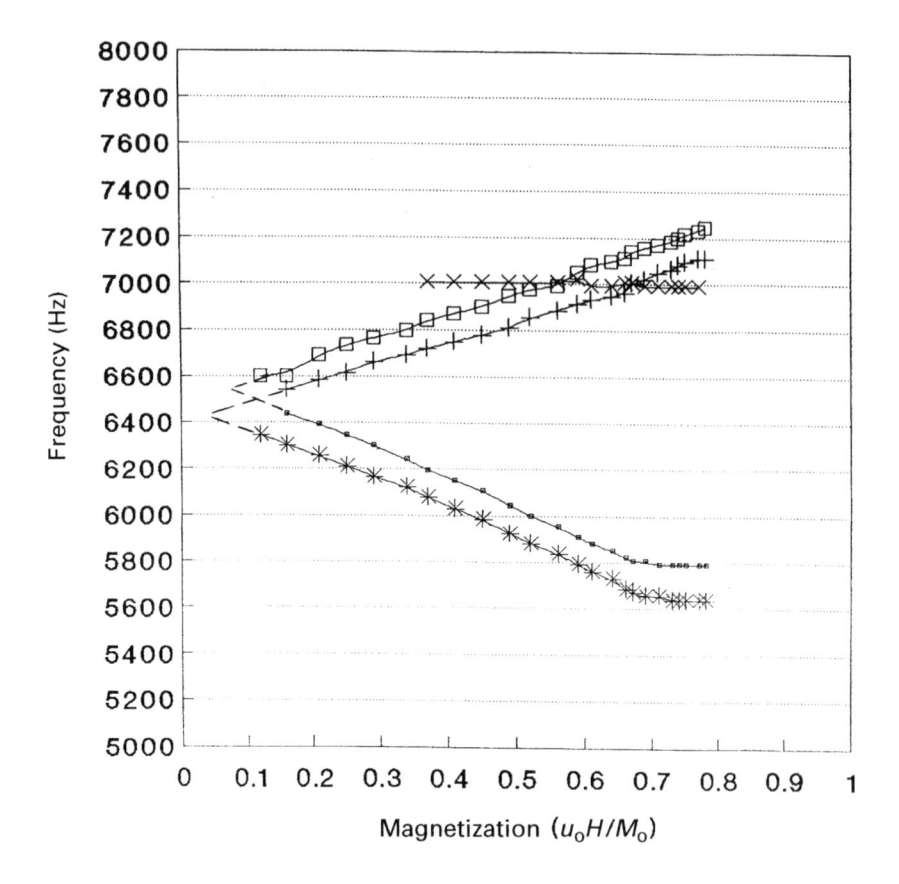

Figure 15.4 Split frequencies versus direct magnetic field ($\mu_0 H_0/M_0$) of evanescent 2-chamber Okada circulator ($R_i = 0.389''$, $R_i/R_0 = 0.82$) (reprinted with permission, see Fig. 15.3)

width of the device to that defined by the upper branch of the lower pair of split frequencies and the lower branch of the upper pair of split frequencies but has otherwise no other obvious bearing on the other parameters which enter into its description.

The existence of spurious modes in the pass-band of many a microwave circuit is often a problem and the Okada resonator is unfortunately not exempt from this shortcoming. The locations of the spurious modes for the geometries under consideration are separately superimposed on these two illustrations. Scrutiny of these results suggests that varying R_i/R_o provides one albeit modest means of perturbing the positions of the undesired modes.

The filling factor (k) employed in obtaining this data is 0.67 and the radius of the ferrite resonator is $R_i=0.389''$. The radius of the cavity formed by the junction of the three standart rectangular waveguide is 0.812''. The direct magnetization of the ferrite material (M_0) is 0.1600 T and it has a dielectric constant (ε_f) of 15.2.

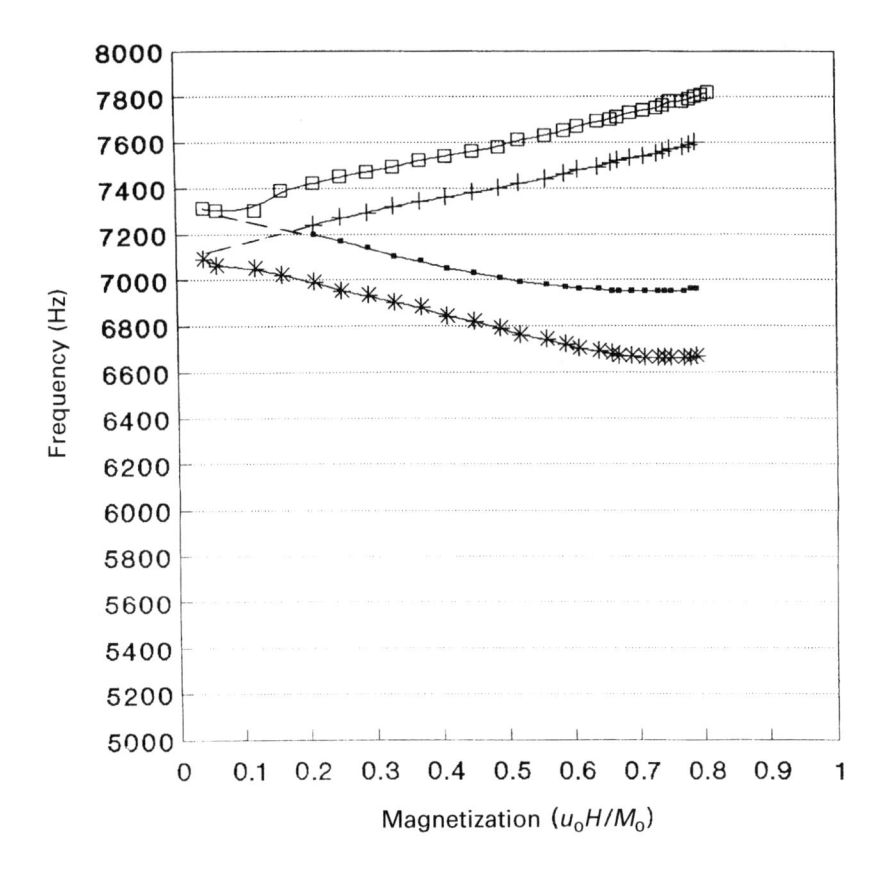

Figure 15.5 Split frequencies versus direct magnetic field ($\mu_0 H_0 / M_0$) of evanescent 2-chamber Okada circulator ($R_i = 0.389''$, $R_i/R_0 = 0.90$) (reprinted with permission, see Fig. 15.3)

15.4 Gyrator Circuit of Evanescent Mode Okada Junction

One important quantity that enters into the description of the complex gyrator of any weakly magnetized junction circulator is its quality factor as already noted. It defines the gain-bandwidth product of the device. This quantity is related to the split frequencies of the gyromagnetic resonator in the usual way by

$$\frac{1}{Q_L} = \sqrt{3}\left(\frac{\omega_+ - \omega_-}{\omega_0}\right) \tag{15.1}$$

Figure 15.6 separately illustrates the relationship between the quality factor and the direct magnetic field for parametric values of R_i/R_o. Scrutiny of this illustration indicates that R_i/R_o is restricted by the required quality factor of the junction. Values between 2 and 2½ are suitable for the design.

Another quantity that enters into the description of the complex gyrator circuit is its real part condition or gyrator conductance (g). It is related to the split frequencies of the gyromagnetic resonator and the normalized susceptance slope parameter (b') of the complex gyrator circuit by

$$g = \sqrt{3}\, b' \left(\frac{\omega_+ - \omega_-}{\omega_0} \right) \tag{15.2}$$

The normalized gyrator conductance is in this sort of device the dependent variable. It may be obtained by direct measurements or by separately measuring the susceptance slope parameter and split frequencies of the junction. One means of directly deducing this quantity is by determining the return loss or *VSWR* at one port with the other two terminated in matched loads and having recourse to the relationships below

$$g = \sqrt{2(VSWR) - 1}\,, \quad g \geq 1 \tag{15.3}$$

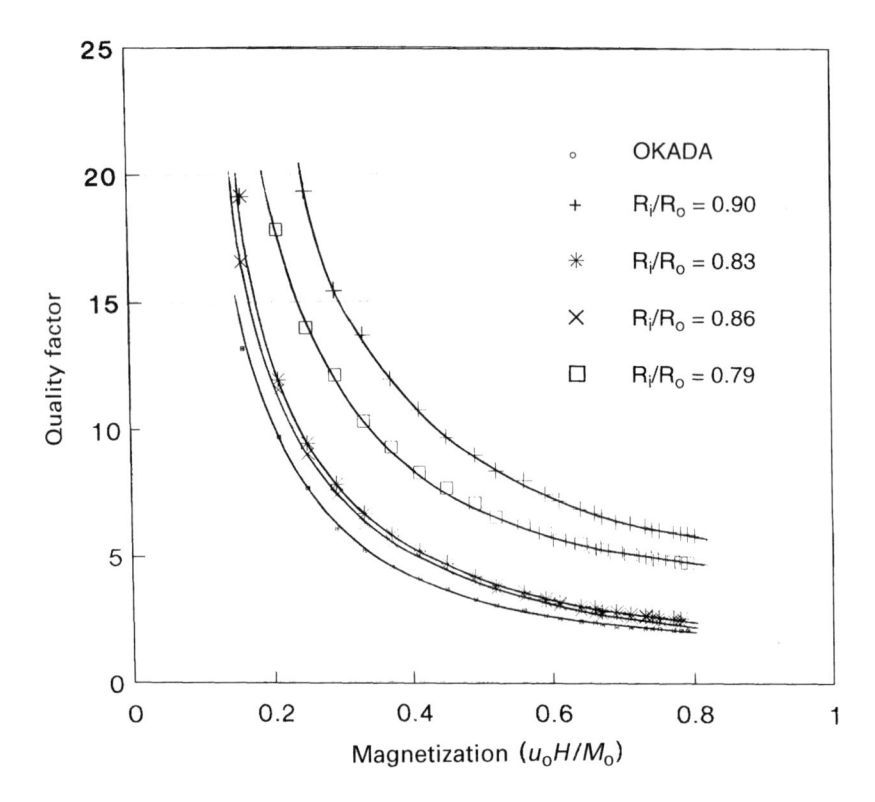

Figure 15.6 Quality factor (Q_L) versus direct magnetic field ($\mu_0 H_0 / M_0$) of evanescent 2-chamber Okada circulator for parametric values of R_i/R_o ($R_i = 0.389''$) (reprinted with permission, see Fig. 15.3.

15.4 GYRATOR CIRCUIT

or

$$g = \sqrt{\frac{2 - (VSWR)}{(VSWR)}}, \quad g \leq 1 \tag{15.4}$$

The third physical variable which enters into the description of the complex gyrator circuit of this sort of problem region, in addition to its quality factor and its gyrator conductance, is the susceptance slope parameter (b') of its tank circuit. One important aspects of the geometry under consideration is also therefore the relationship between the details of the spokes of the cavity region and the susceptance slope parameter of the junction. One means of extracting this quantity is by direct measurement of the frequency response of its return loss, another is by combining the Q-curves and g-curves, and still another is to recognize that the susceptance slope parameter and the quality factor have the same values when the gyrator conductance is equal to unity. The latter two conditions are obtained by having recourse to the standard relationship between these quantitites in a G-STUB load.

$$b' = Q_L g \tag{15.5}$$

Measurements on the gyrator conductance on this sort of circulator are not at this time available. It is, however, understood that it is to some extent fixed by the short evanescent waveguide sections connecting the three standard rectangular waveguides to the junction region. The full characterization of the gyrator circuit of this sort of junction must therefore await a choice of an evanescent mode filter or matching topology.

The midband frequency or radial wavenumber of this sort of resonator is also affected by its radial dimensions. One polynomial representation for this quantity over the field of variable investigated here is

$$k_o R_i = 77.74 - 463.50 \left(\frac{R_i}{R_o}\right) + 1030.95 \left(\frac{R_i}{R_o}\right)^2 - 1001.50 \left(\frac{R_i}{R_o}\right)^3$$

$$+ 359.97 \left(\frac{R_i}{R_o}\right)^4, \quad 0.45 \leq R_i/R_o \leq 0.85, \, R_i = 0.389 \tag{15.6}$$

The relationship between $k_o R_i$ and R_i/R_o specified by the approximation is restricted to the particular filling factor of the Okada resonator and to the dielectric constant and magnetization of the material utilized in the work.

The frequency response of the isolation of one directly coupled arrangement at a direct magnetic field of ($\mu_0 H_0/M_0$) equal to 0.088 is separately depicted in Figure 15.7. Its 20 dB bandwidth is consistent with the quality factor at the same direct magnetic field. The insertion loss is separately superimposed on the illustration for completeness sake.

While the unloaded junction employed here is cut-off it is perhaps necessary to stress that no significant cut-off phenomenon has been observed throughout its loaded arrangement. The description of the junction as an evanescent one is

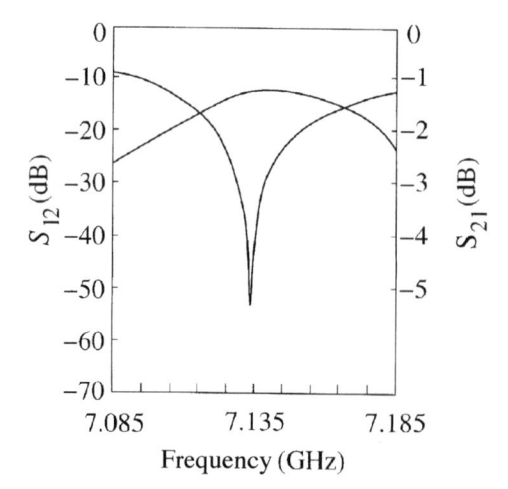

Figure 15.7 Frequency response of Okada junction (reprinted with permission, see Fig. 15.3)

therefore restricted to the understanding that its terminals are compatible with the use of evanescent filter sections.

15.5 Equivalent Circuits of Evanescent Mode Waveguide

As already remarked the evanescent 3-port junction circulator is admirably suited for the design of quarter-wave coupled circulators using evanescent mode filter or transformer sections. The purpose of this section is to deduce the equivalent circuit of this sort of waveguide. One possibility is the π-arrangement comprising three inductors depicted in Figure 15.8a; the other is the dual T-arrangement in Figure 15.8(b). The derivation of the π-topology starts by having recourse to the standard identities for the branch elements of the structure;

$$Y_1 = Y_2 = Y_0 \coth\left(\frac{\gamma l}{2}\right) \tag{15.7}$$

$$Y_3 = Y_0 \sinh(\gamma l) \tag{15.8}$$

Y_0 is the wave admittance of the waveguide, γ is its propagation constant and l is the length of a typical section.

It proceeds by noting that the propagation constant (γ) of a waveguide below cut-off is real for all frequencies from the origin to the cut-off frequency:

$$\gamma^2 = \left(\frac{2\pi}{\lambda_c}\right)^2 - \left(\frac{2\pi}{\lambda_0}\right)^2 \tag{15.9}$$

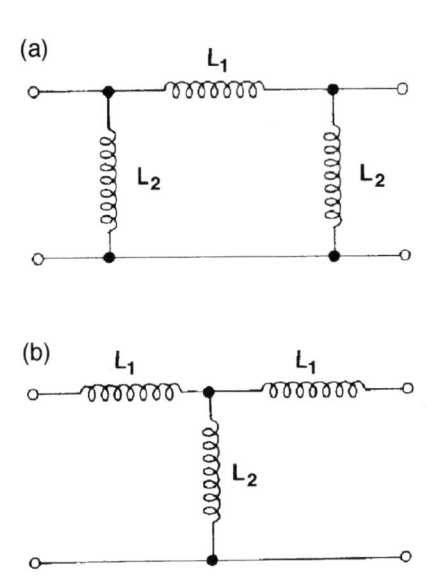

Figure 15.8 (a) π equivalent circuits of evanescent mode waveguide; (b) T equivalent circuits of evanescent mode waveguide

It is also recognized that the waveguide admittance (Y_0) is a pure imaginary quantity under the same circumstances:

$$Y_0 = \frac{\gamma}{-j\omega\mu_0} \qquad (15.10)$$

Introducing these relationships into the branch immittances in equations 7 and 8 of the π prototype indicates that each element is a nearly frequency-independent inductance as asserted. The derivation of the T-equivalent circuit proceeds in a dual manner.

15.6 Practical Immittance Inverters

One widely used concept in the area of filter synthesis is that of the immittance inverter. An impedance inverter K_{ij} is a frequency invariant 2-port network which transforms an impedance Z_j at one plane into an impedance Z_i at another one:

$$Z_i Z_j = K_{ij}^2 \qquad (15.11)$$

An admittance inverter J_{ij} similarly maps an admittance with one value Y_j into another with a value Y_i according to

$$Y_i Y_j = Y_{ij}^2 \qquad (15.12)$$

15 AN EVANESCENT MODE OKADA JUNCTION CIRCULATOR

Four widely used practical realizations are illustrated in Figure 15.9. Although each circuit requires negative elements for its realization these can be absorbed in adjacent positive elements. The equivalence between any of these circuits and an ideal impedance inverter consisting of a UE of characteristic impedance K will now be demonstrated. This may be done by establishing a one-to-one equivalence

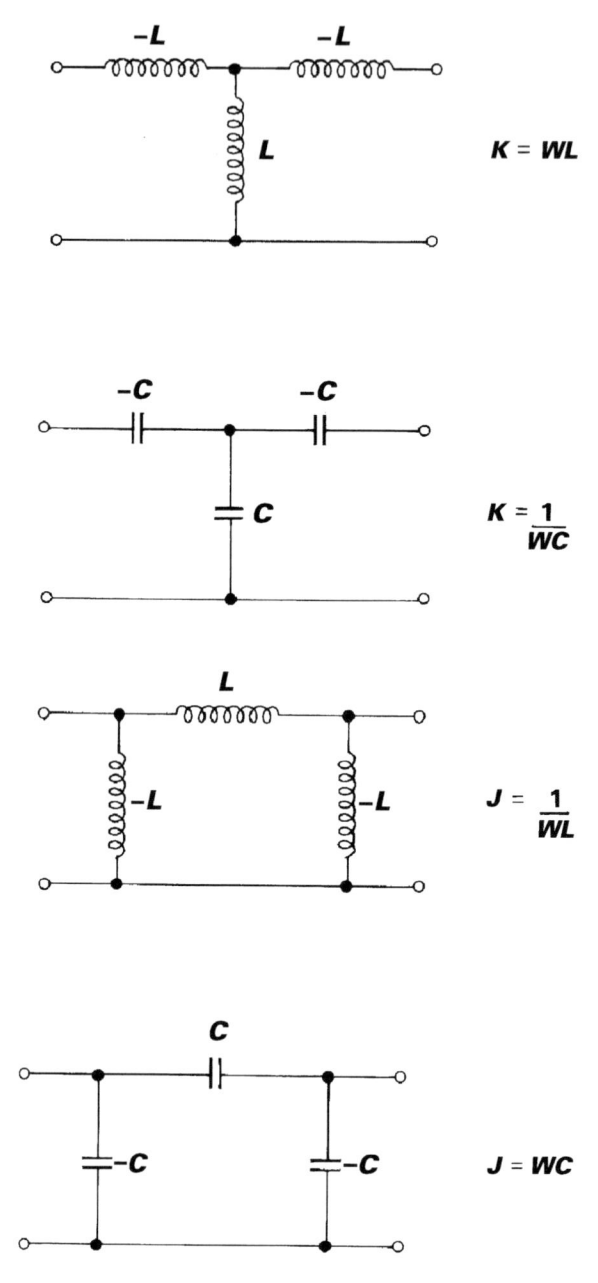

Figure 15.9 Schematic diagrams of immittance inverters using π and T-circuits

15.6 PRACTICAL IMMITTANCE INVERTERS

between the *ABCD* parameters of the two topologies. The derivation begins with the definition of the *ABCD* matrix of a uniform transmission line of characteristic impedance Z_0 and electric length θ.

$$\begin{bmatrix} A & B \\ C & D \end{bmatrix} = \begin{bmatrix} \cos\theta & jZ_0 \sin\theta \\ (j\sin\theta)/Z_0 & \cos\theta \end{bmatrix} \quad (15.13)$$

Evaluating this relationship at $\theta = 90°$ gives:

$$\begin{bmatrix} A & B \\ C & D \end{bmatrix} = \begin{bmatrix} 0 & jK \\ j/K & 0 \end{bmatrix} \quad (15.14)$$

where

$$K = Z_0 \quad (15.15)$$

The derivation continues by forming a one to one equivalence between such a U.E. element and any of the possible arrangements illustrated in Figure 15.9. Taking the T network consisting of series impedances (Z) and a shunt admittance (Y) by way of an example and recalling the overall *ABCD* matrix of such a network gives:

$$A = 1 + ZY \quad (15.16a)$$

$$B = Z(2 + ZY) \quad (15.16b)$$

$$C = Y \quad (15.16c)$$

$$D = 1 + ZY \quad (15.16d)$$

In the situation considered here

$$Z = \frac{1}{-j\omega C} \quad (15.17a)$$

$$Y = j\omega C \quad (15.17b)$$

Evaluating the *ABCD* parameters under these conditions indicates that

$$\begin{bmatrix} A & B \\ C & D \end{bmatrix} = \begin{bmatrix} 0 & j/\omega C \\ j\omega C & 0 \end{bmatrix} \quad (15.18)$$

A comparison between this matrix and that of the ideal impedance inverter suggests that the two are equivalent provided

$$K = \frac{1}{\omega C} \tag{15.19}$$

The equivalences between the topologies of the other possible immittance inverters and a quarter-wave long U.E. follows without ado.

15.7 Quarter-Wave Coupled Device

An outline procedure for the design of quarter-wave coupled circulators using evanescent waveguide sections may be articulated once the equivalent circuit of the waveguide and the notion of an ideal immittance inverter are established. The

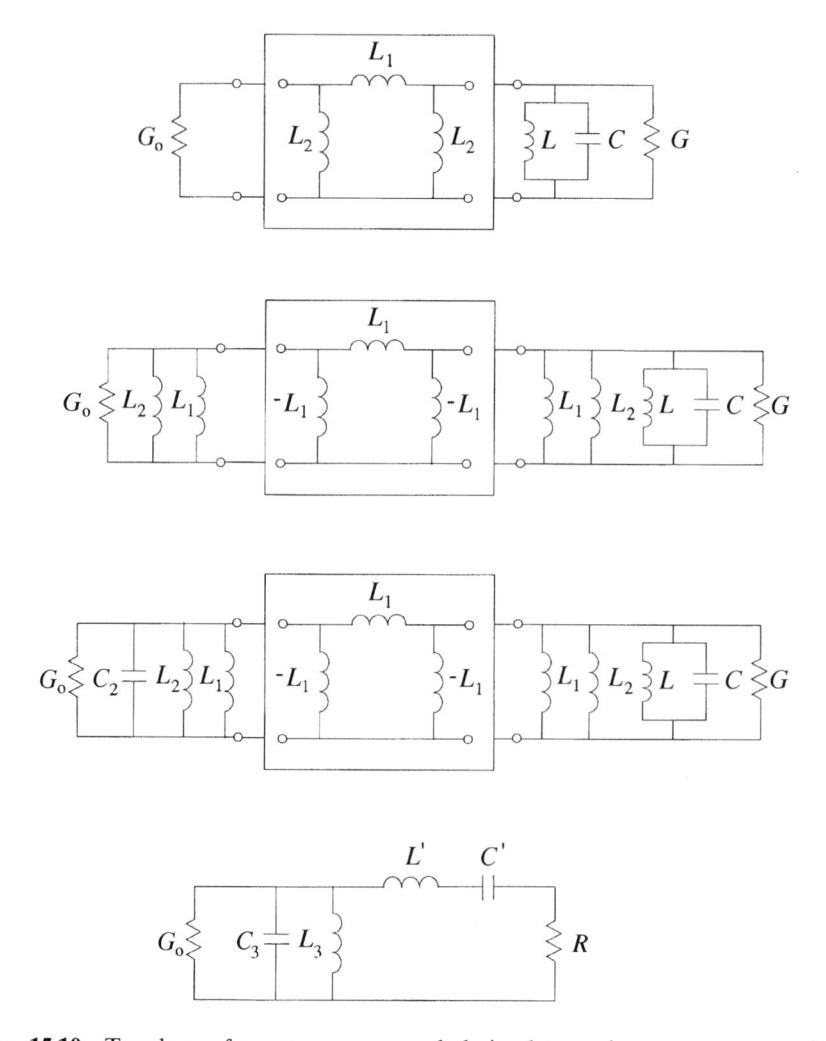

Figure 15.10 Topology of quarter-wave coupled circulator using evanescent mode waveguide

realization of the degree-2 problem is illustrated in Figure 15.10. It begins by of replacing the evanescent waveguide section by an ideal immittance inverter loaded by negative shunt inductances. It continues by symmetrically adding shunt capacitors at its input and output terminals as a preamble to replacing the quarter-wave coupled shunt LCG load by a series LCR one. The final topology is readily recognised as a degree-2 filter circuit. The detailed development of the network problem in terms of a network specification is outside the remit of this work.

References and Further Reading

D'Ambrosio, A. (1978) Slim-guide circulators in C, X, K and U bands, *Proc. Int. Microwave Symp.*, 105.

Craven, G. and Mok, C. K. (1971) The design of evanescent mode waveguide bandpass filters for a prescribed insertion loss, *IEEE Trans. Microwave Theory Tech.*, **MTT-19**(3), 295.

Hauth, W., Lenz, S. and Pivit, E. (1986) Design and realization of very-high-power waveguide junction circulators, *Frequency*, **40**, 2–11.

Helszajn, J. (1967) An H-plane high-power TEM ferrite junction circulator, *Radio and Electron. Eng.*, **33**, 257–261.

Helszajn, J. (1969) Frequency and bandwidth and H-plane TEM junction circulator, *Proc. IEE*, **117**(7), 1235–1238.

Helszajn, J. (1973) Microwave measurement techniques for below resonance junction circulators, *IEEE Trans. Microwave Theory Tech.*, **MTT-29**, 347–351.

Helszajn, J., Guixa, R., Girones, J., Hoyos, E. and Garcia-Taheno, J. (1995) Adjustment of Okada resonator using composite chambers, *IEEE Trans. Microwave Theory Tech.*, **MTT-43**.

Helszajn, J. and Tsounis, V. (1998) Quality factor of an Evanescent Mode Okada Resonator, *IEE Proc., Microwave Propagation and Antennas*, **144**.

Konishi, Y. (1969) A high power UHF circulator, *IEEE Trans. Microwave Theory Tech.*, **MTT-15**, 700–708.

Okada, F. and Ohwit, K. (1975) The development of a high power microwave circulator for use in breaking of concrete and rock, *J. Microwave Power*, **10**(2).

Okada, F., Ohwit, K., Mori, M. and Yasude, M. (1977) A 100 kW waveguide Y-junction circulator for microwave power systems at 915 MHz, *J. Microwave Power*, **12**(3).

Okada, F., Ohwit, K. and Yasude, M. (1978) Design proceedure of high power CW Y-junction waveguide circulators, *IMPI*, 118–120.

Okada, F. and Ohwi, K. (1978) Design of a high power CW Y-junction waveguide circulator, *IEEE Trans. Microwave Theory Tech.*, **MTT-26**, 364–369.

Schieblich, C. and Schünemann, K., (1981) Circulators with evanscent mode resonators, *IEEE Proc. 11th European Microwave Conf.*, 394.

16

COMPLEX GYRATOR CIRCUIT OF AN H-PLANE JUNCTION CIRCULATOR USING E-PLANE TURNSTILE RESONATORS

16.1 Introduction

The most common waveguide circulator met in practice is the H-plane one consisting of a 3-port H-plane junction symmetrically loaded by E-plane open gyromagnetic cavity resonators. Figure 16.1 depicts one practical geometry. It can be visualized as a 7-port network consisting of a 3-port H-plane junction symmetrically coupled along its axis to cylindrical gyromagnetic waveguides each supporting 2-orthogonal ports. The network is reduced to a 3-port one by closing the gyromagnetic waveguides by short-circuit pistons. This sort of circulator has its origin in the turnstile structure indicated in Figure 16.2. The duality between the two arrangements in Figures 16.1 and 16.2 may be achieved by replacing the electric wall in the circular waveguides of the original turnstile junction by open magnetic ones. The first circulation condition in this sort of circulator coincides with that for which the in-phase and degenerate counter-rotating eigenvalues are in anti-phase on the eigenvalue diagram. The second one is established by splitting the degeneracy between the counter-rotating ones by a suitable direct magnetic field. The purpose of this chapter is to outline one solution which relies on a weakly magnetized resonator for the description of its 1-port complex gyrator circuit. This sort of junction is uniquely defined by its quality factor (Q_L), its normalized susceptance slope parameter (b') and its normalized gyrator conductance (g). The description of its circuit assumes that the in-phase eigen-network can be idealized by an ideal electric wall at the terminals of the junction and that its counter-rotating one can be adjusted to impose a magnetic one at the same terminals. It is separately assumed that the frequency variation of the in-phase eigen-network can be neglected compared to that of the degenerate counter-

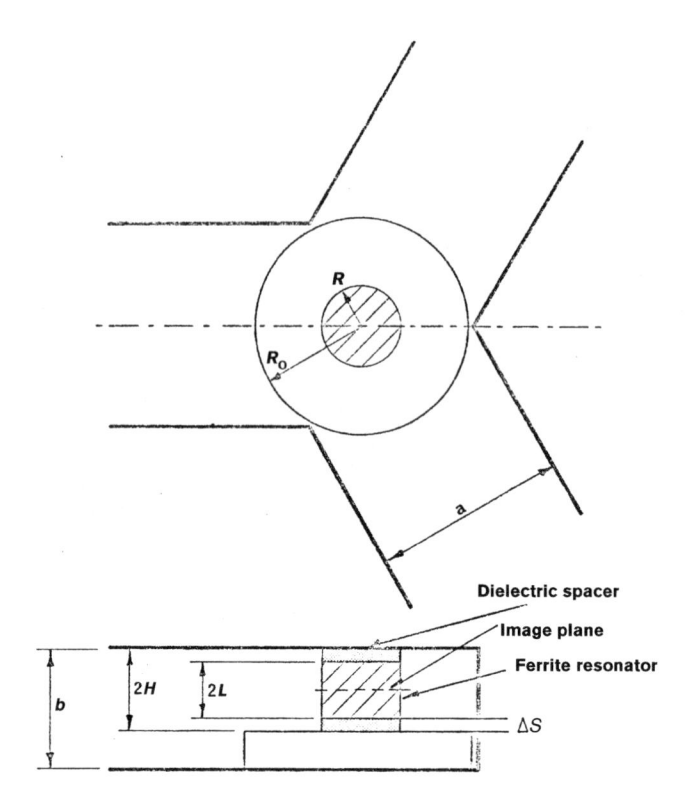

Figure 16.1 Schematic diagram of H-plane waveguide circulators using turnstile junction

rotating ones. The realisable gain-bandwidth of the device is again uniquely fixed by the quality factor of the gyrator circuit. The synthesis problem is completed once the details of the radial or ridge impedance transformer are fixed.

16.2 Image Planes of Turnstile Circulators

The H-plane waveguide circulator illustrated in Figure 16.1 is one of three practical geometries met in practice. The three possible arrangements are depicted in Figure 16.3. The configurations in Figure 16.3(a) and (b) are dual and are described by a single set of variables. The geometries in Figure 16.3(c) is identical to those in Figures 3(a) and 3b, except that its susceptance slope parameter is twice that of the former ones. It may be viewed as a 5-port network instead of a 7-port one comprising a 3-port H-plane junction coupled to a single cylindrical gyromagnetic waveguide supporting two orthogonal ports. The degenerate frequencies of the counter-rotating modes in each of these structures are determined by the odd dominant mode solution of a pair of coupled open demagnetized ferrite circular or triangular waveguides short-circuited at one end. These are then split in the usual way by magnetizing the resonator by a suitable direct magnetic field.

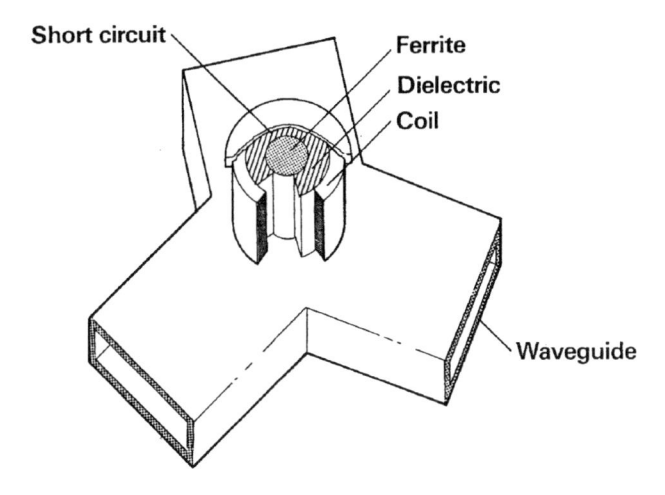

Figure 16.2 Schematic diagram of turnstile circulator

The in-phase mode is fixed by a quasi-planar resonance, whose frequency is determined by the position of the image or waveguide wall. Its eigen-network is common to all three configurations.

16.3 Eigen-networks and Complex Gyrator Circuit

The construction of conventional junction circulators involves the adjustment of two counter-rotating split eigen-networks and one in-phase one. The counter-rotating eigen-networks of the 7-port turnstile circulator consists of two identical eigen-resonators individually coupled by ideal transformers in series with a typical H-plane rectangular waveguide. A typical eigen-resonator in the vicinity of the first pair of degenerate poles consists of a distributed gyromagnetic resonator in shunt with a lumped element LC circuit which represents the effect of the loading of the image wall. The turns ratio of the ideal transformer is not at this time available in analytical form and is usually experimentally deduced. The in-phase eigen-network supports a quasi-planar one made up of a three layer ferrite/dielectric/ferrite region. Its adjustment fixes the position of the image wall or gap of the arrangement. The radial region between the actual junction and the rectangular waveguides is usually described in terms of the $ABCD$ parameters of the region. The eigen-networks under consideration are illustrated in Figure 16.4. The eigen-networks of the 5-port turnstile junction are identical in every way to those of the 7-port arrangement except that its counter-rotating ones only contain a single eigen-resonator and that its susceptance slope parameter is twice that of the original one. This family of eigen-networks is illustrated in Figure 16.5.

The first circulation condition in a 3-port circulator coincides with either the minimum insertion loss between any two of its ports or the minimum return loss at any of its port.

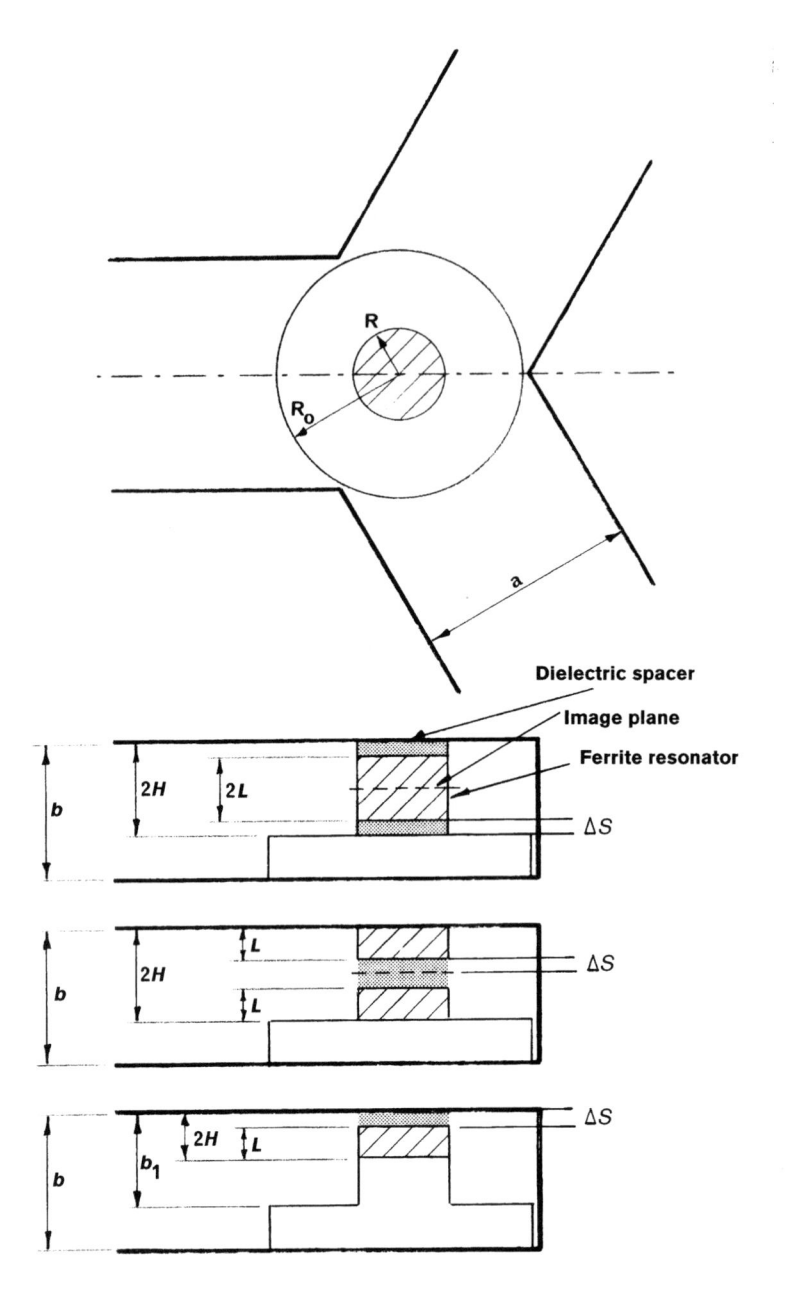

Figure 16.3 Schematic diagrams of H-plane waveguide circulators using turnstile junctions

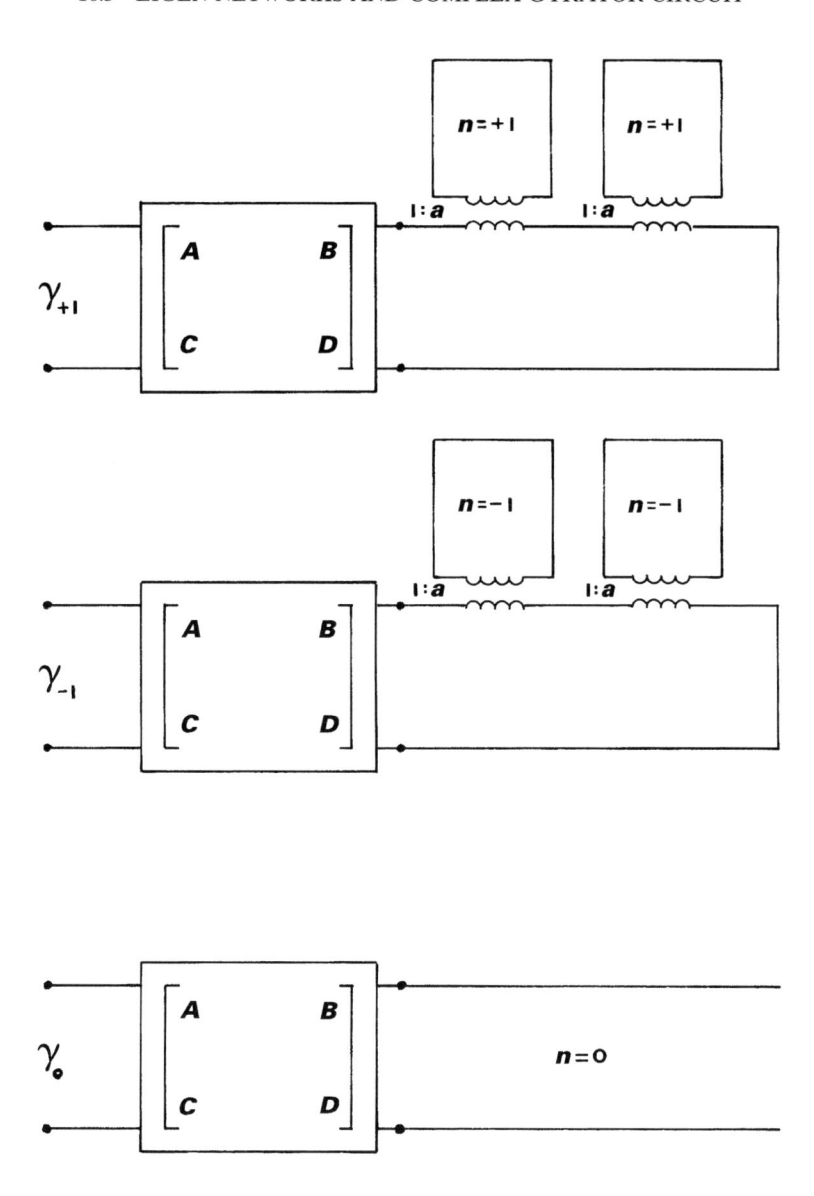

16.3 EIGEN-NETWORKS AND COMPLEX GYRATOR CIRCUIT 315

Figure 16.4 Eigen-networks of waveguide circulators using coupled quarter-wave-long resonators or single half-wave resonators

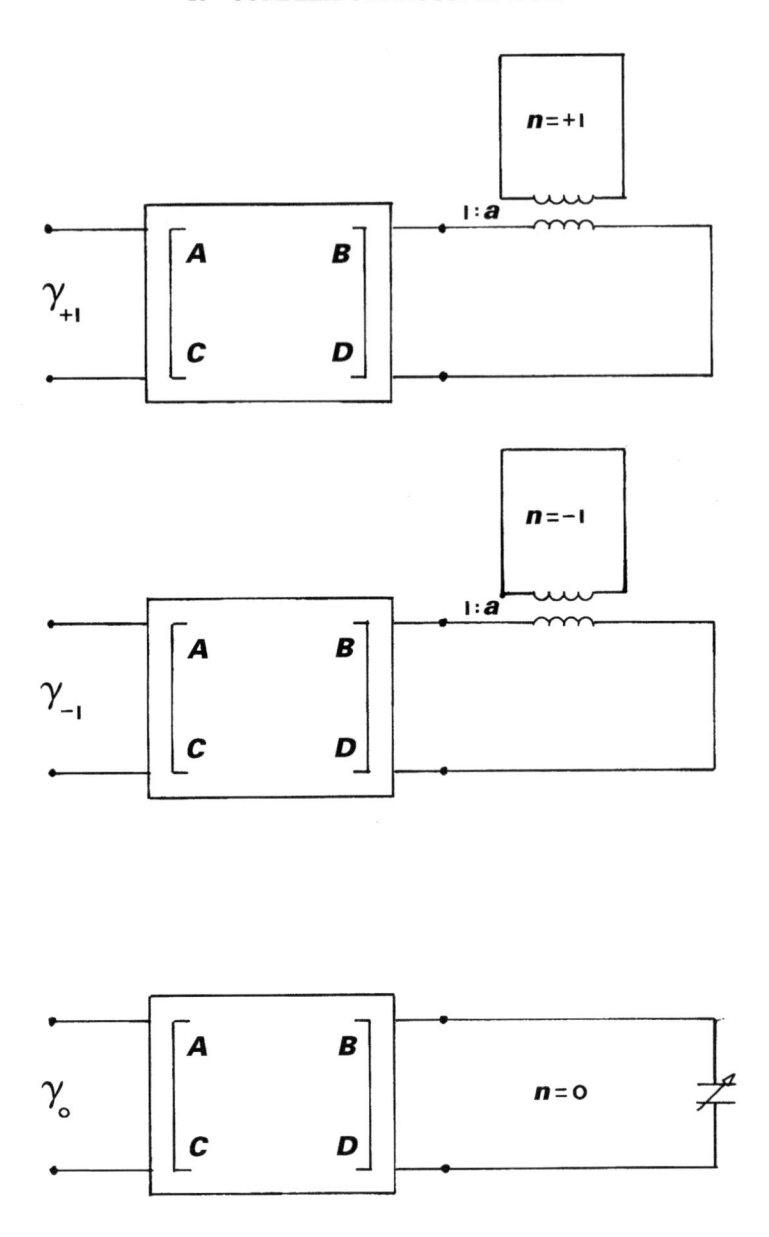

Figure 16.5 Eigen-networks of waveguide circulators using single quarter-wave-long resonator

16.3 EIGEN-NETWORKS AND COMPLEX GYRATOR CIRCUIT

These two quantities are related to the appropriate reflection eigenvalues by

$$S_{11} = \frac{s_0 + 2s_1}{3} \tag{16.1}$$

$$S_{21} = S_{31} = \frac{s_0 - s_1}{3} \tag{16.2}$$

where

$$s_0 = 1.0 \exp(-j2\theta_0) \tag{16.3}$$

$$s_1 = 1.0 \exp[-j2(\theta_1 + \pi/2)] \tag{16.4}$$

One solution at $r = R$ is

$$s_0 = -1 \tag{16.5}$$

$$s_1 = +1 \tag{16.6}$$

The eigenvalue diagram of the first circulation condition is therefore defined by

$$\theta_0 = \pi/2 \tag{16.7}$$

$$\theta_1 = \pi/2 \tag{16.8}$$

Since the determination of the phase θ_0 also requires a knowledge of θ_1, the evaluation of the former quantity affords the experimental means for the latter to be obtained as well, and furthermore, the necessary and sufficient conditions of the first circulation condition to be formed. Figure 16.6 indicates the frequency responses of s_0, s_1, and S_{11} for a turnstile junction using a single resonator with a radial wavenumber $(k_0 R)$ equal to 0.816 and a gap factor (k) equal to 0.80. The condition $s_1 = 1$ provides a simple statement of the resonant frequency of the degenerate counter-rotating eigen-networks. Means by which the phase angles θ_0 and θ_1 may be experimentally obtained are described in some detail in Chapter 24.

The second circulation condition is then satisfied without ado by removing the degeneracy between the counter-rotating eigenvalues with a suitable direct magnetic field. This gives

$$\theta_+ = \theta_1 + \pi/3 \tag{16.9}$$

$$\theta_- = \theta_1 - \pi/3 \tag{16.10}$$

The eigenvalue diagrams under consideration are indicated in Figure 16.7.

318   16 COMPLEX GYRATOR CIRCUIT

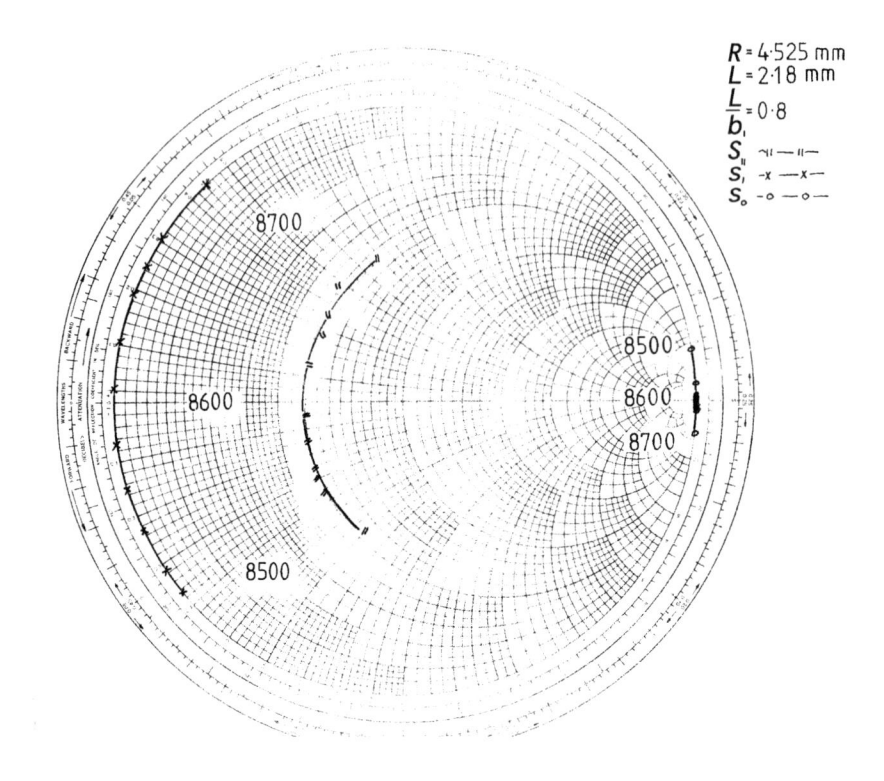

$R = 4.525$ mm
$L = 2.18$ mm
$\dfrac{L}{b_1} = 0.8$
$S_{\shortparallel}$ —ıı—ıı—
S_1 —x—x—
S_0 —o—o—

Figure 16.6 Smith Chart of s_0, s_1 and S_{11} variables (reprinted with permission, Helszajn, J. and Sharp, J., (1985) Adjustment of in-phase mode in turnstile circulator, *IEEE Trans. Microwave Theory Tech.*, **MTT-33**, April) (© 1985 IEEE)

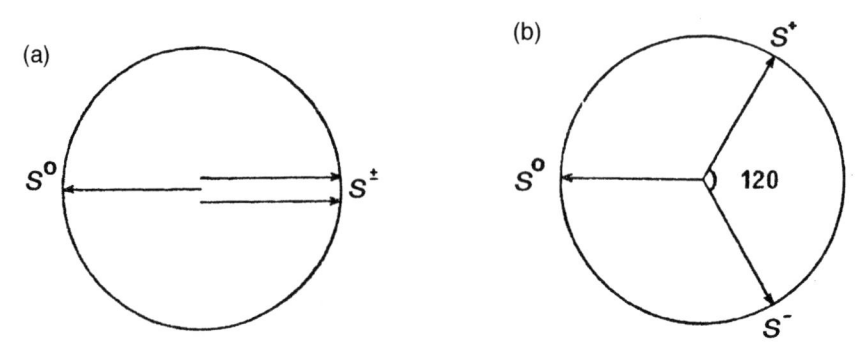

Figure 16.7 (a) Eigenvalue diagram of 3-port junction with $s_0 = 1$ and $s_1 = -1$, (b) Eigenvalue diagram of ideal 3-port junction circulator

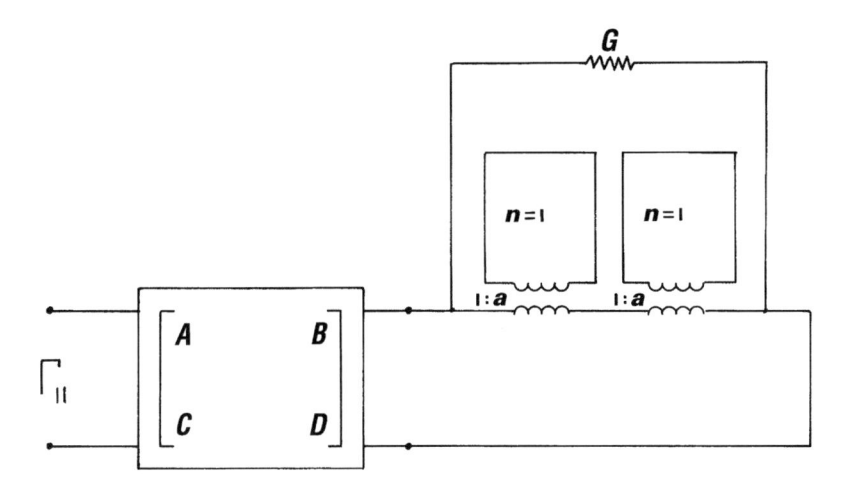

16.3 EIGEN-NETWORKS AND COMPLEX GYRATOR CIRCUIT

Figure 16.8 1-port equivalent circuit of waveguide circulator using coupled quarter-wave-long resonators or single half-wave-long resonators

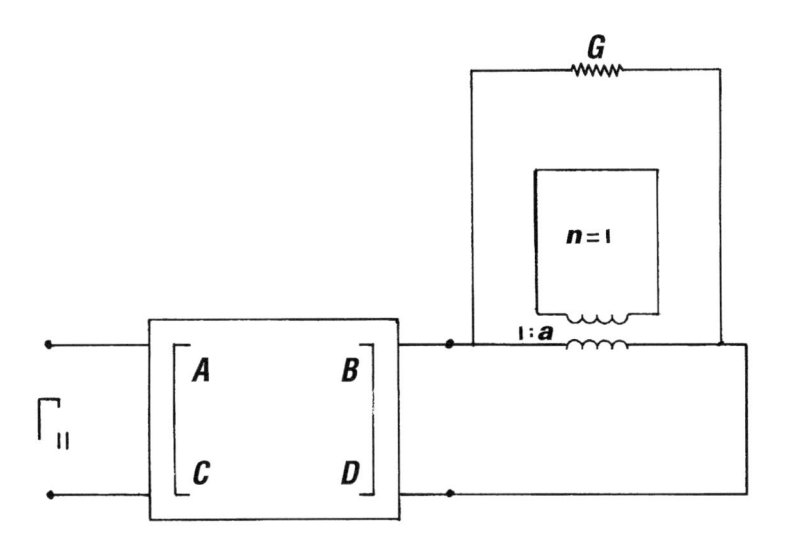

Figure 16.9 1-port equivalent circuit of waveguide circulator using single quarter-wave-long resonator

Once the in-phase and counter-rotating modes of a junction have been correctly adjusted, its eigen-network model may be replaced by a 1-port equivalent circuit. The 1-port complex gyrator circuits for the two devices considered here are illustrated in Figures 16.8 and 16.9. For circulators using weakly magnetized resonators the only case considered here, these 1-port networks exhibit gyrator conductances described by the following classic relationship:

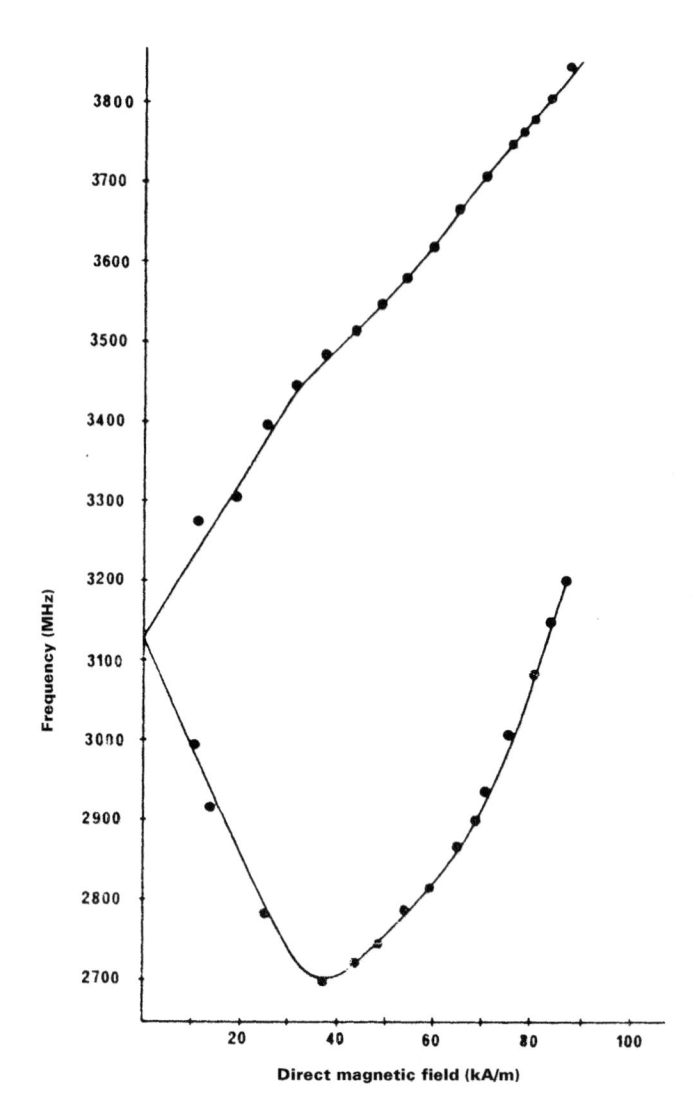

Figure 16.10 Split frequencies of loosely coupled quarter-wave-long resonator versus direct magnetic field (reprinted with permission, Helszajn, J. and Sharp, J. (1983) Resonant frequencies, Q-factor, and susceptance slope parameter of waveguide circulators using weakly magnetized open resonators, *IEEE Trans. Microwave Theory Tech.*, **MTT-31**, 434–441, June) (© 1985 IEEE)

$$g = \sqrt{3}\, b' \left(\frac{\omega_+ - \omega_-}{\omega_0} \right) \qquad (16.11)$$

$\omega_\pm$ are the split resonant frequencies of the magnetized open resonator, ω_0 is that of the demagnetized one, g is the normalized gyrator conductance, and b' is the normalized susceptance slope parameter. Any two of these three quantities are sufficient to describe the gyrator equation. The quality factor (Q_L) is again defined by

$$\frac{1}{Q_L} = \sqrt{3} \left(\frac{\omega_+ - \omega_-}{\omega_0} \right) \qquad (16.12)$$

The gyromagnetic cavity resonator has been dealt with in some detail in Chapter 12. Figure 16.10 shows on experimental plot for completeness.

16.4 Adjustment of Degenerate Counter-rotating Eigen-networks

The operating frequency of any junction circulator using a weakly magnetized gyromagnetic resonator is usually determined by that of the degenerate counter-rotating eigen-networks of the device. The required solution in the case of the H-plane junction circulator coincides with the odd mode solution of two coupled demagnetized open gyromagnetic resonators.

$$\varepsilon_{\text{eff}} \left(\frac{k_0}{\beta_0} \right) \cot \left[\left(\frac{\beta_0}{k_0} \right) \left(\frac{L}{R} \right) k_0 R \right] - \varepsilon_d \left(\frac{k_0}{\alpha_0} \right) \coth \left[\left(\frac{\alpha_0}{k_0} \right) \left(\frac{L}{R} \right) \left(\frac{1-k}{k} \right) k_0 R \right] = 0 \quad (16.13)$$

where

$$k = \frac{L}{L + S} \qquad (16.14)$$

The solution to this problem region is the topic of Chapter 10 and will not be repeated here. Figure 16.11 gives the experimental relationship between the gap or filling factor (k) and the radial wavenumber ($k_0 R$) for some typical arrangements. Good agreement between theory and experiment is usually achieved for engineering purposes by replacing the open gyromagnetic waveguide by a closed one and replacing the actual dielectric constant of the material by an effective one.

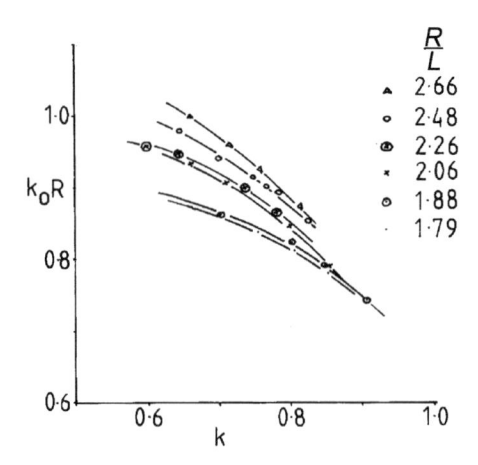

Figure 16.11 Experimental mode chart for counter-rotating modes of turnstile junctions using single disk resonator (reprinted with permission, , Helszajn, J. and Sharp, J. (1983) Resonant frequencies, Q-factor, and susceptance slope parameter of waveguide circulators using weakly magnetized open resonators, *IEEE Trans. Microwave Theory Tech.*, **MTT-31**, 434–441, June)

16.5 Adjustment of In-Phase Eigen-network

While the frequency variation of the in-phase eigen-network is usually neglected in the 1-port approximation of the circulator compared to that of the counter-rotating ones, it is still necessary to establish the proper phase angle for this network. This in-phase quasi-planar network determines the spacing between the open resonator and image wall. For an ideal circulator, the reflection coefficients s_0 for the in-phase mode at $r = R$ and R_0 are

$$s_0 = -1, r = R \tag{16.15a}$$

$$s_0 = +1, r = R_0 \tag{16.15b}$$

respectively.

Since these two conditions cannot be simultaneously satisfied for a radial junction with the value of R_0 obtained from the boundary condition of the counter-rotating modes, these have in the past placed upper and lower bounds on the filling factor (k) of the quasi-planar resonator. However, experience indicates that the former condition is to be preferred, so that the boundary condition $s_0 = -1$ will be adopted here. Adopting this condition gives

$$J_0(k_{\text{eff}}R) = 0 \tag{16.16a}$$

16.5 ADJUSTMENT OF IN-PHASE EIGEN-NETWORK

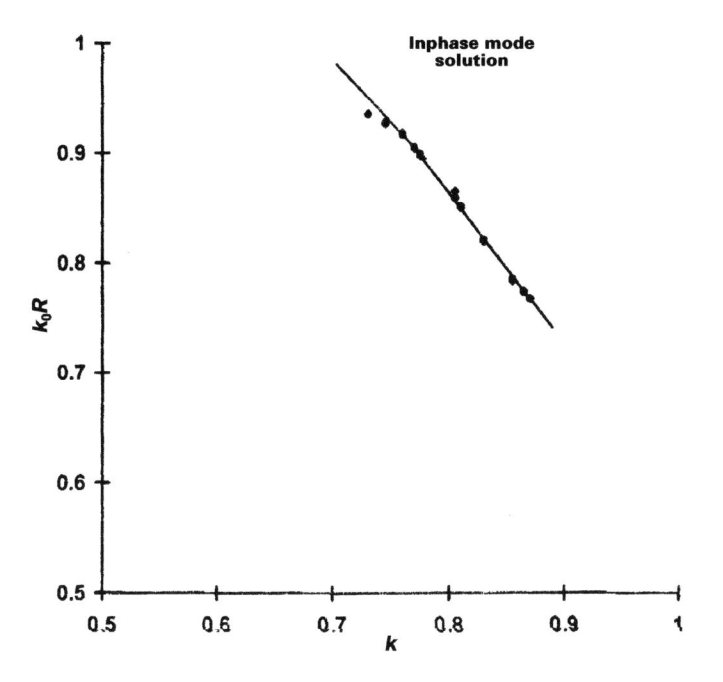

Figure 16.12 Experimental radial-wave number of in-phase eigen-network

or

$$k_{eff}R = 2.405 \tag{16.16b}$$

where

$$k_{eff} = k_0\sqrt{\varepsilon_{eff}} \tag{16.17a}$$

and

$$\varepsilon_{eff} = \frac{\varepsilon_f \varepsilon_d}{[(1-k)\varepsilon_f + k\varepsilon_d]} \tag{16.17b}$$

One polynomial approximation for the filling factor based on the data in Figure 16.12 which is applicable to either single and coupled resonator junctions with $\varepsilon_d = 1$ is

$$k \approx 0.2196 + 2.204(k_0R) - 1.785(k_0R)^2,$$

$$\varepsilon_f = 15, \ 0.75 < k_0R < 0.95, \tag{16.18}$$

or

$$k_{\text{eff}}R \approx 0.6748 + 2.9827(k_0R) - 2.093(k_0R)^2,$$

$$\varepsilon_\text{f} = 15, \, 0.75 < k_0R < 0.95 \tag{16.19}$$

The discrepancy between the two values of the radial variables in (16.16) and (16.19) may be understood by recognizing that the former value is based on a resonator model with ideal electric (magnetic) walls, whereas in practive the boundary conditions consist of imperfect electric (magnetic) walls. This EM problem is outside the remit of this work.

16.6 Adjustment of First Circulation Condition

The necessary and sufficient conditions for the adjustment of this class circulator is now met by superimposing the data in Figures 16.11 and 16.12 in the manner illustrated in Figure 16.13. The intersection of these two relationships gives the required one between θ_1 and θ_0. This result, strictly speaking, applies to a garnet resonator with a magnetization of 0.1600 T and a relative dielectric constant of 15.0. If k_0R is taken as the independent variable, then k and R/L are the dependent ones. The effect of mistuning the in-phase eigen-network is separately illustrated in Figure 16.14. Obviously, the reference plane of S_{11} is in this situation incompatible with the synthesis of high-quality quarter-wave coupled junction circulators.

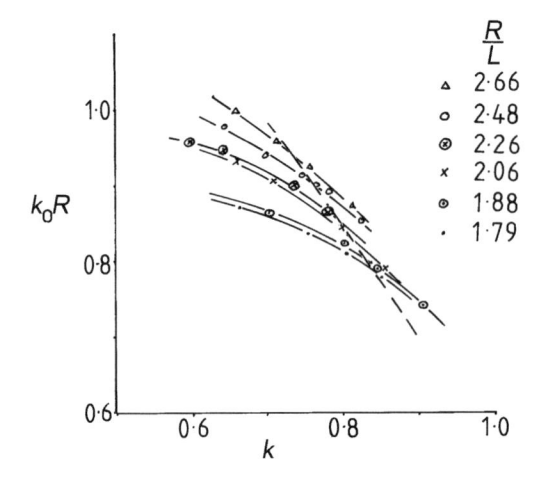

Figure 16.13 First circulation solution for turnstile junctions using single and coupled disk resonator (reprinted with permission, Helszajn, J. and Sharp, J. (1985) Adjustment of in-phase mode in turnstile circulator, *IEEE Trans. Microwave Theory Tech.*, **MTT-33**, April)

16.7 SUSCEPTANCE SLOPE PARAMETER 325

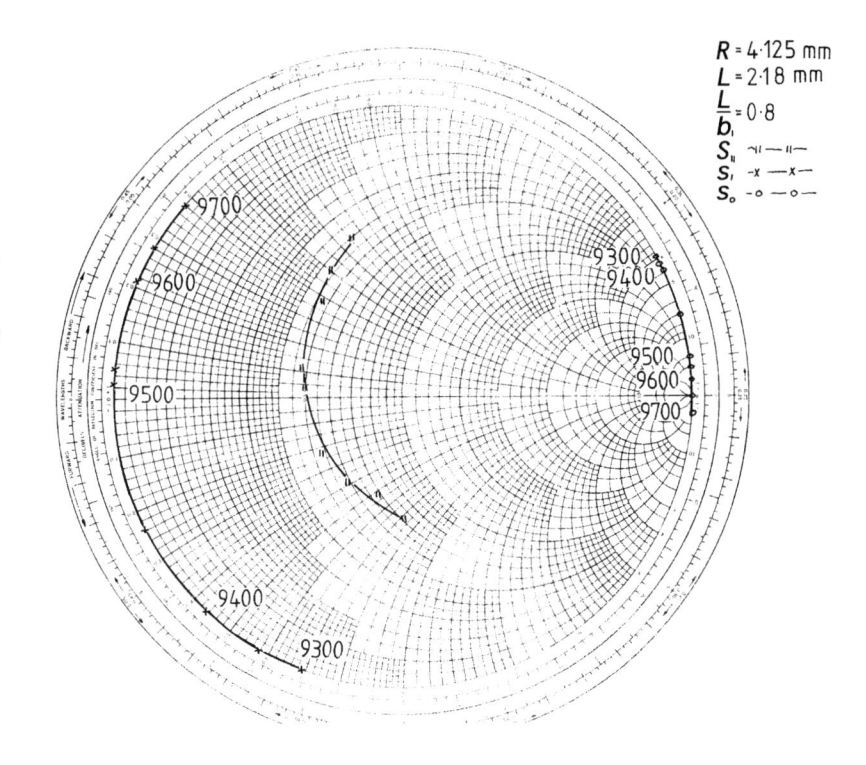

Figure 16.14 Effect of mistuning in-phase eigen-network (reprinted with permission, Helszajn, J. and Sharp, J. (1985) Adjustment of in-phase mode in turnstile circulator, *IEEE Trans. Microwave Theory Tech.*, **MTT-33**, April)

The situation in Figure 16.14 is sometimes displayed on a unit circle in the manner indicated in Figure 16.7(a). Since for this class of junction it is now merely necessary to split the degeneracy between the counter-rotating eigenvalues in the manner indicated in Figure 16.7(b) in order to realize an ideal circulator, this condition is sometimes referred to as the first circulation condition. However, this is not a general result but merely a special case of the more general boundary condition obtained by setting the imaginary part of the complex gyrator immittance to zero. It is obviously an adequate boundary condition for the problem at hand. It is also noted from the data in Figure 16.14 that the frequency variation of the in-phase eigen-network may indeed be neglected compared to those of the degenerate counter-rotating ones as is often assumed.

16.7 Susceptance Slope Parameter

While the circulation solution defined by Figure 16.13 is adequate at a single frequency it is not sufficient for the design of quarter-wave coupled devices.

A knowledge of the susceptance slope parameter of this sort of resonator is mandatory for design. While an exact analytical description of this problem awaits a solution, an approximate one can be formulated without any particular difficulty. The derivation of the required result starts by forming the susceptance of the gyrator circuit in Figure 16.9. This circuit consists of an ideal transformer of turns ratio which represents the coupling between the eigen-resonator of the complex gyrator circuit and the gyrator terminals, a lumped element LC circuit which represents the effect of the image wall and an ideal quarter-wave long gyromagnetic waveguide short-circuited at one end and of open-circuited at the other which coincides with a typical decoupled resonator. The susceptance of the complex gyrator circuits is

$$y_1 = n^2 \zeta_1 \qquad (16.20)$$

where

$$\zeta_1 = j\zeta_0 \left[\frac{\varepsilon_f k_0}{\beta} \cot(\beta L_0) + \frac{\varepsilon_f k_0}{\beta} \tan(\beta \Delta L) - \frac{\varepsilon_d k_0}{\alpha} \coth(\alpha \Delta S) \right] \qquad (16.21)$$

The coefficients in this equation correspond to the wave admittances met in connection with propagating and cut-off waveguide sections supporting TM like waves

$$\zeta_{TM} = \frac{j\omega\varepsilon_0\varepsilon_r}{\gamma} \qquad (16.22)$$

For a propagating waveguide

$$\gamma = j\beta \qquad (16.23a)$$

and

$$\zeta_{TM} = \frac{\varepsilon_r \zeta_0 k_0}{\beta} \qquad (16.23b)$$

For a cut-off one

$$\gamma = \alpha \qquad (16.24a)$$

and

$$\zeta_{TM} = \frac{-j\varepsilon_r \zeta_0 k_0}{\alpha} \qquad (16.24b)$$

where

16.7 SUSCEPTANCE SLOPE PARAMETER

$$\beta^2 = k_0^2 \varepsilon_f - k_c^2 \tag{16.25a}$$

$$\alpha^2 = k_c^2 - k_0^2 \varepsilon \tag{16.25b}$$

n^2 represents the turns ratio of an ideal transformer which embodies the absolute definition of impedance used to describe the change in admittance between the eigen-resonator and eigen-network at the terminals of the junction. One empirical relationship that appears to satisfy the experimental work with $\varepsilon_d = 1$ is

$$n^2 = \frac{1}{3} \frac{ab}{(\pi R)^2} \tag{16.26}$$

ab is the cross-sectional area of the rectangular waveguide, and πR^2 is that of the open demagnetized ferrite wave-guide.

The solution to the quarter-wave long triangular resonator is obtained by duality by replacing

$$k_c R = 1.84 \tag{16.27}$$

with

$$k_c A = \frac{4\pi}{3} \tag{16.28}$$

and writing n^2 on the basis of experiment as

$$n^2 = 0.12 \left(\frac{4ab}{\sqrt{3} \, A^2} \right), \quad \text{apex coupled triangle} \tag{16.29}$$

$$n^2 = 0.18 \left(\frac{4ab}{\sqrt{3} \, A^2} \right), \quad \text{side coupled triangle} \tag{16.30}$$

ab is the cross-sectional area of the rectangular waveguide and $\sqrt{3}A^2/4$ is that of the triangular ferrite resonator.

The susceptance slope parameter B' of the open demagnetized ferrite resonator can now be evaluated in terms of the physical variables of the eigen-network by making use of its definition below:

$$B' = \frac{\omega_0}{2} \frac{\delta y_1}{\delta \omega} \bigg|_{\omega_0} \tag{16.31}$$

ω_0 is obtained by satisfying (16.21) with $\zeta_1 = 0$. Evaluating the preceding equation leads to the required result

$$B' = n^2 \zeta_0 \left\{ \left[\frac{\pi \varepsilon_f^2}{4} \left(\frac{k_0}{\beta_0}\right)^3 + \frac{\varepsilon_f^2}{2} \left(\frac{k_0}{\beta_0}\right)^3 \right] \cdot \left[\tan(\beta_0 \Delta L) - \frac{\beta_0 \Delta L}{\cos^2(\beta_0 \Delta L)} \right] \right.$$

$$\left. + \frac{1}{2} \left(\frac{k_0}{\alpha}\right)^3 \left[\coth\left(\frac{k_0}{\alpha}\right)^3 \left[\coth(\alpha S) + \frac{\alpha S}{\sinh^2(\alpha S)} \right] \right] \right\} \quad (16.32)$$

A normalized susceptance slope parameter is now defined as

$$b' = \frac{B'}{\zeta_{TE}} \quad (16.33)$$

ζ_{TE} is the wave admittance of the rectangular waveguide propagating a TE_{10} mode

$$\zeta_{TE} = \zeta_0 \left(\frac{\beta}{k_0}\right) \quad (16.34)$$

and

$$\beta^2 = \left(\frac{2\pi}{\lambda_0}\right)^2 - \left(\frac{2\pi}{\lambda_c}\right)^2 \quad (16.35)$$

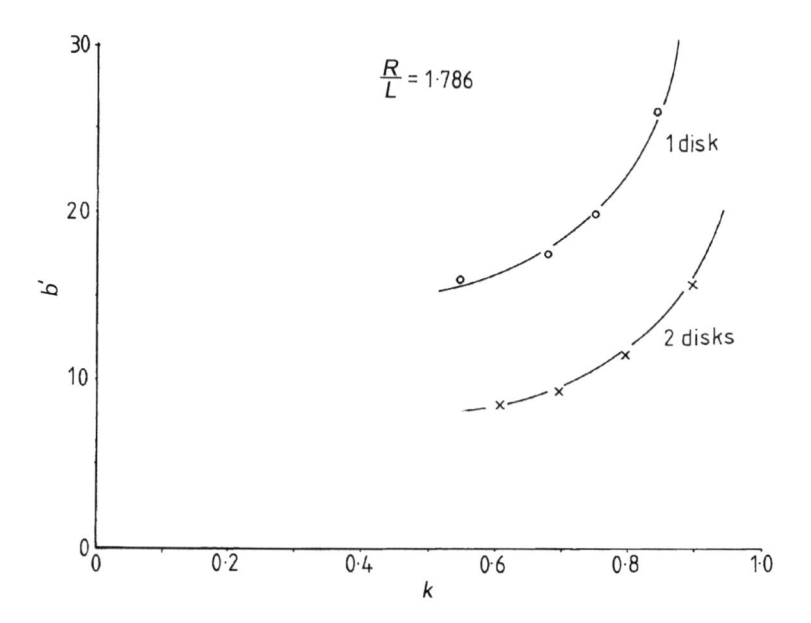

Figure 16.15 Susceptance slope parameter of degenerate counter-rotating eigen-networks for single and coupled disk resonators (reprinted with permission, Helszajn, J. and Sharp, J. (1983) Resonant frequencies, Q-factor, and susceptance slope parameter of waveguide circulators using weakly magnetized open resonators, *IEEE Trans. Microwave Theory Tech.*, **MTT-31**, 434–441, June)

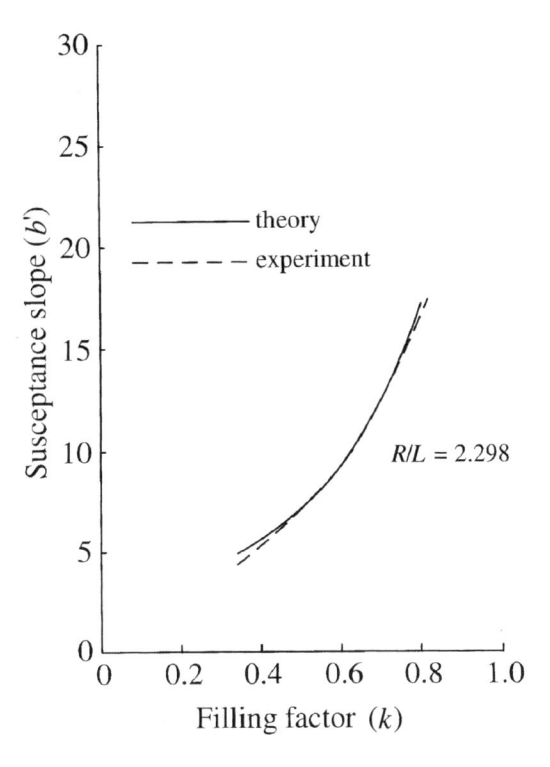

Figure 16.16 Normalized susceptance slope parameter of single quarter-wave-long resonator (reprinted with permission, Helszajn, J. and Tan, F. C. F. (1975) Susceptance slope parameter of waveguide partial-height ferrite circulators, *Proc. Inst. Elec. Eng.*, **122**(72), 1329–1332, December)

It is experimentally observed that the susceptance slope parameter for the coupled disks arrangement is half that of the single disk resonator. This feature is indicated in Figure 16.15. Figure 16.16 compares theory and practice in the case of one junction. The experimental result applies equally well to the power voltage definition of admittance in standard rectangular waveguide

$$Y_{PI} = \left(\frac{a}{2b}\right) \zeta_{TE} \tag{16.36}$$

It is therefore appropriate in the design of quarter-wave coupled devices based on the power-voltage definition of impedance.

The susceptance of the complex gyrator circuit (B) may be recovered from a knowledge of the susceptance slope parameter (B') by forming

$$B = j2\delta B' \tag{16.37}$$

where δ is the normalized frequency variable, $\delta = (\omega - \omega_0)/\omega_0$.

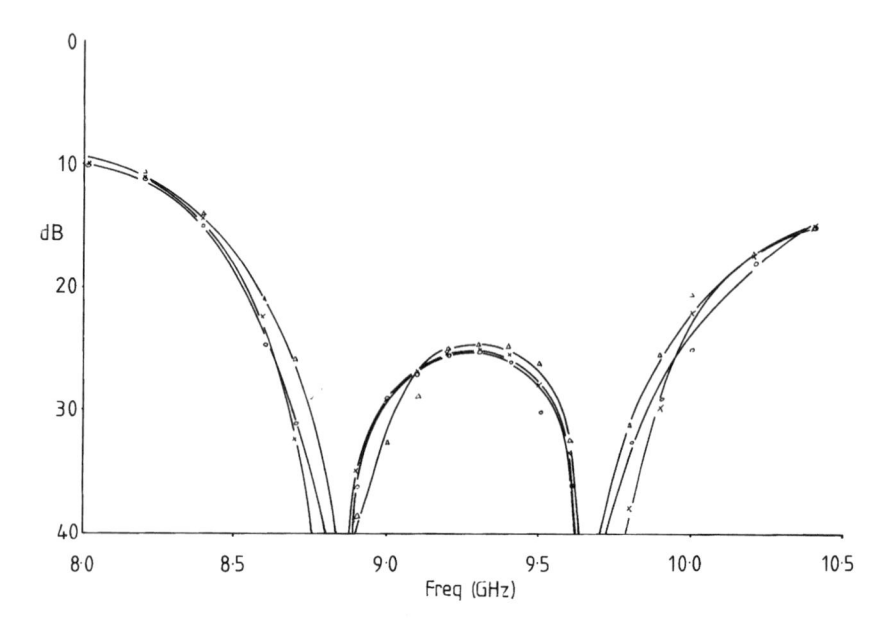

Figure 16.17 Frequency response of radial line coupled turnstile junction using single resonator (reprinted with permission, Helszajn, J. and Sharp, J. (1985) Adjustment of in-phase mode in turnstile circulator, *IEEE Trans. Microwave Theory Tech.*, **MTT-33**, April)

The first circulation adjustment described here has been incorporated in the experimental construction of one circulator in WR90 waveguide using a radial coupled turnstile junction employing a single disk resonator. Figure 16.17 depicts the frequency response of the device. the details of the second circulation condition (radial transformer, magnetic variables, susceptance slope parameters, etc.) are outside the remit of this work. Those of the first circulation condition are in very close agreement with the data developed here. No tuning whatsoever was utilized in obtaining this result.

16.8 Synthesis of Waveguide Circulators using Partial-Height Resonators

In the synthesis problem, the variables of the complex gyrator circuit (g, b', Q_L) are usually fixed by the maximum and minimum values of the $VSWR$, $S(\max)$ and $S(\min)$ and the bandwidth W of the circulator specification. This problem is fully discussed in Chapter 23 and will not be repeated here. However, it is recalled that varying $S(\min)$ has a significant influence on the absolute values of g and b'.

In the case of circulators using partial-height resonators, the susceptance slope parameter and the lower bound on the loaded Q-factor are fixed by the choice of the radial wavenumber of the resonator. If the lower bound for Q_L is adopted for design, then the gyrator conductance is also fixed. The latter quantities as well as $S(\max)$ may then be taken as the independent variables and $S(\min)$ and the

16.8 SYNTHESIS OF WAVEGUIDE CIRCULATORS

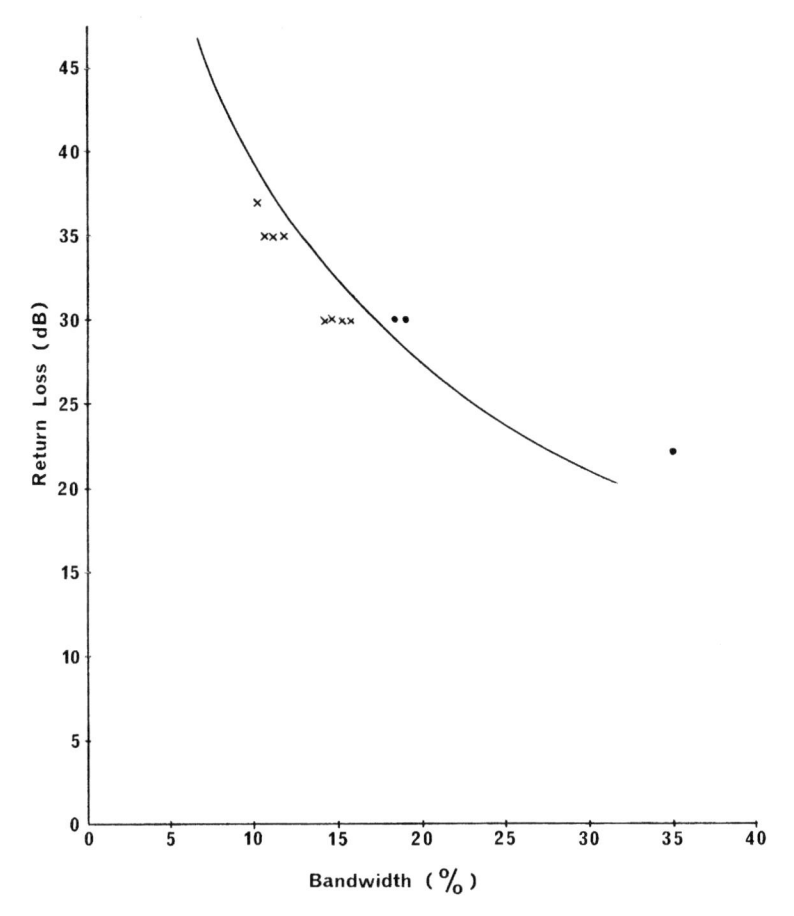

Figure 16.18 Relationship between return-loss and bandwidth (with $S(\text{min}) = 1$ and $Q_L = 2.0$) (Experimental data courtesy Ferranti Ltd. Thompson-CSF, Microwave Associates, Marconi.) (Reprinted with permission, Helszajn, J. (1984) Design of waveguide circulators with Chebyshev characteristics using partial-height resonators, *IEEE Trans. Microwave Theory Tech.*, **MTT-32**)

normalized bandwidth W may be taken as the dependent ones and the ensuing network problem may be tested for realizability. If no solution is possible, or if $S(\text{min})$ or W are unacceptable, the value of b' or Q_L must be modified by varying either the radial wavenumber of the resonator or its shape.

Figure 16.18 depicts the relationship between the theoretical return loss (with $S(\text{min})=1$ and $Q_L=2$) and the bandwidth W and some experimental results on circulators using single cylinders and coupled disks resonators obtained by a number of industrial organizations. This result suggests that the weakly magnetized theory outlined here is suitable for the realization of devices with 30% bandwidth at the 20 dB return-loss frequencies.

16.9 Overall Frequency Response of Circulator

The overall frequency response of a circulator can be constructed once the physical variables of the junction are fixed. The admittance at the input terminals is now formed in terms of the radial waveguide $ABCD$ matrix and the complex gyrator admittance. The result is:

$$Y_{in} = a^2 \cdot \frac{jC + DY_L}{A + jBY_L} \tag{16.38}$$

where

$$Y_L = G_L + jB_L \tag{16.39}$$

The entries of the $ABCD$ matrix of a radial waveguide are summarized in Chapter 11.

The turns ratio of an ideal transformer is a^2 which represents the transition between the rectangular and radial waveguides. It is calculated assuming that the terminals of the transformer coincide with those of the radial waveguide formed by the junction of the rectangular waveguides. Although this is not quite the case, the two are sufficiently close to permit the region between the terminals of the circulator and the radial cavity to be neglected. Taking a 9-GHz circulator in WR90 waveguide as an example indicates that $k_0R_0 = 2.48$ at the terminals of the radial cavity, whereas it is typically 2.36 in practice.

Applying the boundary conditions at the terminals of the radial and rectangular waveguides gives the turns ratio between the radial and rectangular waveguides as

$$a^2 \approx \frac{3 \sin(n\phi)}{2\pi n\phi} \left[\frac{\sin(3/2 - n)\phi}{(3/2 - n)} + \frac{\sin(3/2 + n)\phi}{(3/2 + n)} \right] \tag{16.40}$$

This equation is evaluated with n=1 and

$$\phi = \pi/3 \tag{16.41}$$

The 1-port equivalent circuits which apply here are depicted in Figures 16.8 and 16.9.

16.10 Input Terminals of Radial Waveguide Circulator

For a radial waveguide junction using a magnetized resonator, the notion of characteristic plane already introduced elsewhere in this text must be modified to account for the fact that the output terminals of the line are no longer an open circuit but a variable conductance which is determined by the gyrator level of the junction. The input terminals of such a junction now coincide with those at which the input admittance is real. Since the radial line is nonuniform, this plane is a function of the gyrator level specializing the relationship in (16.38) gives.

$$Y_{in} = a^2 \frac{jC + DG}{A + jBG} \qquad (16.42)$$

where it is assumed that the gyrator admittance is real at the centre frequency. Rearranging this quantity leads to

$$Y_{in} = a^2 \cdot \frac{(AD + BC)G + j(AC - BDG^2)}{A^2 + B^2G^2} \qquad (16.43)$$

Setting the imaginary part to zero fixes the input terminals of the device as

$$AC - BDG^2 = 0 \qquad (16.44)$$

The real part is separately determined by

$$a^2 \frac{(AD + BC)G}{A^2 + B^2G^2} \qquad (16.45)$$

16.11 Computation

The step-by-step design of a 2.85 GHz circulator using a single disk junction in WR284 waveguide based on the topology in Figure 16.19 will now be illustrated. It starts by choosing a trial value for k_0R (0.89) and proceeds to evaluate the radius of the resonator (R), the length of the resonator (L) and the location of the image wall (S). Once the junction configuration is fixed, its susceptance slope parameter and the lower bound on the loaded Q-factor may be evaluated. The results are $b' = 11.60$ and $Q_L = 1.63$ respectively. This calculation also, incidentally, fixes the lower bound for the magnetization of the ferrite resonator. Using these gyrator parameters in conjunction with a return-loss specification of $S(\min) = 1$, $S(\max) = 1.03$, fixes the gyrator level as $g = 7.26$ and the required quality factor Q_L as 1.9. Since this value of Q-factor exceeds the lower bound estimate, the calculation may proceed; otherwise a different radial wavenumber must be selected or the specification must be modified. The corresponding bandwidth is now determined from the network problem as $W = 12.0\%$. The radius R_0 of the radial waveguide is next calculated in terms of the gyrator level. The result is $k_0R_0 = 2.36$ for which $R_0 = 40.5$ mm. Finally, equation (16.45) gives $b_2/b = 0.51$, which completes the design procedure.

The calculated and experimental quantities are listed below:

$$k_0R = 0.89 \qquad (0.89)$$

$$k_0R_0 = 2.36 \qquad (2.42)$$

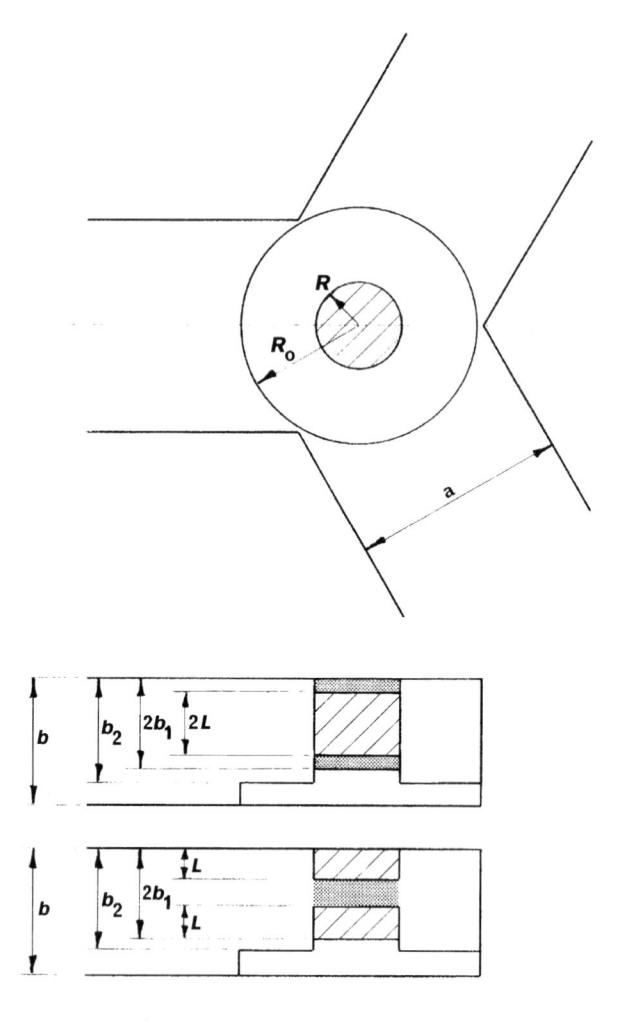

Figure 16.19 Schematic diagrams of waveguide circulators using partial-height resonators with $b_2 \neq 2b_1$

$R/L = 2.13$ (2.10)

$k = 0.76$ (0.70)

$b_2/b = 0.51$ (0.63)

$\varepsilon_d = 1$ (1)

$M_0 = 0.0618$ T   (0.0680 T)

The overall frequency response of the device obtained by fine tuning each port is depicted in Figure 16.20.

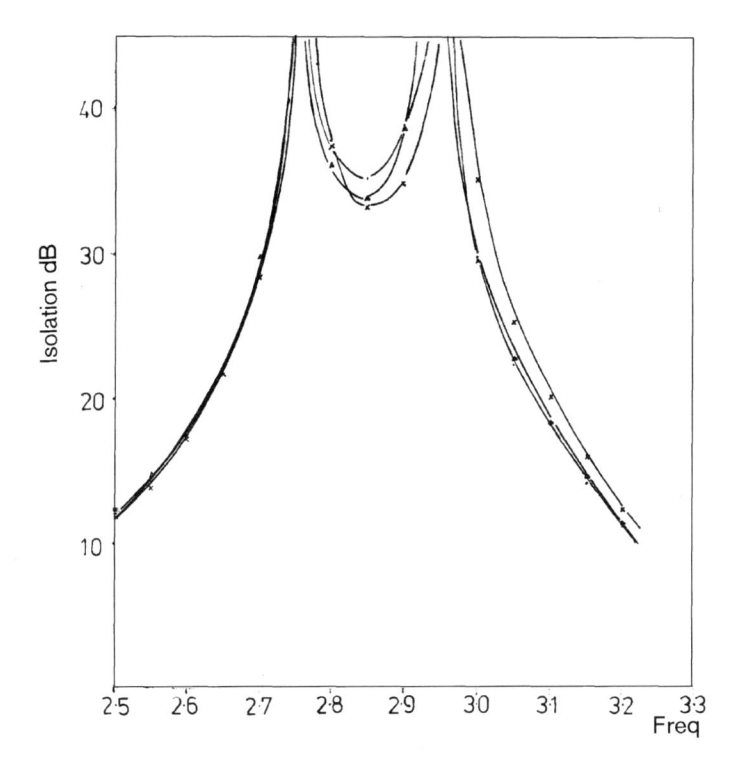

Figure 16.20 Experimental frequency response of quarter-wave coupled waveguide circulator in WR284 using single quarter-wave-long resonator (reprinted with permission, Helszajn, J. (1981) High-power waveguide circulators using quarter-wave long composite ferrite/dielectric resonators, *Proc. Inst. Elec. Eng., part H*, **128**, 268–273)

The agreement between theory and practice is good for the first four quantities (which define the phase angles of the eigen-networks) but is less good for the quantity b_2/b (which determines the impedance level of the eigen-networks). This lack of correlation is partly in keeping with the fact that the exact direct magnetic field of the experimental junction was not established as a preamble to quarter-wave coupling it, so that some experimental trade-off between the gyrator conductance and the impedance transformer was necessary.

References and Further Reading

Akaiwa, Y. (1974) Operation modes of a waveguide Y-circulator, *IEEE Trans. Microwave Theory Tech.*, **MTT-22**, 954–959.

Akaiwa, Y. (1977) Mode classification of triangular ferrite post for Y-circulator, *IEEE Trans. Microwave Theory Tech.*, **MTT-25**, 59–61.

Akaiwa, Y. (1978) A numerical analysis of waveguide H-plane Y-junction circulators with circular partial-height ferrite post, *J. Inst. Electron Commun. Eng. Jap.*, **E61**, 609–617.

Anderson, L. K. (1967) An analysis of broadband circulators with external tuning elements, *IEEE Trans. Microwave Theory Tech.*, **MTT-15**, 42–47.

Auld, B. S. (1959) The synthesis of symmetrical waveguide circulators, *IRE Trans. Microwave Theory Tech.*, **MTT-7**, 238–246.

Denlinger, E. J. (1974) Design of partial-height ferrite waveguide circulators, *IEEE Trans. Microwave Theory Tech.*, **MTT-22**, 810–813.

Dou, W. B. and Li, S. F. (1988) On volume modes and surface modes in partial-height ferrite circulators and their bandwidth expansion at millimeter wave band, *Microwave and Optical Tech. Letter.*, **1**, 200–208.

Hauth, W. (1981) Analysis of circular waveguide cavities with partial-height ferrite insert, *Proc. Eur. Microwave Conf.*, 383–388.

Helszajn, J. (1972) The synthesis of quarter-wave coupled circulators with Chebyshev characteristics, *IEEE Trans. Microwave Theory Tech.*, **MTT-20**, 764–769.

Helszajn, J. (1973) Frequency response of quarter-wave coupled reciprocal stripline junctions, *IEEE Trans. Microwave Theory Tech.*, **MTT-21**, 533–537.

Helszajn, J. (1973) Microwave measurement techniques for below resonance junction circulators, *IEEE Trans. Microwave Theory Tech.*, **MTT-21**, 347–351.

Helszajn, J. (1974) Common waveguide circulator configurations, *Electron Eng.*, 66–68.

Helszajn, J. (1975) Scattering matrices of junction circulators with Chebyshev characteristics, *IEEE Trans. Microwave Theory Tech.*, **MTT-23**, 548–551.

Helszajn, J. (1976) A unified approach to lumped element, stripline and waveguide junction circulators, *Proc. Inst. Elec. Eng., part H* (1), September, 18–26.

Helszajn, J. (1981) High-power waveguide circulators using quarter-wave long composite ferrite/dielectric resonators, *Proc. Inst. Elec. Eng., part H*, **128**, 268–273.

Helszajn, J. (1984) Design of waveguide circulators with Chebyshev characteristics using partial-height ferrite resonators, *IEEE Trans. Microwave Theory Tech.*, **MTT-32**, 908–917.

Helszajn, J. and Sharp, J. (1983) Resonant frequencies, Q-factor, and susceptance slope parameter of waveguide circulators using weakly magnetized open resonators, *IEEE Trans. Microwave Theory Tech.*, **MTT-31**, 434–441.

Helszajn, J. and Sharp, J., (1985) Adjustment of in-phase mode in turnstile circulator, *IEEE Trans. Microwave Theory Tech.*, **MTT-33**, April.

Helszajn, J. and Tan, F. C. F. (1975) Design data for radial waveguide circulators using partial-height ferrite resonators, *IEEE Trans. Microwave Theory Tech.*, **MTT-23**, 288–298.

Helszajn, J. and Tan, F. C. F. (1975) Mode charts for partial-height ferrite waveguide circulators, *Proc. Inst. Elec. Eng.*, **122**(1).

Helszajn, J. and Tan, F. C. F. (1975) Susceptance slope parameter of waveguide partial-height ferrite circulators, *Proc. Inst. Elec. Eng.*, **122**(72), 1329–1332.

Levy, R. and Helszajn, J. (1982) Specific equations for one and two section quarter-wave matching networks for stub-resistor loads, *IEEE Trans. Microwave Theory Tech.*, **MTT-30**, 55–62.

Levy, R. and Helszajn, J. (1983) Short-line matching networks for circulators and resonant loads, *Proc. Inst. Elec. Eng., Part. H*, **130**, 385–390.

Milano, U., Saunders, J. H. and Davis, L. E. Jr. (1960) A Y-junction stripline circulator, *IRE Trans. Microwave Theory Tech.*, **MTT-8**, 346–351.

Montgomery, C. G., Dicke, R. H. and Purcell, E. M. (1948) *Principles of Microwave Circuits*. McGraw-Hill, New York.

Owen B. and Barnes, C. E. (1970) The compact turnstile circulator, *IEEE Trans. Microwave Theory Tech.*, **MTT-18**, 1096–1100.

Owen, B. (1972) The identification of modal resonances in ferrite loaded waveguide Y-junction and their adjustment for circulation, *Bell Syst. Tech. J.*, **51**(3).

Piotrowski, W. S. and Raul, J. E. (1976) Low-loss broad-band EHF circulator, *IEEE Trans. Microwave Theory Tech.*, **MTT-24**, 863–866.

Riblet, G., Helszajn, J. and O'Donnell, B. (1979) Loaded Q-factors of partial height and full-height triangular resonators for use in waveguide circulators. In: *Proc. Eur. Microwave Conf.*, 420–424.

Schaug-Patterson, T. (1958) Novel design of a 3-port circulator, Norwegian Defence Research Establishment, January.

17

COMPLEX GYRATOR CIRCUIT OF AN EVANESCENT-MODE E-PLANE JUNCTION CIRCULATOR USING H-PLANE TURNSTILE RESONATORS

17.1 Introduction

An important class of 3-port waveguide circulators is that comprising an evanescent-mode E-plane junction loaded by one or two re-entrant H-plane turnstile resonators with quasi-magnetic walls. The purpose of this chapter is to examine in some detail the eigenvalue problem and gyrator circuit of this sort of arrangement as a prerequisite to design. Figure 17.1(a) depicts the physical details of the symmetrical E-plane junction considered here. The radial wavenumber of each H-plane turnstile ferrite resonator need not coincide with that of the junction. The other physical variables entering into its design are the axial thickness of each resonator, the spacing between the two and the gyrotropy. The gap between the two resonators is such that the junction is cut off when these are removed. The use of some other shape (triangular or semispherical) for the resonator is another possibility. Figure 17.1(b) illustrates the asymmetrical geometry. Latched prototypes come readily to mind. Such circulators have gyrator circuits akin to those met, in connection with the classic H-plane arrangements. The design of quarter wave coupled device with Chebyshev frequency characteristics does not therefore pose any particular difficulty.

Some appreciation of the operation of the symmetrical E-plane junction employing H-plane turnstile resonators may be gained by recognizing that it is essentially a 7-port circuit comprising three rectangular waveguide ports and two orthogonal ports for each round turnstile waveguide. Since the output terminals of the round waveguides are short-circuited, its overall matrix description reduces in the usual way to a symmetrical 3-port network. By the same token, the conventional H-plane waveguide circulator is a 7-port junction comprising an H-plane

17 COMPLEX GYRATOR CIRCUIT OF AN E-PLANE JUNCTION

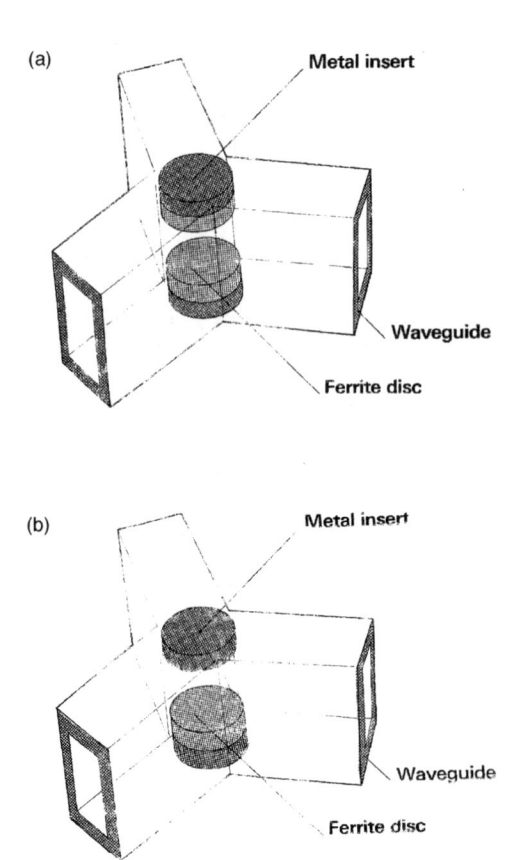

Figure 17.1 (a) Schematic diagram of E-plane junction using coupled quasi-H-plane turnstile resonators, (b) Schematic diagram of E-plane junction using single quasi-H-plane turnstile resonator

junction and E-plane turnstile resonators. The asymmetric E-plane circulator is a 5-port circuit. In the E-plane device, with the direction of propagation taken along the z-axis, H_z rather than H_x is perpendicular to the symmetry axis of the device. The magnetic fields corresponding to the counter-rotating eigenvectors are therefore circularly polarized at the side instead of the top and bottom walls of the waveguide. Figure 17.2 illustrates one possible turnstile arrangement. The value of the susceptance slope parameter met in the description of the complex gyrator circuit of the double turnstile junction, once the first circulation condition is met, is consistent with the synthesis of quarter-wave coupled devices. Furthermore, that associated with the latter structure is half that of the single turnstile one, again in keeping with the situation encountered in the description of the H-plane geometry.

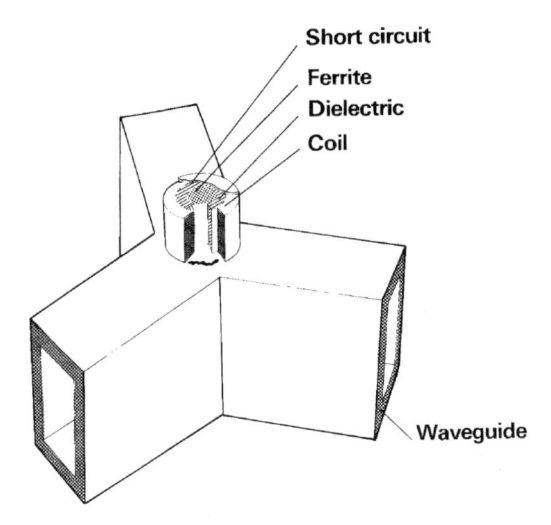

Figure 17.2 Schematic diagram of 3-port E-plane junction using single-turnstile resonator

17.2 Operation of E-Plane Circulator

The adjustment of any circulator requires the perturbation of two counter-rotating and one in-phase eigen-network. The solution investigated in this chapter experimentally displays, in every instance, a passband in the demagnetized state. Devices with stopbands have, however, also been described. The degenerate counter-rotating eigenvalues are in the present geometry determined by that of a quarter-wave-long open circular demagnetized ferrite or dielectric waveguide propagating the HE_{11} mode with one flat face short-circuited and the other flat face loaded by a magnetic wall. This situation differs from that of the more familiar H-plane configuration in that its open flat face is loaded by an electric wall. The frequency of the former arrangement increases as the spacing between the two resonators is reduced; in the latter case it decreases. The corresponding eigen-networks experimentally display magnetic walls at the terminals of the junction. If the junction is propagating for the in-phase eigenvector, symmetry indicates that the corresponding eigen-network has an electric wall boundary condition on its symmetry axis. The dimensions of the junction can then in principle be adjusted so that this eigen-network has a magnetic wall boundary condition at its input terminals. In the arrangement considered here, however, measurements indicate that the junction is evanescent for this excitation and that the in-phase eigen-network has an approximate electric wall boundary condition at its terminals which is independent of the physical geometry of the device. This condition is of course not encountered in the classic H-plane device. Symmetry would otherwise dictate, contrary to the experimental data, that the demagnetized junction would exhibit a bandstop instead of a bandpass response as is the case here.

17.3 Adjustment of E-Plane Circulator

An understanding of the nature of the in-phase and counter-rotating eigenvalue problems is a prerequisite to the adjustment of any junction circulator. The E-plane circulator described in this chapter has a bandpass response in its demagnetized state. Scrutiny of the eigenvalue diagrams in Chapter 19 indicates that this condition is satisfied if its in-phase and counter-rotating eigen-networks form either an *e,2m* or an *m,2e* eigenvalue diagram. Whether one or the other

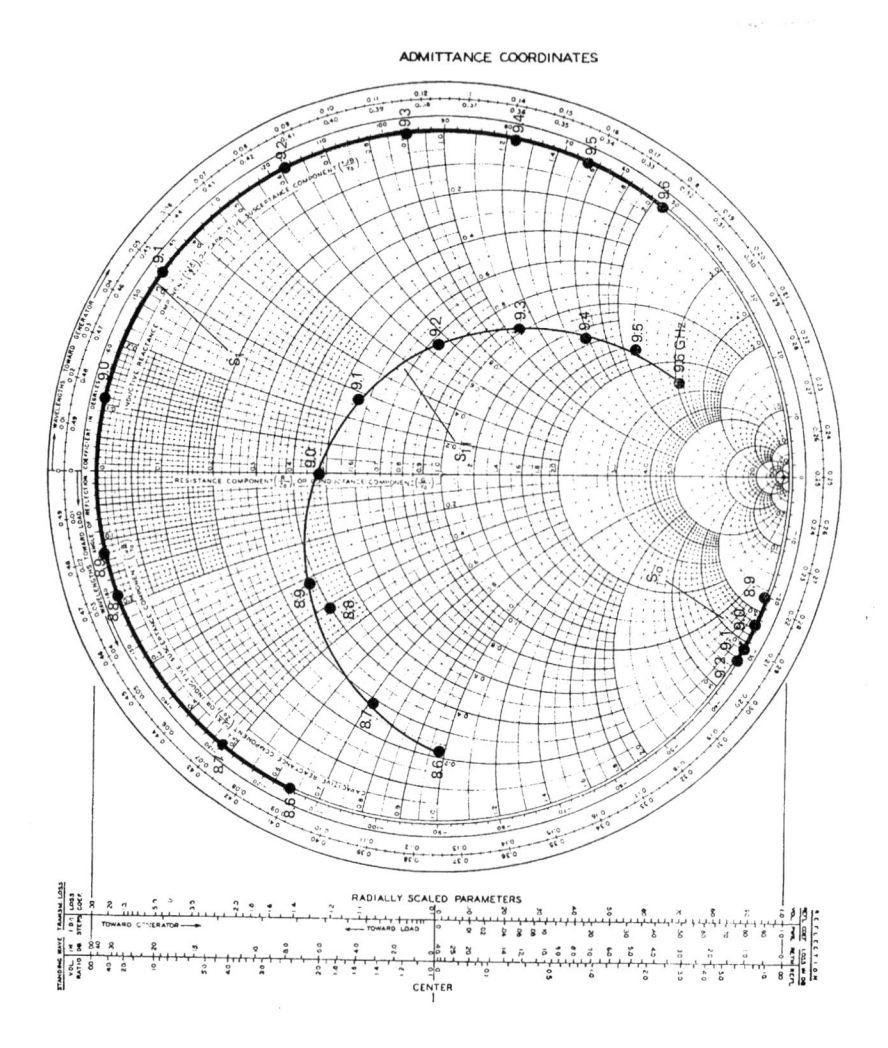

Figure 17.3 Smith chart (admittance coordinates) of in-phase and degenerate counter-rotating eigenvalues and demagnetized frequency response of E-plane junction using coupled H-plane $HE_{111/2}$ turnstile resonators ($\varepsilon_f = 15.0$, $R = 6.0$ mm, $R/R_0 = 1.0$, $R/L = 2.49$, $k = 0.44$) (reprinted with permission, Helszajn, J. (1987) Complex gyrator circuit of an evanescent mode E-plane junction circulator using H-plane turnstile resonator, *IEEE Trans.*, **MTT-35**, 797–806) (© 1987 IEEE)

17.3 ADJUSTMENT OF E-PLANE CIRCULATOR

case prevails requires a theoretical or experimental appreciation of the eigenvalue problem. Of course, whether an eigenvalue exhibits an electric or a magnetic wall depends on the choice of the reference terminals. The ones adopted in this work coincide with those defined by the three waveguides used to form the junction. The experimental in-phase and degenerate admittance eigenvalues, as well as the frequency response of the demagnetized junction, are depicted in Figure 17.3 for one situation. The configuration investigated may therefore with some adjustment of the in-phase eigenvalue be characterized as a degree 1 $e,2m$ R-STUB circuit; the topology in fact associated with the more common H-plane device

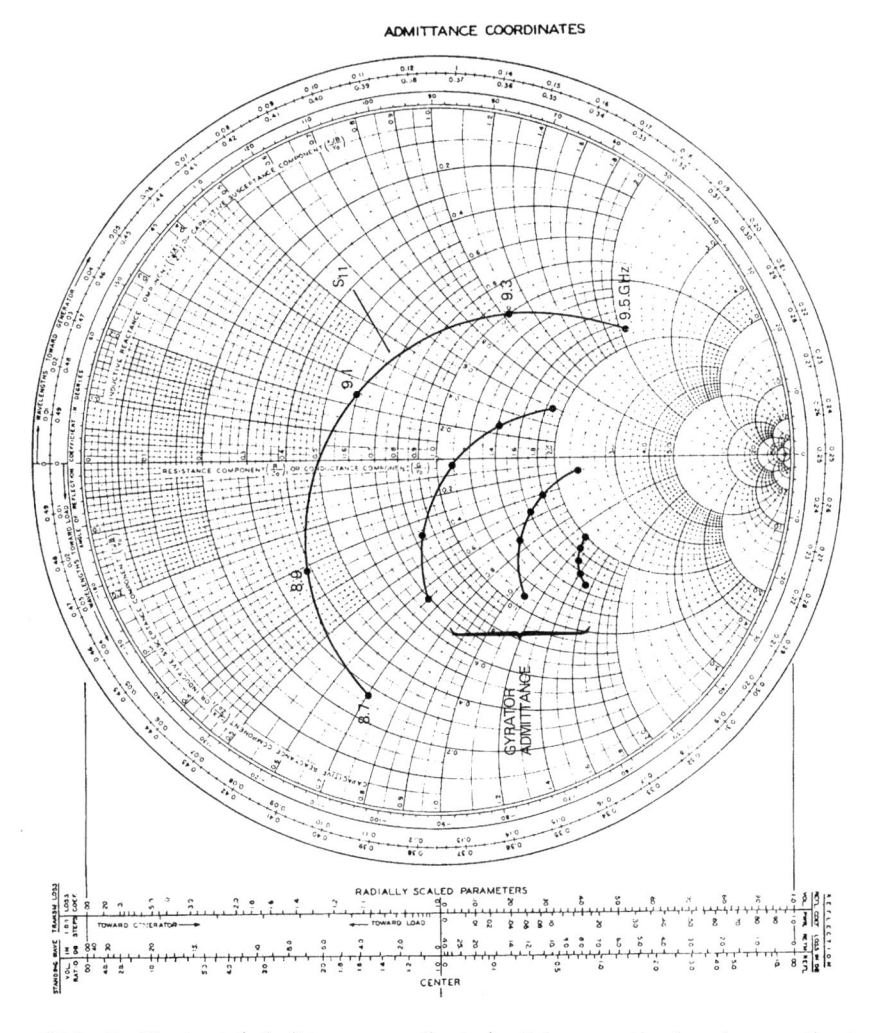

Figure 17.4 Smith chart (admittance coordinates) of demagnetized and magnetized complex gyrator immittance of E-plane junction using coupled H-plane $HE_{111/2}$ turnstile resonators ($\varepsilon_f = 15.0$, $R = 6.0$ mm, $R/R_0 = 1.0$, $R/L = 2.49$, $k = 0.44$) (reprinted with permission, Helszajn, J. (1987) Complex gyrator circuit of an evanescent mode E-plane junction circulator using H-plane turnstile resonator, *IEEE Trans.*, **MTT-35**, 797–806) (© 1987 IEEE)

using E-plane re-entrant turnstile resonators. The type of measurement described here also provides a precise method of determining the operating frequencies and susceptance slope parameters of the eigen-networks of the demagnetized junction of this sort of junction.

The nature of the eigenvalue problem may also be understood from a knowledge of the immittance of the complex gyrator circuit at different direct magnetic fields. If the junction has a passband when demagnetized and if the real part of its complex gyrator immittance is asymptotic to an electric wall as it is magnetized, then it is associated with an $e,2m$ eigenvalue diagram and a degree 1 $e,2m$ R-STUB gyrator circuit. If it is asymptotic to a magnetic wall, it coincides with an $m,2e$ eigenvalue diagram and an $m,2e$ R-STUB circuit. Figure 17.4 shows this result for the eigenvalue solution in Figure 17.3. These immittances do not move along constant-conductance circles on this diagram because the in-phase eigen-network is in this instance some-what overlength.

The ferrite material utilized in this work was a garnet one with a saturation magnetization (M_0) equal to 0.1600 T and a relative dielectric constant (ε_f) of 15.0. The relative dielectric constant (ε_d) of the region between the two resonators was unity. The radius of the resonator (R_i) was arbitrarily made equal to that of the junction formed by the three WR90 waveguides ($R_0 = 6$ mm).

17.4 Counter-Rotating Eigen-networks

The resonator mode associated with the degenerate eigen-networks met in the design of this class of device has not always been identified but is usually the limit $TM_{1,1,1/2}$ or, more strictly speaking, the hybrid $HE_{1,1,1/2}$ one with its open face loaded by a magnetic wall. The $TM_{2,1,1/2}$ is another possibility. Modes in cylindrical cavities with ideal magnetic walls are either TE or TM in type and are described by a three-digit notation as $TE_{l,m,n}$ or $TM_{l,m,n}$. Here, l is the number of full-period variations of E_r with respect to θ, m is the number of half-period variations of E_θ with respect to r, n is 1 for a half-wavelong resonator open-circuited at each end, and n is ½ for a quarter-wave-long resonator open-circuited at one end and short-circuited at the other end. The characteristic equation for the frequency of the degenerate counter-rotating modes of the E-plane junction using H-plane turnstile resonators coincides with the even eigenvalue of two $HE_{1,1,1/2}$ resonators coupled by a section of round waveguide with a contiguous magnetic wall below cutoff; the frequency of the conventional H-plane junction using E-plane turnstile resonator corresponds to the odd eigenvalue of the same geometry. The required arrangement is illustrated in Figure 17.5.

The characteristic equation of the even-mode solution of two coupled resonators from which the length of each resonator (L) may be calculated from a knowledge of its radius (R), the spacing ($2S$) between the two, the propagation constants (α and β), and the constitutive parameters (ε_{eff} and ε_d) of the two regions is

$$\frac{\varepsilon_{eff}}{\beta} \cot(\beta L) - \frac{\varepsilon_d}{\alpha} \tanh(\alpha S) = 0 \qquad (17.1)$$

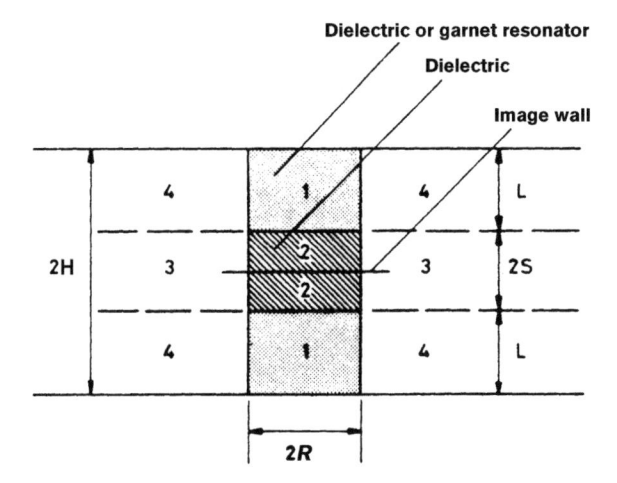

Figure 17.5 Even-mode geometry of coupled turnstile resonators (magnetic wall arrangement)

where

$$\beta^2 = k_0^2 \varepsilon_{\text{eff}} - \left(\frac{1.84}{R}\right)^2 \tag{17.2}$$

$$\alpha^2 = \left(\frac{1.84}{R}\right)^2 - k_0^2 \varepsilon_d \tag{17.3}$$

The propagation constant β is determined by solving the characteristic equation for the HE_{11} mode of the open demagnetized ferrite or dielectric waveguide from a knowledge of ε_f and R. The parameter ε_{eff} is the effective dielectric constant of an equivalent round waveguide with an ideal magnetic wall; it is obtained from a knowledge of β and $k_0 R$ in Chapter 7. The solution adopted here satisfies the boundary conditions between regions 1 and 4 and 1 and 2 in Figure 17.5 but neglects those between 3 and 4 and 2 and 3. Some typical results are summarized in Chapter 9.

17.5 In-Phase Eigen-network

Provisions for the adjustment of this eigenvalue may be at first sight made by noting that the physical variables of the degenerate counter-rotating ones investigated in Chapter 7 are not unique. Figure 17.6 indicates the variation of the admittance associated with S_{11}, with frequency for three different combinations of aspect ratio (R/L) and filling factor (k) for which $k_0 R \sqrt{\varepsilon_f} = 4.38$ at 9.0 GHz. While the frequency dispersion of each arrangement is in keeping with the values of the susceptance slope parameter noted in the next section, the reference terminals of the in-phase eigen-network appears in each instance to be independent of the details of the junction. One explanation for this situation, in keeping with the transition between the stopband and passband solutions noted by Omori (1968), is that the junction is evanescent for the in-phase eigenvector with the mounts in place. In the absence of the resonator mounts, on the other hand, it is propagating, and its eigen-network exhibits an electric wall at the symmetry axis or, equivalently, a magnetic wall at the reference terminals. In the former case the junction has a bandpass characteristic; in the latter instance it has a stopband one.

The effect of a nonideal in-phase eigenvalue may be incorporated in the gyrator circuit without too much difficulty in the manner indicated in Figure 17.7. The

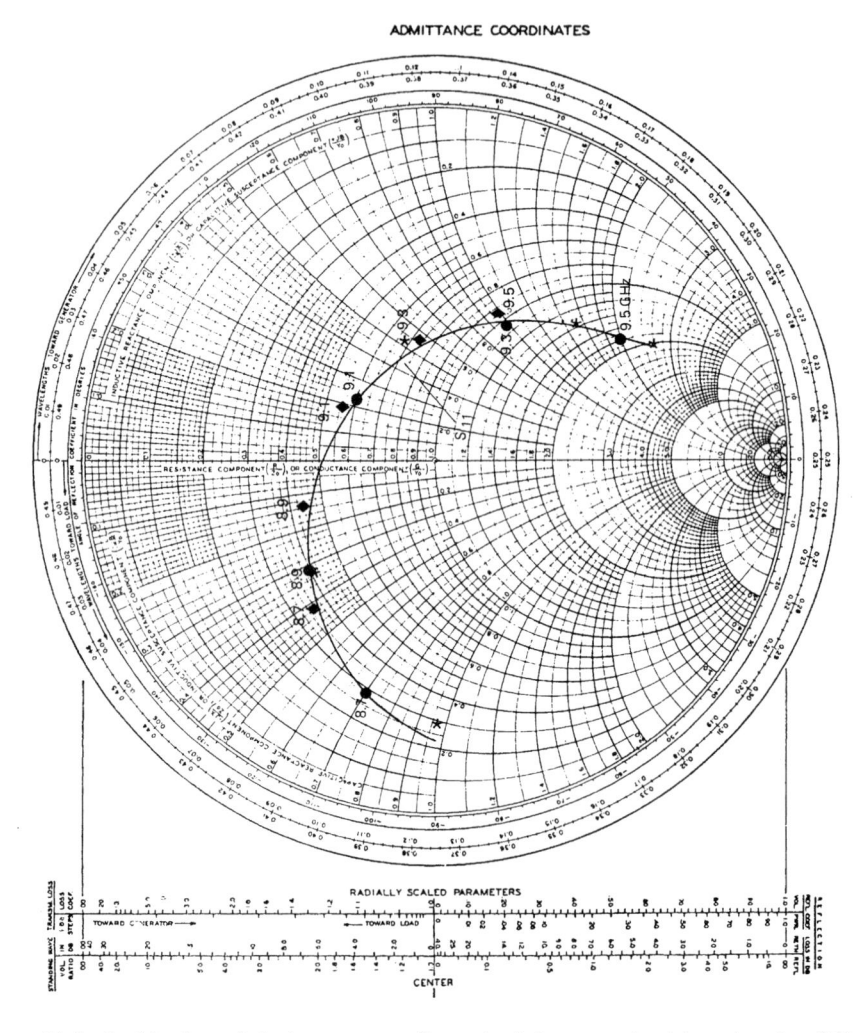

Figure 17.6 Smith chart (admittance coordinates) of demagnetized junction for different values of R/L and k ($R/L = 2.43$, $k = 0.48$, $R/L = 2.49$, $k = 0.44$, $R/L = 2.55$, $k = 0.36$, $R/R_0 = 1.0$) (reprinted with permission, Helszajn, J. (1987) Complex gyrator circuit of an evanescent mode E-plane junction circulator using H-plane turnstile resonator, *IEEE Trans.*, **MTT-35**, 797–806) (© 1987 IEEE)

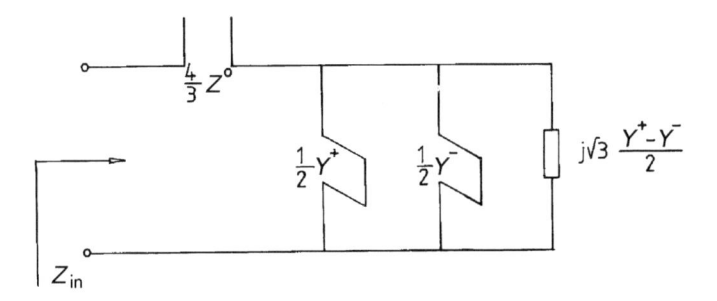

Figure 17.7 Complex gyrator circuit of 3-port junction circulator in terms of immittance eigenvalues (reprinted with permission, Helszajn, J. (1987) Complex gyrator circuit of an evanescent mode E-plane junction circulator using H-plane turnstile resonator, *IEEE Trans.*, **MTT-35**, 797–806) (© 1987 IEEE)

topology of this circuit is reproduced for completeness from the material in Chapter 19; it involves the in-phase impedance eigenvalue (Z^0) and counter-rotating split admittance eigenvalues ($Y^\pm$). Although it is possible to absorb the effect of a nonideal in-phase immittance in the matching network, this should be avoided if at all possible. Scrutiny of the data in the Smith chart in Figure 17.4 indicates that the in-phase eigen-network displays an approximate electric wall at the terminals of the junction, more strictly speaking, a short section of a short-circuited stub or series inductance. It also indicates that the frequency variation of the in-phase eigen-network may be neglected compared to that of the degenerate ones. The parameters of the in-phase eigen-network do not therefore appear in the complex gyrator circuit of this class of device provided it is commensurate with that of the degenerate counter-rotating ones.

17.6 Complex Gyrator Circuit

The complex gyrator circuit of any junction is defined as that exhibited by the device at port 1 with port 3 decoupled from port 2 by terminating the latter port by its complex conjugate load. The data outlined in Figures 17.3 and 17.4 indicate that it may in an evanescent mode E-plane junction, as already noted, be approximated by an $e,2m$ R-STUB network. The elements of this circuit are usually specified in terms of a normalized susceptance slope parameter (b'), a normalized gyrator conductance (g), and loaded quality factor (Q_L) as is by now understood. If the frequency variation of the in-phase eigen-network may be neglected compared to that of the degenerate counter-rotating ones then these parameters are related in a simple way by the split frequencies of the magnetized resonator by

$$g = \sqrt{3}\, b' \left(\frac{\omega_+ - \omega_-}{\omega_0} \right) \tag{17.4}$$

provided again that the in-phase and degenerate eigen-networks are commensurate.

This relationship may be utilized to define the Q-factor of the gyrator circuit in terms of the split frequencies

$$\frac{1}{Q_L} = \sqrt{3}\left(\frac{\omega_+ - \omega_-}{\omega_0}\right) \qquad (17.5)$$

While these two classic equations have now been mentioned more than once in this text these are reproduced here once more the sake of completeness.

Figure 17.8 illustrate the relationships between the normalized susceptance slope parameter and the physical variables of one symmetrical and one asymmetric circuit. This quantity may be evaluated from a knowledge of the frequency response of the directly coupled device in terms of the frequencies ($f_{1,2}$) on either side of the circulation one (f_0) at which the VSWR has some convenient value:

$$b' = \frac{(VSWR - 1)}{2\delta_0 \sqrt{VSWR}} \qquad (17.6)$$

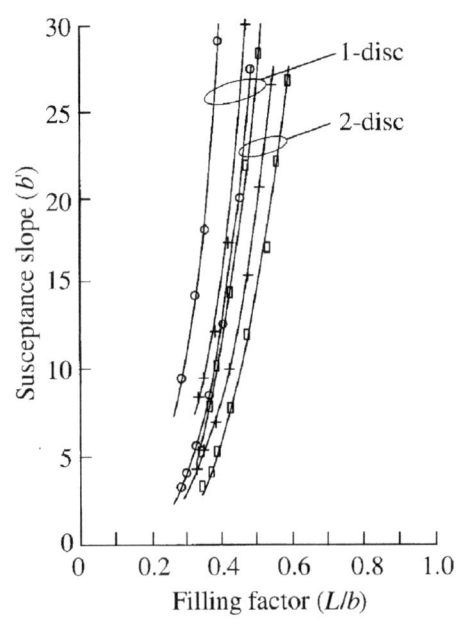

○ $R/L = 3.06$, + $R/L = 2.62$, □ $R/L = 2.36$

Figure 17.8 Susceptance slope parameter of E-plane junction using single and coupled H-plane $HE_{111/2}$ turnstile resonators ($\varepsilon_f = 15.0$, $R = 6.0$ mm, $R/L = 2.36$, 262 and 3.06, $R/R_0 = 1$) (reprinted with permission, Helszajn, J. (1987) Complex gyrator circuit of an evanescent mode E-plane junction circulator using H-plane turnstile resonator, *IEEE Trans.*, **MTT-35**, 797–806) (© 1987 IEEE)

17.6 COMPLEX GYRATOR CIRCUIT

Although this measurement assumes that the in-phase and degenerate counter-rotating eigen-networks are commensurate, a condition not altogether satisfied here, it is on past experience good enough for engineering purposes. A comparison between the value obtained in this manner for the specific case treated in Figure 17.3 and that derived using its more formal definition in terms of the frequency variation of the susceptance (b') of the degenerate counter-rotating immittance eigenvalues.

$$b' = \frac{\omega}{2} \frac{\partial b_1}{\partial \omega} \bigg|_{\omega = \omega_0} \qquad (17.7)$$

indicates that this is indeed so.

Interestingly enough, the susceptance slope parameter of the one-disk configuration is twice that of the coupled-disk geometry (Figure 17.8); a feature also met in the description of the H-plane structure. Figures 17.9 and 17.10 summarize some

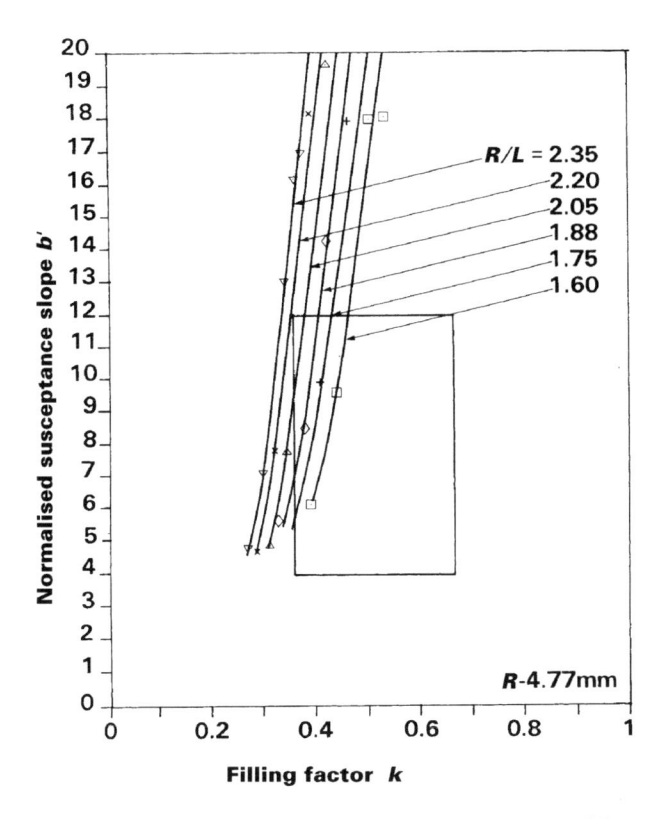

Figure 17.9 Susceptance slope parameter of E-plane circulator for different aspect ratios. ($R = 4.77$ mm, $R/R_0 = 0.795$, $\varepsilon_f = 15.1$) (reprinted with permission, Helszajn, J. and Cheng, S. (1990) Aspect ratio of open resonators in the design of evanescent mode E-plane circulators, *IEE Proc. Microwaves, Antennas and Propagation*, **137**, 55–60)

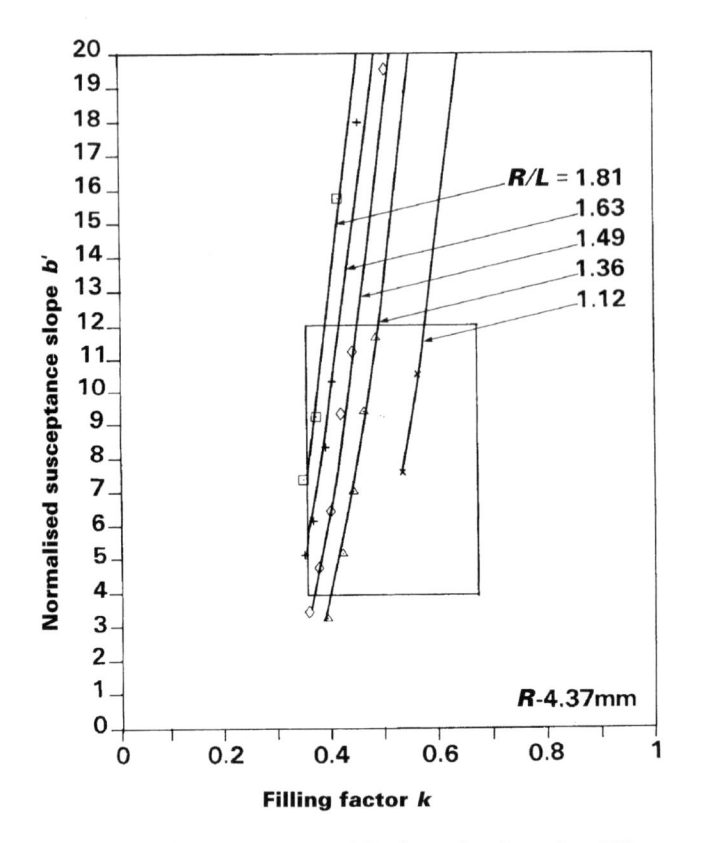

Figure 17.10 Susceptance slope parameter of E-plane circulator for different aspect ratios. $R = 4.37$ mm, $R/R_0 = 0.728$, $\varepsilon_f = 15.1$) (reprinted with permission, Helszajn, J. and Cheng, S. (1990) Aspect ratio of open resonators in the design of evanescent mode E-plane circulators, *IEE Proc. Microwaves, Antennas and Propagation*, **137**, 55–60)

additional results on coupled disk geometries for R/R_0 not equal to unity. Values of susceptance slope parameter between 4 and 12 are suitable for the design of quarter-wave coupled devices. The field of useful solutions lies in the cross hatched area on these illustrations. The susceptance slope parameter is, however, not an independent parameter since it relies upon the same physical parameters employed to satisfy the first circulation condition. Since the radial wavenumber ($k_0 R$) is a flat function of the filling factor (k), it follows that the former may be employed to set the frequency and the latter may be utilized to fix the susceptance slope parameter.

Scrutiny of equation (17.4) or of the data in Figure 17.4 indicates that the gyrator conductance may be adjusted in the usual way by the gyrotropy of the resonator. This parameter may be evaluated from a knowledge of the quantities in (17.4) or by utilizing one or the other of the methods described in Chapter 24. Figure 17.11 gives the same additional data on the radial wavenumber of this sort of circuit and Figure 17.12 some further data on its susceptance slope parameter. The quality factor of this sort of resonator is given in Chapter 10.

17.6 COMPLEX GYRATOR CIRCUIT 351

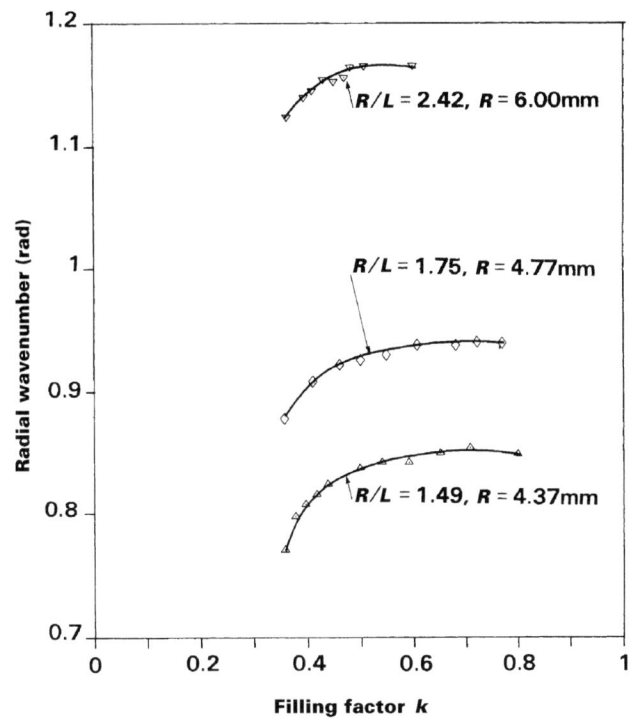

Figure 17.11 Normalized radial wavenumbers parameters of E-plane circulator for different aspect ratios for three resonator geometries displaying a solution at 9 GHz at a filling factor of approximately 0.40

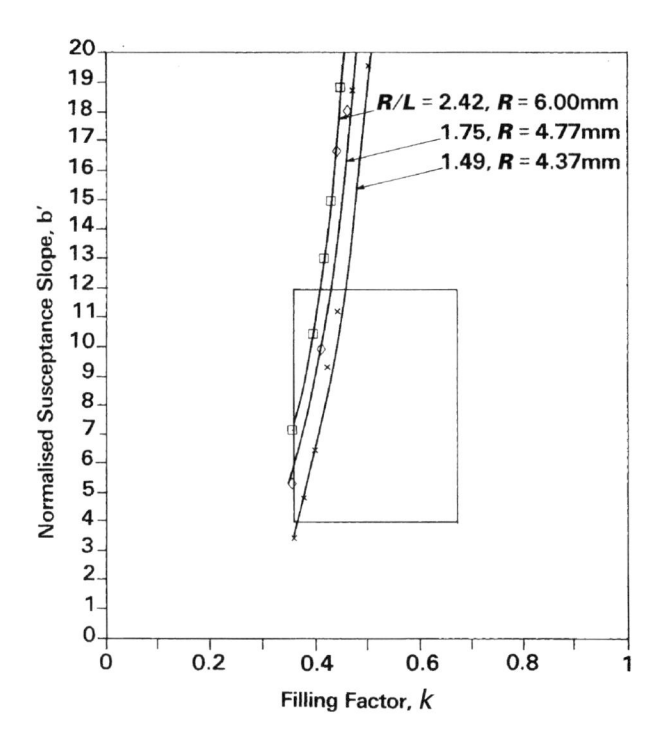

Figure 17.12 Normalized susceptance slope parameters of E-plane circulator for different aspect ratios for three resonator geometries displaying a solution at 9 GHz at a filling factor of approximately 0.40

17.7 Quarter-Wave Coupled E-Plane Junction Circulator

Scrutiny of the element values of the gyrator circuit of the E-plane device examined in this work indicates that these are in keeping with those associated with the more conventional H-plane one. It is therefore expected that, if the in-phase eigen-network in the complex gyrator circuit in Figure 17.5 is absorbed in the matching network, network specifications akin to those realizable with the latter junction should also be achievable with the former device. Figure 17.13 gives one frequency response.

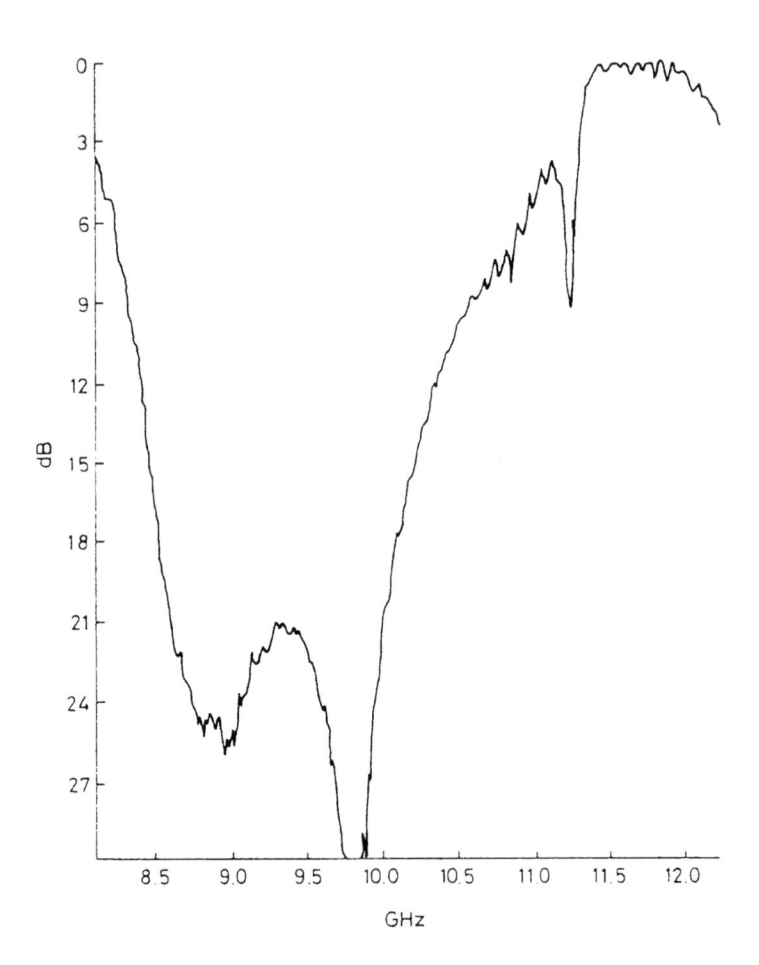

Figure 17.13 Frequency response of E-plane circulator, $R = 4.77$ mm, $R/L = 1.75$ (reprinted with permission, see Fig. 17.10)

17.8 E-Plane Finline Circulator Using H-Plane Turnstile Resonators

Another circulator geometry is an evanescent E-plane finline junction symmetrically loaded in the H-plane by open ferrite resonators. The arrangement is therefore a seven-port circuit, comprising 3-symmetrical finline E-plane ports and 2-orthogonal ports for each round open waveguide port in the H-plane. Figures 17.14(a) and (b) depict arrangements using bilateral and unilateral structures. The linear dimensions of the junction are defined by its overall axial dimension ($2H$), by the radius (R) and length (L) of each ferrite or garnet resonator. Figures 17.15 illustrates one typical finline circuit

The operating frequency of this sort of device corresponds again to the even solution of two coupled $TM_{1,1,1/2}$ resonators with quasi-magnetic walls. Its counter-rotating eigen-networks may therefore be represented in terms of quarter-wave long open resonators, with one flat face open-circuited and the other short-circuited. The axial details of the junction are separately adjusted so that it is cut-off for the in-phase eigen-network. Its eigen-network may therefore be approximately idealized by an electric wall at the terminals of the junction, provided its cut-off frequency is sufficiently below that of the operating frequency of the device. Symmetry would otherwise require that it exhibits a magnetic wall there. This arrangement separately ensures that the overall device acts as a bandpass circuit in its demagnetized state in keeping with experiment. The frequency variation of

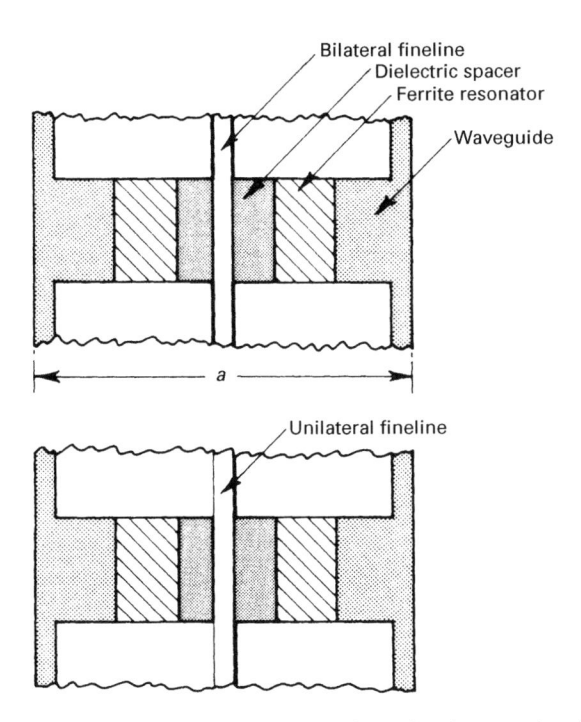

Figure 17.14 Schematic diagrams of E-plane finline circulator using H-plane turnstile resonators

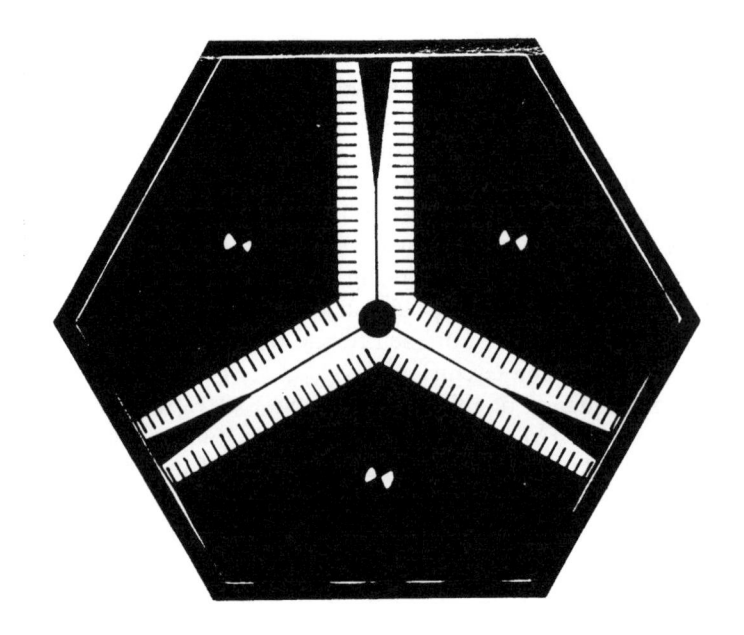

Figure 17.15 Bilateral finline E-plane junction

the in-phase eigenvalue is again neglected, compared to those of the split or degenerate counter-rotating ones. The dielectric spacer which is usually utilized to adjust the in-phase eigen-network in the more common H-plane device may therefore, in this instance, be used as an independent variable to establish the open face of the counter-rotating modes and its coupling to the terminal planes of the junction. The coupling between the terminals of the eigen-resonators and those of the eigen-networks is less well understood, and is usually described by an ideal transformer; the dielectric spacer provides one independent means of varying this quantity. The eigenvalue diagram of the device has therefore the same nature as that of the E-plane circulator already dealt with in this chapter.

17.9 Experimental Adjustment of Finline Circulator

The relationship between the filling factor L/H or k and the radial wavenumber $k_0 R$ of one experimental device is depicted in Figure 17.16 for four different values of the aspects ratio R/L of the resonator. The even-mode solution of two coupled $TM_{1,1,1/2}$ resonators, using open and closed walls, are separately superimposed on this illustration. The open resonator model adopted in this calculation consists of a pair of quarter-wave long resonators, with an ideal magnetic sidewall and an effective dielectric constant to cater for the open wall condition, separated by a continuous section of cut-off waveguide with a magnetic sidewall. The discrepancy between theory and experiment is, in part, due to the fact that the operating frequency of circulators for which the in-phase eigen-network has not been idealized coincides with the frequency at which the phase angles of the counter-rotating

17.9 EXPERIMENTAL ADJUSTMENT OF FINLINE CIRCULATOR 355

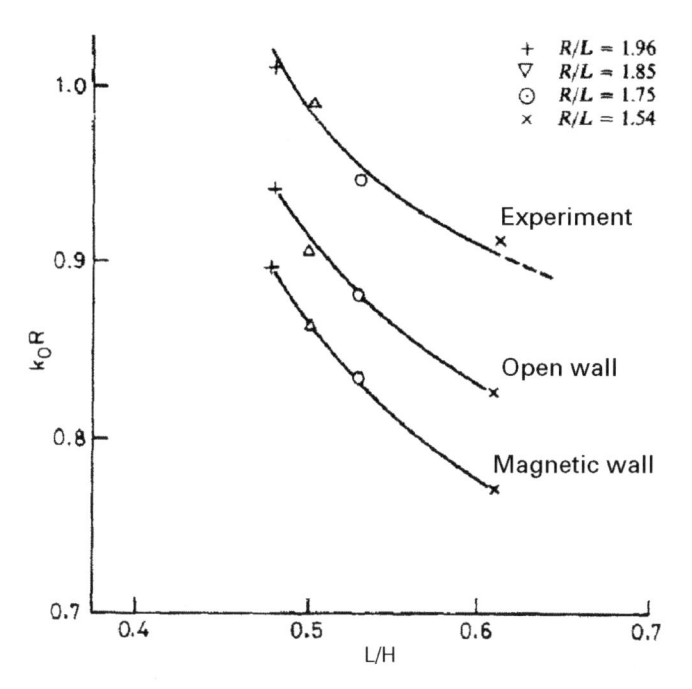

Figure 17.16 Relationship between filling factor and radial wavenumber for different aspect ratios of the resonator (reprinted with permission, Helszajn, J., McDonald, G. and Sutherland, D. (1988) Design of finline circulators using turnstile functions, *IEE Proc., Microwaves, Antennas and Propagation*, 279–281)

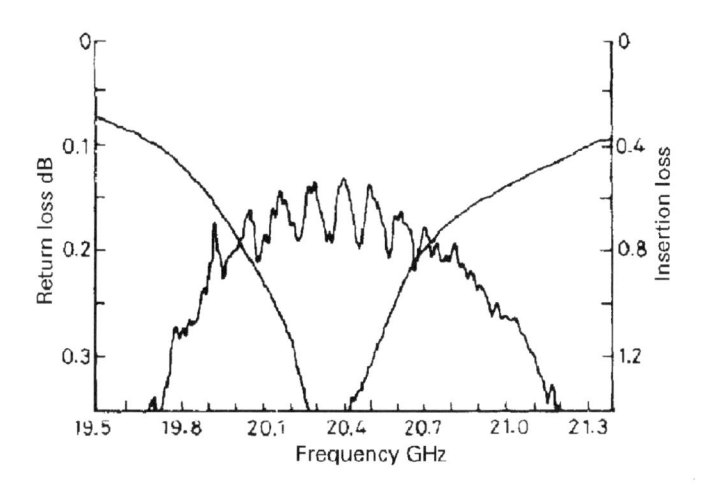

Figure 17.17 Frequency response of directly coupled E-plane finline circulator using H-plane turnstile resonators (reprinted with permission, Helszajn, J., McDonald, G. and Sutherland, D. (1988) Design of finline circulators using turnstile functions, *IEE Proc., Microwaves, Antennas and Propagation*, 279–281)

and in-phase eigen-networks are in antiphase, rather than at that for which these are commensurate.

The radius R of the junction employed in obtaining the result in Figure 17.16 was fixed at 2.15 mm and its overall axial dimension 2H was kept constant at 4.60 mm. This radius corresponds to that of the junction of the three E-plane WR 42 waveguides employed to form the device. The length of each resonator was separately varied between 1.10 and 1.40 mm; each experimental point in this illustration is therefore associated with a resonator with a different aspect ratio R/L. The material used in this work was a lithium ferrite with a magnetization M_0 of 0.4100 T, and a relative permittivity ε_f of 14.8.

Figure 17.17 depicts the insertion and return losses of one directly coupled circulator. The insertion loss of the device includes that of the finline circuits at each port. It is of note that its susceptance slope parameter is typically of the order encountered in the related E-plane problem region.

References and Further Reading

Alexander, I. M. and Flett, I. (1977) Design and development of high-power, Ku-band junction circulators, *Trans. IEEE Magn.*, **MAG-13**(5), 1255–1257.

Auld, B. A. (1959) The synthesis of symmetrical waveguide circulators, *IRE Trans. Microwave Theory Tech.*, **MTT-7**, 238–246.

Beyer, A. and Solbach, K. (1981) A new finline ferrite isolator for integrated millimeter-wave circuit, *IEEE Trans.*, **MTT-29**, 1344–1348.

Beyer, A. and Wolff, I. (1981) Finline ferrite isolator and circulator for K-band, In: *Proc. of 11th Eur. Microwave Conf.*, 321–326.

Braas, M. and Schieblich, C. (1981) E-type circulator for finlines, *Electron. Letters*, **17**, 701–702.

Buchta, G. (1966) Miniaturized broad-band E Tee Circulator at X-band, *IEEE Trans. Microwave Theory Tech.*, **MTT-14**, 1607–1608.

Darwent, A. (1967) Symmetry in high-power circulators for 35 GHz, *IEEE Trans. Microwave Theory Tech.*, **MTT-15**.

Davis, L. E. and Longley, S. R. (1963) E-plane 3-port X-band waveguide circulators, *IEEE Trans. Microwave Theory Tech.*, **MTT-11**, 443–445.

Goebel, U. and Schieblich, C. (1982) Broadband finline circulators, *Proc. IEEE Symp. Digest*, 249–251.

Goebel, U. and Schieblock, C. (1983) A unified equivalent circuit representation of H and E-plane junction circulators. In: *European Microwave Conf.*, 803–808.

Helszajn, J. (1974) Common waveguide circulator configuration, *Electron. Eng.*, 66–68.

Helszajn, J. (1985) Synthesis of quarter-wave coupled junction circulators with degrees 1 and 2 complex gyrator circuits, *IEEE Trans. Microwave Theory Tech.*, **MTT-33**, 382–390.

Helszajn, J. (1987) Complex gyrator circuit of an evanescent mode E-plane junction circulator using H-plane turnstile resonator, *IEEE Trans.*, **MTT-35**, 797–806.

Helszajn, J. and McDermott, M. (1972) Mode charts for E-plane circulators, *IEEE Trans. Microwave Theory Tech.*, **MTT-20**, 187–188.

Helszajn, J. and Thorpe, W. (1985) 18–26 GHz finline resonance isolators using hexagonal ferrites, *IEE Proc. H, Microwaves, Antennas & Propag.*, **132**, 73–76.

Helszajn, J., McDonald, G. and Sutherland, D. (1988) Design of finline circulators using turnstile functions, *IEE Proc., Microwaves, Antennas and Propagation*, **135**, 279–281.

Helszajn, J. and Cheng, S. (1990) Aspect ratio of open resonators in the design of evanescent mode E-plane circulators, *IEE Proc. Microwaves, Antennas and Propagation*, **137**, 55–60.

Meier, P. J. (1976) Integrated finline/A versatile and proven millimeter transmission line, *Microwave J.*, 24–25.

Meier, P. J. (1978) Millimeter integrated circuits suspended in the E-plane of rectangular waveguide, *IEEE Trans.*, **MTT-26**, 726–733.

Omori, M. (1968) An improved E-plane waveguide circulator, *G-MTT, Symp. Dig.*, 228–236.

Solbach, K. (1982) Equivalent circuit of the E-plane Y-junction circulator, *IEEE Trans. Microwave Theory Tech.*, **MTT-30**, 806–809.

Solbach, K. (1983) Status of printed millimeter-wave E-plane circuits, *IEEE Trans.*, **MTT-31**, 107–121.

Wright, W. and McGowan, J. (1968) High-power, Y-junction E-plane circulators, *IEEE Trans. Microwave Theory Tech.*, **MTT-16**, 557–559.

Yoshida, S. (1959) E-type T-circulator, *Proc. IRE*, **47**, 208.

18

WAVEGUIDE CIRCULATORS USING TRIANGULAR AND PRISM RESONATORS

18.1 Introduction

Two important resonator geometries not dealt with so far which have the symmetry of the 3-port junction circulator are the triangular post and the prism or re-entrant turnstile arrangements. The chapter includes the solution of the isotropic triangular resonator with top and bottom electric walls and ideal magnetic side-walls. The standing wave solution of an ideal circulator is obtained with this sort of resonator by taking suitable linear combinations of the isotropic ones. One symmetry met with either the post or prism resonator is illustrated in Figure 18.1. Another is obtained by rotating the resonator by 120°. The odd and even mode solutions of a pair of prism resonators which enter into the descriptions of H and E-plane circulators are separately described.

The chapter includes some experimental data on the mode chart, quality factor and susceptance slope parameter of these sorts of circulators. The quality factor of the prism resonator is a property of its geometry. In practice both E and H-plane circulators have similar values of gain-bandwidth products.

18.2 TM Field Patterns of Triangular Planar Resonator

The TM-mode field patterns in a triangular-shaped demagnetized ferrite or dielectric resonator having no variation of the field patterns along the thickness of the resonator are given by

$$E_z = A_{m,n,l} T(x, y) \tag{18.1a}$$

$$H_x = \frac{j}{\omega \mu u_0 \mu_e} \frac{\partial E_z}{\partial y} \tag{18.1b}$$

18 WAVEGUIDE CIRCULATORS

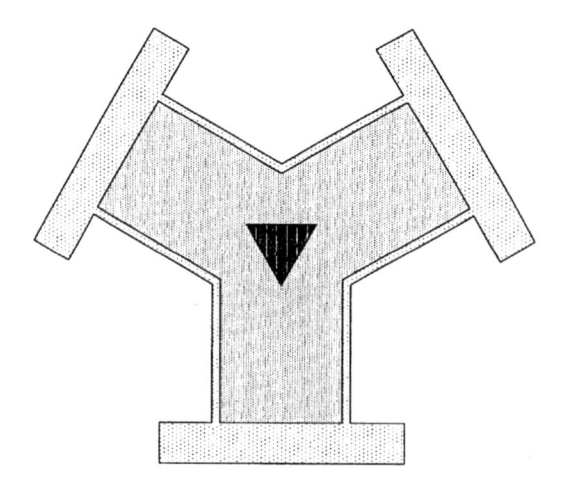

Figure 18.1 Schematic diagrams of waveguide circulator using prism resonators

$$H_y = \frac{-j}{\omega\mu_0\mu_e} \frac{\partial E_z}{\partial x} \tag{18.1c}$$

$$H_z = E_x = E_y = 0 \tag{18.1d}$$

where $A_{m,n,l}$ is a constant. Figure 18.2 shows the geometry of the planar resonator discussed in this text.

For magnetic boundary conditions, $T(x,y)$ may be obtained by duality from that of the TE mode with electric boundary conditions

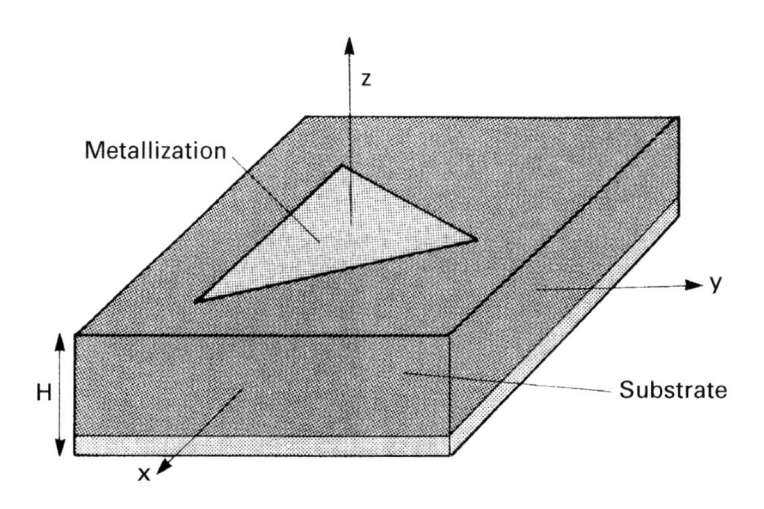

Figure 18.2 Schematic of planar triangular resonator

18.2 TM FIELD PATTERNS

$$T(x,y) = \cos\left[\left(\frac{2\pi x}{\sqrt{3}\,A} + \frac{2\pi}{3}\right)l\right]\cos\left[\frac{2\pi(m-n)y}{3A}\right]$$

$$+ \cos\left[\left(\frac{2\pi x}{\sqrt{3}\,A} + \frac{2\pi}{3}\right)m\right]\cos\left[\frac{2\pi(n-l)y}{3A}\right]$$

$$+ \cos\left[\left(\frac{2\pi x}{\sqrt{3}\,A} + \frac{2\pi}{3}\right)n\right]\cos\left[\frac{2\pi(l-m)y}{3A}\right] \quad (18.2)$$

where A is the length of the triangle side and

$$m + n + l = 0 \quad (18.3)$$

$T(x, y)$ satisfies the wave equation below

$$\left(\frac{\partial^2}{\partial x^2} + \frac{\partial^2}{\partial y^2} + k_{m,n,l}^2\right)E_z = 0 \quad (18.4)$$

where

$$k_{m,n,l} = \left(\frac{4\pi}{3A}\right)\cdot\sqrt{m^2 + mn + n^2} \quad (18.5)$$

The complete standing wave solution is

$$E_z = A_{m,n,l}\, T(x,y) \quad (18.6a)$$

$$H_x = \frac{-jA_{m,n,l}}{\omega\mu_0\mu_e}\Bigg\{\frac{2\pi(m-n)}{3A}\cos\left[\left(\frac{2\pi x}{\sqrt{3}\,A} + \frac{2\pi}{3}\right)l\right]\cdot\sin\left[\frac{2\pi(m-n)y}{3A}\right]$$

$$+ \frac{2\pi(n-l)}{3A}\cdot\cos\left[\left(\frac{2\pi x}{\sqrt{3}\,A} + \frac{2\pi}{3}\right)m\right]\cdot\sin\left[\frac{2\pi(n-l)y}{3A}\right]$$

$$+ \frac{2\pi(l-m)}{3A}\cdot\cos\left[\left(\frac{2\pi x}{\sqrt{3}\,A} + \frac{2\pi}{3}\right)n\right]\cdot\sin\left[\frac{2\pi(l-m)y}{3A}\right]\Bigg\} \quad (18.6b)$$

$$H_y = \frac{-jA_{m,n,l}}{\omega\mu_0\mu_e}\Bigg\{\frac{2\pi l}{\sqrt{3}\,A}\sin\left[\left(\frac{2\pi x}{\sqrt{3}\,A} + \frac{2\pi}{3}\right)l\right]\cdot\cos\left[\frac{2\pi(m-n)y}{3A}\right]$$

$$+ \frac{2\pi m}{\sqrt{3}\,A}\cdot\sin\left[\left(\frac{2\pi x}{\sqrt{3}\,A} + \frac{2\pi}{3}\right)m\right]\cdot\cos\left[\frac{2\pi(n-l)y}{3A}\right]$$

$$+ \frac{2\pi n}{\sqrt{3}\,A}\cdot\sin\left[\left(\frac{2\pi x}{\sqrt{3}\,A} + \frac{2\pi}{3}\right)n\right]\cdot\cos\left[\frac{2\pi(l-m)y}{3A}\right]\Bigg\} \quad (18.6c)$$

It is observed that the interchange of the three digits m,n,l leaves the cut-off number $k_{m,n,l}$ unchanged; similarly, the field patterns are retained, without rotation.

18.3 TM$_{1,0,-1}$ Field Components of Triangular Planar Resonator

The field patterns of the dominant mode in a planar triangular resonator are given by (6) with $m = 1$, $n = 0$, $l = -1$. The result is

$$E_z = A_{1,0,-1}\left[2\cos\left(\frac{2\pi x}{\sqrt{3}\,A} + \frac{2\pi}{3}\right)\cdot\cos\left(\frac{2\pi x}{3A}\right) + \cos\left(\frac{4\pi y}{3A}\right)\right] \quad (18.7a)$$

$$H_x = -jA_{1,0,-1}\zeta_e\left[\cos\left(\frac{2\pi x}{\sqrt{3}\,A} + \frac{2\pi}{3}\right)\cdot\sin\left(\frac{2\pi x}{3A}\right) + \sin\left(\frac{4\pi y}{3A}\right)\right] \quad (18.7b)$$

$$H_y = j\sqrt{3}\,A_{1,0,-1}\zeta_e\left[\sin\left(\frac{2\pi x}{\sqrt{3}\,A} + \frac{2\pi}{3}\right)\cdot\cos\left(\frac{2\pi y}{3A}\right)\right] \quad (18.7c)$$

where

$$k_{1,0,-1} = \frac{4\pi}{3A} \quad (18.8)$$

$$\zeta_e = \sqrt{\frac{\varepsilon_0\varepsilon_r}{\mu_0\mu_e}} \quad (18.9)$$

Figure 18.3 is a sketch of the magnetic and electric fields for the dominant TM$_{1,0,-1}$ mode in a triangular resonator, and Figure 18.4 (a) and (b) indicate the corresponding magnetic field and equipotential lines.

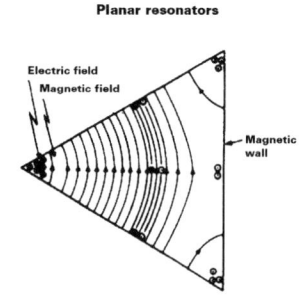

Figure 18.3 TM$_{1,0,-1}$ dominant mode field pattern in triangular resonator with magnetic walls (reprinted with permission, Helszajn, J. and James, D. S. (1978) Planar triangular resonators with magnetic walls, *IEEE Trans. Microwave Theory Tech.*, **MTT-26**, 95–100) (© 1978 IEEE)

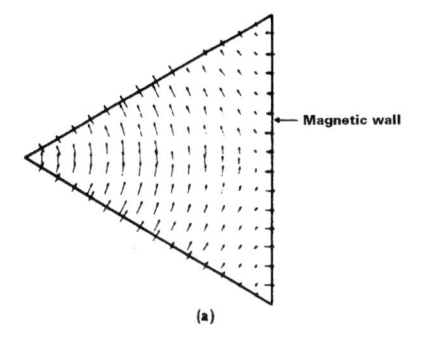

Figure 18.4 (a) Magnetic field pattern for TM$_{1,0,-1}$ dominant mode; (b) lines of equipotential. for TM$_{1,0,-1}$ mode (reprinted with permission, Helszajn, J. and James, D. S. (1978) Planar triangular resonators with magnetic walls, *IEEE Trans. Microwave Theory Tech.*, **MTT-26**, 95–100) (© 1978 IEEE)

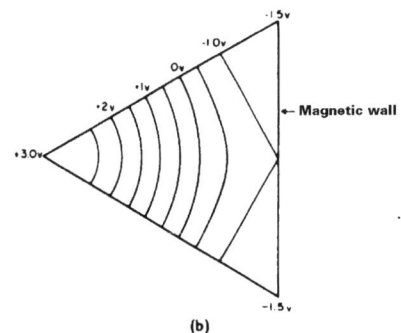

18.4 TM$_{1,1,-2}$ Field Components of Triangular Planar Resonator

In addition to the dominant mode the next higher order mode in a planar triangular resonator has also been investigated. This mode is a symmetrical or in-phase one for which the field pattern is given in Figure 18.5. It is obtained with $m = 1$, $n = 1$, $l = -2$; $T(x,y)_{m,n,l}$, and $k_{m,n,l}$ are therefore described by

$$T(x,y)_{1,1,-2} = \cos 2\left(\frac{2\pi x}{\sqrt{3}\,A} + \frac{2\pi}{3}\right) + 2\cos\left(\frac{2\pi x}{\sqrt{3}\,A} + \frac{2\pi}{3}\right) \cdot \cos\left(\frac{2\pi y}{A}\right) \quad (18.10)$$

$$k_{1,1,-2} = \frac{4\pi}{\sqrt{3}\,A} \quad (18.11)$$

A property of this mode is that it has a maximum at the origin, unlike the $T(x,y)_{1,0,-1}$ one which has a zero there. This maximum is given by

$$T(0,0)_{1,1,-2} = \frac{-3}{2} \quad (18.12)$$

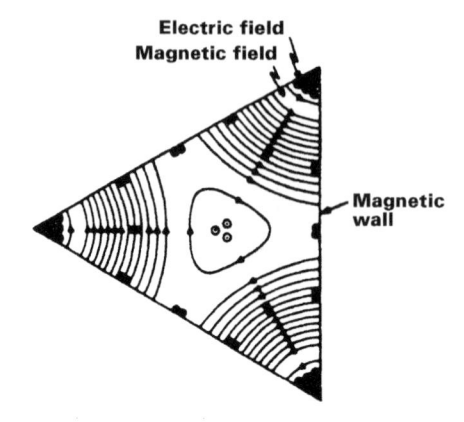

Figure 18.5 TM$_{1,1,-2}$ field pattern in triangular resonator with magnetic walls (reprinted with permission, Helszajn, J. and James, D. S. (1978) Planar triangular resonators with magnetic walls, *IEEE Trans. Microwave Theory Tech.*, **MTT-26**, 95–100) (© 1978 IEEE)

It is easily verified that this mode is symmetrical,

$$T\left(0, \frac{A}{2\sqrt{3}}\right)_{1,1,-2} = T\left(\frac{-A}{4\sqrt{3}}, \frac{A}{4}\right)_{1,1,-2} = T\left(\frac{-A}{4\sqrt{3}}, \frac{-A}{4}\right)_{1,1,-2} \quad (18.13)$$

18.5 Triangular Post Circulator

The most simple waveguide circulator arrangement based on a triangular resonator is the post one with top and bottom electric walls and magnetic sidewalls. The operating frequency of this geometry is approximately given by

$$k_{1,0,-1} A = \frac{4\pi}{3} \quad (18.14)$$

where

$$k_{1,0,-1} = k_0 \sqrt{\mu_r \varepsilon_r} \quad (18.15)$$

and k_0 is the wave-number at its centre frequency.

The construction of the dominant mode standing wave solutions is separately depicted in Figure 18.6.

The quality factor of such a gyromagnetic resonator may be shown to be given by

$$\frac{1}{Q_L} = \left(\frac{3}{\pi}\right) \kappa \quad (18.16)$$

This sort of resonator may also be quarter-wave coupled as is readily appreciated. Figure 18.7 summarizes some possibilities.

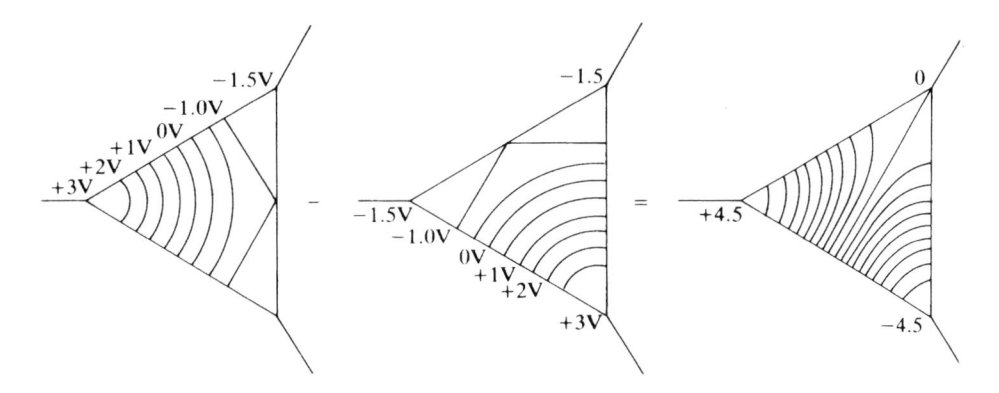

Figure 18.6 Standing wave solutions of waveguide circulator using a triangular resonator (reprinted with permission, Helszajn, J. and James, D. S. (1978) Planar triangular resonators with magnetic walls, *IEEE Trans. Microwave Theory Tech.*, **MTT-26**, 95–100) (© 1978 IEEE)

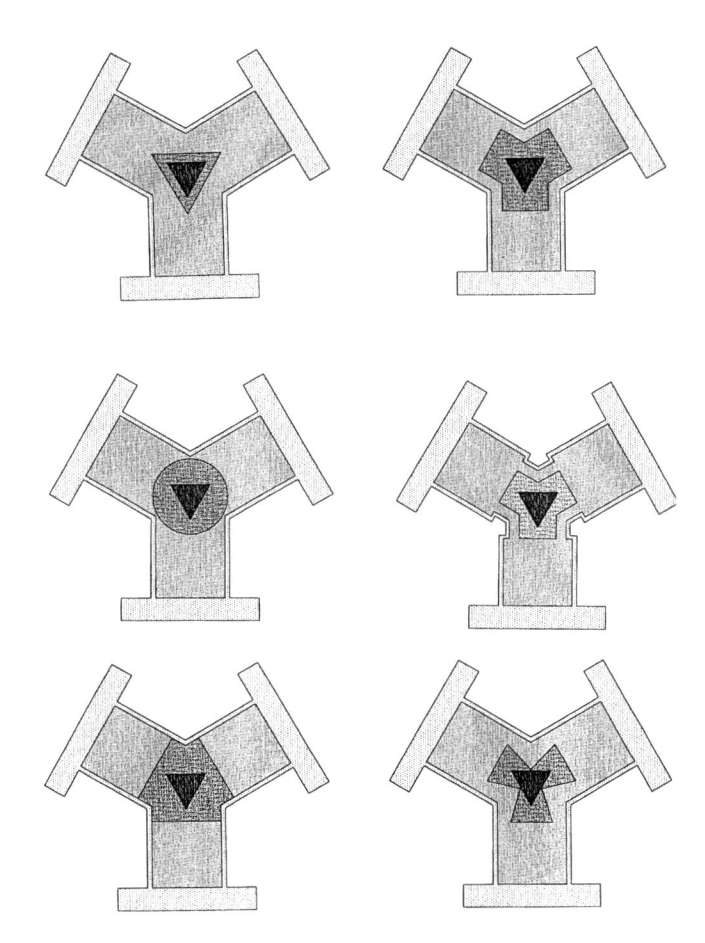

Figure 18.7 Schematic diagrams of waveguide circulators using quarter-wave coupled triangular resonators

18.6 Triangular Waveguide with Magnetic Wall

A waveguide of some interest is the equilateral one with either an electric or a magnetic wall. The problem region of interest in the design of waveguide circulators using prism resonators, however, is the magnetic one. The mode nomenclature employed in the description of this waveguide is that met in connection with the cut-off space of the related planar problem region.

$$k^2_{n,m,l} = \left(\frac{4\pi}{3A}\right)(m^2 + mn + n^2) \tag{18.17}$$

where

$$m + n + l = 0 \tag{18.18}$$

and A is the side dimension of the waveguide.

The waveguide wavelength is defined in the usual way by

$$\left(\frac{2\pi}{\lambda_g}\right)^2 = \left(\frac{2\pi}{\lambda_0}\right)^2 \varepsilon_r - (k_{n,m,l})^2 \tag{18.19}$$

The dominant solution is the $TM_{1,0,-1}$ one for which $m = 1$; $n = -1$; $l = 0$ and $k_{1,0,-1}$ satisfies (18.14) and (18.15).

18.7 Perturbation Theory of Closed Gyromagnetic Prism Resonator

Another resonator of some interest is a quarter-wave long prism arrangement open-circuited at one end and short-circuited at the other. The perturbation solution for the phase constants of such an arrangement with ideal magnetic side-walls is reproduced below.

$$\beta^2_\pm = k_0^2 \varepsilon_{\text{eff}} (\mu \mp C\kappa) - \left(\frac{4\pi}{3}\right)^2 \tag{18.20}$$

The constant C is given by

$$C = \frac{\sqrt{3}}{2\pi} \tag{18.21}$$

κ is the off-diagonal element of the tensor permeability and μ is equal to unity. The cut-off number of the dominant mode in this sort of waveguide is given by (18.20) and (18.21). The degenerate frequencies employed in this approximation are given approximately by

18.7 PERTURBATION THEORY

$$(k_0 A)^2 = \frac{\left(\frac{\pi}{2}\right)^2 \left(\frac{A}{L}\right)^2 + \left(\frac{4\pi}{3}\right)^2}{\mu_{\text{eff}} \varepsilon_r} \tag{18.22}$$

A property of this sort of resonator is that both its degenerate cut-off numbers and its phase constants are split by the gyrotropy. The split frequencies are

$$(k_- A)^2 = \frac{\left(\frac{\pi}{2}\right)^2 \left(\frac{A}{L}\right)^2 + \left(\frac{4\pi}{3}\right)^2}{(\mu + C\kappa)\,\varepsilon_r} \tag{18.23}$$

$$(k_+ A)^2 = \frac{\left(\frac{\pi}{2}\right)^2 \left(\frac{A}{L}\right)^2 + \left(\frac{4\pi}{3}\right)^2}{(\mu - C\kappa)\,\varepsilon_r} \tag{18.24}$$

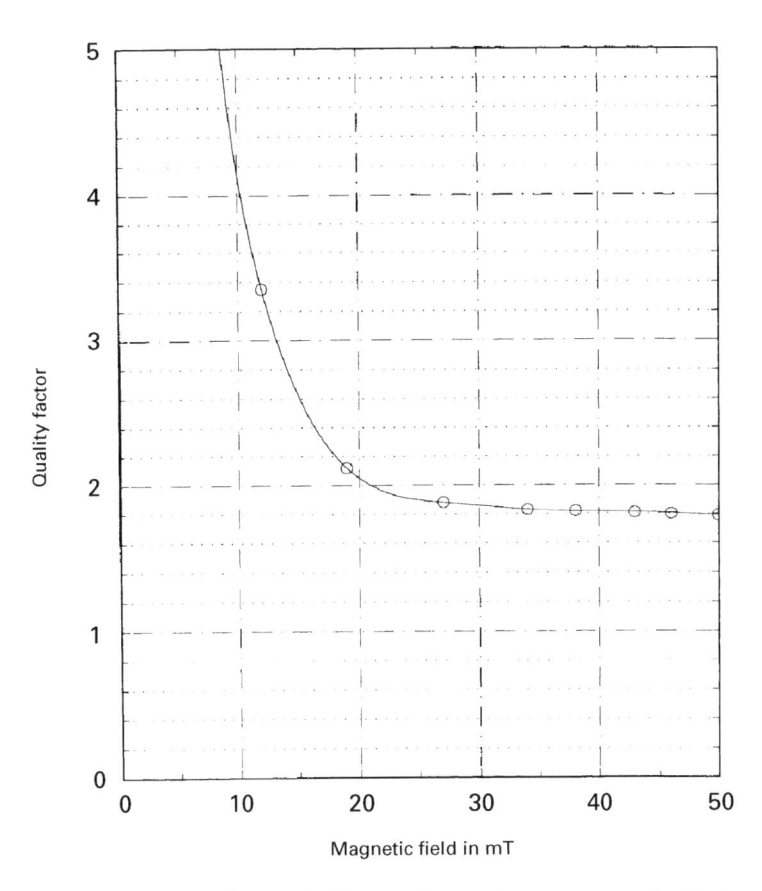

Figure 18.8 Quality factor of open half-wave long prism resonator in H-plane junction (A/L) = 4.5, ε_d = 2.1, k_f = 0.725

The difference between the split frequencies is separately given by

$$\frac{\omega_+ - \omega_-}{\omega_0} = \left(\frac{\sqrt{3}}{\pi}\right)\kappa \qquad (18.25)$$

The quality-factor of the complex gyrator circuit is defined in the usual way by the split frequencies of the gyromagnetic resonator. The required result is the same as that met with the planar resonator

$$\frac{1}{Q_L} = \left(\frac{3}{\pi}\right)\kappa \qquad (18.26)$$

Figure 18.8 indicates the experimental quality factor of one such an arrangement in a 3-port H-plane junction in terms of the normalized direct magnetic field. It suggests that the theoretical result provides a lower bound on the actual one. Figure 18.9 depicts a similar result for a 3-port E-plane one.

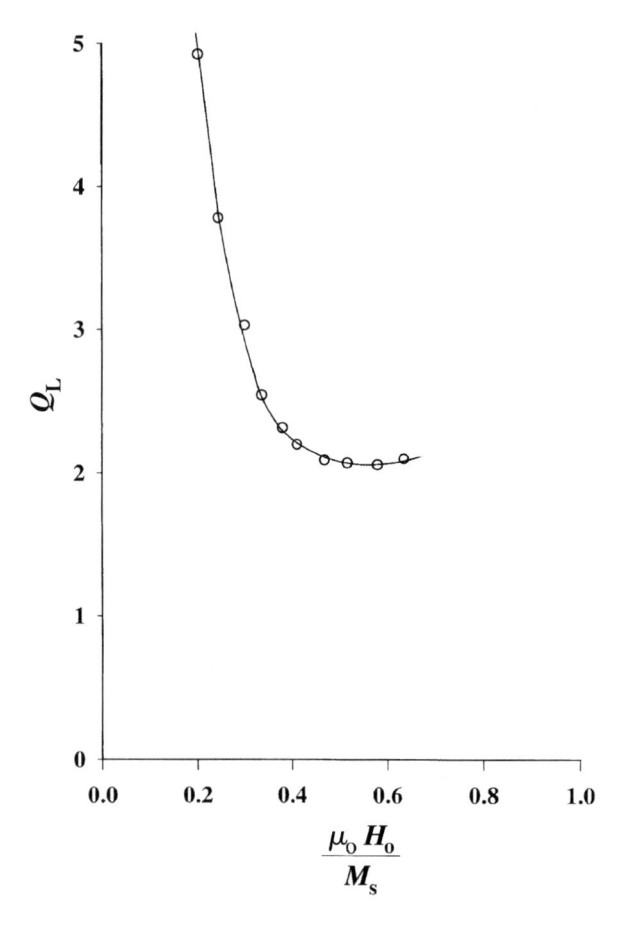

Figure 18.9 Quality factor of open quarter-wave long prism resonator in E-plane junction

18.8 Even and Odd Mode Solutions of Coupled Prism Resonators

The even and odd mode solutions of a pair of coupled prism resonators, in keeping with the development of the related cylindrical problem region, correspond to the operating frequencies of weakly magnetized H and E-plane waveguide circulators respectively. The purpose of this section is to summarize the solutions associated with this class of resonators. The two problem regions considered here are indicated in Figure 18.10. A scrutiny of the cylindrical and prism problem regions suggests that, except for the cut-off numbers, the two formulations are identical.

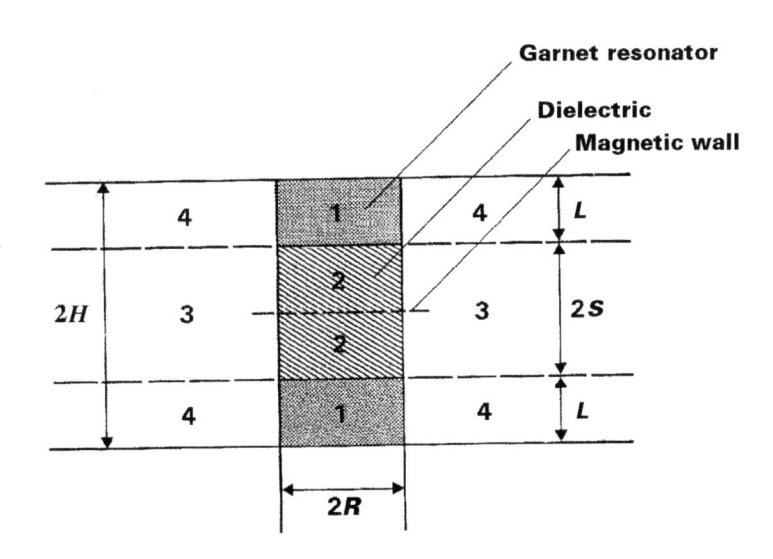

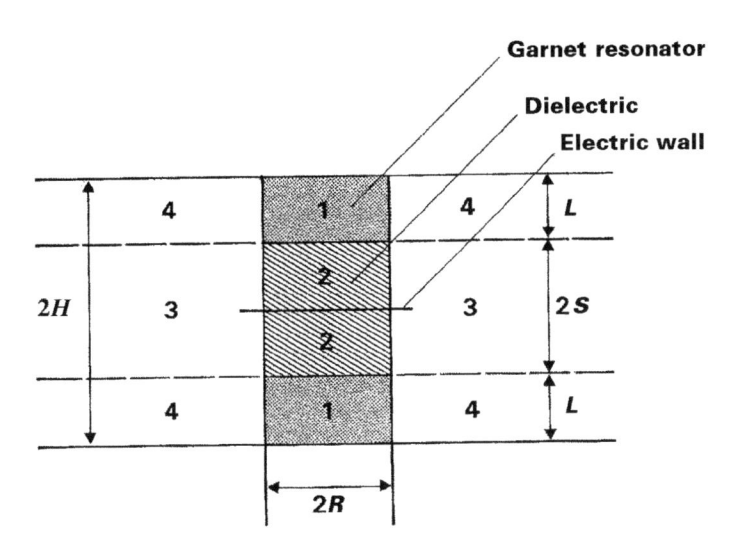

Figure 18.10 Even and odd modes problem regions

The odd mode characteristic equation is

$$\frac{\varepsilon_{\text{eff}} k_0}{\beta_0} \cot(\beta_0 L) - \frac{\varepsilon_d k_0}{\alpha_0} \coth(\alpha_0 S) = 0 \tag{18.27a}$$

where

$$\left(\frac{\beta_0}{k_0}\right)^2 = \varepsilon_{\text{eff}} \mu_d - \left(\frac{k_{m,n,l}}{k_0}\right)^2 \tag{18.27b}$$

$$\left(\frac{\alpha_0}{k_0}\right)^2 = \left(\frac{k_{m,n,l}}{k_0}\right)^2 - \varepsilon_d \tag{18.27c}$$

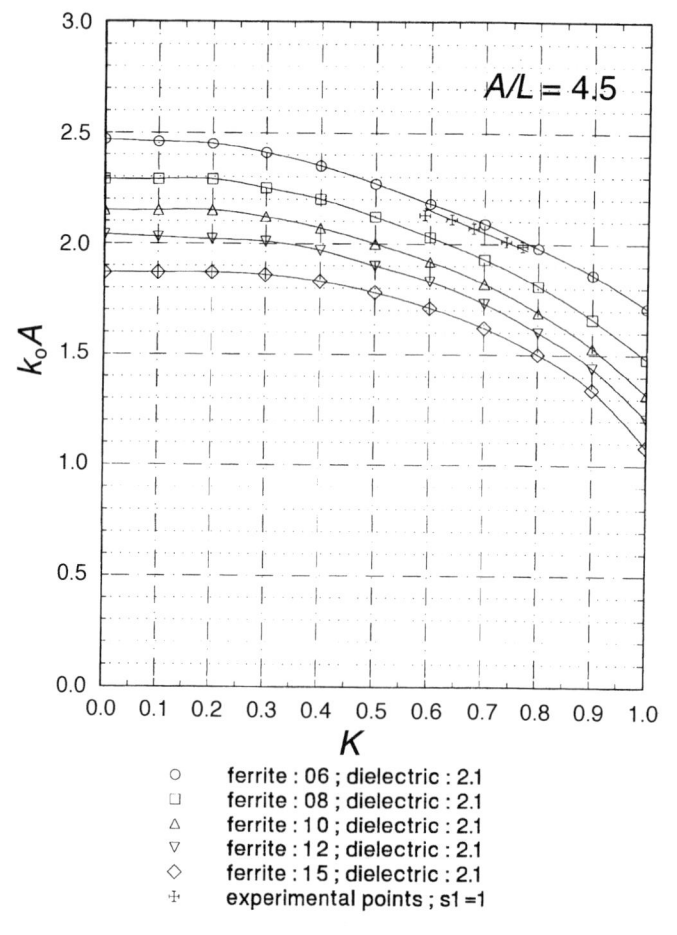

Figure 18.11 Odd mode chart of open half-wave prism resonator for parametric values of effective dielectric constant (A/L) = 4.5, $\varepsilon_d = 1.6$

18.8 EVEN AND ODD MODE SOLUTIONS

ε_{eff} is the effective dielectric constant of the gyromagnetic resonator; it is obtained from either measurement or calculation. ε_d is that of the dielectric constant of the spacers. μ_d is the demagnetized permeability

$$\mu_d = \tfrac{1}{3} + \tfrac{2}{3}(1 - p^2)^{1/2} \tag{18.28}$$

p is the normalized magnetization

$$p = \frac{\gamma M_0}{\mu_0 \omega} \tag{18.29}$$

γ is the gyromagnetic ratio (2.21×10^5 (rad/m)/(A/m)), M_0 is the saturation magnetization (T), μ_0 is the free space permeability ($4\pi \times 10^{-7}$ H/m) and ω is the radian frequency (rad/sec)

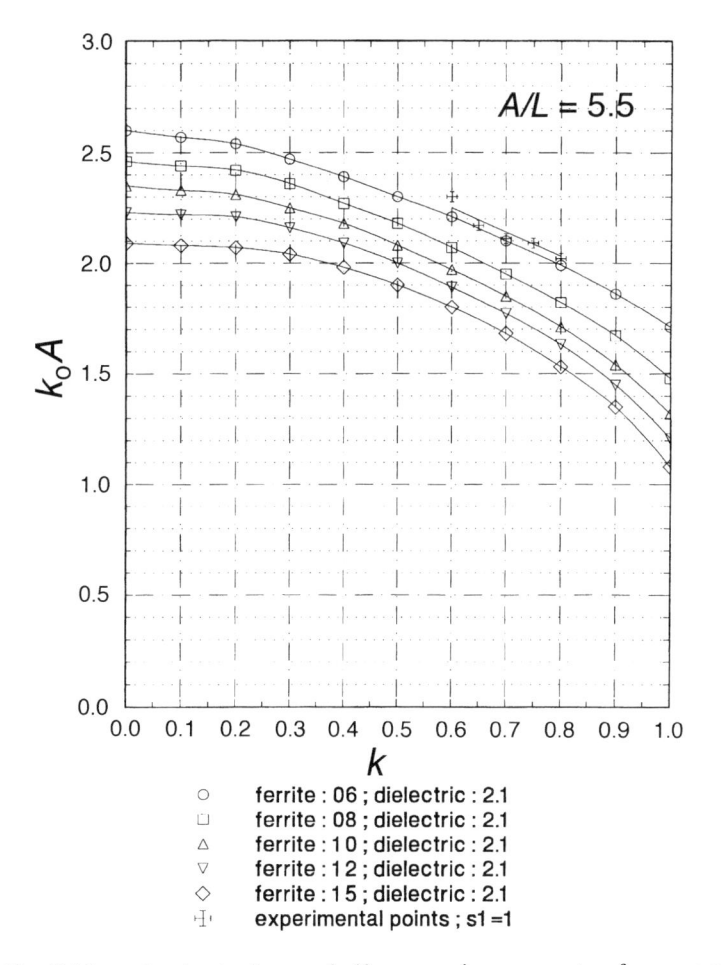

Figure 18.12 Odd mode chart of open half-wave prism resonator for parametric values of effective dielectric constant (A/L) = 5.5, $\varepsilon_d = 1.6$

The separation constant of the dominant mode of a prism waveguide with an ideal magnetic wall is given by equation (18.14).

The free space cut-off number (k_c) of the equivalent waveguide model is separately given by

$$k_c = \frac{k_{1,0,-1}}{\sqrt{\varepsilon_{\text{eff}}}} \tag{18.30}$$

The condition in equation (18.27) may for computational purposes be written as

$$\varepsilon_{\text{eff}} \left(\frac{k_0}{\beta_0}\right) \cot\left[\left(\frac{\beta_0}{k_0}\right)\left(\frac{L}{A}\right) k_0 A\right]$$
$$- \varepsilon_d \left(\frac{k_0}{\alpha_0}\right) \coth\left[\left(\frac{\alpha_0}{k_0}\right)\left(\frac{L}{A}\right)\left(\frac{1-k}{k}\right) k_0 A\right] = 0 \tag{18.31}$$

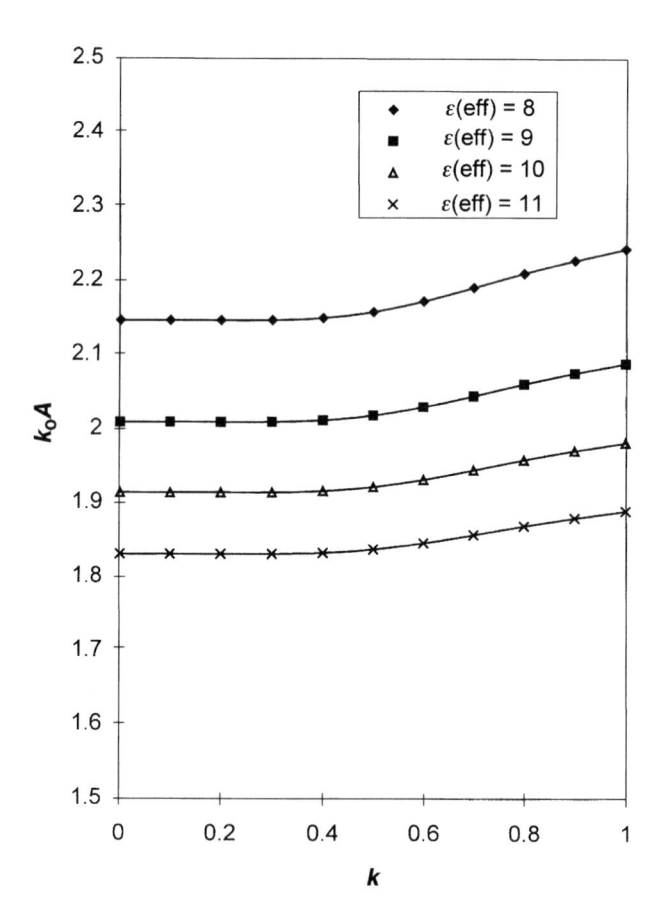

Figure 18.13 Even mode chart of coupled pair of prism resonators with ideal magnetic side walls ($k_0 A$ versus k for parametric values of $\epsilon_f(\text{eff})$ with $A/L = 2.73$)

18.8 EVEN AND ODD MODE SOLUTIONS

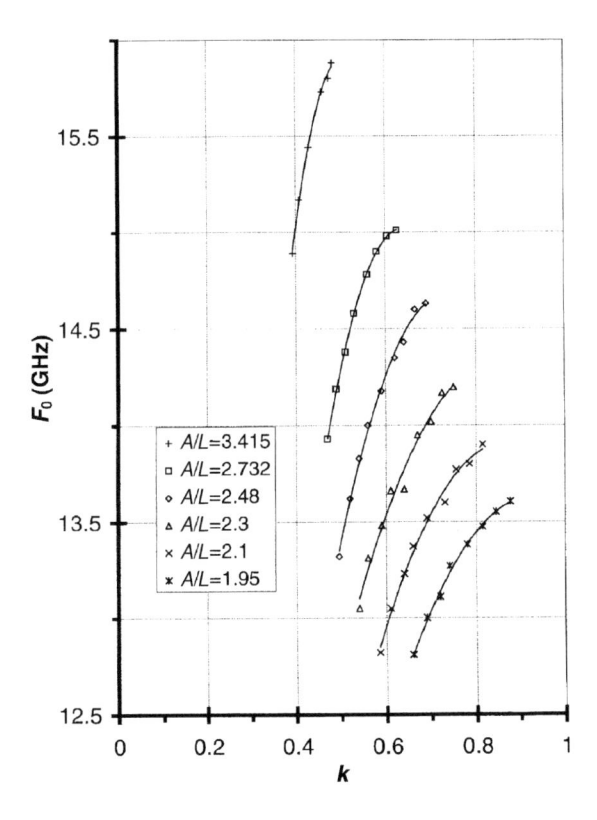

Figure 18.14 Gap factor versus frequency of E-plane waveguide circulator in WR75 waveguide for parametric values of A/L ($A = 6.83$ mm)

k is the gap factor previously defined in connection with the cylindrical problem region.

Figures 18.11 and 18.12 indicate two typical mode charts.

The even mode characteristic equation is identical to that of the odd one except that the hyperbolic cotangent term is replaced by a hyperbolic tangent one. The result is written for computational purposes as

$$\varepsilon_{\text{eff}} \left(\frac{k_0}{\beta_0}\right) \cot\left[\left(\frac{\beta_0}{k_0}\right)\left(\frac{L}{A}\right) k_0 A\right]$$
$$- \varepsilon_d \left(\frac{k_0}{\alpha_0}\right) \tanh\left[\left(\frac{\alpha_0}{k_0}\right)\left(\frac{L}{A}\right)\left(\frac{1-k}{k}\right) k_0 A\right] = 0 \quad (18.32)$$

The theoretical mode chart of this arrangement is indicated in Figure 18.13.

Figure 18.14 gives one experimental relationship between the frequency and the filling factor in this sort of device. Figure 18.15 separately shows the result in normalized of form.

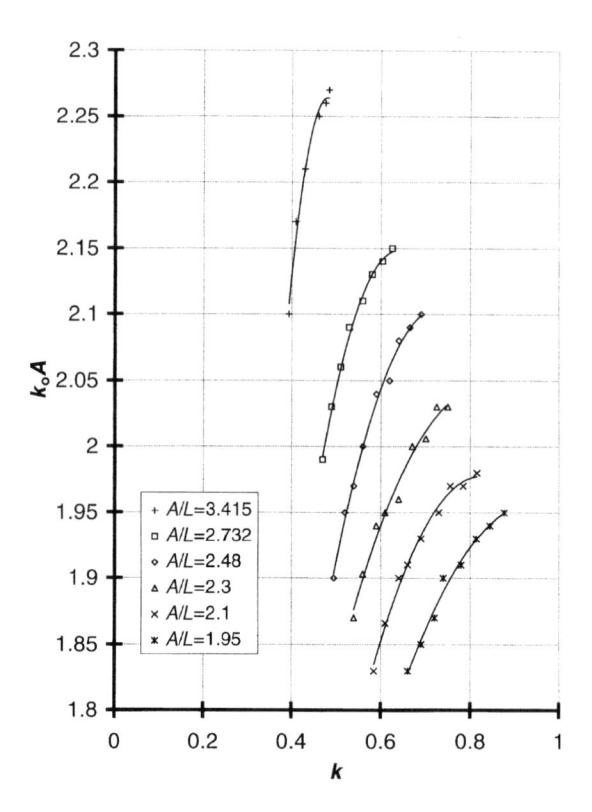

Figure 18.15 Gap factor versus radian side dimension of E-plane circulator in WR75 waveguide for parametric values of A/L ($A = 6.83$ mm)

18.9 The H-plane Junction Circulator Using Prism Resonators

The operation of the H-plane turnstile waveguide circulator relies on either cylindrical or prism resonators. The purpose of this section is to briefly summarize some experimental data on the gyrator circuit of one arrangement employing a half-wave prism resonator open-circuited at each end supported by suitable dielectric spacers. Figure 18.16 shows the arrangement. This sort of circulator may be visualized as a 7-port junction with four of its ports closed by suitable electric walls. The quantities of interest in its description are its mode chart, its split frequencies or quality factor and its susceptance slope parameter. Figures 18.11 and 18.12 give the mode charts of two arrangements. These correspond to the odd mode frequency of a pair of quarter wave long prism resonators open-circuited at one end and short-circuited at the other. The quality factor is separately illustrated in Figure 18.8. Figure 18.17 indicates its susceptance slope parameter.

The gyrator conductance which is also of interest in the adjustment of this sort of device is the dependent variable. It is related to the susceptance slope parameter and quality factor of the junction in the usual way by

18.9 THE H-PLANE JUNCTION CIRCULATOR

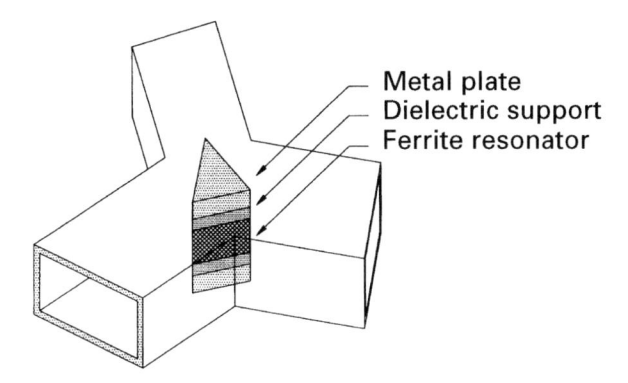

Figure 18.16 Schematic diagram of H-plane waveguide turnstile circulator using open half-wave long prism resonator

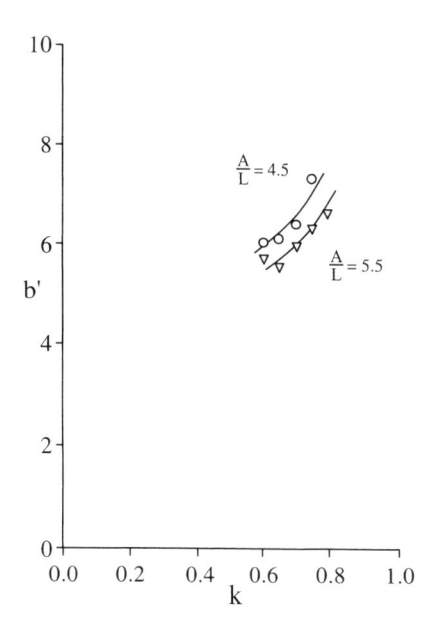

Figure 18.17 Susceptance slope parameter versus gap factor for parametric values of (A/L) for half-wave long prism resonator in H-plane junction ($\varepsilon_d = 2.1$)

$$g = \frac{b'}{Q_L} \tag{18.33}$$

or

$$g = \sqrt{3}\, b' \left(\frac{\omega_+ - \omega_-}{\omega_0} \right) \tag{18.34}$$

Figure 18.18 shows the frequency response of one quarter-wave coupled device.

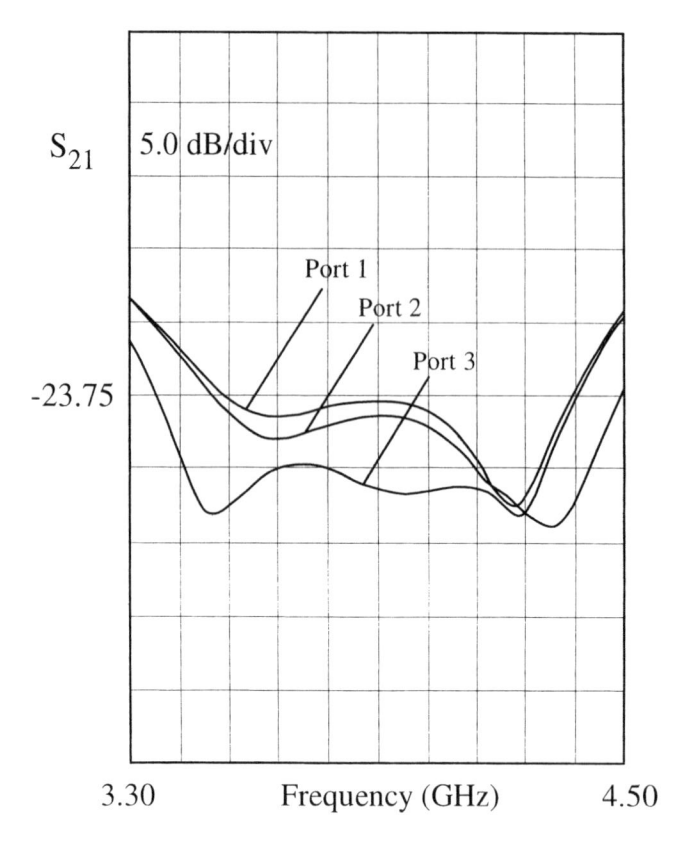

Figure 18.18 Frequency response of quarter-wave long coupled H-plane circulator

18.10 The E-Plane Junction Circulator using Prism Resonators

The geometry of the E-plane circulator considered here is illustrated in Figure 18.19. Its topology, in keeping with that using cylindrical gyromagnetic resonators, consists of an E-plane junction loaded by two quarter-wave long prism resonators. It may therefore be considered as a 7-port turnstile junction. The single prism or re-entrant turnstile version may be viewed as a 5-port junction without ado. The quantities entering into its description are again its mode chart, its susceptance slope parameter (b') and its quality factor (Q_L). The gyrator conductance, which is also of interest in it adjustment is the dependent variable. It is related to the susceptance slope parameter and quality factor in the usual way by the relationship in equation (18.34).

The susceptance slope parameter of one typical resonator is given in Figure 18.20. Some experimental data on it mode chart and quality factor are separately summarized in Figures 18.12, 18.13 and 18.15. Figure 18.21 indicates a single resonator version.

18.10 THE E-PLANE JUNCTION CIRCULATOR

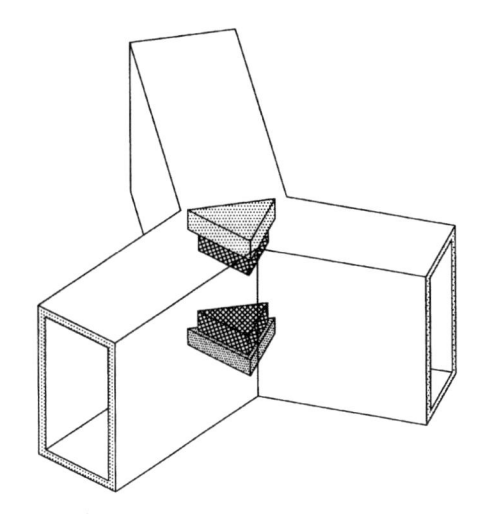

Figure 18.19 Schematic diagram of E-plane circulator using coupled quarter-wave long prism resonators

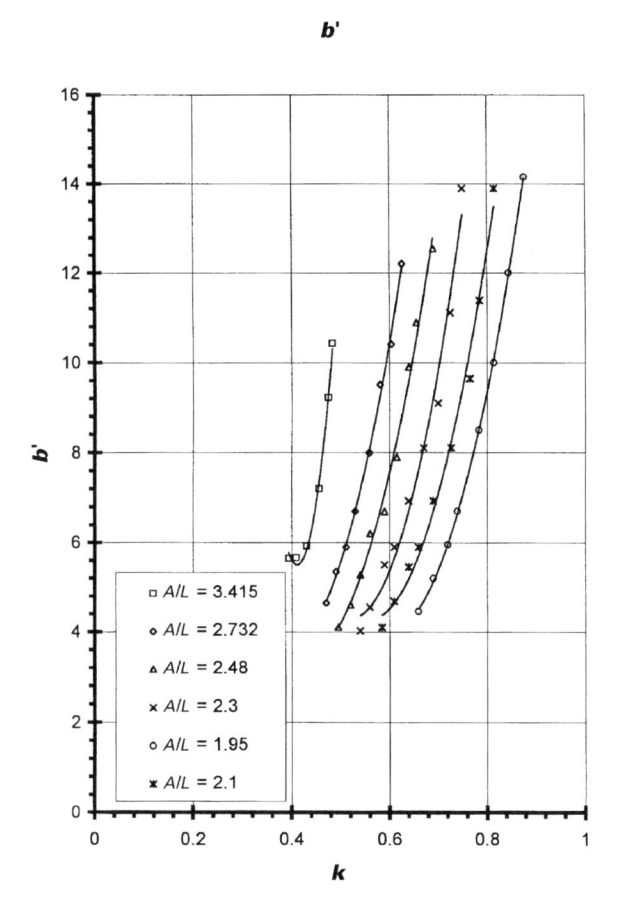

Figure 18.20 Susceptance slope parameter of E-plane junction circulator using coupled quarter-wave long prism resonators for parametric values of (A/L) ($A = 6.83$ mm)

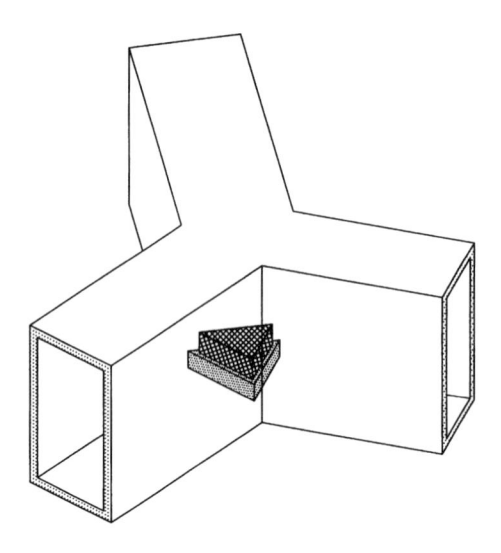

Figure 18.21 Schematic diagram of E-plane circulator using single quarter-wave long prism resonator

References

Akaiwa, Y. (1977) Mode classification of triangular ferrite post for Y-circulator, *IEEE Trans. Microwave Theory Tech.*, **MTT-25**, 59–61.

Chadha, R. and Gupta, K. C. (1980) Green's functions for triangular segments in planar microwave circuits, *IEEE Trans. Microwave Theory Tech.*, **MTT-28**, 1139–1143.

Helszajn, J. and James, D. S. (1978) Planar triangular resonators with magnetic walls, *IEEE Trans. Microwave Theory Tech.*, **MTT-26**, 95–100.

Helszajn, J., James, D. S. and Nisbet, W. T. (1979) Circulators using planar triangular resonators, *IEEE Trans. Microwave Theory Tech.*, **MTT-27**, 188–193.

Helszajn, J. and Sharp, J. (1983) Resonant frequencies, Q-factor, and susceptance slope parameter of waveguide circulators usng weakly magnetized open resonators, *IEEE Trans. Microwave Theory Tech.*, **MTT-31**, 434–441.

Schelkunoff, M. (1943) *Electromagnetic Waves*, Van c Rostrand N.Y.

Riblet, G., Helszajn, J. and O'Donnell, B. (1979) Loaded Q-factors of partial-height and full-height triangular resonators for use in waveguide circulators, In: *Proc. Eur. Microwave Conf.*, 420–424.

19

SYNTHESIS OF QUARTER-WAVE COUPLED JUNCTION CIRCULATORS WITH DEGREES 1 AND 2 COMPLEX GYRATOR CIRCUITS

19.1 Introduction

The construction of the 3-port circulator involves the adjustment of one in-phase and two split counter-rotating eigen-networks. The usual arrangement is that for which the in-phase eigen-network displays an electric wall at the terminals of the junction and the degenerate ones a magnetic wall there. The in-phase eigen-network may be commensurate with those of the demagnetized ones or it may coincide with the frequency at which the split eigen-networks exhibit complex conjugate immittances, or it may, in general, be noncommensurate. If it is idealized by a frequency-independent electric wall boundary condition, then the 1-port complex gyrator immittance of the junction is a STUB R-circuit of degree 1, otherwise it is a STUB R-circuit of degree 2. While this represents the common situation it is not the only one, since it is also possible for the in-phase eigen-network to exhibit a magnetic wall at the terminals of the junction and the degenerate counter-rotating ones an electric one. Indeed it is also possible for all three eigen-networks to exhibit either an electric or magnetic at the same terminals, so that it is in fact possible to realize four 1-port gyrator circuits for each class of solution. The main purpose of this chapter is to summarize the four possible complex gyrator networks of each degree and to form the network problem for the degree 2 topologies. A knowledge of the appropriate eigen-networks and complex gyrator circuit in any given situation is, of course, an essential prerequisite for design. The 1-port complex gyrator circuits of degree 2, interestingly enough, explicitly exhibit both the in-phase and split counter-rotating eigen-networks of the magnetized junction and thus permit the synthesis problem to be directly posed in terms of the microwave problem. The chapter includes the adjustment of one degree-2 waveguide circulator.

19.2 Degrees 1 and 2 Circulators with Open- and Short-Circuited In-Phase Eigen-networks

The complex gyrator circuit of a 3-port circulator may be classified in one of four possible ways according to whether the reflection eigenvalues associated with the first circulation condition satisfy one of the four solutions below

$$s_0 = -1, \; s_\pm = 1 \tag{19.1a}$$

$$s_0 = 1, \; s_\pm = -1 \tag{19.1b}$$

$$s_0 = s_{\pm 1} = 1 \tag{19.1c}$$

$$s_0 = s_\pm = -1 \tag{19.1d}$$

The required relationship between the reflection eigenvalues of the ideal circulator can then be obtained in a systematic way by making use of that between the scattering coefficients and the eigenvalues. If the adjustment is only to be made in terms of the reflection coefficient at a typical port, then it is not necessary to diagonalize the scattering matrix S. In this instance, the required relationship between S_{11} and the eigenvalues can be obtained quite simply by equating the spur of the scattering matrix to the sum of the eigenvalues:

$$S_{11} = \frac{s_0 + s_{+1} + s_{-1}}{3} \tag{19.2}$$

A scrutiny of this relationship suggests that the demagnetized or reciprocal junction exhibits a passband response in the first two instances and a stop band in the second two situations. Figure 19.1 depicts the complete family of solutions.

The in-phase and counter-rotating eigenvalues are as is now understood the reflection variables of 1-port reactive networks known as the eigen-networks of the junction. These may be realized in terms of the poles of the immittance eigenvalues in either a first or second Foster form, in the manner illustrated in Figure 19.2(a) and (b). Counter-rotating poles, in this expansion, that have the symmetry of the junction are associated with the in-phase eigen-network. Whether a pole of an eigenvalue exhibits an electric or magnetic wall at the symmetry plane is readily established by application of the appropriate in-phase or counter-rotating eigenvectors at the terminals of the junction. Although the lowest order in-phase pole is usually associated with a magnetic-wall boundary condition at the symmetry plane of the junction, it may also exhibit an electric wall there. One such instance occurs in the case in an E-plane junction, another is met if a thin metal wall is introduced through the plane of symmetry of an H-plane one. Likewise, although the lowest order counter-rotating poles of the junction are usually associated with electric-wall boundary conditions at the symmetry plane of the junction, these may still exhibit electric-wall boundary conditions at its terminals by realizing the eigen-networks by half-wave-long transmission lines instead of quarter-wave ones.

19.2 DEGREES 1 AND 2 CIRCULATORS

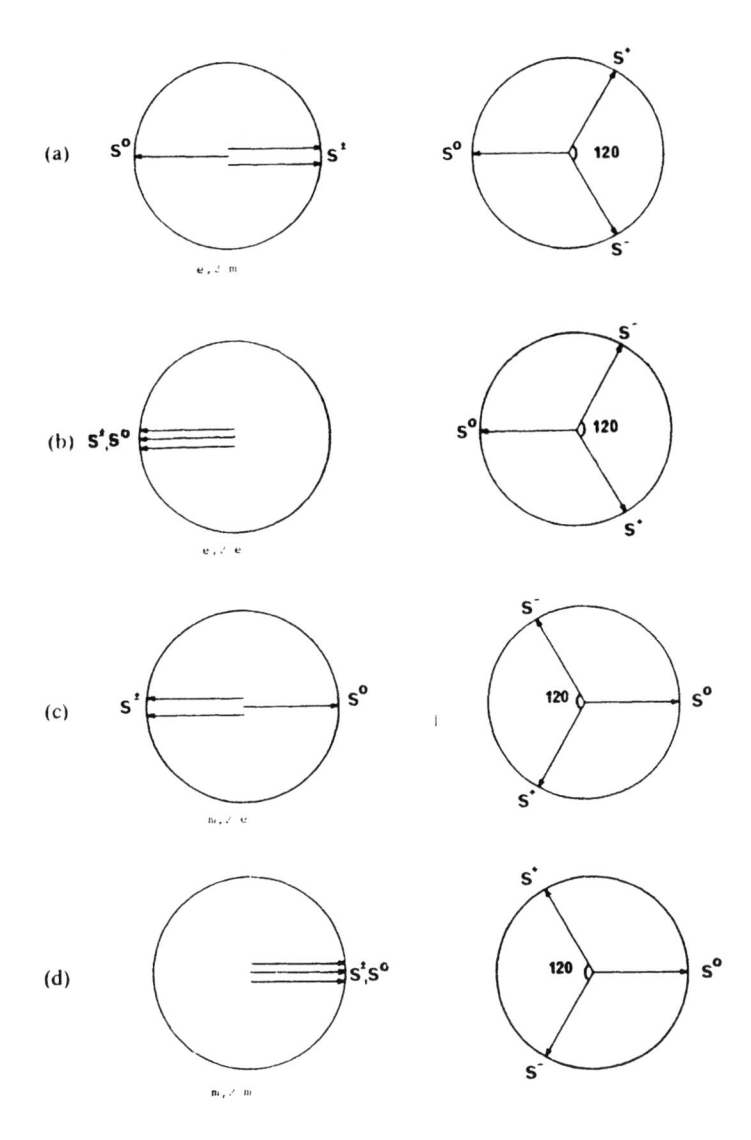

Figure 19.1 (a) First and second circulation adjustments for junction circulatior with $S^0 = -1$ and $S^\pm = 1$, (b) First and second circulation adjustments for junction circulator with $S^0 = -1$ and $S^\pm = -1$, (c) First and second circulation adjustment for junction circulator with $S^0 = 1$ and $S^\pm = -1$, (d) First and second circulation adjustment for junction circulator with $S^0 = 1$ and $S^\pm = 1$ (reprinted with permission, Helszajn, J. (1985) Synthesis of quarter-wave coupled junction circulators with degree 1 and 2 complex gyrator circuits, *IEEE Trans. Microwave Theory Tech.*, **MTT-32**, 382–390)

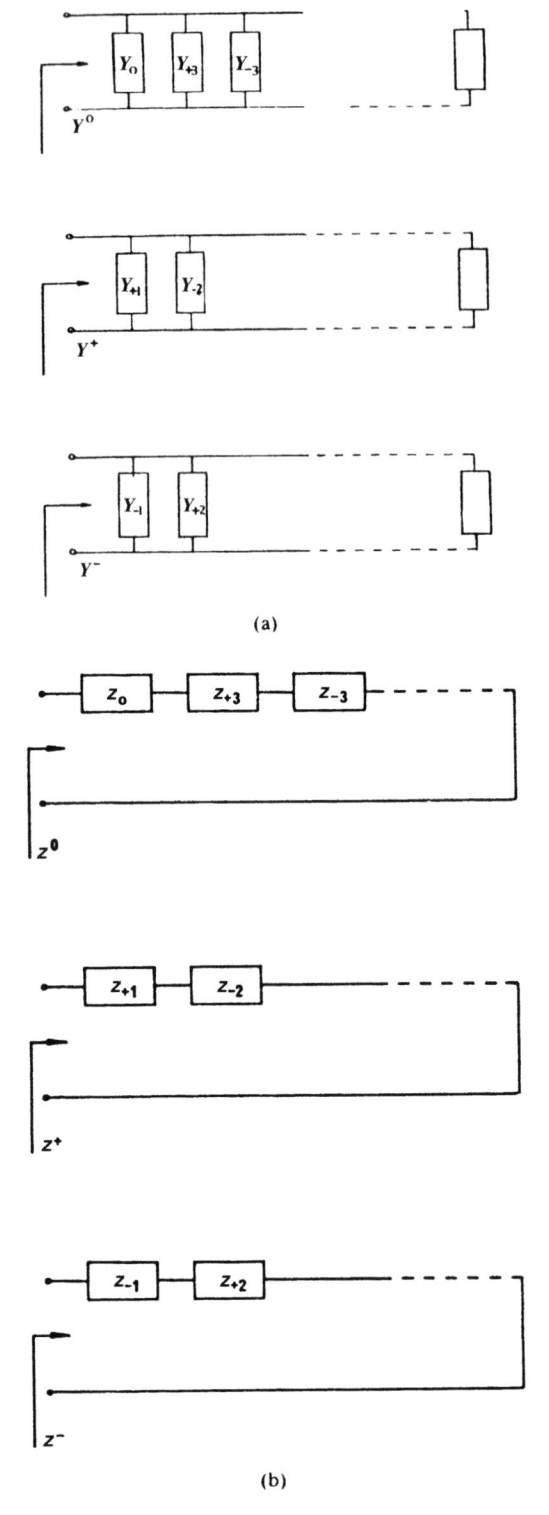

Figure 19.2 First and second Foster forms of in-phase and counter-rotating eigennetworks of junction circulator

19.2 DEGREES 1 AND 2 CIRCULATORS 383

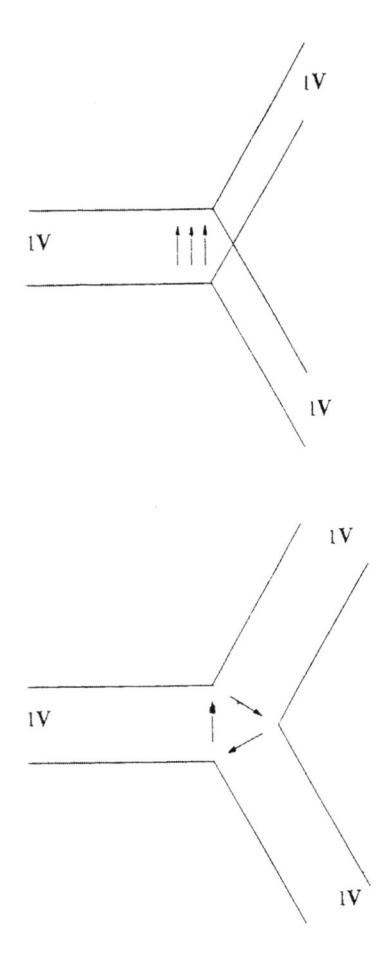

Figure 19.3 In-phase excitations of E- and H-plane junctions

Figure 19.3 illustrates the distinction between the in-phase eigen-networks for E- and H-plane junctions. Substitution of an electric wall for a magnetic one, for either of the eigenvalues, leads to a reversal in the direction of circulation of the junction, if the splitting between the eigenvalues is correctly reset, as is readily verified. This is also the case if $s^{\pm}$ are interchanged.

The open-circuit parameters at the terminals of the junction do not exist if the in-phase eigen-network exhibits a magnetic-wall boundary condition, and conversely the short-circuit parameters do not exist if it displays an electric-wall boundary condition. If the in-phase and degenerate eigen-networks exhibit dual walls, then the demagnetized junction has neither open- nor short-circuited parameters. The scattering matrix is of course always realizable. If the in-phase eigen-network is idealized by a frequency-independent electric or magnetic wall at the terminals of the junction then the corresponding eigenvalue diagram is of degree 1, otherwise it is of degree 2. A number of practical examples of the latter class have been mentioned in the open literature.

The eigenvalue diagrams in Figure 19.1(a)–(d) may be labelled according to whether the eigen-networks exhibit electric or magnetic walls at the terminals of the junction and according to whether the in-phase eigen-network is idealized by a frequency-independent electric or magnetic wall or not as *e,2m*, *e,2e*, *m,2e*, and *m,2m* of degree 1 or 2. If the in-phase eigen-network is idealized by either a frequency-independent open- or short-circuited stub at the terminals of the junction, then the 1-port complex gyrator circuit is a STUB-R load of degree 1, otherwise it is a STUB-R load of degree 2. The 1-port gyrator circuits of degrees 1 and 2 discussed here may be realized in terms of the in-phase (Z^0, Y^0) and counter-rotating ($Z^{\pm}, Y^{\pm}$) immittance eigenvalues; both the in-phase and counter-rotating eigen-networks are explicitly exhibited by the gyrator circuits of degree 2.

19.3 Complex Gyrator Circuits of Degree 1 Junction Circulators with Open- and Short-Circuited In-Phase Eigen-networks

The complex gyrator circuits of degree 1, obtained by idealizing the in-phase eigen-network by a frequency-independent electric wall at the terminals of the junction and those of the degenerate counter-rotating ones by either magnetic or electric walls there will now be examined as a preamble to summarizing the dual problem for which the in-phase eigen-network exhibits a magnetic wall at the same terminals. Figure 19.4 gives the complex gyrator topologies for the four possible degree 1 situations, and Figure 19.5 illustrates the corresponding lumped-element versions. Since the short-circuit parameters do not in the former instance exist, the derivation of the required equivalent circuit starts by forming the classic 1-port gyrator impedance in terms of the open-circuit parameters of the junction with $V_3 = I_3 = 0$

$$Z_{in} = Z_{11} - \frac{Z_{12}^2}{Z_{13}} \tag{19.3}$$

The open-circuit parameters Z_{11}, Z_{12}, and Z_{13} are given in the usual way in terms of the impedance eigenvalues Z^0 and $Z^{\pm}$ by

$$Z_{11} = \frac{Z^0 + Z^+ + Z^-}{3} \tag{19.4a}$$

$$Z_{12} = \frac{Z^0 + \alpha Z^+ + \alpha^2 Z^-}{3} \tag{19.4b}$$

$$Z_{13} = \frac{Z^0 + \alpha^2 Z^+ + \alpha Z^-}{3} \tag{19.4c}$$

Exact complex gyrator circuits of degree 1 may now be formed at the frequencies at which the counter-rotating eigen-networks exhibit either electric or magnetic walls.

19.3 COMPLEX GYRATOR CIRCUITS

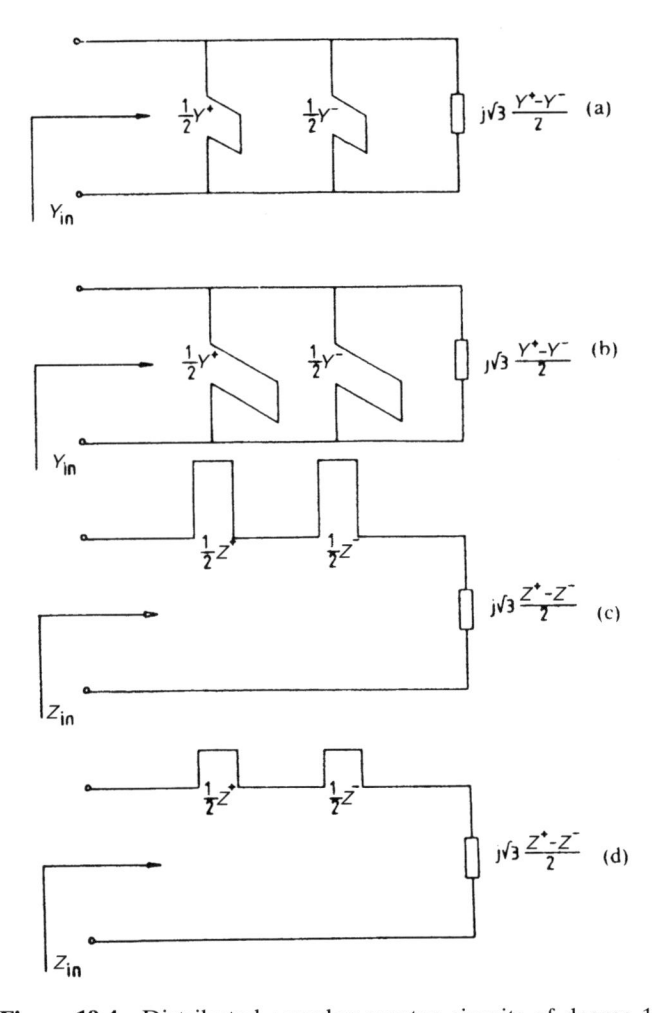

Figure 19.4 Distributed complex gyrator circuits of degree 1

Idealizing the in-plane eigen-network at the terminals of the junction by an electric wall gives

$$Z^0 = 0. \tag{19.5}$$

The required result is given in Chapter 4 by

$$Y_{in} = \frac{1}{Z_{in}} = \frac{(Y^+ + Y^-)}{2} - j\sqrt{3}\,\frac{(Y^+ - Y^-)}{2} \tag{19.6}$$

This result has historically been given with an approximation sign but is exact as can be readily verified by tracing (19.3) and (19.6). The case where the degenerate eigenvalues exhibit magnetic walls may now be distinguished from that for

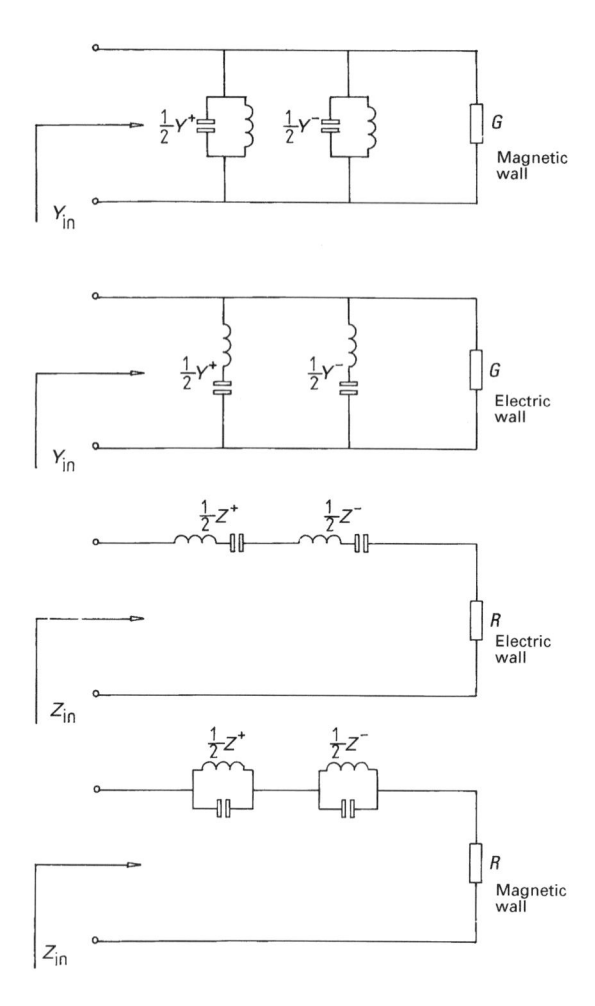

Figure 19.5 Lumped-element complex gyrator circuits of degree 1

which these have electric walls. In the first situation, the counter-rotating eigen-networks may be realized using quarter-wave-long short-circuited stubs, and the complex gyrator circuit has the topology in Figure 19.4(a). In the second case, the counter-rotating networks may be realized using half-wave-long short-circuited stubs, and the corresponding complex gyrator circuit has the layout indicated in Figure 19.4(b). The counter-rotating eigenvalues are also, in this latter instance, interchanged on the eigenvalue diagram so that the junction now circulates in the opposite direction.

Some additional distinguishable properties of these two solutions are that for the eigenvalue diagram in Figure 19.1(a) the real part of the gyrator immittance tends to a magnetic wall as the junction is demagnetized, while that in Figure 19.1(b) tends to an electric wall under the same condition. Whether one or the other situation applies is determined by whether the degenerate eigen-networks have magnetic or electric walls. Furthermore, the required angular splitting between the degenerate eigenvalues is in the former case, half that of the latter one. Surprisingly enough, the susceptance slope parameter of the second solution is now a function of the splitting. All of these aspects may be readily verified by scrutinizing Figures 19.1(a) and (b) and Figures 19.4(a) and (b). These properties may also be directly demonstrated by assuming that $Y^\pm$ in Figure 19.2 may phenomenologically be written as

$$Y^\pm = -ja_1^2 Y_1 \cot(\theta_1 \pm \Delta\theta_1) \tag{19.7}$$

θ_1 is the electrical length of the degenerate counter-rotating eigen-networks, ($\Delta\theta_1$ represents the perturbation in the demagnetized eigen-networks by the gyrotropy, Y_1 is the characteristic admittance of the counter-rotating eigen-networks which, for simplicity, are assumed to coincide with those of the demagnetized junction, and a_1 is the turns ratio of an ideal transformer that represents the coupling between the three transmission lines and the resonator.

The derivation of the 1-port complex gyrator circuits of degree 1 for the case where the in-phase eigen-network is idealized by a frequency independent magnetic wall and those of the counter-rotating ones by either magnetic or electric walls proceeds in a similar fashion except that short-circuit parameters are employed to derive the complex gyrator immittances, and that Y^0 instead of Z^0 is assumed to be zero in the approximation problem

$$Z_{in} = \frac{(Z^+ + Z^-)}{2} + j\sqrt{3}\frac{(Z^+ - Z^-)}{2} \tag{19.8}$$

The appropriate equivalent circuits are illustrated in Figure 19.4(c) and (d). The gyrator resistance is asymptotic to an electric wall in the first instance and to a magnetic one in the second case. The solution in Figure 19.4(c) is well behaved in the vicinity of its midband, but that in Figure 19.4(d) exhibits stopbands on either side of its passband and a stopband in its demagnetized state.

19.4 Complex Gyrator Circuits of Degree 2 Circulators with Open- and Short-Circuited In-Phase Eigen-networks

The derivation of the complex gyrator circuits of degree 2 for the two eigenvalue diagrams in Figure 19.1(a) and (b), as well as for those in Figure 19.1(c) and (d) will now be undertaken. Figure 19.6 indicates the distributed topologies associated with this class of solutions and Figure 19.7 the corresponding equivalent lumped-element layouts. In realizing these circuits, it has been assumed that the real part of the complex gyrator immittance may be formed by idealizing the in-phase eigen-networks by either an electric or magnetic wall. It is readily appreciated that the split shunt parallel and series resonators in the vicinities of the passband frequencies, reduce to shunt resonators and that, likewise, the series combinations of the series and parallel resonators reduce to series resonators at

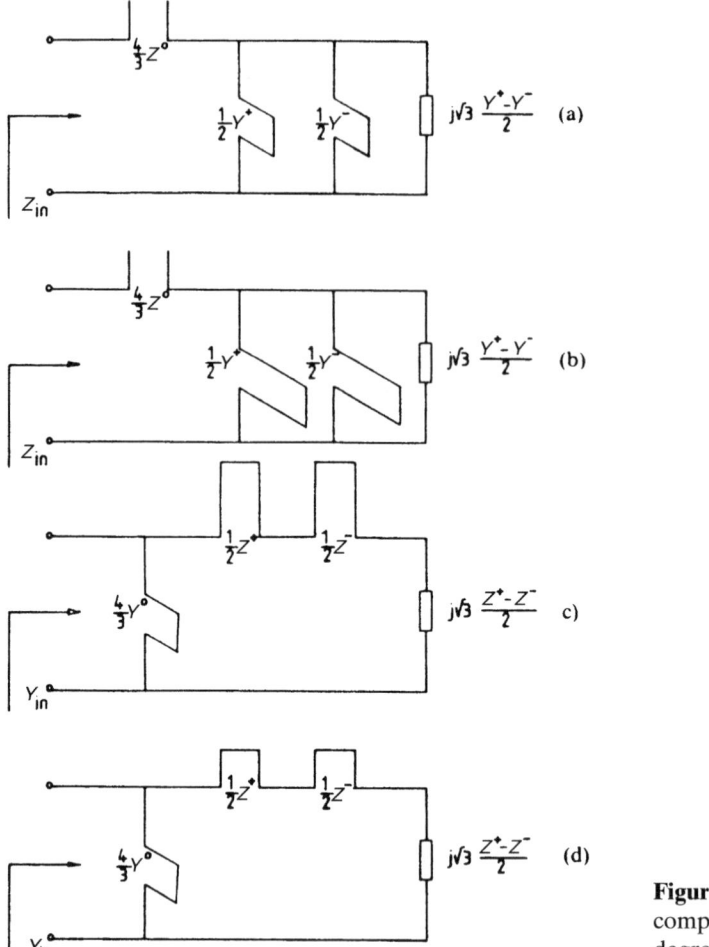

19 SYNTHESIS OF QUARTER-WAVE COUPLED JUNCTION CIRCULATORS

Figure 19.6 Distributed complex gyrator circuits of degree 2

the same frequencies. Each of the gyrator circuits of degree 2 has the same transmission zeros and thus each can be arranged to exhibit the same transmission characteristic. The insertion-loss function for this class of network is akin to that realizable with a complex gyrator circuit of degree 1 coupled by a single U.E.

The realization of a typical result starts again with a statement of the 1-port gyrator immittance in terms of the immittance parameters of the overall 3-port circuit. In the case for which $s_0 = -1$, it is again appropriate to employ open-circuit parameters in forming the complex gyrator immittance. The required result is

$$Z_{in} \approx \frac{8Z^0 - (Z^+ + Z^-)}{6} + j\frac{(Z^+ - Z^-)}{2\sqrt{3}} \qquad (19.9)$$

In obtaining this result, the in-phase eigenvalue Z^0 has been idealized by a short-circuit boundary condition in forming the real part of the gyrator immittance. This

19.4 COMPLEX GYRATOR CIRCUITS

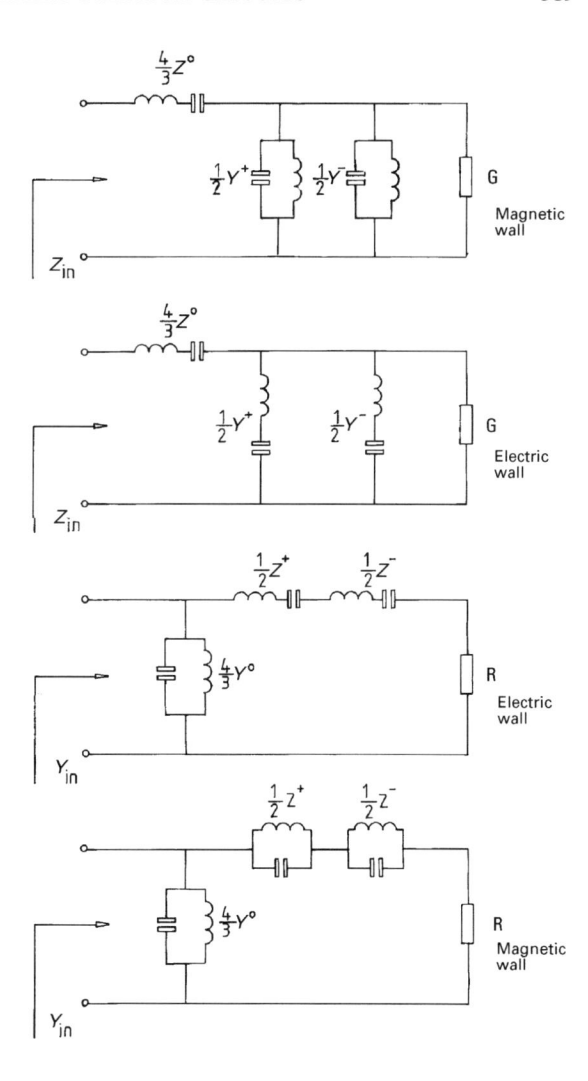

Figure 19.7 Lumped-element complex gyrator circuits of degree 2

impedance is readily synthesized in the form indicated in Figure 19.6 by writing Z_{in} as

$$Z_{in} = Z_1 + \frac{1}{Y_1} \tag{19.10}$$

where

$$Z_1 \approx \frac{4Z^0}{3} \tag{19.11}$$

$$Y_1 \approx \frac{(Y^+ + Y^-)}{2} + j\sqrt{3}\,\frac{(Y^+ - Y^-)}{2} \tag{19.12}$$

The realization of the gyrator circuit of a junction in the situation where the in-phase eigen-networks exhibits a magnetic wall at its terminals ($s_0 = +1$) proceeds in a dual fashion to the preceding case except that short-circuit, instead of open-circuit, parameters are employed.

$$Y_{in} = \frac{8Y^0 - (Y^+ + Y^-)}{6} + j\frac{(Y^+ - Y^-)}{2\sqrt{3}} \qquad (19.13)$$

This admittance may be synthesized in the form illustrated in Figure 19.6 by expressing Y_{in} as

$$Y_{in} = Y_1 + \frac{1}{Z_1} \qquad (19.14)$$

where

$$Y_1 \approx \frac{4Y^0}{3} \qquad (19.15)$$

$$Z_1 \approx \frac{(Z^+ + Z^-)}{2} + j\sqrt{3}\frac{(Z^+ - Z^-)}{2} \qquad (19.16)$$

19.5 Adjustment of Degree-2 Circulator

The phenomenological adjustment of one degree-2 junction circulator using a post gyromagnetic resonator will now be outlined. Such an adjustment may be understood by considering the $TM_{\pm1,1,0}$ and TM_{010} standing wave solutions in the simple planar gyromagnetic resonator, illustrated in Figure 19.8(a). Since the two solutions have different cut-off numbers some means of bringing these together has of course to be introduced. One possibility is the introduction of a metal post at the centre of the junction. A scrutiny of the field patterns in Figure 19.8(a) suggests that this adjustment should leave the frequency of the $TM_{\pm1,1,0}$ modes unperturbed but that it will in all likelihood tune that of the TM_{010} mode down in frequency. This is in practice the case. Once this condition is met the degeneracy between the counter-rotating modes may be removed without ado by the application of a suitable direct magnetic field. An ideal circulator is realized by rotating this standing wave solution by 60° in the manner indicated in Figure 19.8(b). Figure 19.9 indicates one practical arrangement and Figure 19.10 the steps in the required eigenvalue diagram.

19.6 Experimental Adjustment of Degree-1 Circulator

The degree-1 adjustment of a junction comprising a simple gyromagnetic post resonator will first be described as a preamble to discussing that of the degree-2

19.5 EXPERIMENTAL ADJUSTMENT OF DEGREE-1 CIRCULATOR

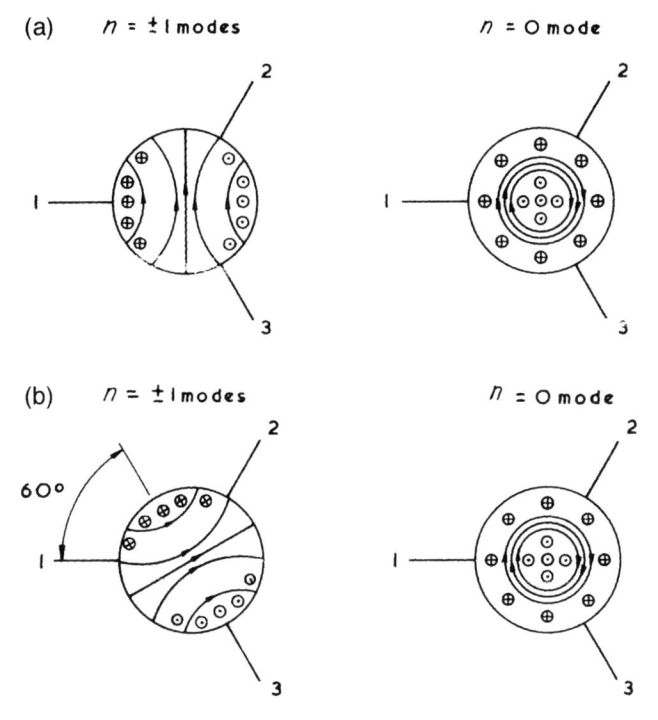

Figure 19.8 (a) $n = 1$ and $n = 0$ field patterns in unmagnetized ferrite disk, (b) $n = 1$ and $n = 0$ field patterns in magnetized ferrite disk (reprinted with permission, Helszajn, J. (1972) Three resonant mode adjustment of the waveguide circulator, *Radio Electron. Eng.*, **42**, 1–4)

arrangement. In this adjustment the two usual variables are the diameter of the ferrite disk and the magnitude of the direct magnetic field or the gyrotropy of the ferrite material The geometry under consideration here is indicated in Figure 19.9.

The first circulation condition is established by determining the frequency at which the reflection coefficient is a minimum. For an ideal circulator this condition coincides with $S_{11} = 1/3$. Figure 19.11 shows the variation of the return loss as a function of frequency. The operating frequency of the device coincides

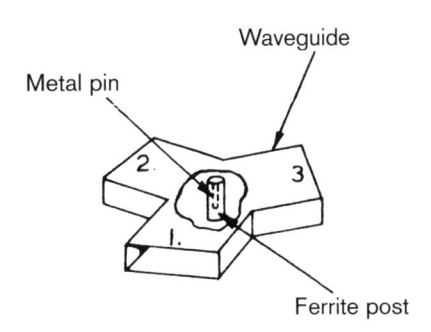

Figure 19.9 Configuration of experimental waveguide circulator

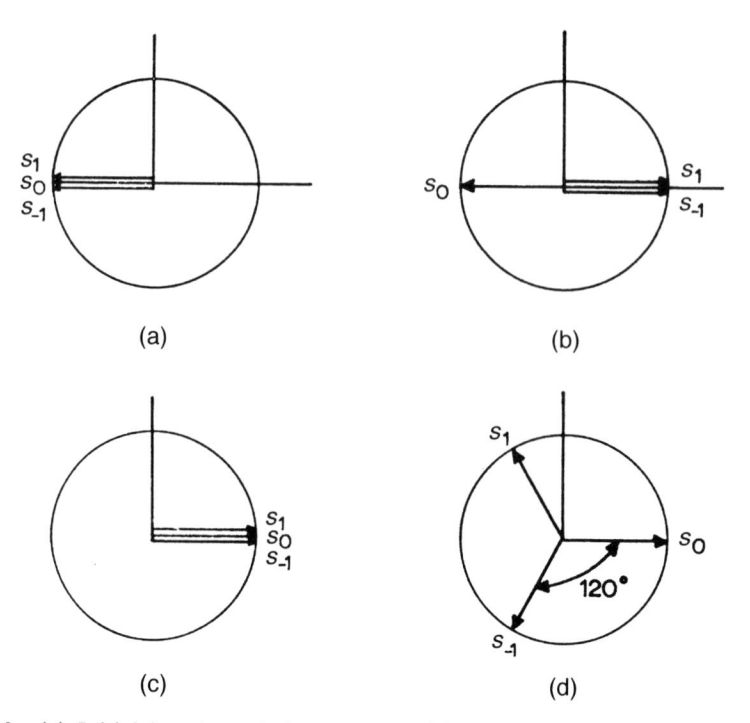

Figure 19.10 (a) Initial location of eigenvectors: (b) Location of eigenvalues after first circulation adjustment for wideband junction; (c) Location of eigenvalues after second circulation adjustment for wideband junction; (d) Location of eigenvalues after third circulation adjustment for wideband junction (ideal circulator)

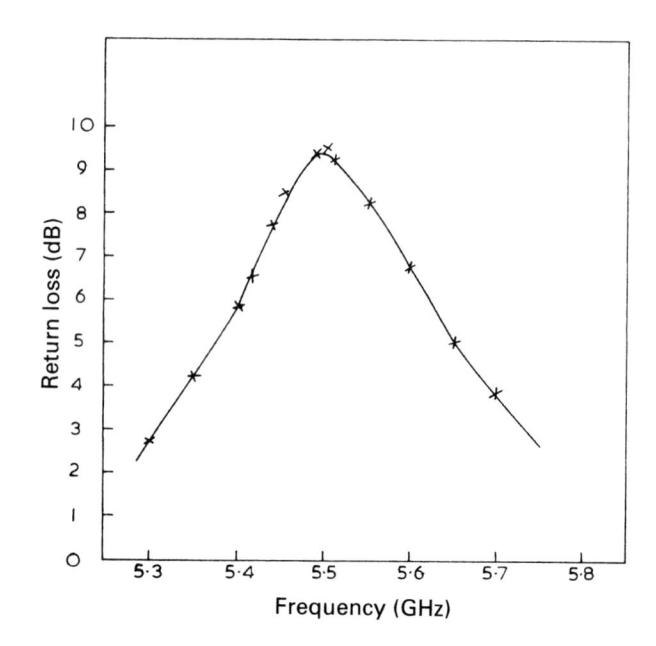

Figure 19.11 Variation of return loss with frequency for 12.2 mm diameter garnet disk in WR187 waveguide (reprinted with permission, Helszajn, J. (1972) Three-resonant mode adjustment of the waveguide circulator, *Radio Electron. Eng.*, **42**, 1–4)

19.6 EXPERIMENTAL ADJUSTMENT OF DEGREE-1 CIRCULATOR

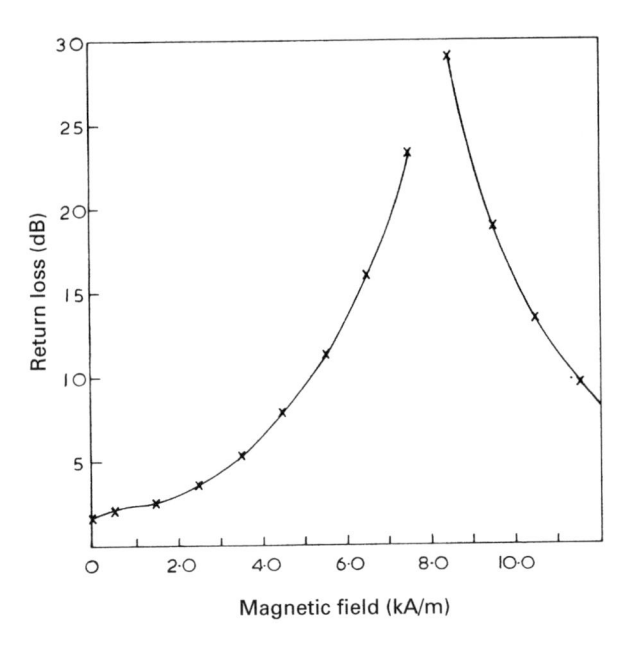

Figure 19.12 Variation of return loss with direct magnetic field for 12.2 mm diameter garnet at 5.47 GHz in WR187 waveguide (reprinted with permission, Helszajn, J. (1972) Three-resonant mode adjustment of the waveguide circulator, *Radio Electron. Eng.*, **42**, 1–4)

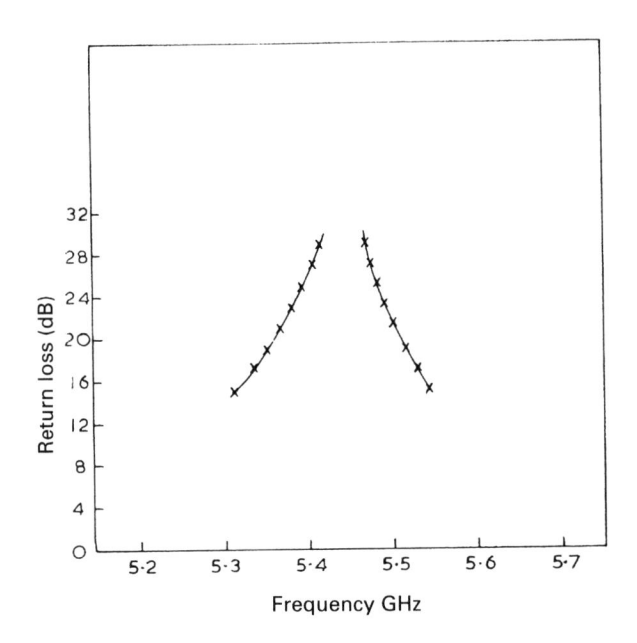

Figure 19.13 Variation of return loss with frequency for two-resonant mode circulator in WR187 waveguide (reprinted with permission, Helszajn, J. (1972) Three-resonant mode adjustment of the waveguide circulator, *Radio Electron. Eng.*, **42**, 1–4)

with the maximum return loss of 9.5 dB. The second circulation condition is now simply realized by adjusting the magnitude of the direct magnetic field until $S_{11} = 0$. Figure 19.12 indicates the variation of the return loss as a function of direct magnetic field. An ideal circulator is obtained at a direct magnetic field given by $H_0 = 6.7$ kA/m.

The overall bandwidth of the device is shown in Figure 19.13. The bandwidth of the circulator here is 2.9%, which is typical for a directly coupled junction of this type.

19.7 Experimental Adjustment of Degree-2 Circulator

One experimental adjustment of a degree-2 junction circulator can be realized by making use of the arrangement depicted in Figure 19.9. The first circulation adjustment establishes the $TM_{\pm1,1,0}$ field patterns within the junction and is common to both the degree -1 and 2 circuits. The second circulation adjustment is now realized by introducing one additional resonant field pattern within the junction. One suitable mode is the TM_{010} one which can be tuned by introducing a thin metal post through the centre of the resonator. This perturbation of the junction is satisfied when S_{11} passes through unity. The variation of the return loss, as the length of the tuning post is varied, shown in Figure 19.14, indicates that the return loss indeed passes through a minimum when the length of the metal post is suitably adjusted. This perturbation of the junction therefore establishes the TM_{010} field pattern within it as asserted. The electrical length of the pin utilized here is the same as that used to establish the TM_{010} mode in the construction of the single junction 4-port circulator described in Chapter 22. The final perturbation of the

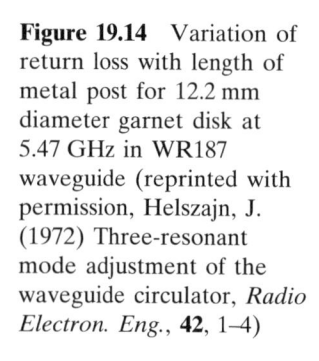

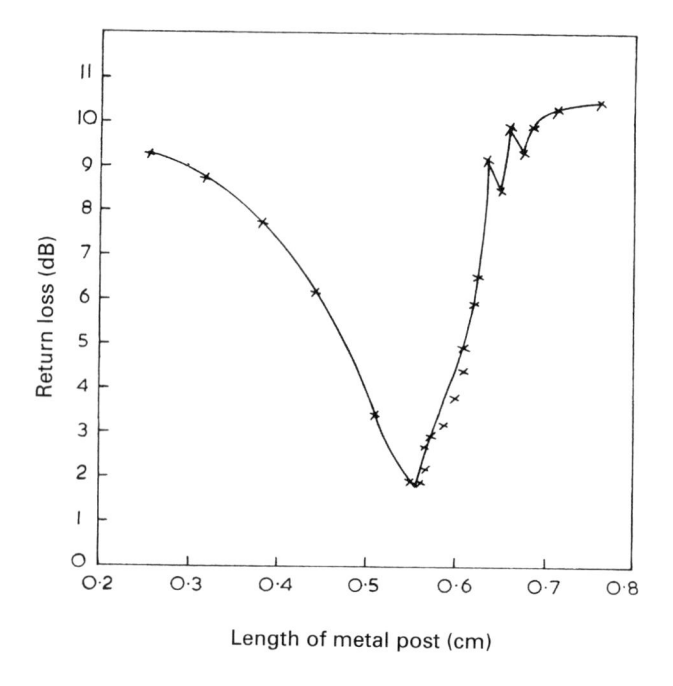

Figure 19.14 Variation of return loss with length of metal post for 12.2 mm diameter garnet disk at 5.47 GHz in WR187 waveguide (reprinted with permission, Helszajn, J. (1972) Three-resonant mode adjustment of the waveguide circulator, *Radio Electron. Eng.*, **42**, 1–4)

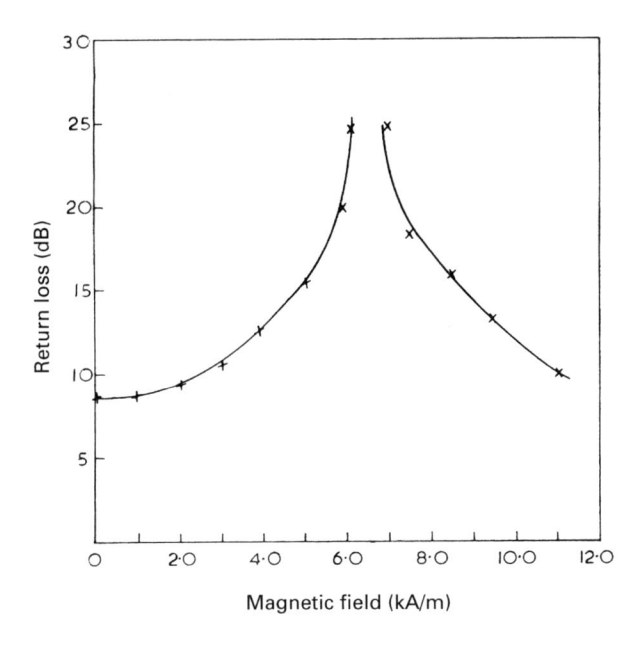

Figure 19.15 Variation of return loss with direct magnetic field for 12.2 mm diameter garnet post at 5.47 GHz with 5.5 mm long metal post in WR187 waveguide (reprinted with permission, Helszajn, J. (1972) Three-resonant mode adjustment of the waveguide circulator, *Radio Electron. Eng.*, **42**, 1–4)

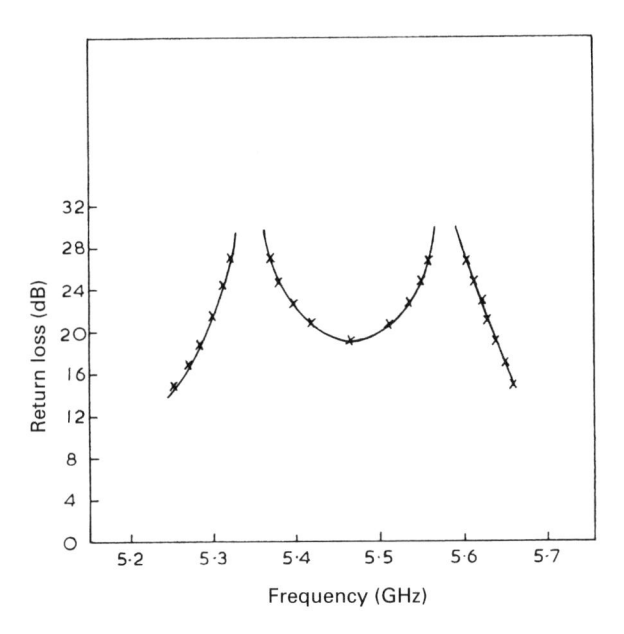

Figure 19.16 Variation of return loss with frequency for 3 resonant mode circulator in WR187 waveguide (reprinted with permission, Helszajn, J. (1972) Three-resonant mode adjustment of the waveguide circulator, *Radio Electron. Eng.*, **42**, 1–4)

junction is now met by applying a static magnetic field to the device. Figure 19.15 depicts the variation of the return loss as a function of the direct magnetic field. The condition associated with an ideal circulator is met with $H_0 = 8.5$ kA/m without ado. Figure 19.16 indicates the variation of the return loss as a function of the frequency. The bandwidth achieved here is about 7%. In obtaining this illustration a slight perturbation of the length of metal post from about 5.5 mm to 5.3 mm was employed.

The material used was a garnet with saturation magnetization of 0.1750 T. Its relative dielectric constant was $\varepsilon_r = 14.5$. The waveguide size used was WR 187 and the operating frequency was about 5.5 GHz. The garnet disk was 12 mm in diameter with at 2.5 mm diameter hole through its centre to take the tuning post. The trimming of the tuning post was done with a micrometer. Spring finger were used to make contact to the tuning post where the latter entered the waveguide. The direct magnetic field was applied using a coil.

References and Further Reading

Akaiwa, Y. (1973) Input impedance of a circulator with an in-phase eigen-excitation resonator, *Electron Lett.*, **9**.

Anderson, L. K. (1967) An analysis of broadband circulators with external tuning elements, *IEEE Trans. Microwave Theory Tech.*, **MTT-15**, 42–47.

Auld, B. A. (1959) The synthesis of symmetrical waveguide circulator, *IRE Trans. Microwave Theory Tech.*, **MTT-7**, 238–246.

Bergman, J. O. and Christensen, C. (1968) Equivalent circuit for a lumped element Y circulator, *IEEE Trans. Microwave Theory Tech.*, **MTT-16**, 308–310.

Bittar, J. and Verszely, J. (1980) A general equivalent network of the input impedance of symmetric three-port circulators, *IEEE Trans. Microwave Theory Tech.*, **MTT-28**, 807–808.

Bosma, H. (1968) A general model for junction circulators: choice of magnetization and bias field, *IEEE Trans. Mag.*, **MAG-4**, 587–596.

Butterweck, H.-J. (1963) Der Y-Zirkulator, *Arc. Elek. Ubertragung.*, **17**, 163–176.

Fay, C. E. and Comstock, R. L. (1965) Operation of the ferrite junction circulator, *IEEE Trans. Microwave Theory Tech.*, **MTT-13**, 15–27.

Goebel, U. and Schieblich, C. (1983) A unified equivalent circuit representation of H and E-plane junction circulators. In: *Eur. Microwave Conf.*, 803–808.

Helszajn, J. (1970) Wideband circulator adjustment using $n = 1$ and $n = 0$ electromagnetic-field patterns, *Electron Lett.*, **6**, 729–731.

Helszajn, J. (1972) Three-resonant mode adjustment of the waveguide circulator, *Radio Electron. Eng.*, **42**, 1–4.

Helszajn, J. (1972) The synthesis of quarter-wave coupled circulators with Chebyshev characteristics, *IEEE Trans. Microwave Theory Tech.*, **MTT-20**, 764–769.

Helszajn, J. (1983) Complex gyrator circuits of planar circulators using higher order modes in a disk resonator, *IEEE Trans. Microwave Theory Tech.*, **MTT-31**, 931–937.

Helszajn, J. (1981) Operation of tracking circulator, *IEEE Trans. Microwave Theory Tech.*, **MTT-29**, 700–707.

Helszajn, J. (1981) Standing wave solution of planar irregular and hexagonal resonators, *IEEE Tans. Microwave Theory Tech.*, **MTT-29**.

Helszajn, J. (1985) Synthesis of quarter-wave coupled junction circulators with degree 1 and 2 complex gyrator circuits, *IEEE Trans. Microwave Theory Tech.*, **MTT-32**, 382–390.

Humphreys, B. L. and Davies, J. B. (1962) The synthesis of n-port circulators, *IRE Trans. Microwave Theory Tech.*, **MTT-10**, 551–554.

Konishi, Y. (1967) A high power u.h.f. circulator, *IEEE Trans. Microwave Theory Tech.*, **MTT-15**, 700–708.

Milano, U., Saunders, J. U. and Davis, L. E. Jr. (1960) A Y-junction strip-line circulator, *IRE Trans. Microwave Theory Tech.*, **MTT-8**, 346–350.

Naito, Y. and Tanaka, N. (1971) Broad-banding and changing operation frequency of circulator, *IEEE Trans. Microwave Theory Tech.*, **MTT-19**, 367–372.

Omori, M. (1968) An improved E-plane waveguide circulator. In: *G-MTT Int. Microwave Symp. Dig.*, 228. IEEE, New York.

Piotrowsky, W. S. and Raul, J. E. (1976) Low-loss broad-band EHF circulators, *IEEE Trans. Microwave Theory Tech.*, **MTT-24**, 863–866.

Salay, S. J. and Peppiatt, H. J. (1972) An accurate junction circulator design procedure, *IEEE Trans. Microwave Theory Tech.*, **MTT-20**, 192–193.

Solbach, K. (1982) Equivalent circuit representation for the E-plane circulator, *IEEE Trans. Microwave Theory Tech.*, **MTT-30**, 806–809.

20

THE 4-PORT SINGLE JUNCTION WAVEGUIDE CIRCULATOR

20.1 Introduction

While the 3-port single junction circulator is the most common arrangement met in practice 4-port ones may also be realized without too much difficulty. The purpose of this chapter is to describe a number of 4-port single junction H-plane waveguide circulators. Such junctions, in common with 3-port devices exhibit some of the properties of transmission line cavity resonators between ports 1 and 2, and a definite standing wave pattern exists within the junction with nulls at ports 3 and 4 also. However, an important difference between the two is that the 4-port device cannot be adjusted with external tuning elements only. This remark may be understood by recognizing that a 4-port device can be matched without being a circulator. Scrutiny of its eigenvalue problem region indicates that its adjustment requires three independent variables that may be established in a systematic way by perturbing the scattering matrix eigenvalues one at a time on the unit circle until these coincide with those of an ideal circulator. This procedure has been fully discussed in Chapter 3. It is facilitated provided the nature of the field patterns used to construct the device are known. The present chapter describes four possible H-plane waveguide junctions one of which relies on the use of a higher order mode for its operation. The E-plane 4-port single junction circulators has been separately described in the literature.

The possibility of realizing 6 or 8-port circuits based on turnstile models is separately understood.

20.2 The H-Plane 4-Port Single Junction Circulator

While both E and H-plane 4-port topologies have been described in the literature the attention of the chapter is restricted to the H-plane one. One possible 4-port H-plane waveguide circulator consists of a single magnetized ferrite disk on one side of the waveguide and a metal post on the other wall. The three independent variables used in its adjustment are a pair of $HE_{\pm1,1,1/2}$ open dielectric resonances in an unmagnetized ferrite disk, a $TM_{0,1,1/2}$ resonance along a metal post, and the magnitude of a direct field to remove the degeneracy between the $HE_{\pm1,1,1/2}$ modes. Since the overall length of the ferrite disk and the quarter-wavelong cylinder is approximately equal to the height of the waveguide, the two are allowed to make firm flat contact with each other and with the waveguide walls. These two conditions satisfy two of the three independent variables of the junction. The third and last one is met by rotating the standing wave formed by the hybrid $HE_{\pm1,1,1/2}$ modes by the required 45° by the application of a suitable direct magnetic field. The procedure used to adjust this circulator is described in Chapter 3. It allows each of the three independent variables for this type of junction to be established one at a time in a systematic way.

Another H-plane junction is one based upon the use of $TM_{\pm1,1,0}$ counter-rotating modes in a radial resonator and relies again upon the $TM_{0,1,1/2}$ mode associated with a quarterwave-long metal post for the symmetrical mode.

Still another possible waveguide realization rests altogether on radial modes in a post resonator for both its counter-rotating and symmetrical modes. This example uses a $TM_{0,1,0}$ in conjunction with a capacitor reactance on its axis and $TM_{\pm1,1,0}$ mode in a simple ferrite post with no variation of the electric field along the axis of the resonator.

Still another version relies on its operation on a pair of radial $TM_{\pm3,1,0}$ radial degenerate resonances, a radial $TM_{0,1,0}$ resonance, and the amplitude of a suitable direct field to split the degeneracy between the former modes. Since the modes in this construction have nearly equal radial wave numbers, a simple ferrite disk immediately meets the first two of the three circulation conditions. The third and last condition is met in the usual way by removing the degeneracy between the degenerate modes by the application of a suitable direct magnetic field.

20.3 4-Ports Single Junction Circulator using Axial Modes

The 4-port single junction circulator described in this section relies for its operation on a linear combination of $HE_{\pm1,1,1/2}$ and $TM_{0,1,1/2}$ axial modes. Once these modes are established within the junction, the $HE_{\pm1,1,1/2}$ one is rotated by the application of a direct magnetic field until two output ports are decoupled from one input port. The geometry under consideration is illustrated in Figure 20.1.

The electric fields employed in this arrangement are given in Figures 20.2(a)–(d). The illustrations in Figures 20.2(a) and (b) show the standing wave solutions of the $HE_{\pm1,1,1/2}$ and $TM_{0,1,1/2}$ modes of the unmagnetized junction. Figures 20.2(c) and (d) give the same field patterns with the $HE_{\pm1,1,1/2}$ mode rotated through 45°. Adding the amplitudes of the E-fields at the four ports indicates that circulation takes place between ports 1 and 2 while ports 3 and 4 are decoupled.

The angle through which the $HE_{\pm1,1,1/2}$ hybrid mode is rotated is established by taking a linear combination of the electric fields around the periphery of the ferrite disk.

20.3 4-PORTS SINGLE JUNCTION CIRCULATOR USING AXIAL MODES

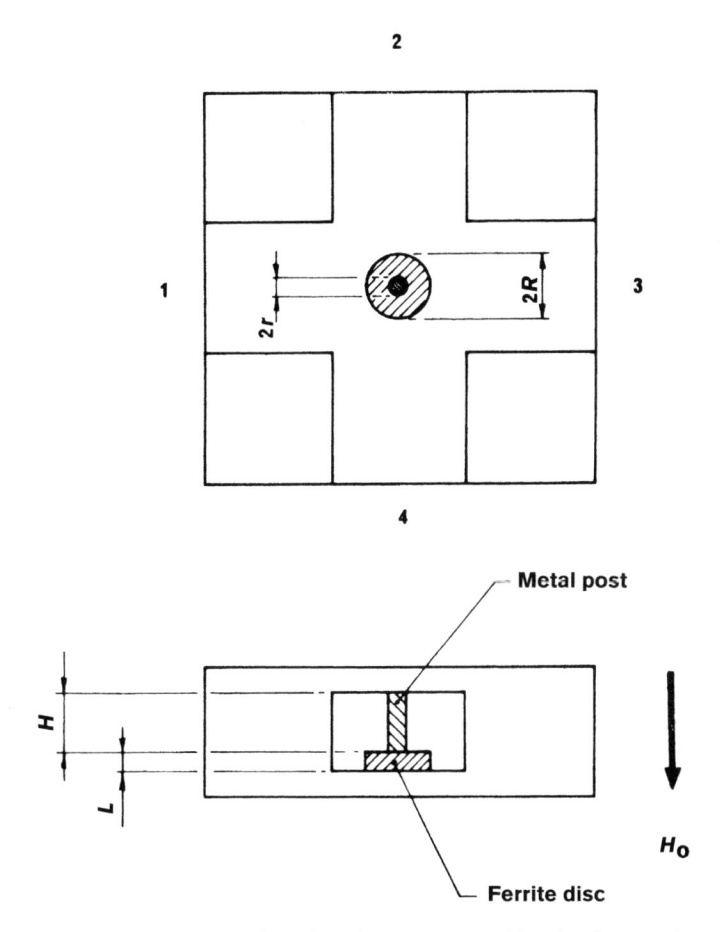

Figure 20.1 Schematic of single junction 4-port waveguide circulator using axial modes

$$E(R,\phi) = a_{01} + a_{11} \cos(\tau_{11} + \phi) \tag{20.1}$$

where a_{01} and a_{11} are arbitrary constants and τ_{11} is the phase angle through which the $HE_{\pm 1,1,1/2}$ mode is rotated. Applying the boundary conditions of an ideal circulator at the four ports gives

$$E(R,0) = a_{01} + a_{11} \cos \tau_{11} = +1 \tag{20.2a}$$

$$E(R,\pi/2) = a_{01} + a_{11} \cos(\tau_{11} + \pi/2) = +1 \tag{20.2b}$$

$$E(R,\pi) = a_{01} + a_{11} \cos(\tau_{11} + \pi) = 0 \tag{20.2c}$$

$$E(R,3\pi/2) = a_{01} + a_{11} \cos(\tau_{11} + 3\pi/2) = 0 \tag{20.2d}$$

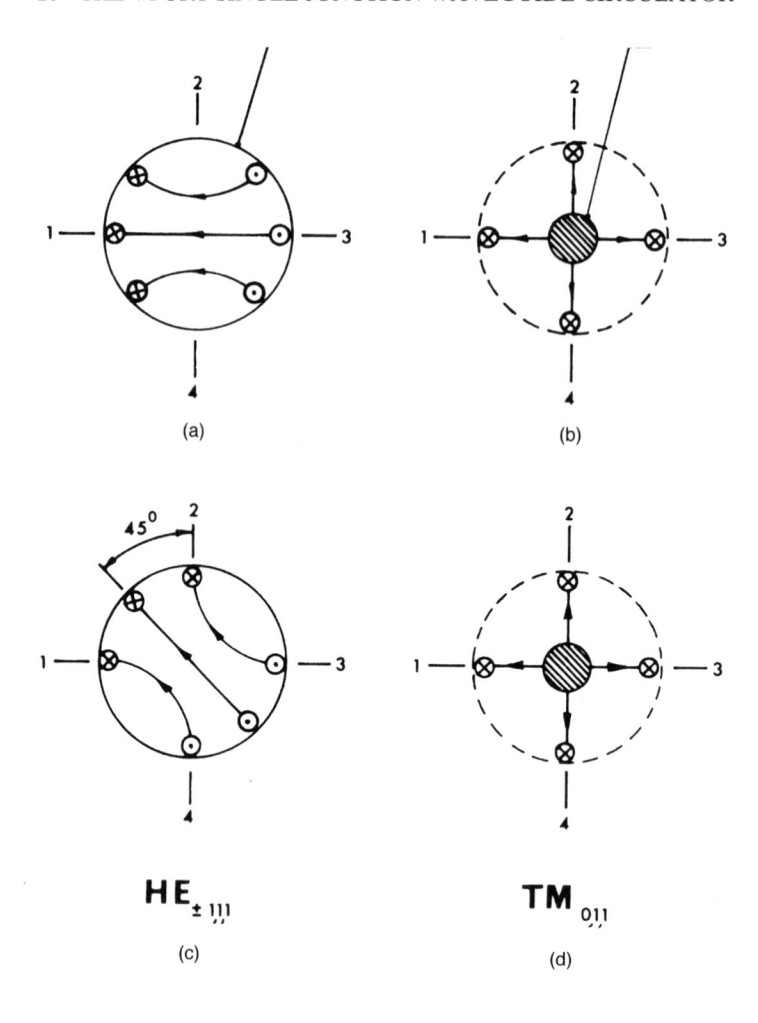

Figure 20.2 (a) $HE_{\pm 1,1,1/2}$ E-field pattern in unmagnetized disk, (b) $TM_{0,1,1/2}$ E-field pattern on metal post, (c) $HE_{\pm 1,1,1/2}$ E-field pattern in magnetized disk, (d) $TM_{0,1,1/2}$ E-field pattern on metal post (reprinted with permission, Helszajn, J. (1973) Waveguide and stripline 4-port single junction circulators, *IEEE Trans. Microwave Theory Tech.*, **21**, **MTT-20**, 630–633) (© 1973 IEEE)

The result is

$$\tau_{11} = 45° \qquad (20.3a)$$

$$a_{01} = \tfrac{1}{2} \qquad (20.3b)$$

$$a_{11} = \tfrac{1}{\sqrt{2}} \qquad (20.3c)$$

20.3.1 Adjustment of Counter-Rotating Modes of Junction

The center frequency of the 4-port single junction circulator resides in the vicinity of a suitable pair of counter-rotating modes of the junction geometry. In the geometry considered here these are the $HE_{\pm1,1,1/2}$ hybrid modes associated with an open quarter wave long dielectric disk resonator open-circuited at one flat face and short-circuited at the other. The mode chart of this problem region has been dealt with in Chapter 9. It gives the aspect ratio R/L of the disk as a function of the radial wavenumber k_0R. For a circulator at 9.30 GHz one has $R = 5.0$ mm, $L = 3.0$ mm, with $\varepsilon_r = 12.4$ and $4\pi M_0 = 0.2400$ T. Here, R is the radius of the disk, L is its length, and the other variables have the usual meaning.

The first circulation adjustment is now met provided

$$S_{11} = S_{21} = S_{41} = 0 \qquad (20.4a)$$

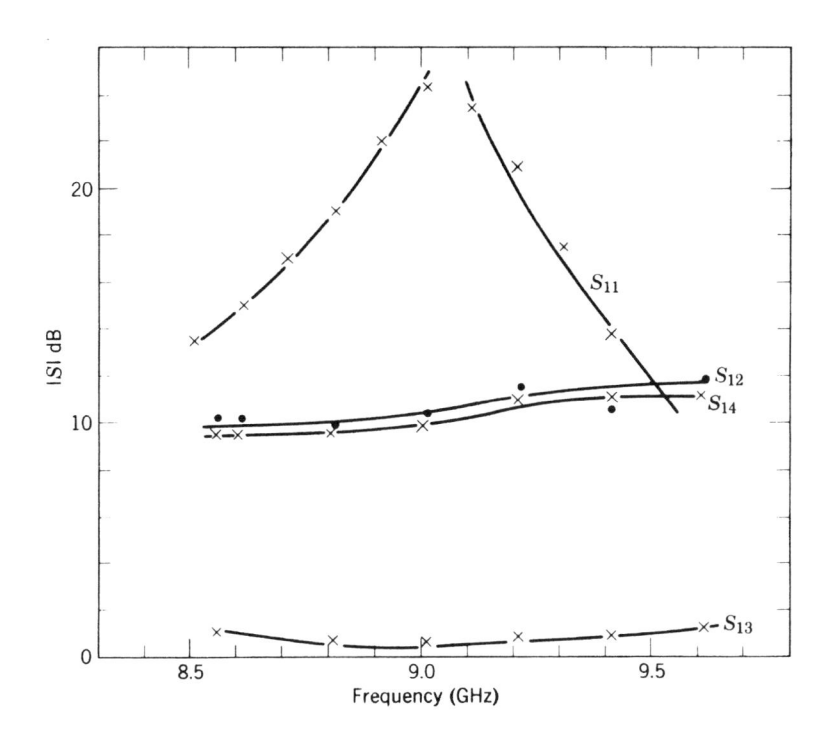

$$S_{31} = 1 \qquad (20.4b)$$

Figure 20.3 Scattering parameters of junction after first circulation adjustment (reprinted with permission, Helszajn, J. (1973) Waveguide and stripline 4-port single junction circulators, *IEEE Trans. Microwave Theory Tech.*, **21**, **MTT-20**, 630–633) (© 1973 IEEE)

The scattering parameters of the junction as a function of frequency supporting such a $HE_{\pm1,1,1/2}$ mode in a ferrite disk resonator is indicated in Figure 20.3. The first circulation condition of this arrangement is satisfied at a frequency of 9.10 GHz, which is in good agreement with the calculated value of 9.30 GHz.

20.3.2 Adjustment of Symmetrical Mode of Junction

The second circulation condition is satisfied with the scattering coefficients given by

$$S_{11} = S_{21} = S_{31} = S_{41} = \tfrac{1}{2} \tag{20.5}$$

This condition is obtained by establishing a suitable symmetrical mode within the junction. One possibility is to employ the axial $TM_{0,1,1/2}$ mode which exists along a quarterwave-long thin metal post. The condition for resonance on such a post is approximately given by

$$l = \frac{\lambda_0}{4} - \frac{(a-r)}{2 \log_e (a/r)} \tag{20.6}$$

Equation (20.6) is asymptotic to $\lambda_0/4$ for an infinitely thin post. This equation is obtained by having the incident tangential electric field set up a current in the post the net effect of which is to cancel the incident field at the post. The current I_x along the wire has the nature of a standing wave

$$I_x = \sin\left[\frac{2\pi(l-x)}{\lambda_0}\right] \tag{20.7}$$

where x lies between 0 and l.

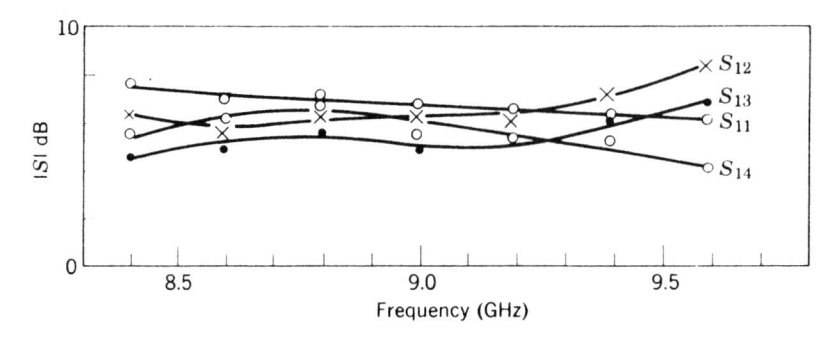

Figure 20.4 Scattering parameters of junction after second circulation adjustment (reprinted with permission, Helszajn, J. (1973) Waveguide and stripline 4-port single junction circulators, *IEEE Trans. Microwave Theory Tech.*, **21**, **MTT-20**, 630–633) (© 1973 IEEE)

20.3 4-PORTS SINGLE JUNCTION CIRCULATOR USING AXIAL MODES   405

Figure 20.4 indicates the scattering variables of one junction with a metal post of 1 mm radius. Since the electrical length of the post at 9.10 GHz is 0.22 λ_0 it has been allowed to make contact with the ferrite disk to maximize its length. No effort was made to improve on the geometry used here.

20.3.3 Adjustment of Direct Magnetic Field

The third and last adjustment of the junction is the amplitude of the direct magnetic field required to rotate the standing wave associated with the counter-rotating $HE_{\pm 1,1,1/2}$ modes by 45°. For an ideal lossless junction the scattering parameters are

$$S_{11} = S_{31} = S_{41} = 0 \tag{20.8a}$$

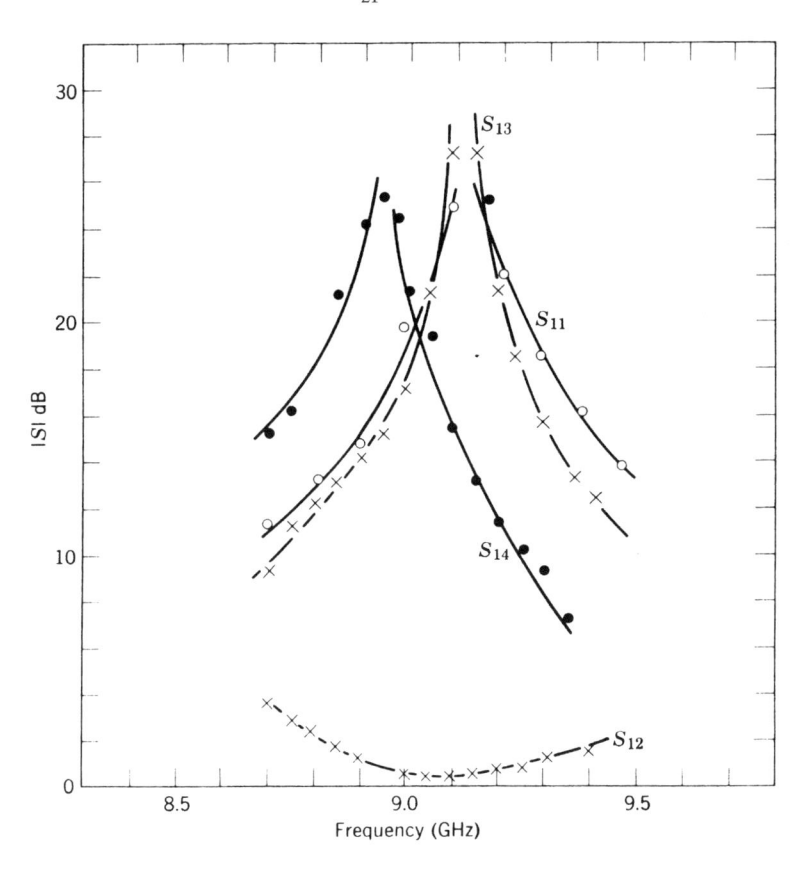

$$S_{21} = 1 \tag{20.8b}$$

Figure 20.5 Frequency variation of scattering parameters of junction after third circulation adjustment (reprinted with permission, Helszajn, J. (1973) Waveguide and stripline 4-port single junction circulators, *IEEE Trans. Microwave Theory Tech.*, **21**, **MTT-20**, 630–633) (© 1973 IEEE)

Figure 20.5 depicts the various scattering coefficients of the junction as a function of frequency, This illustration also indicates a property of the 4-port single junction circulator, which is that S_{11} and S_{31} are interdependent. These two parameters may be wide-banded with the help of external matching but S_{41} is a property of the basic junction alone.

The overall bandwidth of the device is of course determined by the quality factors of the individual modes used in its construction. Since the resonant frequency of an axially resonating mode does not require a unique shape, some adjustment of the Q factor is possible in such resonators by varying the geometry. Another way of widebanding the return loss of the junction is by means of a radial line transformer as in the case of the 3-port junction.

20.4 4-Port Single Junction Circulator Using Mixed Radial and Axial Modes

The purpose of this section is to describe the adjustment of a 4-port single junction circulator based on the geometry illustrated in Figure 20.6. Its arrangement consists of a full-height ferrite post between the top and bottom walls of the waveguide at the junction of four waveguides with a partial-height metal post through its centre. The modes used in this structure are the $TM_{\pm 1,1,0}$ radial ones associated with a post resonator with no variation of the fields along its axis and the axial $TM_{0,1,1/2}$ mode which exists along a quarterwave-long metal post. The patterns are shown in Figures 20.7(a) and (b). The first shows the $TM_{\pm 1,1,0}$ radial modes of the simple ferrite post and the $TM_{0,1,1/2}$ mode associated with a quarterwave-long metal post in a waveguide. The second illustration gives the same patterns but with the $TM_{\pm 1,1,0}$ modes rotated through 45°. A linear combination of the latter standing wave patterns at the four ports again coincides once again with the boundary conditions of an ideal circulator. The three circulation adjustments are obtained one at a time in the same way as in the previous case as is easily verified.

20.4.1 Adjustment of counter-rotating radial modes

The counter-rotating radial modes used for the construction of this geometry are the $TM_{\pm 1,1,0}$ modes with $kR = 1.84$. This circulation condition is satisfied as before with

$$S_{11} = S_{21} = S_{41} = 0 \tag{20.9a}$$

$$S_{31} = 1 \tag{20.9b}$$

Figure 20.8 depicts the variation of S_{31} with frequency for this arrangement. The radius of the ferrite post is 9.8 mm, its magnetization is 0.1200 T and its relative dielectric constant is 14.4.

20.4 4-PORT SINGLE JUNCTION CIRCULATOR

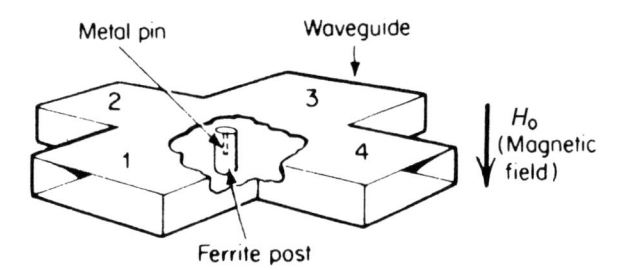

Figure 20.6 Schematic of 4-port single junction circulator using mixed radial and axial modes (reprinted with permission, Helszajn, J. and Buffler, C.R. (1968) Adjustment of the 4-port single junction circulator, *Radio Electron. Eng.*, **35**(6))

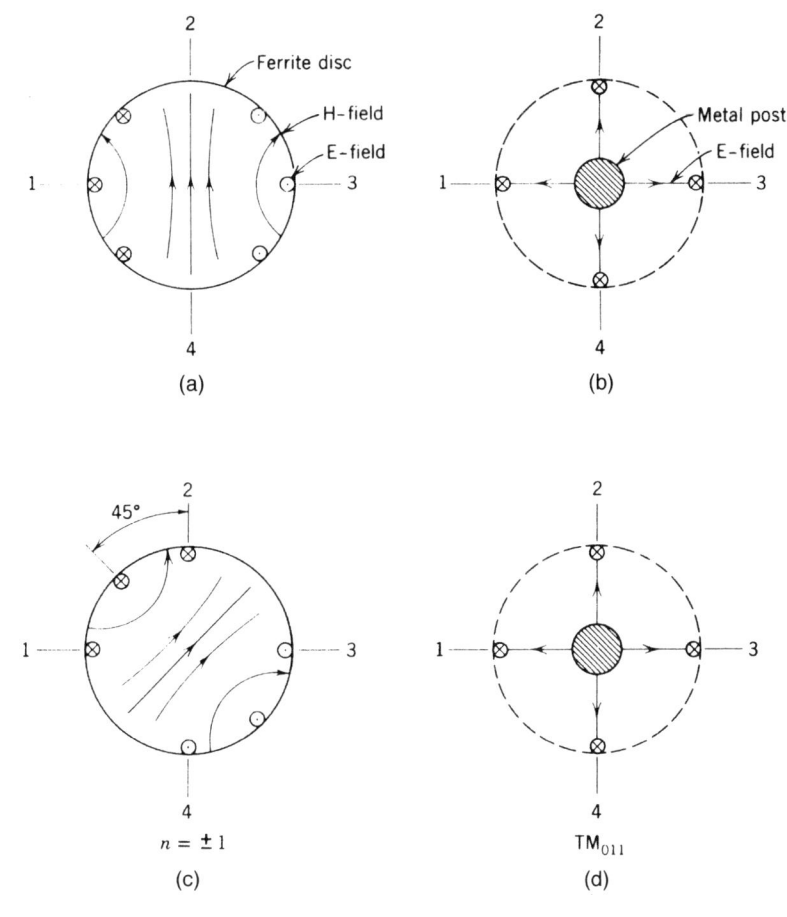

Figure 20.7 (a) $TM_{\pm,1,1,0}$ field patterns for unmagnetized ferrite post, (b) $TM_{0,1,1/2}$ E-field pattern on metal post, (c) $TM_{\pm,1,1,0}$ modes in magnetized ferrite post, (d) $TM_{0,1,1/2}$ E-field on metal post (reprinted with permission, Helszajn, J. and Buffler, C.R. (1968) Adjustment of the 4-port single junction circulator, *Radio Electron. Eng.*, **35**(6))

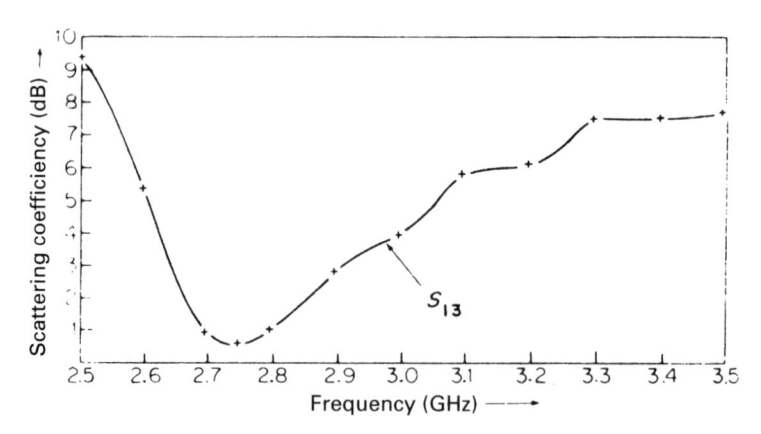

Figure 20.8 Scattering coefficient S_{31} versus frequency for fixed ferrite diameter for 4-port radial mode junction (reprinted with permission, Helszajn, J. and Buffler, C.R. (1968) Adjustment of the 4-port single junction circulator, *Radio Electron. Eng.*, **35**(6))

20.4.2 Adjustment of symmetrical axial mode of junction

Once the $TM_{\pm1,1,0}$ radial modes have been established within the junction the $TM_{0,1,1/2}$ axial mode is tuned by varying the length of the metal post. This circulation condition is again met provided

$$S_{11} = S_{21} = S_{31} = S_{41} = \tfrac{1}{2} \tag{20.10}$$

Figure 20.9 indicates the scattering coefficients of the junction as a function of the length of the metal post. The length of the metal post is a quarterwave in the ferrite medium.

20.4.3 Splitting of degenerate radial $TM_{\pm1,1,0}$ modes

The final and third adjustment of the circulator is now met by rotating the standing wave formed by the $TM_{\pm1,1,0}$ modes through 45° by magnetizing its junction. The variation of the scattering coefficients with the direct magnetic field is illustrated in Figure 20.10. An ideal circulator is obtained provided

$$S_{11} = S_{31} = S_{41} = 0 \tag{20.11a}$$

$$S_{21} = 1 \tag{20.11b}$$

This condition is satisfied at a direct field equal to 2780 A/m.

20.4 4-PORT SINGLE JUNCTION CIRCULATOR

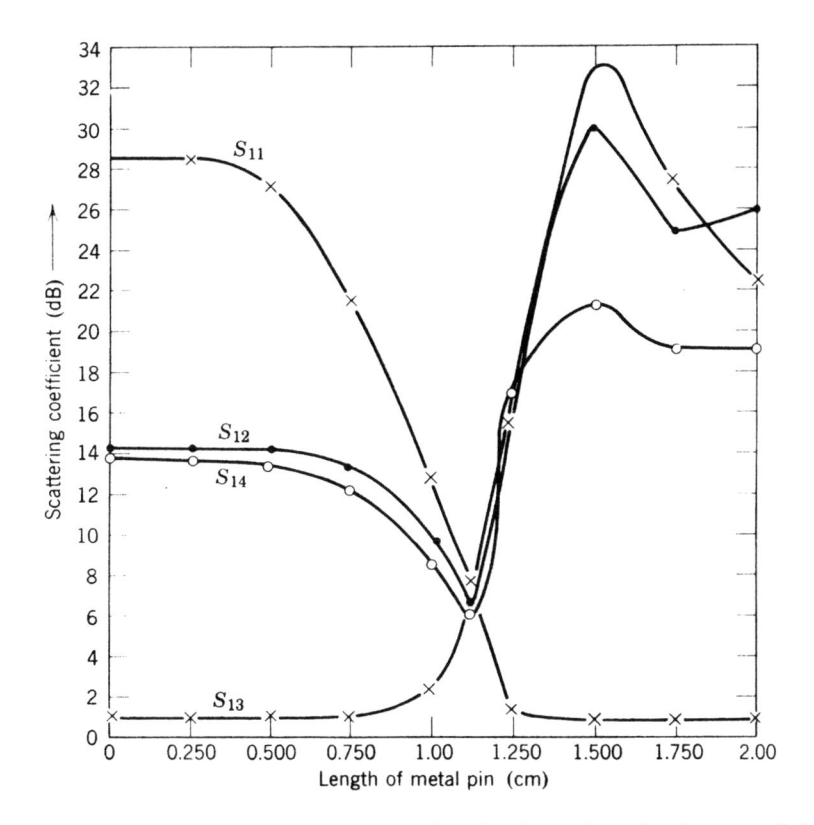

Figure 20.9 Scattering coefficients versus length of metal pin for 4-port radial mode junction (reprinted with permission, Helszajn, J. and Buffler, C.R. (1968) Adjustment of the 4-port single junction circulator, *Radio Electron. Eng.*, **35**(6))

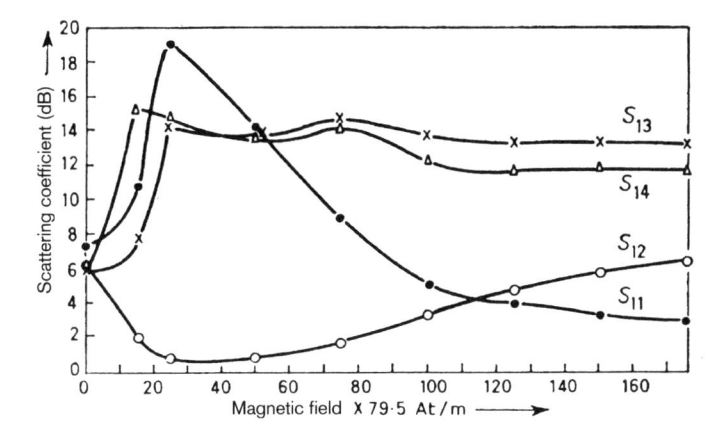

Figure 20.10 Scattering coefficients as a function of magnetic field for 4-port radial mode junction (reprinted with permission, Helszajn, J. and Buffler, C.R. (1968) Adjustment of the 4-port single junction circulator, *Radio Electron, Eng.*, **35**(6))

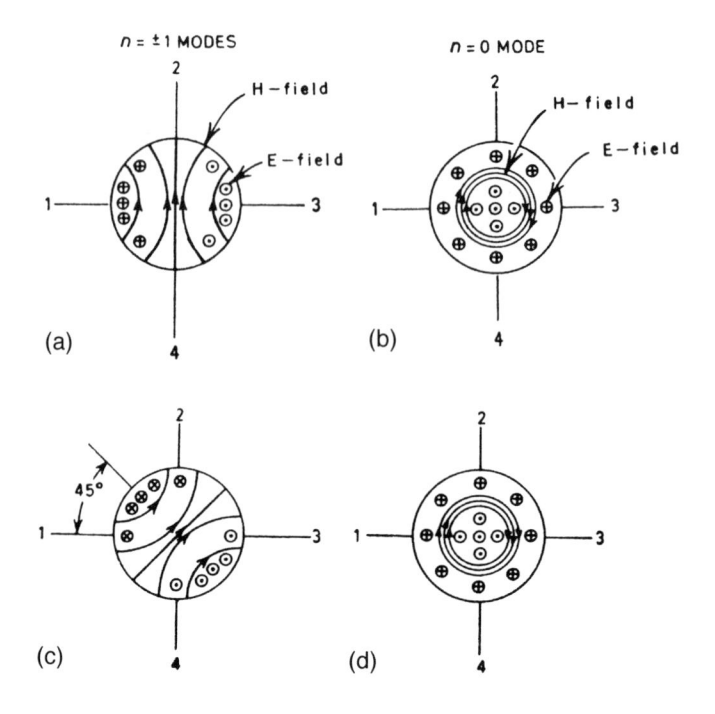

Figure 20.11 (a) $TM_{\pm1,1,0}$ field patterns for unmagnetized ferrite post, (b) TM_{010} field patterns for unmagnetized ferrite post, (c) $TM_{\pm1,1,0}$ field patterns for magnetized ferrite post, (d) TM_{010} field patterns for magnetized ferrite post. Scattering coefficients versus length of metal pin for 4-port radial mode junction (reprinted with permission, Fay, C. E. and Comstock, R. L. (1965) Operation of the ferrite junction circulator, *IEEE Trans. Microwave Theory Tech.*, **MTT-13**, 15–27)

20.5 4-Port Single Junction Circulator using Radial Modes

A 4-port single junction circulator may also be realized by having recourse to a linear combination of radial $TM_{\pm1,1,0}$ and $TM_{0,1,0}$ modes. The field patterns in this case are shown in Figures 20.11(a) and (b). The adjustment of this solution follows closely that of the first two geometries already described except that the $TM_{0,1,0}$ mode is tuned to the frequency of the $TM_{\pm1,1,0}$ modes with the help of a thin nonresonant capacitive post at the center of junction instead of a resonant one. Figures 20.11(c) and (d) indicates the field patterns of the magnetized junction.

20.6 4-Port Single Junction Circulator Using Higher Order Modes

The adjustment of one solution based on the use of degenerate higher order modes in a post resonator for its operation will now be described. Its schematic diagram

20.6 4-PORT SINGLE JUNCTION CIRCULATOR USING HIGHER ORDER MODES

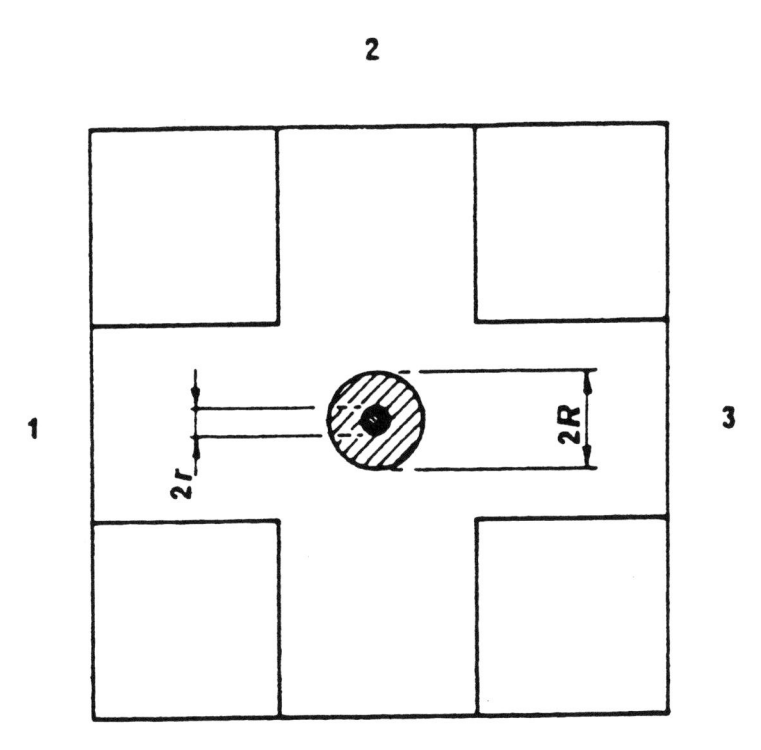

Figure 20.12 Schematic of single junction 4-port using a higher order gyromagnetic resonator

is illustrated in Figure 20.12. It consists of a simple ferrite post at the junction of four waveguides. The direct magnetic field is applied perpendicular to the plane of the ferrite disk in the usual way.

The modes used in this arrangement are the $TM_{0,1,0}$ and $TM_{\pm,3,1,0}$ radial ones in a simple ferrite post with no variations of the field patterns along its axis. Its experimental adjustment relies on the fact that the $TM_{0,1,0}$ mode already lies approximately midway between the split $TM_{\pm,3,1,0}$ ones. It is, therefore, possible in this instance to omit the second circulation adjustment. Figures 20.13(a) and (b) depict the field patterns of the modes in question. The radial wavenumbers for such a demagnetized ferrite disk are $kR = 3.83$ for $TM_{\pm,3,1,0}$ and $kR = 4.20$ for $TM_{\pm,3,1,0}$. A radial wavenumber of between $3.86 < kR < 4.10$ has been used on one experimental waveguide junction. The third circulation adjustment is now met by removing the degeneracy between the counter-rotating modes. Figures 20.13(c) and (d) indicate the field patterns of the modes with the $TM_{\pm,3,1,0}$ ones rotated by 15° by magnetizing the junction with a direct magnetic field. If the magnitudes of the electric fields for the two patterns are equal (after rotation) at the four ports, transmission occurs between ports 1 and 2 and ports 3 and 4 are isolated.

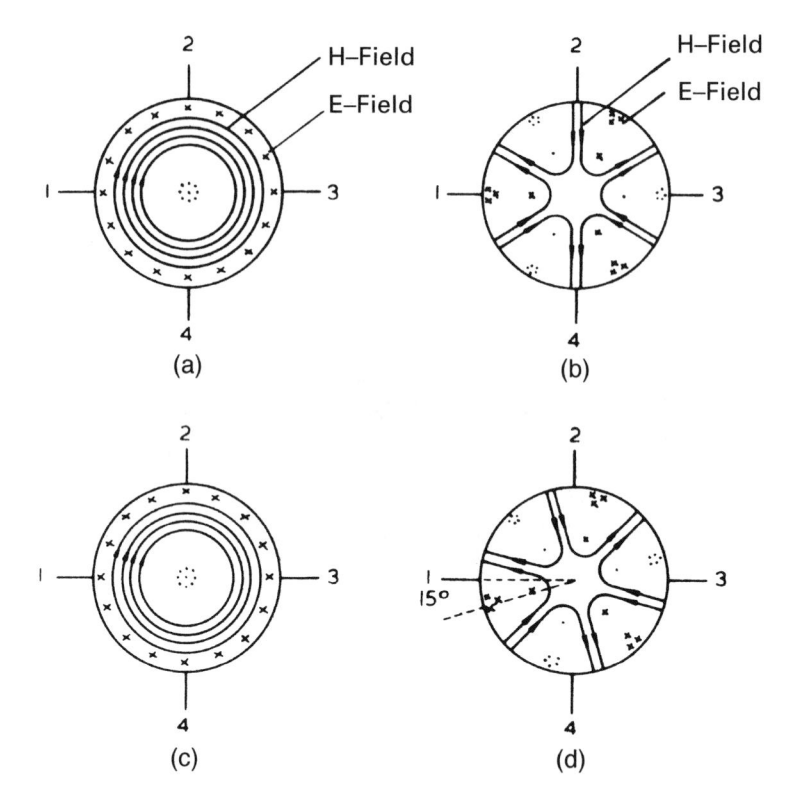

Figure 20.13 (a) $TM_{0,1,0}$ field patterns for unmagnetized disk, (b) $TM_{\pm 3,1,0}$ field patterns for unmagnetized disk, (c) $TM_{0,1,0}$ field patterns for magnetized disk, (d) $TM_{\pm 3,1,0}$ field patterns for magnetized disk (reprinted with permission, Helszajn, J. (1973) Waveguide and stripline 4-port single junction circulators, *IEEE Trans. Microwave Theory Tech.*, **21**, **MTT-20**, 630–633) (© 1973 IEEE)

The distribution of the electric field around the periphery of the ferrite disk for the 4-port circulator described here is

$$E_z = a_0 + a_3 \cos 3(\tau_3 + \phi) \qquad (20.12)$$

a_0 and a_3 are arbitrary constants and τ_3 is the phase angle through which the standing wave solution formed by the degenerate modes is rotated.

$$E_z(\phi=0) = a_0 + a_3 \cos 3\tau_3 = +1 \qquad (20.13a)$$

$$E_z(\phi=\pi/2) = a_0 + a_3 \cos 3(\tau_3 + \pi/2) = +1 \qquad (20.13b)$$

$$E_z(\phi=\pi) = a_0 + a_3 \cos 3(\tau_3 + \pi) = 0 \qquad (20.13c)$$

$$E_z(\phi=3\pi/2) = a_0 + a_3 \cos 3(\tau_3 + 3\pi/2) = 0 \quad (20.13d)$$

The result is

$$\tau_3 = 15° \quad (20.14a)$$

$$a_0 = \tfrac{1}{2} \quad (20.14b)$$

$$a_3 = \tfrac{1}{\sqrt{2}} \quad (20.14c)$$

References and Further Reading

Bogdanov, A. G. (1969) Design of waveguide x-circulators, *Radio Eng. Electron. Phys.*, **14**(4).

Bosma, H. (1964) On stripline circulation at U.H.F., *Trans. IEEE Microwave Theory Tech.*, **MTT-12**, 61–72.

Courtney, W. E. (1970) Analysis and evaluation of a method of measuring the complex permittivity and permeability of microwave insulators, *IEEE Trans. Microwave Theory Tech.*, **MTT-18**(8), 476–485.

Davies, J. B. and Cohen, P. (1963) Theoretical design of symmetrical junction stripline circulators, *IEEE Trans. Microwave Theory Tech.*, **MTT-11**, 506–512.

Davis, L. E., Coleman, M. D. and Cotter, J. J. (1964) Four-port crossed waveguide junction circulators, (corresp.) *IEEE Trans. Microwave Theory Tech.*, **MTT-12**, 43–47.

Fay, C. E. and Dean, W. A. (1966) The four-port single junction circulator in stripline, *PG-MTT, Digest Int. Symp.*, IEEE, New York.

Fay, C. E. and Comstock, R. L. (1965) Operation of the ferrite junction circulator, *IEEE Trans. Microwave Theory Tech.*, **MTT-13**, 15–27.

Helszajn, J. (1970) The adjustment of the m-port single junction circulator, *IEEE Trans. Microwave Theory Tech.*, **MTT-18**, 705–711.

Helszajn, J. (1972) Three resonant mode adjustment of the waveguide circulator, *Radio Electron. Eng.*, **42**(5), 357–360.

Helszajn, J. (1973) Waveguide and stripline 4-port single junction circulators, *IEEE Trans. Microwave Theory Tech.*, **MTT-20**, 630–633.

Helszajn, J. and Buffler, C. R. (1968) Adjustment of the 4-port single junction circulator, *Radio Electron, Eng.*, **35**(6) 357–360.

Ku, W. H. and Wu, Y. S. (1973) On stripline four port circulator' *IEEE Microwave Theory Tech. Int. Microwave Symp. Digest*, IEEE Cat. No. 73, chap. 736–9 MTT, 86–88.

Landry, D. N. (1963) A single junction four-port coaxial circulator, *Wescon*, Part 5, Session 4.2 IEEE, New York.

Lewin, L. (1951) *Advanced Theory of Waveguides*, Iliffe and Sons, Ltd., London.

Longley, S. R. (1967) Experimental 4-port E-plane junction circulator, *IEEE Trans. Microwave Theory Tech.*, (corresp.), **MTT-15**, 378–380.

Markuvitz, *Microwave Handbook*, McGraw-Hill, New York.

Owen, B. (1972) The identification of modal resonances in ferrite loaded waveguide Y-junctions and their adjustments for circulation, *Bell Syst. Tech J.*, **51**(3), 595–627.

Watkins, J. (1969) Circulator resonant structures on microstrip, *Electron. Tech. Lett.*, **5**, 524–525.

Yoshida, S. (1959) X-circulator, *Proc. Inst. Radio Eng.*, **47**, 1150.

21

MICROWAVE SWITCHING USING JUNCTION CIRCULATORS

21.1 Introduction

Ferrite control devices, as is by now appreciated, may be actuated using electromagnets, or may be controlled by externally or internally latched circuits. Important parameters in the description of such devices include switching speed, hysteresis effects, temperature behaviour, shielding effects due to induced eddy currents in waveguide walls in externally actuated devices, influence of air gaps on the properties of the hysteresis loop, stored magnetic energy in switching circuit, switching power, demagnetizing factors and resettability. The purpose of this chapter is to qualitatively and quantitatively discuss some of these different design aspects. One drawback of the use of an electromagnet to actuate ferrite control devices is that the switching power is moderately large and that the switching speed is restricted because of demagnetizing fields and eddy currents to about 10 µs. The demagnetizing fields are determined by the shape of the ferrite; the eddy currents are induced in the waveguide walls due to the changing magnetic field, which results in a certain degree of shielding. One way of removing both limitations is to use a ferrite shape in the microwave circuit that can be biased by a single-wire loop and that preferably has a square hysteresis loop. The magnetic energy is here determined by the inductance of a single-wire turn and that required to move between the two states of the hysteresis loop and is extremely small. Switching times of less than 0.5 µs are possible with this type of operation because the switching wire is located inside the waveguide. Another advantage of this technique is, of course, that no holding current is required. If the magnitude of the magnetic field is changed, energy must be provided to, or removed from, the magnetic field. The magnetic energy stored in the circuit is therefore an important quantity in rating different switches. This energy, divided by the switching time, is related to the instantaneous power required from the driver. The switching power is inversely proportional to the switching time.

416 21 MICROWAVE SWITCHING USING JUNCTION CIRCULATORS

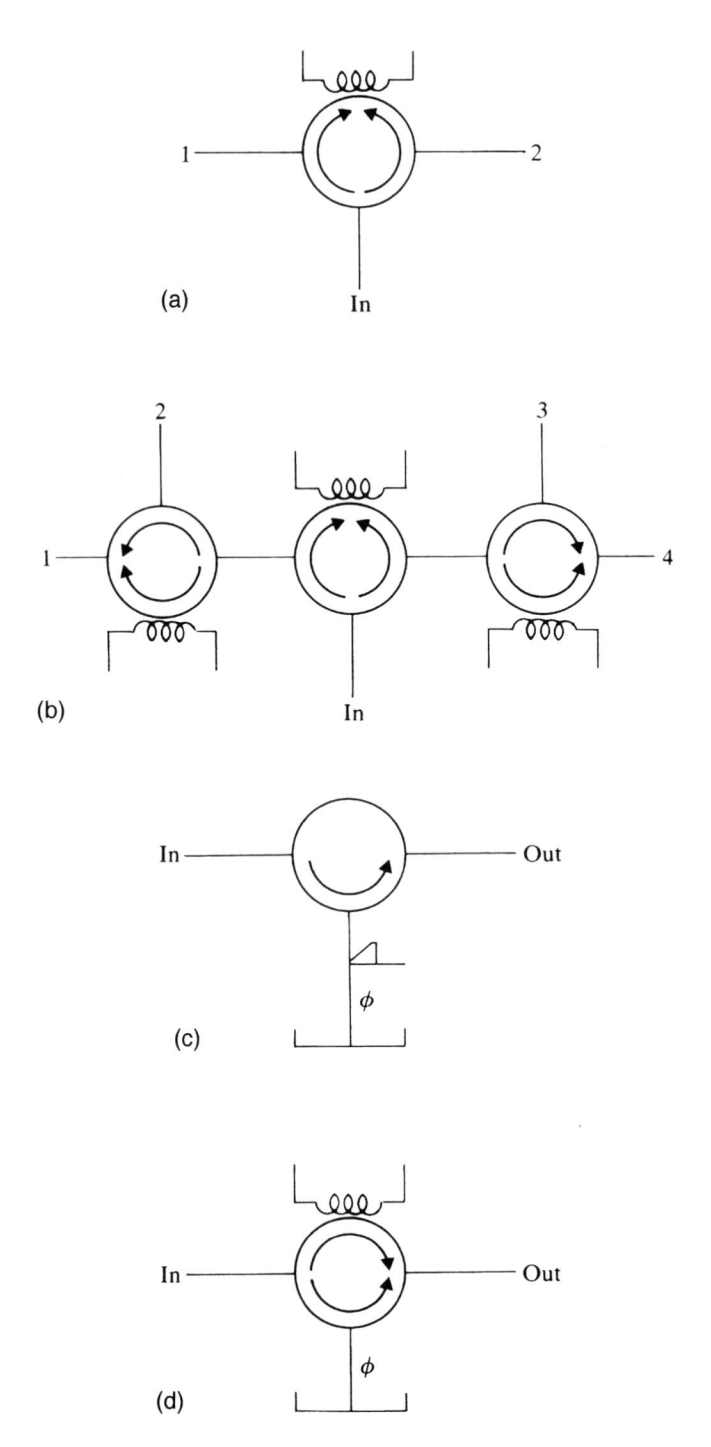

Figure 21.1 Microwave phase shifter using (a) Schematic of circulator switch (b) SP4T Butler switch using circulators (c) p–i–n dioded switch and fixed circulator and (d) switched circulator

21.2 Microwave Switching Using Circulators

Since the direction of circulation of a circulator is determined by that of the direct magnetic field it may be employed to switch an input signal at one port to either one of the other two. Switching is achieved by replacing the permanent magnet by an electromagnet or by latching the microwave ferrite resonator directly by embedding a current-carrying wire loop within the resonator. The schematic diagram of a switched junction is shown in Figure 21.1(a). It is particularly useful in the construction of Butler-type matrices in phased array systems. A single-pole three throw version is depicted in Figure 21.1(b).

Two common arrangements in which ferrite circulators may be employed to obtain microwave phase shifting are separately illustrated in Figure 21.1(c) and (d). The first uses a fixed circulator in conjunction with a *p–i–n* diode switch to vary the short-circuit plane terminating port 2. A transmission analogue phase shifter is therefore obtained between ports 1 and 3 with this mode of operation. The second version is also a transmission configuration but now a switchable circulator is used to control the path between ports 1 and 3 of the circulator. The switching speed of the *p–i–n* device is normally the faster one.

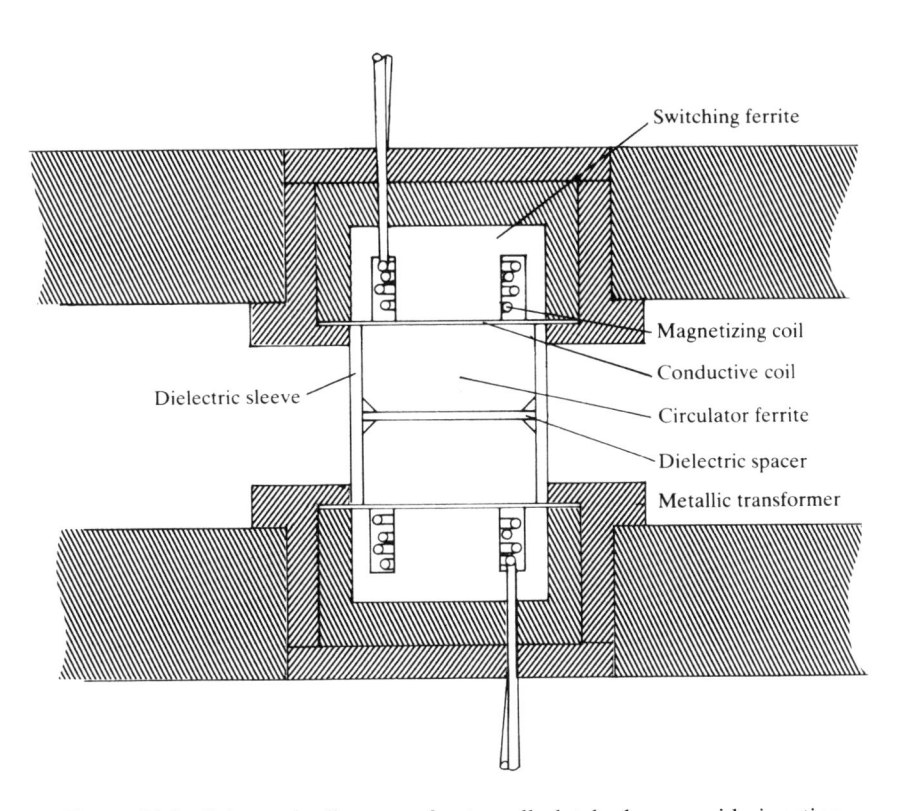

Figure 21.2 Schematic diagram of externally latched waveguide junction

21.3 Externally and Internally Latched Junction Circulators

Circulators may be either actuated by an electromagnet or they may be operated by internally or externally latching the ferrite resonator. Figure 21.2 illustrates one externally latched arrangement. Figure 21.3(a) and (b) depict internally latched waveguide devices using half-wave long resonators. Figure 21.4 summarizes some resonators with the required three-fold symmetry for the construction of this class

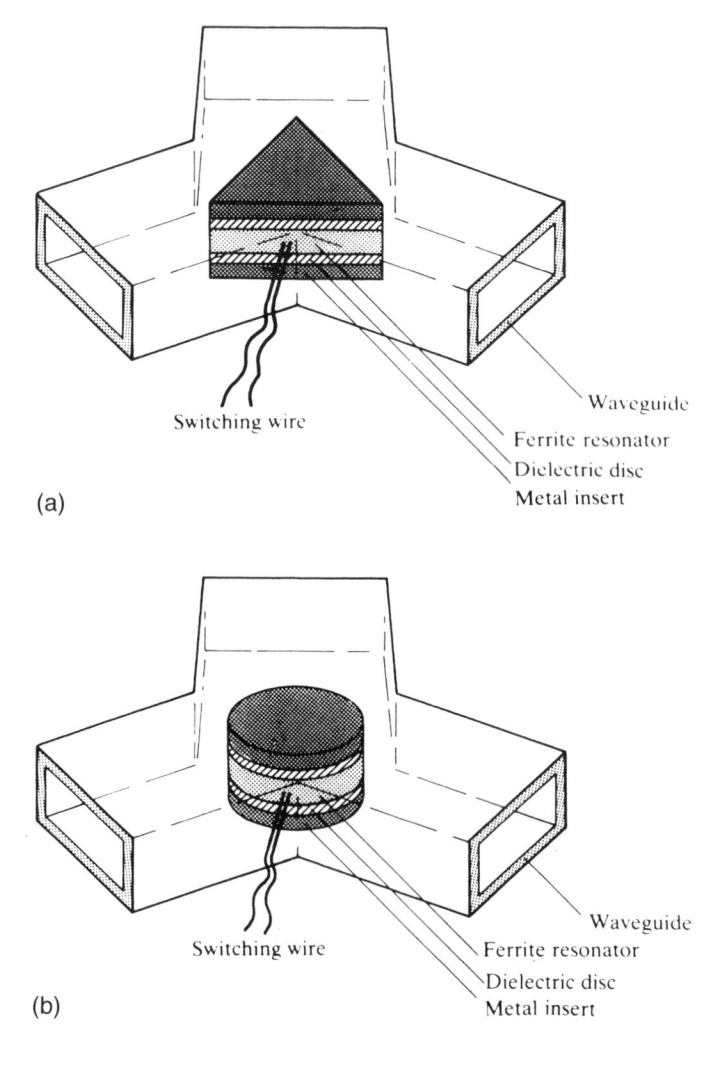

Figure 21.3 Schematic diagram of waveguide junction circulator using a partial height (a) triangular and (b) circular resonator with a wire loop

21.3 EXTERNALLY AND INTERNALLY LATCHED JUNCTION CIRCULATORS

of device. The derivation of the two circulation conditions of this sort of device is outside the remit of this work but the split frequencies of the latched post resonator, an important parameter in the description of any circulator, may be derived without too much difficulty, as will be demonstrated in the next section

The standing wave solution of the wye resonator is also of some interest. Figure 21.5 indicates its solution.

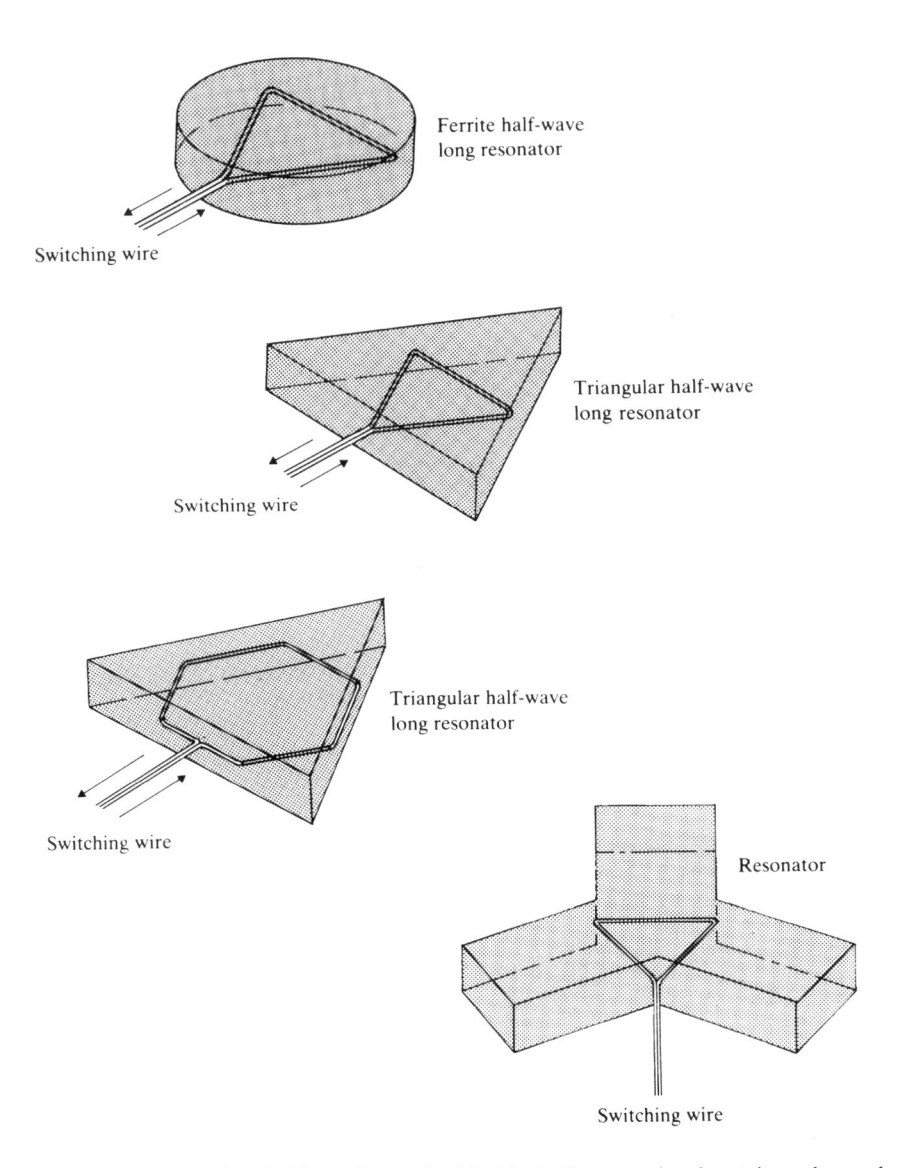

Figure 21.4 Details of switching wire embedded in half-wave circular, triangular and wye resonator

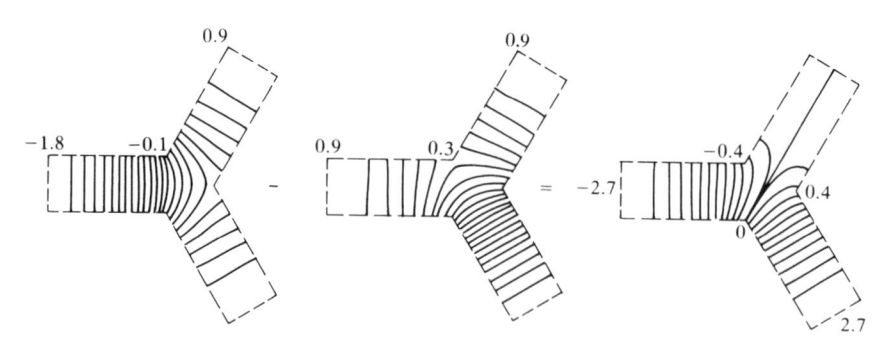

Figure 21.5 Standing wave solution of wye post resonator

21.4 Quality Factor of Latched Post Resonator

The splitting of the degenerate frequencies of a latched resonator is of some import in the design of any junction circulator as is by now understood. The purpose of this section is to establish this quantity in the case of a post resonator bounded by electric walls on its two flat faces and by a magnetic wall on its sidewall. The direct magnetic field is along the positive z-axis in the inner region and the opposite direction in the outer region. The switching wire is indicated in Figure 21.6. The derivation starts by noting that for a post or planar resonator

$$\frac{\partial}{\partial z} = 0 \tag{21.1}$$

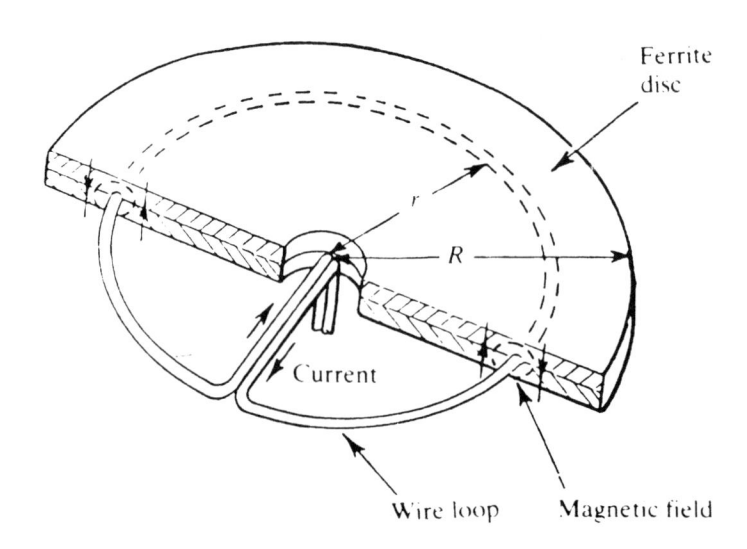

Figure 21.6 Current and magnetic field in ferrite disc

21.4 QUALITY FACTOR OF LATCHED POST RESONATOR

The preceding equation implies

$$E_\theta = E_r = 0 \tag{21.2}$$

and

$$H_z = 0 \tag{21.3}$$

The transverse fields H_θ and H_r in each region are obtained from a knowledge of E_z there in the usual way with the aid of Maxwell's equations. The characteristic equation for the split frequencies of the magnetized resonator may thereafter be constructed from a knowledge of the field patterns and the boundary conditions at $r = b$ and $r = a$. If only the frequencies of the resonator are required, a knowledge of E_z and H_θ in each region is sufficient.

In the inner region of the ferrite resonator the electric field may be described in terms of Bessel functions of the first kind of order n by

$$E_z = \sum_{n=-\infty}^{\infty} A_n J_n(kr) \exp(-jn\theta) \tag{21.4}$$

The θ component H_θ of the magnetic field in a magnetized ferrite medium is related to E_z by

$$H_\theta = \frac{-j}{\omega\mu_0\mu_e} \left[\frac{\partial E_z}{\partial r} + j\left(\frac{\kappa}{\mu}\right) \frac{1}{r} \frac{\partial E_z}{\partial \theta} \right] \tag{21.5}$$

This gives

$$H_\theta = \sum_{n=-\infty}^{\infty} -jA_n \zeta_e \left[J'_n(kr) - \left(\frac{\kappa}{\mu}\right) \frac{nJ_n(kr)}{kr} \right] \exp(-jn\theta) \tag{21.6}$$

where k is the wave number:

$$k = \omega\sqrt{\mu_0\mu_e\varepsilon_0\varepsilon_f} \tag{21.7}$$

and ζ_e is the wave admittance:

$$\zeta_e = \sqrt{\frac{\varepsilon_0\varepsilon_f}{\mu_0\mu_e}} \tag{21.8}$$

The relative effective permeability μ_e is described in terms of the diagonal and off-diagonal entries μ and κ of the tensor permeability by

$$\mu_e = \frac{\mu^2 - \kappa^2}{\mu} \tag{21.9}$$

and ε_f is the relative dielectric constant of the ferrite or garnet material.

The electric field in the outer region is described in terms of Bessel functions $J_n(x)$ and $Y_n(x)$ of the first and second kind by

$$E_z = \sum_{n=-\infty}^{\infty} A_n [B_n J_n(kr) + C_n Y_n(kr)] \exp(-jn\theta) \tag{21.10}$$

where B_n and C_n are constants to be determined.

The azimuthal magnetic field in the outer region of magnetized resonator is now formed with the aid of equation 21.5 but with the sign κ reversed: This gives

$$H_\theta = \sum_{n=-\infty}^{\infty} -j A_n \zeta_e \left\{ B_n \left[J'_n(kr) + \left(\frac{\kappa}{\mu}\right) \frac{n J_n(kr)}{kr} \right] \right.$$

$$\left. + C_n \left[Y'_n(kr) + \left(\frac{\kappa}{\mu}\right) \frac{n Y_n(kr)}{kr} \right] \right\} \exp(-jn\theta) \tag{21.11}$$

The unknown constants B_n and C_n can be deduced in terms of the physical variables by using the conditions that E_z and H_θ are continuous at $r = b$:

$$B_n J_n(kb) + C_n Y_n(kb) = J_n(kb) \tag{21.12}$$

$$B_n \left[J'_n(kb) + \left(\frac{\kappa}{\mu}\right) \frac{n J_n(kb)}{(kb)} \right] + C_n \left[Y'_n(kb) + \left(\frac{\kappa}{\mu}\right) \frac{n Y_n(kb)}{kb} \right]$$

$$= J'_n(kb) - \left(\frac{\kappa}{\mu}\right) \frac{n J_n(kb)}{kb} \tag{21.13}$$

The required result is

$$B_n = 1 - \frac{\pi \kappa}{\mu} n J_n(kb) Y_n(kb) \tag{21.14}$$

$$C_n = \frac{\pi \kappa}{\mu} n J_n^2(kb) \tag{21.15}$$

The two roots for the uncoupled resonator are given by applying the boundary conditions at the sidewall:

$$H_\theta(r = a) = 0 \tag{21.16}$$

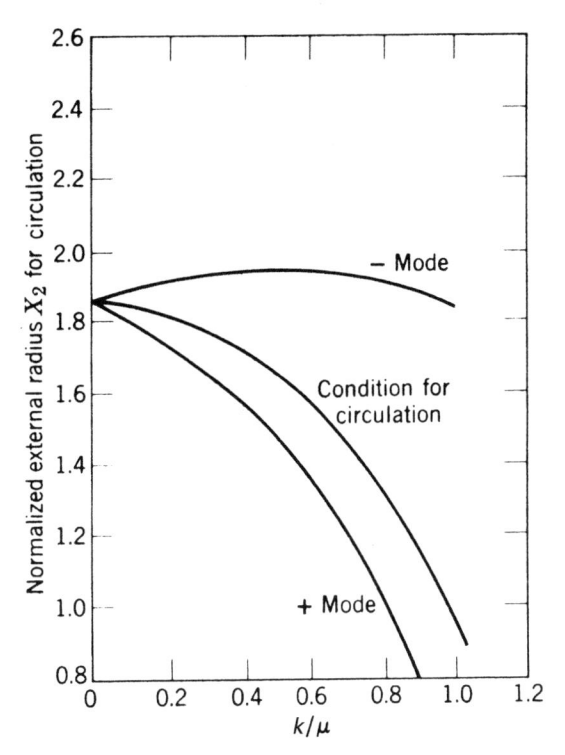

Figure 21.7 Normalized outside radius for circulation of oppositely magnetized ferrite elements for the $n = 1$ mode as function of κ/μ for $a = \sqrt{2}\,b$ (reprinted with permission, Siekanowicz, W. W. and Schilling, W. A. (1968) A new type of latching switchable ferrite junction circulator, *IEEE Trans. on MTT*, **MTT-16**, 177–183)

Figure 21.7 gives the roots of ka for $n = \pm 1$ as a function of κ/μ for $a = \sqrt{2}\,b$. This indicates that for a given value of κ/μ the splitting between the resonator modes of the latched configuration is smaller than that of the standard junction. The maximum bandwidth, which is proportional to the splitting between the two normal modes, is in this case approximately reduced by a factor of 0.70. The operating frequency is also now a function of κ/μ.

21.5 Magnetic Circuit Using Major Hysteresis Loop

The direct magnetic field in a junction circulator can be established using either an external electromagnet or it can be switched by current pulses through a magnetizing wire between the two remanent states of the major or indeed of a minor hysteresis loop of a closed magnetic circuit. The former arrangement requires a holding current to hold the device in a given state. In the latter one, however, no such current is necessary; the device remains latched in a given state until another switching operation is required. The advantages and disadvantages of each type of circuit is understood.

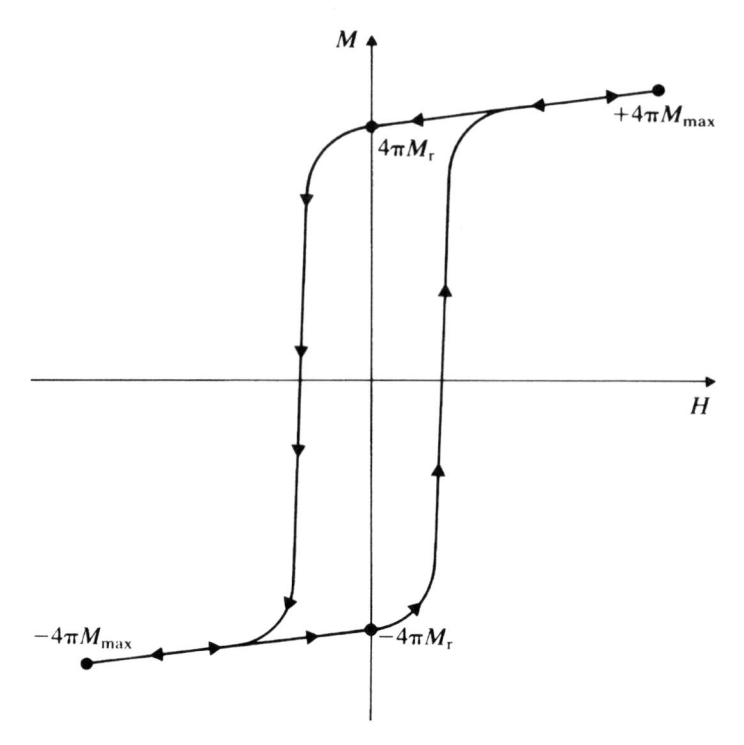

Figure 21.8 Typical hysteresis loop of a latching phase shifter operating with a major loop switching

Operation on the major hysteresis loop may be understood by scrutinizing the hysteresis loop in Figure 21.8, providing it is recognized that the size and shape of this loop may vary with the speed of the switching process. In this situation, the magnetization of the toroid is driven between two remanent states ($\pm 4\pi M_r$) equidistant from the origin by the application of a current pulse sufficiently large to produce a field perhaps three or five times that of the coercive force. After this point is reached, the current pulse is removed and the magnetization will move to the remanent value ($\pm 4\pi M_r$) and remain there until another switching operation is desired. This sort of electronic driver circuit is relatively simple since it is only required that the toroids be driven back and forth between the major remanent states of the hysteresis loop.

21.6 Display of Hysteresis Loop

The magnetic properties and parameters of a magnetic core or toroid under different operating conditions, such as temperature, say, are best discussed in terms of the details of its hysteresis loop. Some experimental quantities that are of particular interest include the saturation magnetization (M_0), the remanent magnetization (M_r) and the coercive force (H_c). The experimental display of such

21.6 DISPLAY OF HYSTERESIS LOOP

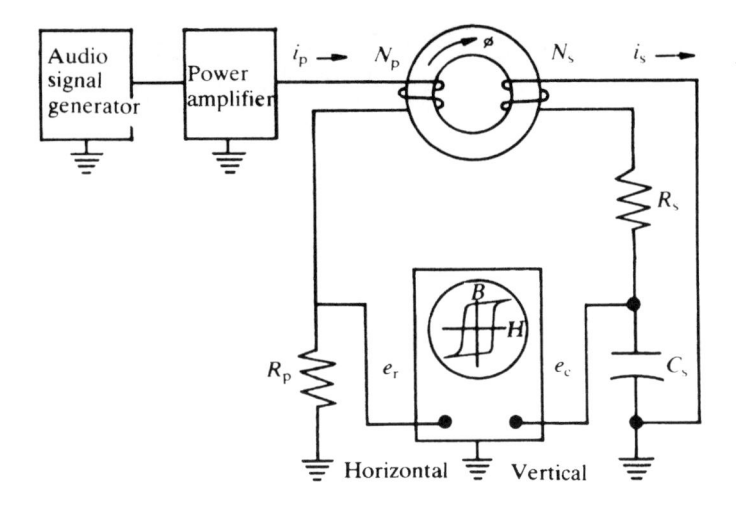

Figure 21.9 Schematic diagram of hysteresis display

loops is therefore of some interest. One circuit that may be used for this purpose is outlined in Figure 21.9. This arrangement develops voltages V_p and V_i that are proportional to B and H respectively.

The magnetic field (H) in the core is monitored by measuring the voltage (V_p) across a resistor in series with the primary winding:

$$H = \frac{N_p}{I_p}\left(\frac{V_p}{R_p}\right), \quad \text{A/m} \tag{21.17}$$

where I_p is the current in the primary winding, N_p is the number of turns of the primary winding (10 to 30) and R_p is the resistor in series with the primary coil (10 Ω). The magnetization (B) is likewise evaluated by forming the voltage (V_i) across the capacitance of the RC integrator in the secondary circuit:

$$B \approx -\frac{V_i R_i C_i}{N_s A} \tag{21.18}$$

where R_i is the series resistance of the integrator (100 kΩ), C_i is the capacitance of the integrator (0.10 μF), N_s is the number of turns of the secondary winding (10 to 30) and A is the cross-sectional area of the core.

The data shown in Figure 21.10 on the effect of small air gaps on the squareness of the hysteresis loop has been obtained using the arrangement outlined here.

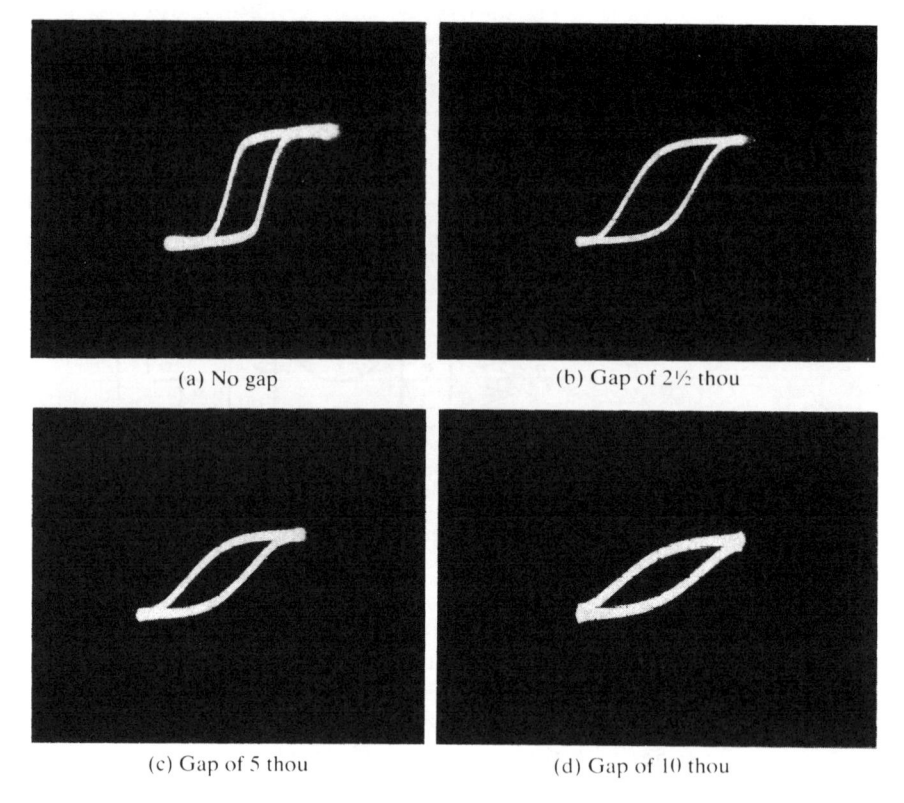

426 21 MICROWAVE SWITCHING USING JUNCTION CIRCULATORS

(a) No gap

(b) Gap of 2½ thou

(c) Gap of 5 thou

(d) Gap of 10 thou

Figure 21.10 Photographs of hysteresis loops showing effect of gap in magnetic circuit

21.7 Switching Coefficient of Magnetization

The change of magnetization in a ferrite core consists usually in reversal of the magnetization, e.g. from negative remanence to the positive remanence corresponding to the magnetic field applied. The ultimate state of magnetization that is set up (after the passage of the current pulse) is here always symmetrical with respect to zero. It is observed that in most cases the change in the magnetization produced by this field cannot follow the increase in the current. The general situation is quite complicated but for an applied magnetic field slightly in excess of the coercive force, H_c, domain wall motion will, in general, be the predominant reversal mechanism. In this case, the flux change is accomplished by the motion of Bloch walls, which separate the domains of differently oriented magnetization. For suitable oriented single crystals of ferrite, a very simple domain configuration may be achieved, which makes it possible to obtain information on the behaviour of a moving domain wall. Studies on single crystals of ferrite have demonstrated that, under this condition, the wall velocity depends linearly on the applied magnetic field. This leads to a linear relation between the direct field and the reciprocal of the switching time. Such a relationship is also noted experimentally for polycrystalline ferrites although the actual domain configuration is not known.

21.7 SWITCHING COEFFICIENT OF MAGNETIZATION

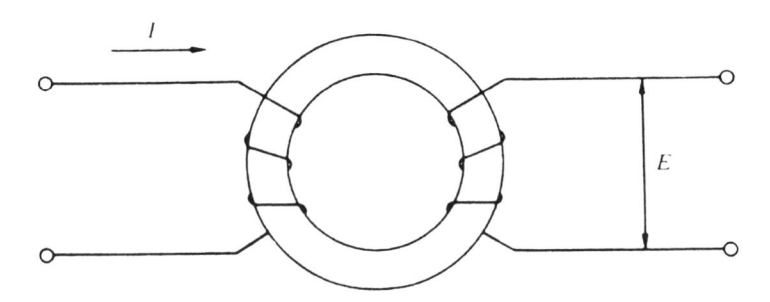

Figure 21.11 Ferrite core with two windings for measuring switching time

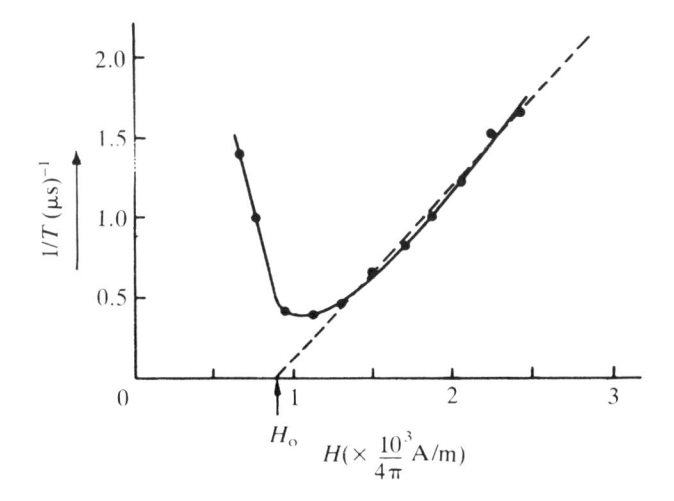

Figure 21.12 Reciprocal reversal time $1/T$ as a function of direct magnetic H for ferroxcube (reprinted with permission Van der Heide, Bruijning, H. G. and Wijn, H. P. J. (1956) Switching time of ferrites with rectangular hysteresis loop, *Philips Tech. Rev.*, **18**, 339)

The switching time is usually measured by using a core with two windings as shown in Figure 21.11. The output voltage pulse appearing at the terminals of the secondary winding exhibits a characteristic shape with two separate maxima when a current pulse is passed through the magnetization winding. For ferrites with rectangular hysteresis loops the first maximum in the output voltage represents a small percentage of the total area under the curve and hence of the change in magnetization. The duration T of a voltage pulse is defined as the time (counted from the beginning of the current pulse) that elapses before the voltage has dropped to 10% of the maximum value; for the maximum value the second peak is considered. The dependence of the switching time on the magnetizing producing flux reversal is most clearly represented by plotting $1/T$ as a function of the magnetic strength H, in the manner indicated in Figure 21.12. Over the majority of the range shown, $1/T$ has a linear dependence on H, which may be adequately represented by $T(H - H_0) = S$

In this expression S is known as the switching coefficient, and H_0, which is of the same order of magnitude as the coercive force H_c of the material, may be termed the threshold field for irreversible magnetization. It should be noted that, although the curve is continued to values of H less than H_0, the switching of the core under these conditions produces a smaller hysteresis loop; i.e., the material is not driven to magnetic saturation and such operation is not desirable. The optimum squareness ratio R_s occurs for values of H very little different from H_0, but it is usual to adopt magnetizing fields of between $2H_0$ and $5H_0$, the slight deterioration in squareness being accepted in the interests of faster switching.

It is found that the great majority of ferrites with the squareness and the coercive force usual for switching elements all have a 'reversal constant' $S = (H - H_0)T$ of the same order of magnitude. In the case of ferroxcube 6A considered above, the resultant value is $H_0 = 80$ μs (A/m).

References and Further Reading

Betts, J., Temme, D. H., and Weiss, J. A. (1966) A switching circulator s-band; stripline; 15 kilowatts; 10 microseconds; temperature-stable, *IEEE Trans. Microwave Theory Tech.*, **MTT-14**, 665–669.

Clavin, A. (1963) Reciprocal and nonreciprocal switches utilizing ferrite junction circulators, *IEEE Trans. Microwave Theory Tech.*, **MTT-11**, 217–218.

Davis, L. E. (1966) Computed phase-shift and performance of a latching 3-port waveguide circulator, *NEREM Rec.*, 96.

Freiberg, L. (1961) Pulse operated circulator switch, *I.R.E. Trans. Microwave Theory Tech.*

Goodman, P. C. (1965) A latching ferrite junction circulator for phased array-switching applications, *IEEE G-MTT Symposium*.

Helszajn, J. (1960) Switching criteria for waveguide ferrite devices, *Radio Electron. Eng.*, **30**, 289–296.

Helszajn, J. and Hines, M. L. (1968) A high speed TEM junction ferrite modulator using a wire loop, *Radio Electron, Eng.*, **53**, 81–82.

Jelonek, Z. J. and Boomer, G. I. (1958) A correlation between the transient and frequency responses in servomechanisms, *J. Brit. IRE*, **18**, 101–114.

King, L. V. (1933) Electromagnetic shielding at radio frequencies, *Phil. Mag., Ser. 7*, **15**, 201.

Laplume, J. (1945) Shielding effect of a cylindrical tube placed in a uniform magnetic field perpendicular to its axis, *Annales de Radioélectricité*, **1**(1), 65–73.

Levy, L. and Silber, L. M. (1960) A fast switching X-band circulator utilizing ferrite toroids, *IRE Wescon Convention Rec.*, Part 1.

Lyons, W. (1933) Experiments on electromagnetic shielding at frequencies between one and thirty kilocycles, *Proc. IRE*, **21**(4), 574–590.

McCleery, D. K. (1961) *Introduction to Transients*, Wiley, New York.

Siekanowicz, W. W., Schilling, W. A. (1968) A new type of latching switchable ferrite junction circulator, *IEEE Trans. Microwave Theory Tech.*, 177–183.

Siekanowicz, W. W., Paglione, R. W. and Walsh, T. E. (1970) A latching ring-and-post ferrite waveguide circulator, *IEEE Trans. Microwave Theory Tech.*, **MTT-18**, April.

Stern, R. A. and Ince, W. J. (1967) Design of composite magnetic circuits for temperature compensation of microwave ferrite devices, *IEEE Trans. Microwave Theory Tech.*, **MTT-15**, 295–300.

Taft, D. R. and Hodges, L. R. Jr (1965) Square loop materials for digital phase shifter applications, *J. Appl. Phys.*, **36**, 1263–1264.

Treuhaft, M. A. and Silber, L. M. (1958) Use of microwave ferrite toroids to eliminate external magnets and reduce switching power, *Proc. IRE*, **46**(8).

Van der Heide, Bruijning, H. G. and Wijn, H. P. J. (1956) Switching time of ferrites with rectangular hysteresis loop, *Philips Tech, Rev.*, **18**, 339.

22

INSERTION LOSS OF WAVEGUIDE CIRCULATORS

22.1 Introduction

An important quantity in the description of any junction circulator is its insertion loss. One sort of loss, already discussed, is that due to return or reflection loss. Another type of loss is dissipation in magnetic, dielectric and circuits which occur in nonideal materials and metals in waveguides. Such losses not only affect the insertion loss and power rating of any device but also its adjustment in that all existing relationships between the entries of its scattering matrix based on the unitary condition have to be revisited. Specifically, each eigen-network of the problem region may now be associated with a different dielectric, magnetic and circuit dissipation so that the magnitude of each reflection coefficient may now be different from unity. In order to formally cater for such dissipation a dissipation matrix is often introduced. Such a matrix caters for the fact that the energy absorbed in the circuit must be satisfied with an equal or greater than zero sign instead of an equal to unity condition. One consequence of this feature is that if the ports of a circulator are adjusted to have perfect isolation than this adjustment no longer ensures a perfect return loss at a typical port. Since an ideal circulator has many of the properties of a 2-port transmission cavity it is appropriate to discuss its insertion loss in terms of unloaded, external and loaded quality factors and this is the approach usually employed. Another type of loss encountered in most ferrite devices is a nonlinear one that manifests itself under perpendicular pumping as a power sensitive subsidiary resonance at a dirct magnetic field below that required to establish the main Kittel line. This process involves the excitation of spinwaves which propagate at about 45° with respect to the direct magnetic field and which are synchronous with half the frequency of the radio frequency magnetic field.

22.2 Unloaded, External, and Loaded Q-Factors

The insertion loss of a 2-port resonant network is often expressed in terms of an unloaded, an external and a loaded quality factor. Since an ideal circulator has some of the features of a transmission line cavity between two of its ports it is appropriate to do so here also.

In defining the Q-factor, the power that is dissipated within the circuit is normally separately expressed from that within the external circuit. The power dissipated within the circuit is described by an unloaded Q-factor, Q_u as

$$Q_u = \frac{\omega_0 \text{ (energy stored in the circuit)}}{\text{power dissipated in circuit}} \tag{22.1}$$

If coupling ports, loops or probes, are introduced into the resonator, the power dissipated in each load may be separately expressed by a single external quality factor, $Q_{ex,n}$, defined by

$$Q_{ex,n} = \frac{\omega_0 \text{ (energy stored in the circuit)}}{\text{power reflected at port } n} \tag{22.2}$$

The total dissipated power is described by a loaded Q-factor, Q_L, given by

$$Q_L = \frac{\omega_0 \text{ (energy stored in the circuit)}}{\text{power dissipated} + \text{power reflected at all ports}} \tag{22.3}$$

It is readily demonstrated that

$$\frac{1}{Q_L} = \frac{1}{Q_u} + \frac{1}{Q_{ex,1}} + \frac{1}{Q_{ex,2}} + \ldots \tag{22.4}$$

Equations (22.1) and (22.2) are usually expressed in abbreviated form as

$$Q_u = \frac{\omega_0 U_0}{P_d} \tag{22.5}$$

$$Q_{ex,n} = \frac{\omega_0 U_0}{P_{r,n}} \tag{22.6}$$

22.3 Unloaded Q-Factor

The unloaded Q-factor in a practical gyromagnetic circuit is made up of terms which are due to dielectric (Q_d), magnetic (Q_{eff}) and circuit losses (Q_c).

22.3 UNLOADED Q-FACTOR

$$\frac{1}{Q_u} = \frac{1}{Q_d} + \frac{1}{Q_{\text{eff}}} + \frac{1}{Q_c} \tag{22.7}$$

In a gyromagnetic cavity the effective or magnetic term may be separately decomposed into the circularly polarized variables Q_+ and Q_- of the problem region

$$\frac{2}{Q_{\text{eff}}} \approx \left(1 - \frac{\kappa}{\mu}\right)\left(\frac{1}{Q_+}\right) + \left(1 + \frac{\kappa}{\mu}\right)\left(\frac{1}{Q_-}\right) \tag{22.8}$$

If $Q_+ \approx Q_-$ or if κ/μ is small then

$$\frac{2}{Q_{\text{eff}}} \approx \frac{1}{Q_+} + \frac{1}{Q_-} \tag{22.9}$$

This result reduces to that one in 2-port reciprocal circuit for $Q_+ = Q_-$

Q_d and Q_{eff} are separately given in terms of the dielectric and magnetic loss tangents of the ferrite substrate in the usual way by

$$Q_d = \frac{1}{\tan \delta_d} \tag{22.10a}$$

$$Q_{\text{eff}} = \frac{1}{\tan \delta_{\text{eff}}} \tag{22.10b}$$

where

$$\tan \delta_d = \frac{\varepsilon_r''}{\varepsilon_r'} \tag{22.10c}$$

$$\tan \delta_{\text{eff}} = \frac{\mu_{\text{eff}}''}{\mu_{\text{eff}}'} \tag{22.10d}$$

and

$$Q_+ = \frac{1}{\tan \delta_+} \tag{22.10e}$$

$$Q_- = \frac{1}{\tan \delta_-} \tag{22.10f}$$

where

$$\tan \delta_+ = \frac{\mu_+''}{\mu_+'} \tag{22.10g}$$

$$\tan \delta_- = \frac{\mu_-''}{\mu_-'} \tag{22.10h}$$

ε'_r and ε''_r are the real and imaginary parts of the relative dielectric constant respectively; the primed permeability terms are defined in a like manner.

For most ferrite materials, $\tan \delta_d \approx 0.0002$, so that the dielectric loss is usually negligible compared to the magnetic ones and so is often disregarded.

22.4 Split Unloaded Q-factors

The derivation of the relationship between the effective and circular quality factors start by forming μ_{eff} in terms of the scalar or circular variable μ_+ and μ_-

$$\frac{2}{\mu_{\text{eff}}} = \frac{1}{\mu_+} + \frac{1}{\mu_-} \tag{22.11}$$

where

$$\mu_+ = \mu - \kappa \tag{22.12}$$

$$\mu_- = \mu + \kappa \tag{22.13}$$

$$\mu_{\text{eff}} = \frac{\mu^2 - \kappa^2}{\mu} \tag{22.14}$$

In order to cater for dissipation complex quantities are introduced into the preceding equation. This gives

$$\frac{2}{\mu'_{\text{eff}} - j\mu''_{\text{eff}}} = \frac{1}{\mu'_+ - j\mu''_+} + \frac{1}{\mu'_- - j\mu''_-} \tag{22.15a}$$

The real part is now given by

$$\frac{2}{\mu'_{\text{eff}}} \approx \frac{1}{\mu'_+} + \frac{1}{\mu'_-} \tag{22.15b}$$

and the imaginary part by

$$\frac{2\mu''_{\text{eff}}}{(\mu'_{\text{eff}})^2} \approx \frac{\mu''_+}{(\mu'_+)^2} + \frac{\mu''_-}{(\mu'_-)^2} \tag{22.15c}$$

This last equation may readily be written in the form of (22.8) as asserted by multiplying both sides of this equation by μ'_{eff}.

The derivation of the split unloaded Q-factors in terms of the original variables starts by replacing σ by $\sigma + j\alpha$ in the definitions of $\mu_\pm$

$$\mu_{\pm} = 1 + \frac{p}{\sigma + j\alpha \mp 1} \qquad (22.16)$$

p is the normalized magnetization

$$p = \frac{\gamma M_0}{\mu_0 \omega} \qquad (22.17)$$

and σ is the normalized internal direct magnetic field

$$\sigma = \gamma \left(H_0 - \frac{N_z M_0}{\mu_0} \right) \Big/ \omega \qquad (22.18)$$

α is the effective normalized linewidth

$$\alpha = \frac{\gamma \Delta H_{\text{eff}}}{2\omega} \qquad (22.19)$$

and ΔH_{eff} is the effective linewidth (A/m). Away from the main resonance, the effective linewidth can be replaced by the spinwave one ΔH_k.

It continues by setting σ to zero and proceeds by forming the real and imaginary parts of $\mu_{\pm}$

$$\mu'_{\pm} = 1 \mp p \qquad (22.20)$$

$$\mu''_{\pm} = \alpha p \qquad (22.21)$$

The required result is

$$Q_{\pm} = \frac{1 \pm p}{\alpha p} \qquad (22.22)$$

The condition $\sigma = 0$ employed here implies that the material is saturated. It is usually associated with a gyromagnetic resonator biased below the Kittel line; the only case specifically treated here.

22.5 Effective Quality Factor of Junction Circulator

The derivation of the effective quality factor associated with the magnetic loss of a gyromagnetic resonator starts with the definition of the effective permeability

$$\mu_{\text{eff}} = \frac{(p + \sigma)^2 - 1}{\sigma^2 + p\sigma - 1} \qquad (22.23)$$

It continues by replacing σ by $\sigma + j\alpha$ as a preamble to reducing σ to zero. μ_{eff} is then given by

$$\mu'_e - j\mu''_e = \frac{(p + j\alpha)^2 - 1}{(j\alpha)^2 + j\alpha p - 1} \tag{23.24}$$

The real and imaginary parts of the effective permeability are now given by

$$\mu'_e \approx 1 - p^2 \tag{22.25a}$$

$$\mu''_e = p(1 + p)\alpha \tag{22.25b}$$

The effective unloaded Q factor is, therefore,

$$Q_{\text{eff}} = \frac{\mu'_e}{\mu''_e} = \frac{1 - p^2}{p(1 + p^2)\alpha} \tag{22.26}$$

The nature of Q_{eff} in equation (22.8) may also be verified in terms of the circular quality factors $Q_\pm$ by having recourse to equation (22.22).

Scrutiny of this relationship indicates that the effective or magnetic unloaded Q-factor reduces to the imaginary part of the counter-rotating scalar susceptibilities in the limit as p^2 becomes negligible compared to unity.

$$Q_{\text{eff}} \approx \frac{1}{\alpha p} \tag{22.7}$$

The origin of this relationship may be deduced by starting with the definition of the circular susceptibilities of the gyromagnetic problem region

$$\chi_\pm = \frac{p}{\sigma \pm 1} \tag{22.28}$$

It again continues by replacing σ by $\sigma + j\alpha$ as a preamble to setting σ equal to zero. This readily gives

$$\chi'_\pm = \mp p \tag{22.29}$$

$$\chi''_\pm = \alpha p \tag{22.30}$$

A comparison between the definitions of Q_{eff} and χ''_{eff} readily gives the equivalence between the two as asserted

$$Q_{\text{eff}} \approx \frac{1}{\chi_\pm''} \tag{22.31}$$

If Q_{eff} is expressed in terms of the original variables then

$$Q_{\text{eff}} \approx \frac{2\omega^2}{(\gamma M_0)(\mu_0 \gamma \Delta H_{\text{eff}})} \tag{22.32}$$

22.6 Insertion Loss of Symmetrical Transmission Cavity

In order to form the insertion loss of a 2-port cavity it is necessary to construct its transmission parameter S_{21} in terms of its circuit elements. One possible arrangement for this purpose is indicated in Figure 22.1(a).

One way of doing so is to have recourse to the eigenvalue approach. The required development starts by constructing the impedance eigenvalues of the circuit in Figure 22.1(a) for the eigen-networks in Figure 22.1(b) and 22.1(c)

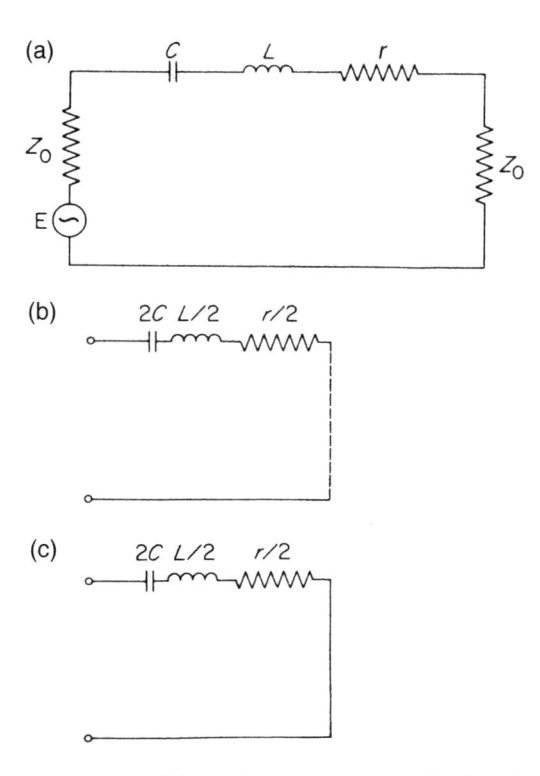

Figure 22.1 (a) Series resonator; (b) In-phase eigen-network of series resonator; (c) Out-of-phase eigen-network of series resonator

$$Z_1 = Z_{o/c} = \infty \qquad (22.33a)$$

$$Z_2 = Z_{s/c} = \frac{R}{2} + \frac{j\omega L}{2} + \frac{1}{j2\omega C} \qquad (22.33b)$$

The corresponding scattering eigenvalues are

$$s_1 = 1 \qquad (22.34a)$$

$$s_2 = \frac{(R/2 + j\omega L/2 + 1/j2\omega C) - Z_0}{(R/2 + j\omega L/2 + 1/j2\omega C) + Z_0} \qquad (22.34b)$$

S_{21} is therefore given by

$$S_{21} = \frac{s_1 - s_2}{2} = \frac{2Z_0/R}{1 + 2Z_0/R + (j\omega L/R)(\omega/\omega_0 + \omega_0/\omega)} \qquad (22.35)$$

The required result is now obtained by rewriting the preceding equation in terms of Q_u, Q_{ex} and Q_L.

$$|S_{21}|^2 = \frac{4Q_L^2}{Q_{ex}^2} \left| \frac{1}{1 + Q_L^2(\omega/\omega_0 - \omega_0/\omega)^2} \right| \qquad (22.36)$$

For a symmetric cavity, the only case considered in this text,

$$\frac{1}{Q_L} = \frac{2}{Q_{ex}} + \frac{1}{Q_u} \qquad (22.37)$$

where

$$Q_u = \frac{\omega_0 L}{R} \qquad (22.38)$$

$$Q_{ex} = \frac{\omega_0 L}{Z_0} \qquad (22.39)$$

and

$$\frac{R}{Z_0} = \frac{Q_{ex}}{Q_u} \qquad (22.40)$$

without ado.

The midband insertion loss is therefore given by

$$S_{21} = \frac{2Q_L}{Q_{ex}} \tag{22.41}$$

The insertion loss in decibels is now defined in terms of the single external Q-factor and the unloaded one by

$$L \text{ (dB)} = 20 \log_{10}\left(1 + \frac{Q_{ex}}{2Q_u}\right) \tag{22.42}$$

Since the specification of a circulator is usually given in terms of its quality factor, Q_L, the insertion is sometimes written as

$$L \text{ (dB)} = 20 \log_{10}\left(1 + \frac{Q_L}{Q_u}\right) \tag{22.43}$$

It is also often useful to decompose Q_u in terms of its individual factors Q_{eff}, Q_d and Q_c. This readily gives

$$L \text{ (dB)} = 20 \log_{10}\left[1 + Q_L\left(\frac{1}{Q_{eff}} + \frac{1}{Q_d} + \frac{1}{Q_c}\right)\right] \tag{22.44}$$

If Q_L/Q_u is small compared to unity then

$$L \text{ (dB)} \approx 8.686 \left(\frac{Q_L}{Q_u}\right) \tag{22.45}$$

The insertion loss of practical waveguide circulators is of the order of 0.10 dB and the loaded quality factor is usually between 1.8 and 2.2. Introducing these assumptions into the description of the insertion loss of a typical device indicates that its unloaded quality factor must be equal to 87 or more.

22.7 Insertion Loss of Waveguide Circulators

Some experimental results on the insertion loss of a number of waveguide circulators using different ferrite materials are summarized in Figure 22.2. The primary variable employed in constructing this quantity is the imaginary part of the scalar susceptibilities term $\chi''_{\pm}$.

The results in Figure 22.2 have been separately summarized by introducing an insertion loss function of the form

$$L \text{ (dB)} = 20 \log_{10}\left[1 - (A\chi''_{\pm} - B \tan \delta_d - C)\right] \tag{22.46}$$

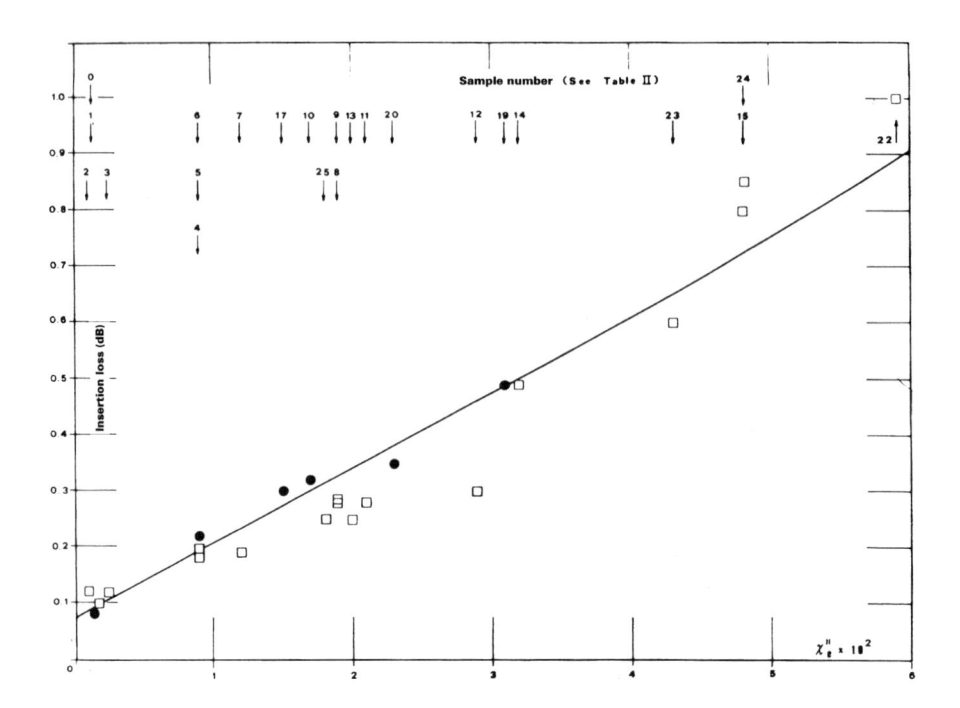

Figure 22.2 Insertion of various waveguide circulator for different material compositions (reprinted with permission, Roveda, R. (1972) Measurement of dissipative parameters of ferrite and insertion losses, *IEEE. Trans Microwave Theory Tech.* **MTT-20**, 89–96) (© 1972 IEEE)

One approximate solution that appears to fit the experimental results in Figure 22.2 is $A = 2.85$; $B = 0$; $C = 0.017$.

The equivalence between the empirical form used here for the insertion loss of a junction circulator and that derived previously may be readily deduced by extracting a constant factor A from the inside brackets in (22.46). The value of this constant is not very different from that associated with loaded quality factor of the class of circulator used to construct this empirical solution.

22.8 Scattering Matrix of Semi-ideal Circulators

Semi-ideal circulators are those for which $S_{11} \neq 0$, $S_{21} \neq 1$, and $S_{31} = 0$ or for which dissipation exists. In such circulators either $S_{11} = 0$, $S_{21} \neq 1$, and $S_{31} \neq 0$.

The first situation is obtained when the angle between s_{+1} and s_{-1} is less than |120°|. The second case is obtained when it is larger the |120°|.

The scattering parameters are given in terms of the dissipation eigenvalues introduced in Chapter 2 by

22.8 SCATTERING MATRIX OF SEMI-IDEAL CIRCULATORS

$$3S_{11} = -1 + \left(1 - \frac{q_{+1}}{2}\right)\exp(-j2\theta_+) + \left(1 - \frac{q_{-1}}{2}\right)\exp(j2\theta_+) \quad (22.47a)$$

$$3S_{21} = -1 + \alpha\left(1 - \frac{q_{+1}}{2}\right)\exp(-j2\theta_+) + \alpha^2\left(1 - \frac{q_{-1}}{2}\right)\exp(j2\theta_+) \quad (22.47b)$$

$$3S_{11} = -1 + \alpha^2\left(1 - \frac{q_{+1}}{2}\right)\exp(-j2\theta_+) + \alpha\left(1 - \frac{q_{-1}}{2}\right)\exp(j2\theta_+) \quad (22.47c)$$

where it has been assumed that the splitting is symmetrical: $\theta_{-1} = -\theta_{+1}$.

In semi-ideal circulators S_{11} and S_{31} are completely determined by S_{21} provided it is assumed that the amplitude of s_{+1} and s_{-1} are equal. This means that the latter quantity can be obtained simply by measuring either S_{11} or S_{31}.

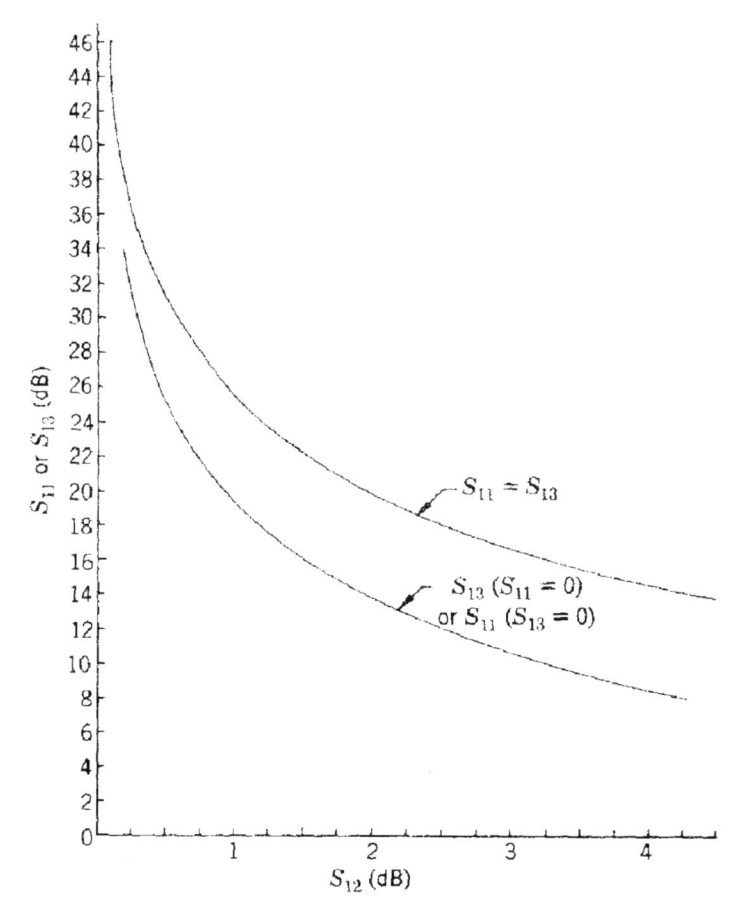

Figure 22.3 Relationship between scattering parameters of semi-ideal circulators (reprinted with permission, Helszajn, J. (1972) Dissipation and scattering matrices of lossy junctions, *IEEE Trans. Microwave Theory Tech.* **MTT-20**, 779–782) (© 1972 IEEE)

The first case to be considered is that in which the angle between s_{+1} and s_{-1} is such that $S_{11} = 0$. This condition is obtained by setting $S_{11} = 0$ in equation (22.47a). The result is

$$-1 + 2\left[1 - 2\left(\frac{Q_L}{Q_u}\right)\right]\cos^2\theta_{+1} = 0 \tag{22.48}$$

S_{21} is now obtained by substituting equation (22.48) into equation (22.47b). This gives

$$|S_{21}| = \frac{1}{2}\left(1 - \frac{\tan^2\theta_{+1}}{\sqrt{3}}\right) \tag{22.49}$$

Similarly the result for S_{31} is

$$|S_{31}| = \frac{1}{2}\left(1 + \frac{\tan^2\theta_{+1}}{\sqrt{3}}\right) \tag{22.50}$$

The relation between S_{21} and S_{31} is given graphically in Figure 22.3.

The second case to be considered here is that for which the angle between s_{+1} and s_{-1} is such that $S_{31} = 0$. This condition is obtained by setting $S_{31} = 0$ in equation (22.47b). The result is

$$-1 + 2\left[1 - 2\left(\frac{Q_1}{Q_u}\right)\cos^2\theta_{+1}\right]\cos(120 + 2\theta_{+1}) = 0 \tag{22.51}$$

S_{11} and S_{21} are now evaluated by substituting the last equation into equations (22.47a) and (22.47c). The results are

$$|S_{11}| = \frac{[1 + (\tan^2\theta_{+1}/\sqrt{3})]}{1 + \sqrt{3}\tan^2\theta_{+1}} \tag{22.52}$$

and

$$|S_{21}| = \frac{2\tan^2\theta_{+1}/\sqrt{3})}{1 + \sqrt{3}\tan^2\theta_{+1}} \tag{22.53}$$

The relation between S_{11} and S_{21} is shown in Figure 22.3.

22.9 Low-Loss Interval between Subsidiary and Main Resonances

A feature of some importance in the design of high power devices is the appearance of a subsidiary resonance at large signal levels at a direct magnetic field below the main Kittel resonance. The subsidiary resonance is a power sensitive phenomenon which manifests itself as a nonlinear insertion loss that limits the peak power rating of ferrite devices. This situation is well understood and is related

22.9 LOW-LOSS INTERVAL

to the transfer of power from the microwave magnetic field to spinwaves at half the frequency of the microwave signal. One standard way of avoiding this instability relies on material technology to widen the spinwave linewidth at the expense of the overall small-signal insertion loss. A second way to avoid this difficulty is to bias the material below the peak of the subsidiary resonance. Still a third solution to this problem is to ensure that the frequency relation between the microwave signal and the spinwaves at half its frequency cannot be satisfied.

One means of meeting this condition is to bias the ferrite material between the subsidiary and main resonances. The region in question is defined in Figure 22.4. Its width may be established without difficulty but is outside the remit of this work. This may be done by forming the difference between the two fields defined by the spinwave dispersion relation and the main resonance. The essential considerations entering into its definition are the direct magnetic field, the separation between the skirts of the subsidiary and main resonances, the band-width of this region and the relation between magnetization and linewidth. It is also necessary

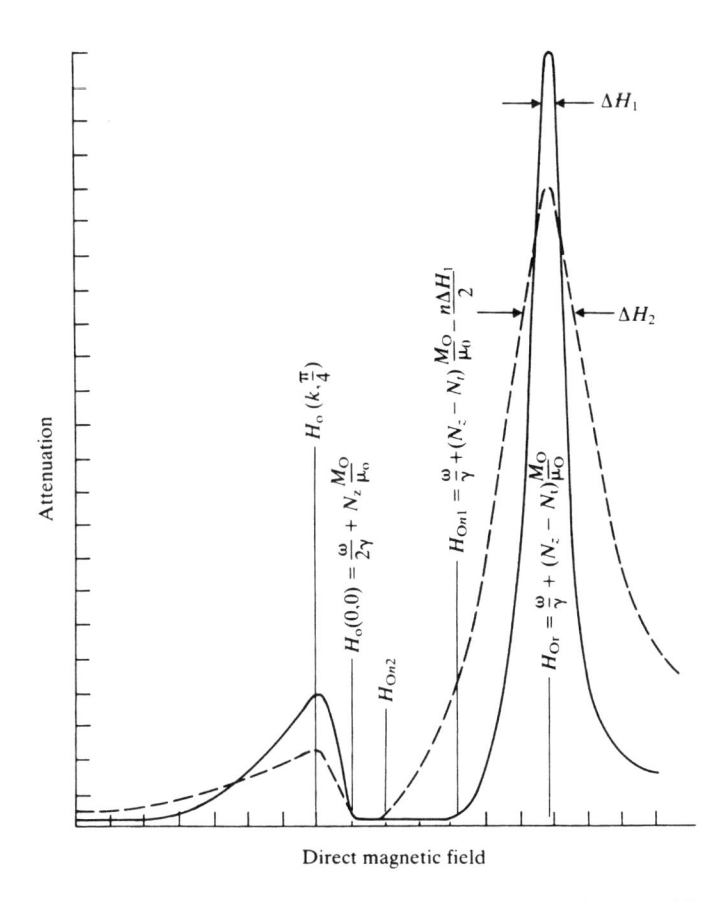

Figure 22.4. Subsidiary and main resonances at large signal showing definition of low loss region (reprinted with permission, Helszajn, J. and Walker, P. N. (1978) Operation of high peak power differential phase shift circulators, *IEEE Trans. Microwave Theory Tech.*, **MTT-26**, 653–658)

to ensure that the effective or scalar permeabilities do not take on negative values over the operating band of the device.

One important advantage of biasing the ferrite material between the subsidiary and main resonances is that the overall magnetic loss is less under this condition than it is below the subsidiary resonance. Since spinwave instability at half the pump frequency is completely suppressed in this situation, the uniform mode linewidth may be selected without regard to that of the spinwave one. The value of the spinwave linewidth has therefore, in this sort of design no bearing on the overall insertion loss of the device.

References and Further Reading

Bex, H. and Schwartz, E. (1971) Performance Limitation of Lossy Circulators, *IEEE Trans. Microwave Theory Tech.*, **MTT-19**, 493–494.

Bosma, H. (1966) Performance of Lossy H-Plane Y-Circulators, *IEEE Trans. Magn.*, **MAG-2**, 273–277.

Damon, R. W. (1953) Relaxation effects in ferromagnetic resonance, *Rev. Mod. Phys.*, **25**, 239–245.

Fay, C. E. and Comstock, R. L. (1965) Operation of the ferrite junction circulator, *IEEE Trans. Microwave Theory Tech.*, **MTT-13**, 15–27.

Hagelin, S. (1969) Analysis of lossy symmetrical three-port networks with circulator properties, *IEEE Trans. Microwave Theory Tech.*, **MTT-17** (6), 328.

Helszajn, J. (1972) Dissipation and scattering matrices of lossy junctions, *IEEE Trans. Microwave Theory Tech.*, **MTT-29**, 779–782.

Helszajn, J. (1979) Standing-wave solution of 3-port junction circulator with 1-port terminated in a variable short circuit, *Proc. IEE, Part H (Microwaves, Optics and Acoustics)*, **126**, 67–69.

Helszajn, J., Riblet, G. P. and Mather, J. P. (1975) Insertion loss of 3-port circulator with one port terminated in variable short circuit, *IEEE Trans. Microwave Theory Tech.*, **MTT-23**, 926–927.

Helszajn, J. and Walker, P. N. (1978) Operation of high peak power differential phase shift circulators, *IEEE Trans. Microwave Theory Tech.*, **MTT-26**, 653–658.

Konishi, Y. (1965) Lumped element circulator, *IEEE Trans. Microwave Theory Tech.*, **MTT-13**, 852–864.

Lagrange, A., Lahmi, H. and Vallatin, B. R. (1973) K-band high-peak-power junction circulator: influence of the static magnetic field, *IEEE Trans. Magnetics*, **MAG-9**, 531–534.

Roveda, R. (1972) Measurement of dissipative parameters of ferrite and insertion lossess, *IEEE. Trans Microwave Theory Tech.*, **MTT-20**, 89–96.

Suhl, H. (1956) The non-linear behaviour of ferrites at high signal level, *Proc. IRE*, **44**, 1270.

23

SYNTHESIS OF STEPPED IMPEDANCE TRANSDUCERS

23.1 Introduction

A property of any 3-port junction circulator is that the isolation between one input port and one output port is infinite and the insertion loss between the same input port and the other output port is zero when the return loss at each port is infinite. The classic equivalent circuit of this type of junction is a complex shunt STUB-G load consisting of a short-circuited quarter-wave long U.E. in shunt with a real conductance. One possible solution met in connection with this topology is the quarter-wave impedance transformer. A second solution based on short line transformers is obtained by replacing the 90° stub by short, typically 30°, open and short-circuited ones. One property of either network is that its gain-bandwidth product is fixed by the quality factor of the load circuit. A shortcoming of both topologies in the design of junction circulators is that the impedance levels encountered with these circuits become unrealizable if the loaded Q-factor of the complex gyrator circuit exceeds about 3. One topology that alleviates this difficulty is to move the reference terminals of the load by a quarter wave to the characteristic plane of the 3-port junction so that is now displays a series STUB-R load consisting of a short-circuited half-wave long U.E. in series with a real resistance and to employ a half-wave long U.E. The t-plane topology obtained in this way is akin to a degree-2 lumped element bandpass filter. The purpose of this chapter is to develop the exact t-plane synthesis of each topology for a degree 2 circuit. This is done both in closed and in tabular forms. The classic solution to any of these types of problems, as in the related filter ones, is based upon an insertion loss specification in the θ plane, the use of the unitary condition to deduce a squared amplitude reflection coefficient from a knowledge of that of the transmission one, the mapping of the θ plane into the t plane, the construction of a bounded real reflection coefficient, the use of the bilinear transformation between immittance and reflection coefficient and the synthesis of a one-port immittance function in terms of U.E.'s and t-plane inductors and capacitors that have the topology of the required two-port network.

A semi-tutorial approach to the actual t-plane synthesis problem has been adopted in this chapter in order to familiarize the non specialist reader with this type of method.

23.2 θ-Plane Insertion Loss Function for Quarter Wave-Long Stepped Impedance Transducers

The most important circuit topology met in the design of commercial junction circulators is that consisting of one or more transformers (U.E.'s) and a complex STUB-G load in terms of a microwave specification. Since the solution of this problem is crucial in the design of practical circulators it will be developed from first principles in closed form.

The first step in the synthesis of any commensurate line network is to establish an appropriate insertion loss specification. One topology that displays a suitable equi-ripple bandpass response centred about 90° which is applicable to a complex STUB-G load and n U.E.'s is indicated in Figure 23.1. The required amplitude-squared frequency response is separately indicated is illustrated in Figure 23.2. It is defined by

$$L(\theta) = 1 + K^2 + \varepsilon^2 \left[\frac{(1+\sin\theta_c)T_{n+1}\left(\frac{\cos\theta}{\cos\theta_c}\right) - (1-\sin\theta_c)T_{n-1}\left(\frac{\cos\theta}{\cos\theta_c}\right)}{2\sin\theta} \right]^2 \quad (23.1)$$

K and ε are related by the minimum and maximum values of the voltage standing wave ratio within the passband,

$$S(\text{max}) = (\sqrt{1 + K^2 + \varepsilon^2} + \sqrt{1 + K^2})^2 \quad (23.2)$$

$$S(\text{min}) = (\sqrt{1 + K^2} + K)^2 \quad (23.3)$$

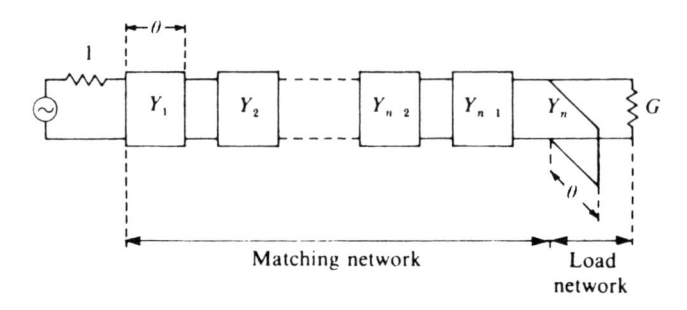

Figure 23.1 Schematic diagram of n U.E.'s terminated in STUB-G load

23.2 θ-PLANE INSERTION LOSS FUNCTION

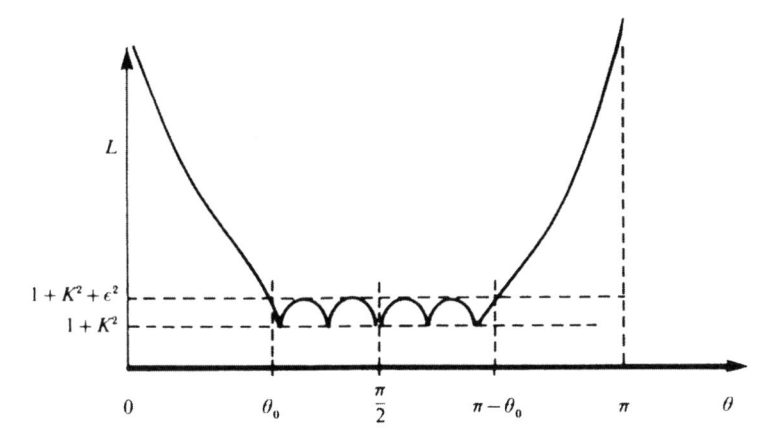

Figure 23.2 Frequency response n U.E.'s loaded by shunt STUB-G load

or

$$K = \frac{S(\min) - 1}{2\sqrt{S(\min)}} \tag{23.4}$$

$$\sqrt{K^2 + \varepsilon^2} = \frac{S(\max) - 1}{2\sqrt{S(\max)}} \tag{23.5}$$

θ and θ_c are the usual radian frequency and the lower bandedge one.

$$\theta = \frac{\pi}{2}\left(\frac{\omega}{\omega_0}\right) \tag{23.6}$$

$$\theta_c = \frac{\pi}{2}\left(\frac{\omega_c}{\omega_0}\right) \tag{23.7}$$

respectively. θ is also sometimes written in terms of a normalized bandwidth parameter as

$$\theta = \frac{\pi}{2}(1 + \delta) \tag{23.8}$$

where

$$\delta = \frac{\omega - \omega_0}{\omega_0} \tag{23.9}$$

ω_0 is the centre frequency of the transformer, ω is the normal frequency variable. θ_c is related to the bandwidth by

$$2 - \frac{4\theta_c}{\pi} \qquad (23.10)$$

$T_n(x)$ is the Chebyshev polynomial of order n and argument x.

The amplitude square of the reflection coefficient is related to the insertion loss function by the unitary condition in the usual way.

$$|\Gamma(\theta)|^2 = \frac{L(\theta) - 1}{L(\theta)} \qquad (23.11)$$

Once the reflection coefficient at either the input or output port in the t-plane is specified, $Z(t)$ at the corresponding port is obtained in the usual way by making use of the classic bilinear transformation between the two.

$$\frac{Y(\theta)}{G} = \frac{1 - \Gamma(\theta)}{1 + \Gamma(\theta)} \qquad (23.12)$$

For a degree-2 network the insertion loss function becomes

$$L(\theta) = 1 + K^2 + \varepsilon^2 \left\{ \frac{(1 + \sin\theta_c)\left[T_2\left(\frac{\cos\theta}{\cos\theta_c}\right) - (1 - \sin\theta_c)\right]}{2\sin\theta} \right\}^2 \qquad (23.13)$$

The frequency response of this insertion loss function is indicated in Figure 23.3.

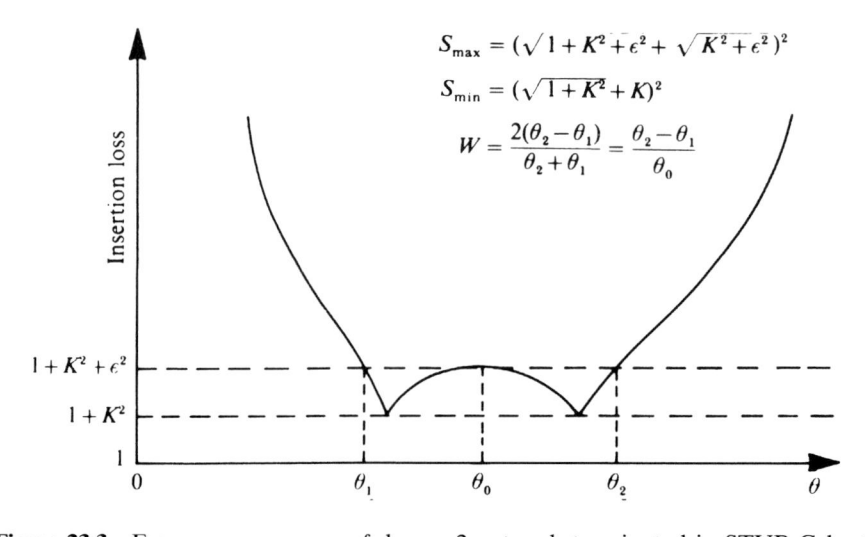

Figure 23.3 Frequency response of degree-2 network terminated in STUB-G load

23.3 t-Plane Synthesis of Quarter Wave Long Stepped Impedance Transducers

The derivation of a realizable immittance proceeds by forming the amplitude square of the reflection coefficient in the t-plane by introducing the t-plane variable throughout.

$$t = j \tan \theta \tag{23.14}$$

If the network is synthesised from the load terminals as is done here then it is necessary to construct $S_{22}(t)S_{22}(-t)$ instead of $S_{11}(t)S_{11}(-t)$. It continues by forming $S_{22}(t)$ from a knowledge of $S_{22}(t)S_{22}(-t)$. Since the numerator polynomial of $S_{22}(t)$ need not be Hurwitz its zeros may be selected from either its left or right half-plane roots. Since the denominator polynomial is Hurwitz its poles are selected from its left half-plane ones. The normalized admittance at the output terminals is then given from a knowledge of $S_{22}(t)$. The result in terms of the original variables is

$$\frac{Y(t)}{G} = \frac{d_2 t^2 + d_1 t + d_0}{n_2 t^2 + n_1 t} \tag{23.15}$$

The coefficients in the numerator and denominator polynomials are given by

$$n_2 = \sqrt{a+1} - \sqrt{a} \tag{23.16a}$$

$$d_2 = \sqrt{a+1} + \sqrt{a} \tag{23.16b}$$

$$n_1 = [2\sqrt{(a+1)}\, c - b + 1]^{1/2} - (2\sqrt{ac} - b)^{1/2} \tag{23.16c}$$

$$d_1 = [2\sqrt{(a+1)}\, c - b + 1]^{1/2} + (2\sqrt{ac} - b)^{1/2} \tag{23.16d}$$

$$n_0 = 0 \tag{23.16e}$$

$$d_0 = 2\sqrt{c} \tag{23.16f}$$

where

$$a = K^2 + \varepsilon^2 \tag{23.17a}$$

$$b = 2\beta\varepsilon^2 - K^2 \tag{23.17b}$$

$$c = \beta^2 \varepsilon^2 \tag{23.17c}$$

and

$$\beta = \tan^2 \theta_c + \frac{\tan \theta_c}{\cos \theta_c} \qquad (23.18)$$

Making use of the relationships between the numerator and denominator coefficients $n_{0,1,2}$ and $d_{0,1,2}$ also indicates that

$$d_1 n_1 - d_0 n_2 = 1 \qquad (23.19a)$$

$$n_2 d_2 = 1 \qquad (23.19b)$$

The synthesis of this immittance commences by extracting a shunt t-plane inductor in a second Cauer form. This step produces a t-plane inductor $L = (Gd_0)/n_1$ and a remainder admittance

$$Y'(t) = G \frac{n_1 d_2 t + 1}{n_1(n_2 t + n_1)}$$

The synthesis of the required circuit continues by extracting a U.E. by replacing t by unity in $Y'(t)$. The result is

$$Y'(1) = G \frac{n_1 d_2 + 1}{n_1(n_2 + n_1)}$$

Making use of the relationships between the various coefficients indicates that $Y'(1)$ may also be written as $Y'(1) = G/(n_1 n_2)$.

The remainder admittance $Y''(t)$ is then given by Richards Theorem as

$$Y''(t) = Y'(1) \frac{Y'(t) - tY'(1)}{Y'(1) - tY'(t)}$$

The remainder admittance after a cancelling a common term $(t^2 - 1)$ is

$$Y''(t) = \frac{G}{n_1^2}$$

This admittance fixes the load at the generator terminals as $G/n_1^2 = 1$.

The t-plane inductor, the U.E. and the load conductance are now specified in terms of the original variables for the purpose of calculations by

$$G = n_1^2 \qquad (23.20)$$

$$Y_1' = \frac{n_1}{n_2} \qquad (23.21)$$

$$L = n_1 d_0 \qquad (23.22)$$

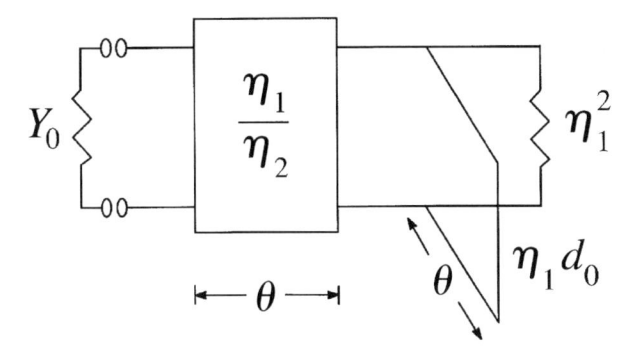

Figure 23.4. Topology of degree-2 STUB-G network

The topology produced by this procedure is now established by realizing the t-plane inductor by a short-circuited transmission line with a characteristic impedance equal to that of the t-plane inductor in shunt with the load conductance G, by realizing the U.E. by a series transmission line with a characteristic admittance equal to that of the U.E., and by fixing the G by the condition at the generator terminals. The required topology is depicted in Figure 23.4.

23.4 Network Parameters of Quarter-Wave Long Impedance Transducers

It is usual, in tabulating solutions to this type of problem, to replace the stub admittance L by its equivalent susceptance slope parameter B':

$$B' = \frac{\pi L}{4} \tag{23.23}$$

This notation has the merit that it permits the Q factor of the load to be also defined without difficulty:

$$Q_L = \frac{B'}{G} \tag{23.24}$$

A typical family of solutions for this sort of degree-2 network is tabulated in Table 23.1. One feature of these tables is that the quality factor is uniquely fixed once the bandwidth and the maximum value of the VSWR in the passband are specified. Another important aspect of this solution is that the immittance level of the complex gyrator circuit can be adjusted by varying the minimum value of the VSWR in the passband. Since the latter parameter is of little interest in the design of any circulator it may be used to absorb any uncertainty in the absolute immittance level of the gyrator circuit.

23 SYNTHESIS OF STEPPED IMPEDANCE TRANSDUCERS

Degree N=2	S(Max)=1.2		S(Min)=1		Degree N=2	S(Max)=1.2		S(Min)=1.06	
W	G	B'	Q	Y₁	W	G	B'	Q	Y₁
0.100	54.899	343.593	6.259	8.117	0.100	45.124	295.209	6.542	7.359
0.150	24.878	102.468	4.119	5.464	0.150	20.502	88.154	4.300	4.960
0.200	14.371	43.608	3.035	4.153	0.200	11.885	37.583	3.162	3.776
0.250	9.507	22.568	2.374	3.378	0.250	7.896	19.492	2.468	3.078
0.300	6.865	13.222	1.926	2.870	0.300	5.730	11.448	1.998	2.622
0.350	5.272	8.440	1.601	2.515	0.350	4.425	7.328	1.656	2.304
0.400	4.239	5.737	1.353	2.255	0.400	3.577	4.994	1.396	2.072
0.450	3.530	4.089	1.158	2.058	0.450	2.997	3.571	1.192	1.896
0.500	3.023	3.026	1.001	1.905	0.500	2.582	2.650	1.026	1.760
0.550	2.648	2.307	0.871	1.782	0.550	2.275	2.026	0.891	1.652
0.600	2.362	1.802	0.763	1.684	0.600	2.042	1.587	0.777	1.565
0.667	2.077	1.336	0.643	1.579	0.667	1.809	1.182	0.653	1.473

Degree N=2	S(Max)=1.2		S(Min)=1.02		Degree N=2	S(Max)=1.2		S(Min)=1.08	
W	G	B'	Q	Y₁	W	G	B'	Q	Y₁
0.100	52.013	332.465	6.392	7.900	0.100	41.195	269.883	6.551	7.031
0.150	23.584	99.178	4.205	5.320	0.150	18.747	80.656	4.302	4.743
0.200	13.634	42.225	3.097	4.045	0.200	10.890	34.423	3.161	3.615
0.250	9.028	21.863	2.422	3.292	0.250	7.255	17.876	2.464	2.951
0.300	6.527	12.816	1.964	2.799	0.300	5.280	10.515	1.991	2.517
0.350	5.019	8.186	1.631	2.454	0.350	4.090	6.741	1.648	2.215
0.400	4.040	5.567	1.378	2.202	0.400	3.318	4.602	1.387	1.995
0.450	3.369	3.971	1.179	2.011	0.450	2.789	3.296	1.182	1.829
0.500	2.889	2.941	1.018	1.862	0.500	2.411	2.450	1.016	1.701
0.550	2.534	2.243	0.885	1.744	0.550	2.132	1.877	0.880	1.599
0.600	2.264	1.754	0.774	1.648	0.600	1.920	1.473	0.767	1.518
0.667	1.995	1.301	0.652	1.547	0.667	1.708	1.098	0.643	1.432

Degree N=2	S(Max)=1.2		S(Min)=1.04		Degree N=2	S(Max)=1.2		S(Min)=1.1	
W	G	B'	Q	Y₁	W	G	B'	Q	Y₁
0.100	48.738	316.180	6.487	7.648	0.100	36.949	240.386	6.506	6.659
0.150	22.118	94.360	4.266	5.152	0.150	16.851	71.919	4.268	4.497
0.200	12.801	40.197	3.140	3.919	0.200	9.818	30.739	3.131	3.432
0.250	8.489	20.827	2.453	3.192	0.250	6.563	15.991	2.437	2.806
0.300	6.147	12.219	1.988	2.716	0.300	4.795	9.424	1.965	2.399
0.350	4.735	7.812	1.650	2.384	0.350	3.730	6.055	1.623	2.116
0.400	3.819	5.318	1.393	2.141	0.400	3.039	4.143	1.363	1.910
0.450	3.191	3.797	1.190	1.957	0.450	2.566	2.974	1.159	1.755
0.500	2.742	2.814	1.027	1.814	0.500	2.228	2.216	0.994	1.635
0.550	2.409	2.149	0.892	1.700	0.550	1.979	1.701	0.859	1.541
0.600	2.157	1.681	0.779	1.609	0.600	1.790	1.337	0.747	1.465
0.667	1.905	1.249	0.656	1.512	0.667	1.601	1.000	0.625	1.386

Table 23.1 Degree-2 network variables (reproduced with permission, Levy, R. and Helszajn, J. (1982) Specific equations for one and two section quarter-wave matching networks for stub-resistor loads, *IEEE Trans. Microwave Theory Tech.*, **MTT-30**, 55–62)

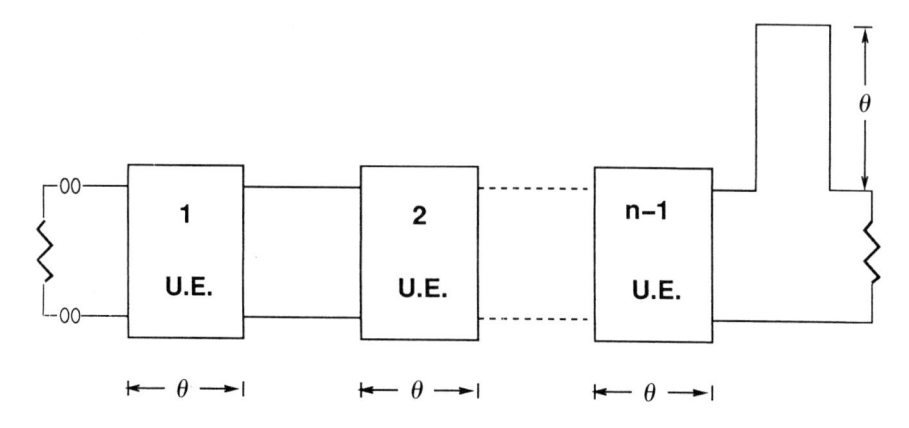

Figure 23.5 Schematic diagram of *n* U.E.'s terminated in series STUB-R load

23.4 NETWORK PARAMETERS

Degree $N = 3$ $S(Max) = 1.15$ $S(Min) = 1$

W	G	B'	Q	Y_1	Y_2
0.200	969.349	3486.515	3.597	4.732	147.312
0.250	399.553	1138.049	2.848	3.818	76.309
0.300	194.299	455.324	2.343	3.214	44.805
0.350	106.011	209.635	1.977	2.789	28.712
0.400	62.989	106.970	1.698	2.474	19.635
0.450	39.988	59.049	1.477	2.233	14.123
0.500	26.777	34.687	1.295	2.044	10.579
0.550	18.741	21.430	1.143	1.893	8.196
0.600	13.620	13.805	1.014	1.770	6.533
0.667	9.356	8.106	0.866	1.639	5.013
0.700	7.902	6.336	0.802	1.584	4.454
0.750	6.265	4.472	0.714	1.514	3.789
0.800	5.083	3.229	0.635	1.454	3.277
0.850	4.211	2.378	0.565	1.403	2.878
0.900	3.555	1.782	0.501	1.359	2.562
0.950	3.054	1.356	0.444	1.321	2.309
1.000	2.665	1.045	0.392	1.289	2.104

Degree $N = 3$ $S(Max) = 1.15$ $S(Min) = 1.08$

W	G	B'	Q	Y_1	Y_2
0.200	580.786	2253.269	3.880	3.881	97.189
0.250	240.220	736.770	3.067	3.144	50.639
0.300	117.341	295.436	2.518	2.660	29.945
0.350	64.382	136.404	2.119	2.321	19.351
0.400	38.519	69.843	1.813	2.071	13.360
0.450	24.657	38.714	1.570	1.882	9.713
0.500	16.672	22.852	1.371	1.735	7.362
0.550	11.801	14.198	1.203	1.618	5.777
0.600	8.687	9.205	1.060	1.524	4.668
0.667	6.083	5.458	0.897	1.425	3.653
0.700	5.193	4.288	0.826	1.384	3.279
0.750	4.187	3.052	0.729	1.332	2.832
0.800	3.458	2.224	0.643	1.288	2.489
0.850	2.918	1.653	0.566	1.251	2.222
0.900	2.511	1.250	0.498	1.220	2.010
0.950	2.200	0.961	0.437	1.194	1.840
1.000	1.957	0.748	0.382	1.171	1.703

Degree $N = 3$ $S(Max) = 1.15$ $S(Min) = 1.02$

W	G	B'	Q	Y_1	Y_2
0.200	893.332	3313.304	3.709	4.559	137.630
0.250	368.379	1081.745	2.936	3.681	71.355
0.300	179.240	432.918	2.415	3.102	41.940
0.350	97.862	199.388	2.037	2.693	26.909
0.400	58.196	101.785	1.749	2.392	18.428
0.450	36.983	56.215	1.520	2.161	13.275
0.500	24.774	33.042	1.333	1.981	9.961
0.550	17.378	20.428	1.176	1.836	7.731
0.600	12.649	13.170	1.041	1.719	6.175
0.667	8.709	7.742	0.889	1.594	4.752
0.700	7.366	6.055	0.822	1.542	4.228
0.750	5.852	4.279	0.731	1.475	3.604
0.800	4.758	3.093	0.650	1.418	3.125
0.850	3.951	2.280	0.577	1.370	2.751
0.900	3.344	1.711	0.512	1.329	2.455
0.950	2.880	1.303	0.453	1.294	2.217
1.000	2.520	1.006	0.399	1.263	2.025

Degree $N = 3$ $S(Max) = 1.15$ $S(Min) = 1.1$

W	G	B'	Q	Y_1	Y_2
0.200	452.332	1742.610	3.853	3.572	79.668
0.250	187.508	570.432	3.042	2.900	41.652
0.300	91.859	229.069	2.494	2.460	24.733
0.350	50.587	105.957	2.095	2.153	16.060
0.400	30.402	54.376	1.789	1.928	11.150
0.450	19.567	30.223	1.545	1.758	8.157
0.500	13.315	17.897	1.344	1.627	6.225
0.550	9.494	11.160	1.175	1.523	4.921
0.600	7.047	7.266	1.031	1.440	4.008
0.667	4.996	4.334	0.868	1.353	3.171
0.700	4.293	3.417	0.796	1.317	2.862
0.750	3.497	2.445	0.699	1.272	2.494
0.800	2.920	1.791	0.613	1.234	2.212
0.850	2.492	1.338	0.537	1.203	1.991
0.900	2.168	1.018	0.470	1.176	1.816
0.950	1.920	0.787	0.410	1.154	1.677
1.000	1.727	0.616	0.356	1.135	1.564

Degree $N = 3$ $S(Max) = 1.15$ $S(Min) = 1.04$

W	G	B'	Q	Y_1	Y_2
0.200	801.796	3043.685	3.796	4.363	125.987
0.250	330.850	994.045	3.005	3.525	65.394
0.300	161.116	397.989	2.470	2.974	38.491
0.350	88.060	183.398	2.083	2.585	24.738
0.400	52.436	93.683	1.787	2.298	16.973
0.450	33.374	51.781	1.552	2.080	12.253
0.500	22.416	30.463	1.359	1.909	9.216
0.550	15.744	18.854	1.198	1.772	7.171
0.600	11.488	12.170	1.059	1.662	5.743
0.667	7.939	7.168	0.903	1.544	4.437
0.700	6.728	5.612	0.834	1.495	3.956
0.750	5.362	3.972	0.741	1.432	3.383
0.800	4.375	2.876	0.657	1.379	2.942
0.850	3.646	2.124	0.583	1.334	2.598
0.900	3.098	1.597	0.515	1.296	2.326
0.950	2.678	1.219	0.455	1.263	2.108
1.000	2.353	0.943	0.401	1.235	1.932

Degree $N = 3$ $S(Max) = 1.15$ $S(Min) = 1.12$

W	G	B'	Q	Y_1	Y_2
0.200	309.011	1152.644	3.730	3.173	59.027
0.250	128.630	378.096	2.939	2.587	31.049
0.300	63.360	152.247	2.403	2.205	18.573
0.350	35.134	70.666	2.011	1.939	12.165
0.400	21.296	36.419	1.710	1.746	8.529
0.450	13.846	20.345	1.469	1.602	6.308
0.500	9.535	12.120	1.271	1.491	4.872
0.550	6.891	7.609	1.104	1.404	3.901
0.600	5.192	4.991	0.961	1.335	3.220
0.667	3.763	3.010	0.800	1.264	2.596
0.700	3.272	2.387	0.730	1.236	2.366
0.750	2.714	1.724	0.635	1.199	2.091
0.800	2.308	1.274	0.552	1.170	1.881
0.850	2.006	0.961	0.479	1.145	1.717
0.900	1.778	0.738	0.415	1.125	1.587
0.950	1.602	0.575	0.359	1.108	1.484
1.000	1.466	0.454	0.310	1.093	1.401

Degree $N = 3$ $S(Max) = 1.15$ $S(Min) = 1.06$

W	G	B'	Q	Y_1	Y_2
0.200	697.205	2688.010	3.855	4.139	112.522
0.250	287.967	878.302	3.050	3.348	58.498
0.300	140.406	351.866	2.506	2.828	34.498
0.350	76.861	162.270	2.111	2.462	22.222
0.400	45.854	82.970	1.809	2.193	15.286
0.450	29.252	45.912	1.570	1.988	11.067
0.500	19.700	27.047	1.373	1.827	8.351
0.550	13.880	16.766	1.208	1.700	6.521
0.600	10.163	10.841	1.067	1.597	5.242
0.667	7.062	6.402	0.907	1.488	4.071
0.700	6.002	5.020	0.836	1.443	3.640
0.750	4.806	3.561	0.741	1.385	3.126
0.800	3.941	2.585	0.656	1.336	2.731
0.850	3.302	1.914	0.580	1.295	2.423
0.900	2.820	1.443	0.512	1.260	2.178
0.950	2.451	1.104	0.450	1.230	1.983
1.000	2.165	0.856	0.396	1.205	1.825

Degree $N = 3$ $S(Max) = 1.15$ $S(Min) = 1.14$

W	G	B'	Q	Y_1	Y_2
0.200	138.408	458.742	3.314	2.528	31.758
0.250	58.354	151.441	2.595	2.084	16.998
0.300	29.231	61.496	2.104	1.799	10.384
0.350	16.559	28.850	1.742	1.604	6.968
0.400	10.302	15.064	1.462	1.465	5.020
0.450	6.908	8.546	1.237	1.363	3.825
0.500	4.928	5.182	1.051	1.287	3.050
0.550	3.704	3.317	0.896	1.229	2.525
0.600	2.911	2.222	0.763	1.184	2.157
0.667	2.238	1.381	0.617	1.139	1.820
0.700	2.005	1.111	0.554	1.122	1.696
0.750	1.739	0.820	0.472	1.100	1.549
0.800	1.545	0.620	0.401	1.083	1.437
0.850	1.400	0.477	0.341	1.069	1.351
0.900	1.290	0.374	0.290	1.058	1.283
0.950	1.523	0.410	0.270	1.084	1.428
1.000	1.738	0.446	0.257	1.105	1.556

Table 23.2 Degree-3 network variables (reproduced with permission, Levy, R. and Helszajn, J. (1982) Specific equations for one and two section quarter-wave matching networks for stub-resistor loads, *IEEE Trans. Microwave Theory Tech.*, **MTT-30**, 55–62)

The synthesis of a degree-3 topology proceeds in a like manner. Table 23.2 summarizes some results.

23.5 Synthesis of Series STUB-R Complex Gyrator Circuits using Half-Wave Long Impedance Transformers

Another network topology that has been employed in the realization of a degree-2 frequency response is obtained by recognizing that the complex gyrator circuit of the junction may, at its characteristic plane, be represented by a half-wave long U.E. in series with a simple resistance. The characteristic planes in a 3-port junction coincide with those at which a short-circuit placed in one port will cause a wave at the input port to be completely reflected, none entering the third one. These planes need not coincide with the terminals of the junction. This circuit can then be embodied in a cascade arrangement of half-wave long U.E.'s. The synthesis of the degree-2 solution is the main task in this section. One possible insertion loss function that displays a suitable equi-ripple bandpass response centred about 180 deg. is illustrated in Figure 23.5. The corresponding frequency response is separately indicated in Figure 23.6. The insertion loss function is defined by

$$L(\theta) = 1 + K^2 + \varepsilon^2 \cos^2\left[(n-1)\cos^{-1}\left(\frac{\sin\theta}{\sin\theta_0}\right) + \cos^{-1}\left(\frac{\tan\theta}{\tan\theta_0}\right)\right] \qquad (23.25)$$

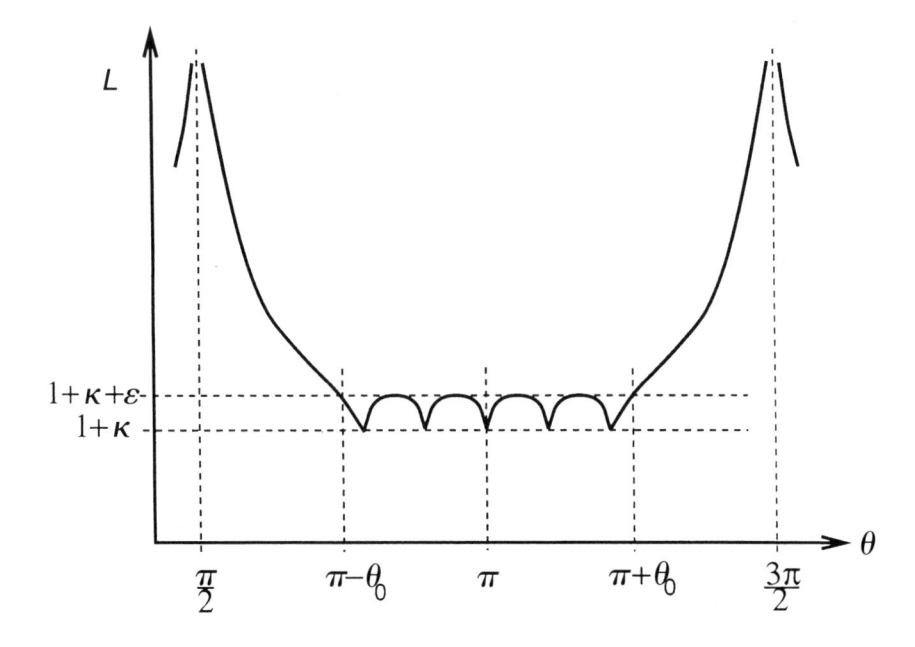

Figure 23.6 Frequency response of n half-wave long U.E.'s loaded by series STUB-R load

23.6 t-PLANE SYNTHESIS OF DEGREE-2 HALF-WAVE LONG TOPOLOGY

K and ε are again related the minimum and maximum values of the voltage standing wave ratio within the passband

θ and θ_0 are the usual radian frequency and the lower band-edge one. θ_0 is related to the bandwidth by

$$w = \frac{2\theta_0}{\pi} \quad (23.26)$$

For a degree-2 network the insertion loss function becomes

$$L(\theta) = 1 + K^2 + \varepsilon^2 \left[\left(\frac{\sin\theta}{\sin\theta_0}\right)\left(\frac{\tan\theta}{\tan\theta_0}\right) - \frac{\sin^2\theta_0 - \sin^2\theta}{\sin^2\theta_0 \cos\theta} \right]^2 \quad (23.27)$$

In order to proceed with synthesis it is again necessary to introduce the t-plane variable. Introducing this transformation into the insertion loss function gives

$$L(t) = 1 + K^2 + \varepsilon^2 \left[\frac{\left(\frac{C_0}{1-C_0}\right)^2 t^4 + \left(\frac{2C_0}{1-C_0}\right)t^2 + 1}{1 - t^2} \right] \quad (23.28)$$

and

$$C_0 = \cos\theta_0 \quad (23.29)$$

23.6 t-Plane Synthesis of Degree-2 Half-Wave Long Topology

The synthesis procedure again begins by forming $\Gamma(t)$ from a knowledge of $\Gamma(t)\Gamma(-t)$. Since the denominator polynominal of $\Gamma(t)$ is Hurwitz its poles are selected from the left half-plane ones of $\Gamma(t)\Gamma(-t)$. Since the numerator polynominal need not be Hurwitz its zeros may be selected from either the left or right half-plane roots of the numerator polynominal. However, the numerator polynominal is also usually constructed from the left half-plane roots. Once the reflection coefficient at either the input or output port in the t-plane is specified, $Z(t)$ at the corresponding port is obtained in the usual way by making use of the classic bilinear transformation between the two.

One possibility at the output port that permits the first extraction to coincide with a series t-plane inductor is

$$\frac{Z(t)}{R} = \frac{n_2 t^2 + n_1 t + n_0}{d_1 t + d_0} \quad (23.30)$$

where

$$n_2 = 2\sqrt{a} \tag{23.31a}$$

$$n_1 = [2\sqrt{a(c+1)} - (b-1)]^{1/2} + [2\sqrt{ac} - b]^{1/2} \tag{23.31b}$$

$$n_0 = \sqrt{c+1} + \sqrt{c} = \frac{1}{\sqrt{S(\max)}} \tag{23.31c}$$

$$d_2 = 0 \tag{23.31d}$$

$$d_1 = [2\sqrt{a(c+1)} - (b-1)]^{1/2} - [2\sqrt{ac} - b]^{1/2} \tag{23.31e}$$

$$d_0 = [2\sqrt{c+1} - \sqrt{c}\,] = \sqrt{S(\max)} \tag{23.31f}$$

and

$$a = \varepsilon^2 \left(\frac{C_0}{1 - C_0}\right)^2 \tag{23.32}$$

$$b = -\left(K^2 - \frac{2C_0\varepsilon^2}{1 - C_0}\right) \tag{23.33}$$

$$c = K^2 + \varepsilon^2 \tag{23.34}$$

Two useful identities between the preceding coefficients are

$$n_0 d_0 = 1 \tag{23.35a}$$

$$n_1 d_1 - n_2 d_0 = 1 \tag{23.35b}$$

The synthesis problem is completed once $Z(t)$ is realized in the required topology.

If the synthesis proceeds from the output port then it starts by extracting a t-plane inductor using a Second Cauer procedure. The value of the inductor obtained in this way is defined without ado by

$$L = \left(\frac{n_2}{d_1}\right) R$$

The remainder impedance is given by $Z'(t) = Z(t) - Lt$. Evaluating this impedance and noting that $n_1 d_1 - n_2 d_0 = 1$, $n_0 d_0 = 1$ gives

$$\frac{Z'(t)}{R} = \frac{t + n_0 d_1}{d_1^2 t + d_0 d_1}$$

The synthesis now proceeds by extracting a U.E. from the preceding function. The value of this U.E. is established by replacing t by 1.

23.6 t-PLANE SYNTHESIS OF DEGREE-2 HALF-WAVE LONG TOPOLOGY

$$\frac{Z'(1)}{R} = \frac{1 + n_0 d_1}{d_1(d_1 + d_0)}$$

This quantity may be further simplified by noting that $n_0 d_0 = 1$. The required result is

$$Z'(1) = \left(\frac{n_0}{d_1}\right) R$$

The remainder impedance is now constructed by having recourse to Richards Theorem.

$$Z''(t) = \left(\frac{n_0}{d_0}\right) R$$

This step reduces the degree of the problem to zero and completes the synthesis proceedure.

If the generator impedance $Z'(1)$ is taken as unity then the elements of the required topology defined by

$$R = \left(\frac{d_0}{n_0}\right) \tag{23.36}$$

$$Z'(1) = \left(\frac{d_0}{d_1}\right) \tag{23.37}$$

$$L = \left(\frac{n_2}{d_1}\right)\left(\frac{d_0}{n_0}\right) \tag{23.38}$$

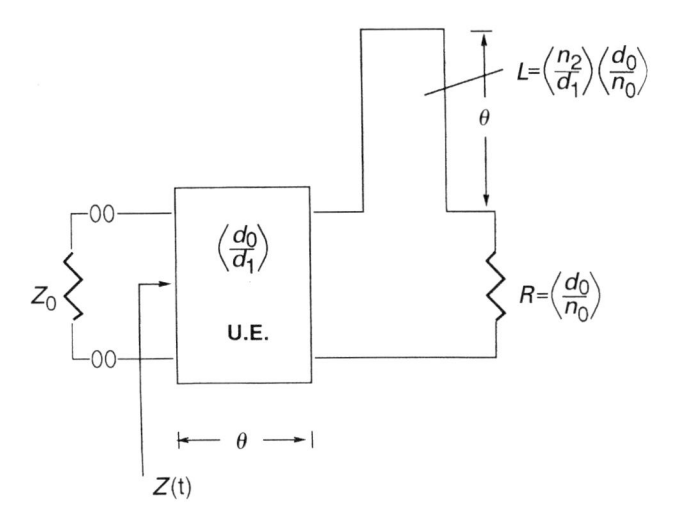

Figure 23.7 Equivalent circuit of degree-2 topology using series t-plane inductor

The equivalent circuit obtained in this manner has the necessary topology for the problem under consideration. It is illustrated in Figure 23.7.

23.7 Network Parameters of Half-Wave Impedance Transducers

The purpose of this section is to summarize some numerical data on a degree-2 half-wave filter topology. Since the effort is dedicated to the practice of circulator work it is done in terms of the loaded Q-factor of the complex R-STUB load and in terms of the reactance slope parameter of the series t-plane inductor. The former quantity is defined in the usual way by

$$x' = \frac{X'}{Z_0} = \left(\frac{\pi}{2}\right)\left(\frac{L}{Z_0}\right) \tag{23.39}$$

and the latter one by

$$Q_L = \frac{x'}{r} \tag{23.40}$$

r is the normalized load resistance

$$r = \frac{R}{Z_0} \tag{23.41}$$

L is the characteristic impedance of the half-wave long short-circuited stub and $Z(1)$ is that of the half wave long U.E. The tabulated data also includes the normalized value of the half-wave U.E.

$$z(1) = \frac{Z(1)}{Z_0} \tag{23.42}$$

One property of the solution outlined in this paper is that the real part of the normalized series STUB-R load is restricted to the reciprocal of the maximum value of the $VSWR$ in the passband of the network.

$$r = \frac{R}{Z_0} = \frac{1}{S(\max)} \tag{23.43}$$

Tables 23.3, 23.4, 23.5 and 23.6 summarize some typical results. Entries in these tables for which the loaded Q-factor is bracketed between 2 and 5 are of interest in the design of practical ciruclators. Such solutions are associated with waveguides with narrow dimensions of between ¼ and ½ that of standard waveguide.

23.7 NETWORK PARAMETERS

S_{max} = 1.05; S_{min} = 1; R = 0.9524					S_{max} = 1.05; S_{min} = 1.025; R = 0.9524				
W	X	Q	L	Z	W	X	Q	L	Z
0.025	11.947	12.544	7.605	0.123	0.025	11.743	12.331	7.476	0.140
0.050	5.829	6.121	3.711	0.241	0.050	5.701	5.986	3.630	0.273
0.075	3.739	3.926	2.381	0.349	0.075	3.630	3.811	2.311	0.393
0.100	2.667	2.801	1.698	0.445	0.100	2.566	2.695	1.634	0.496
0.125	2.012	2.112	1.281	0.527	0.125	1.918	2.013	1.221	0.582
0.150	1.570	1.649	1.000	0.587	0.150	1.483	1.557	0.944	0.653
0.175	1.256	1.319	0.799	0.655	0.175	1.176	1.235	0.749	0.711
0.200	1.022	1.073	0.651	0.703	0.200	0.950	0.997	0.605	0.758
0.225	0.844	0.886	0.537	0.744	0.225	0.779	0.818	0.496	0.795
0.250	0.705	0.741	0.449	0.777	0.250	0.647	0.679	0.412	0.826
0.275	0.595	0.625	0.379	0.805	0.275	0.543	0.570	0.345	0.851
0.300	0.506	0.532	0.322	0.828	0.300	0.459	0.482	0.293	0.871
0.325	0.434	0.455	0.276	0.848	0.325	0.392	0.412	0.250	0.888
0.350	0.374	0.393	0.238	0.864	0.350	0.337	0.354	0.214	0.902
0.400	0.282	0.296	0.180	0.890	0.400	0.252	0.265	0.161	0.924
0.450	0.216	0.227	0.137	0.910	0.450	0.192	0.202	0.122	0.939
0.500	0.167	0.175	0.106	0.924	0.500	0.148	0.155	0.094	0.951
0.550	0.130	0.136	0.083	0.936	0.550	0.115	0.120	0.073	0.959
0.600	0.101	0.106	0.064	0.944	0.600	0.089	0.093	0.057	0.966
0.650	0.078	0.082	0.050	0.951	0.650	0.069	0.072	0.044	0.971

S_{max} = 1.05; S_{min} = 1.0125; R = 0.9524					S_{max} = 1.05; S_{min} = 1.0375; R = 0.9524				
W	X	Q	L	Z	W	X	Q	L	Z
0.025	12.084	12.688	7.693	0.129	0.025	10.561	11.089	6.724	0.166
0.050	5.888	6.183	3.749	0.252	0.050	5.079	5.333	3.233	0.320
0.075	3.769	3.957	2.399	0.364	0.075	3.190	3.349	2.031	0.454
0.100	2.682	2.816	1.707	0.462	0.100	2.221	2.332	1.414	0.564
0.125	2.017	2.118	1.284	0.546	0.125	1.634	1.715	1.040	0.652
0.150	1.570	1.649	1.000	0.617	0.150	1.246	1.308	0.793	0.721
0.175	1.252	1.315	0.797	0.675	0.175	0.975	1.024	0.621	0.775
0.200	1.017	1.068	0.647	0.723	0.200	0.779	0.818	0.496	0.816
0.225	0.838	0.880	0.533	0.763	0.225	0.633	0.664	0.403	0.849
0.250	0.699	0.734	0.445	0.796	0.250	0.521	0.547	0.332	0.874
0.275	0.588	0.618	0.375	0.823	0.275	0.434	0.456	0.276	0.895
0.300	0.500	0.525	0.318	0.845	0.300	0.366	0.384	0.233	0.911
0.325	0.428	0.449	0.272	0.864	0.325	0.310	0.326	0.198	0.924
0.350	0.368	0.387	0.234	0.880	0.350	0.266	0.279	0.169	0.935
0.400	0.277	0.291	0.176	0.904	0.400	0.198	0.208	0.126	0.951
0.450	0.212	0.222	0.135	0.922	0.450	0.150	0.157	0.095	0.962
0.500	0.163	0.172	0.104	0.936	0.500	0.115	0.121	0.073	0.970
0.550	0.127	0.133	0.081	0.946	0.550	0.089	0.093	0.057	0.976
0.600	0.098	0.103	0.063	0.954	0.600	0.069	0.072	0.044	0.980
0.650	0.076	0.080	0.048	0.961	0.650	0.053	0.055	0.034	0.984

Table 23.3 Network variables for $S(\text{max}) = 1.05$ (reproduced with permission, Levy, R. and Helszajn, J. (1993) Synthesis of series STUB-R complex gyrator circuits using half-wave long impedance transformers, *IEE Proc. Microwave Antennas and Propagation*, 426–432)

$S_{max}=1.1$; $S_{min}=1$; $R=0.9091$					$S_{max}=1.1$; $S_{min}=1.05$; $R=0.9091$				
W	X	Q	L	Z	W	X	Q	L	Z
0.025	16.189	17.807	10.306	0.087	0.025	16.300	17.930	10.377	0.103
0.050	7.986	8.785	5.084	0.173	0.050	8.016	8.818	5.103	0.202
0.075	5.209	5.730	3.316	0.255	0.075	5.203	5.723	3.312	0.296
0.100	3.793	4.172	2.415	0.331	0.100	3.765	4.142	2.397	0.383
0.125	2.926	3.219	1.863	0.401	0.125	2.884	3.173	1.836	0.461
0.150	2.338	2.572	1.489	0.465	0.150	2.287	2.516	1.456	0.530
0.175	1.913	2.104	1.218	0.522	0.175	1.856	2.042	1.182	0.590
0.200	1.591	1.750	1.013	0.573	0.200	1.532	1.685	0.975	0.642
0.225	1.340	1.474	0.853	0.618	0.225	1.280	1.408	0.815	0.688
0.250	1.139	1.253	0.725	0.657	0.250	1.081	1.189	0.688	0.726
0.275	0.977	1.075	0.622	0.692	0.275	0.921	1.013	0.586	0.759
0.300	0.843	0.927	0.537	0.722	0.300	0.790	0.869	0.503	0.788
0.325	0.732	0.805	0.466	0.749	0.325	0.682	0.750	0.434	0.812
0.350	0.638	0.702	0.406	0.772	0.350	0.591	0.650	0.376	0.833
0.400	0.490	0.539	0.312	0.810	0.400	0.450	0.495	0.287	0.866
0.450	0.381	0.419	0.242	0.840	0.450	0.347	0.382	0.221	0.891
0.500	0.298	0.327	0.189	0.863	0.500	0.270	0.297	0.172	0.910
0.550	0.233	0.257	0.149	0.882	0.550	0.210	0.231	0.134	0.925
0.600	0.183	0.201	0.116	0.897	0.600	0.164	0.180	0.104	0.937
0.650	0.142	0.156	0.090	0.909	0.650	0.127	0.140	0.081	0.946

$S_{max}=1.1$; $S_{min}=1.025$; $R=0.9091$					$S_{max}=1.1$; $S_{min}=1.075$; $R=0.9091$				
W	X	Q	L	Z	W	X	Q	L	Z
0.025	16.575	18.233	10.552	0.093	0.025	14.860	16.346	9.460	0.123
0.050	8.169	8.986	5.201	0.183	0.050	7.267	7.993	4.626	0.242
0.075	5.321	5.853	3.387	0.269	0.075	4.676	5.144	2.977	0.351
0.100	3.867	4.254	2.462	0.350	0.100	3.348	3.682	2.131	0.449
0.125	2.977	3.275	1.895	0.423	0.125	2.534	2.787	1.613	0.534
0.150	2.373	2.611	1.511	0.489	0.150	1.985	2.183	1.263	0.606
0.175	1.936	2.130	1.233	0.548	0.175	1.591	1.750	1.013	0.667
0.200	1.607	1.767	1.023	0.599	0.200	1.298	1.428	0.826	0.718
0.225	1.350	1.485	0.859	0.645	0.225	1.074	1.181	0.684	0.760
0.250	1.145	1.260	0.729	0.684	0.250	0.898	0.988	0.572	0.795
0.275	0.980	1.078	0.624	0.719	0.275	0.758	0.834	0.483	0.824
0.300	0.844	0.928	0.537	0.748	0.300	0.645	0.710	0.411	0.849
0.325	0.731	0.804	0.465	0.774	0.325	0.553	0.608	0.352	0.869
0.350	0.636	0.700	0.405	0.797	0.350	0.477	0.524	0.303	0.886
0.400	0.487	0.536	0.310	0.834	0.400	0.359	0.395	0.229	0.912
0.450	0.377	0.415	0.240	0.862	0.450	0.275	0.302	0.175	0.931
0.500	0.294	0.324	0.187	0.884	0.500	0.212	0.233	0.135	0.945
0.550	0.230	0.253	0.147	0.901	0.550	0.165	0.181	0.105	0.955
0.600	0.180	0.198	0.115	0.915	0.600	0.128	0.141	0.081	0.963
0.650	0.140	0.154	0.089	0.926	0.650	0.099	0.109	0.063	0.969

Table 23.4 Network variables for $S(\text{max}) = 1.10$ (reproduced with permission, Levy, R. and Helszajn, J. (1993) Synthesis of series STUB-R complex gyrator circuits using half-wave long impedance transformers, *IEE Proc. Microwave Antennas and Propagation*, 426–432)

23.7 NETWORK PARAMETERS

S_{max} = 1.15; S_{min} = 1; R = 0.8696					S_{max} = 1.15; S_{min} = 1.075; R = 0.8696				
W	X	Q	L	Z	W	X	Q	L	Z
0.025	18.989	21.837	12.089	0.072	0.025	19.473	22.394	12.397	0.086
0.050	9.403	10.813	5.986	0.142	0.050	9.619	11.062	6.124	0.170
0.075	6.169	7.094	3.927	0.210	0.075	6.287	7.231	4.003	0.250
0.100	4.527	5.206	2.882	0.275	0.100	4.591	5.279	2.923	0.326
0.125	3.524	4.053	2.244	0.337	0.125	3.553	4.086	2.262	0.397
0.150	2.844	3.270	1.810	0.394	0.150	2.848	3.275	1.813	0.461
0.175	2.350	2.702	1.496	0.447	0.175	2.337	2.687	1.488	0.520
0.200	1.975	2.271	1.257	0.496	0.200	1.950	2.242	1.241	0.572
0.225	1.680	1.932	1.070	0.540	0.225	1.647	1.894	1.048	0.619
0.250	1.443	1.659	0.919	0.580	0.250	1.404	1.615	0.894	0.660
0.275	1.248	1.436	0.795	0.616	0.275	1.207	1.388	0.768	0.696
0.300	1.087	1.250	0.692	0.649	0.300	1.043	1.200	0.664	0.728
0.325	0.950	1.093	0.605	0.678	0.325	0.907	1.043	0.577	0.756
0.350	0.835	0.960	0.531	0.704	0.350	0.792	0.910	0.504	0.780
0.400	0.650	0.747	0.414	0.748	0.400	0.610	0.701	0.388	0.821
0.450	0.510	0.586	0.325	0.784	0.450	0.474	0.545	0.302	0.852
0.500	0.402	0.463	0.256	0.813	0.500	0.371	0.427	0.236	0.877
0.550	0.318	0.365	0.202	0.837	0.550	0.291	0.335	0.185	0.896
0.600	0.250	0.288	0.159	0.856	0.600	0.228	0.262	0.145	0.912
0.650	0.196	0.225	0.125	0.872	0.650	0.177	0.204	0.113	0.924

S_{max} = 1.15; S_{min} = 1.0375; R = 0.8696					S_{max} = 1.15; S_{min} = 1.1125; R = 0.8696				
W	X	Q	L	Z	W	X	Q	L	Z
0.025	19.630	22.575	12.496	0.077	0.025	17.919	20.607	11.408	0.104
0.050	9.713	11.170	6.184	0.152	0.050	8.815	10.137	5.612	0.206
0.075	6.365	7.320	4.052	0.225	0.075	5.723	6.582	3.644	0.301
0.100	4.664	5.363	2.969	0.294	0.100	4.143	4.765	2.638	0.390
0.125	3.624	4.168	2.307	0.359	0.125	3.175	3.651	2.021	0.469
0.150	2.918	3.356	1.858	0.420	0.150	2.518	2.896	1.603	0.540
0.175	2.406	2.767	1.532	0.475	0.175	2.044	2.351	1.301	0.602
0.200	2.017	2.320	1.284	0.525	0.200	1.687	1.940	1.074	0.655
0.225	1.712	1.969	1.090	0.0571	0.225	1.410	1.621	0.898	0.701
0.250	1.467	1.687	0.934	0.611	0.250	1.191	1.369	0.758	0.740
0.275	1.266	1.456	0.806	0.648	0.275	1.014	1.166	0.645	0.773
0.300	1.100	1.265	0.700	0.681	0.300	0.869	0.999	0.553	0.802
0.325	0.960	1.104	0.611	0.710	0.325	0.750	0.862	0.477	0.826
0.350	0.841	0.867	0.536	0.736	0.350	0.650	0.747	0.414	0.847
0.400	0.652	0.750	0.415	0.779	0.400	0.494	0.568	0.315	0.880
0.450	0.510	0.587	0.325	0.814	0.450	0.381	0.438	0.242	0.905
0.500	0.401	0.462	0.256	0.841	0.500	0.295	0.340	0.188	0.923
0.550	0.316	0.364	0.201	0.863	0.550	0.230	0.265	0.147	0.937
0.600	0.248	0.286	0.158	0.881	0.600	0.179	0.206	0.114	0.948
0.650	0.194	0.223	0.123	0.896	0.650	0.139	0.160	0.088	0.956

Table 23.5 Network variables for $S(\max) = 1.15$ (reproduced with permission, Levy, R. and Helszajn, J. (1993) Synthesis of series STUB-R complex gyrator circuits using half-wave long impedance transformers, *IEE Proc. Microwave Antennas and Propagation*, 426–432)

23 SYNTHESIS OF STEPPED IMPEDANCE TRANSDUCERS

Table 4: Network variables for $S_{max} = 1.20$

$S_{max} = 1.2$; $S_{min} = 1$; $R = 0.8333$					$S_{max} = 1.2$; $S_{min} = 1.1$; $R = 0.8333$				
W	X	Q	L	Z	W	X	Q	L	Z
0.025	21.026	25.232	13.386	0.062	0.025	21.901	26.281	13.943	0.076
0.050	10.431	12.517	6.641	0.123	0.050	10.843	13.012	6.903	0.150
0.075	6.865	8.238	4.370	0.183	0.075	7.113	8.536	4.528	0.222
0.100	5.058	6.069	3.220	0.241	0.100	5.218	6.262	3.322	0.291
0.125	3.956	4.748	2.519	0.296	0.125	4.061	4.873	2.585	0.356
0.150	3.210	3.852	2.043	0.348	0.150	3.275	3.930	2.085	0.417
0.175	2.668	3.202	1.699	0.398	0.175	2.705	3.246	1.722	0.473
0.200	2.256	2.707	1.436	0.443	0.200	2.272	2.726	1.446	0.524
0.225	1.931	2.317	1.229	0.486	0.225	1.931	2.317	1.229	0.570
0.250	1.668	2.002	1.062	0.525	0.250	1.657	1.988	1.055	0.612
0.275	1.452	1.742	0.924	0.561	0.275	1.432	1.719	0.912	0.649
0.300	1.271	1.525	0.809	0.594	0.300	1.245	1.495	0.793	0.682
0.325	1.118	1.341	0.712	0.624	0.325	1.088	1.306	0.693	0.712
0.350	0.987	1.184	0.628	0.651	0.350	0.955	1.146	0.608	0.739
0.400	0.775	0.930	0.493	0.699	0.400	0.741	0.890	0.472	0.784
0.450	0.613	0.736	0.391	0.738	0.450	0.581	0.697	0.370	0.820
0.500	0.487	0.585	0.310	0.771	0.500	0.457	0.548	0.291	0.848
0.500	0.387	0.464	0.246	0.798	0.550	0.360	0.432	0.229	0.871
0.600	0.306	0.368	0.195	0.821	0.600	0.283	0.340	0.180	0.889
0.650	0.240	0.289	0.153	0.840	0.650	0.221	0.265	0.141	0.904

$S_{max} = 1.2$; $S_{min} = 1.05$; $R = 0.8333$					$S_{max} = 1.2$; $S_{min} = 1.15$; $R = 0.8333$				
W	X	Q	L	Z	W	X	Q	L	Z
0.025	21.918	26.302	13.954	0.067	0.025	20.304	24.364	12.926	0.093
0.050	10.867	13.040	6.918	0.133	0.050	10.018	12.021	6.377	0.184
0.075	7.144	8.573	4.548	0.198	0.075	6.535	7.842	4.161	0.271
0.100	5.257	6.308	3.346	0.260	0.100	4.760	5.712	3.030	0.353
0.125	4.105	4.926	2.614	0.319	0.125	3.673	4.407	2.338	0.428
0.150	3.325	3.989	2.116	0.374	0.150	2.934	3.521	1.868	0.496
0.175	2.758	3.309	1.756	0.426	0.175	2.400	2.879	1.528	0.556
0.200	2.326	2.792	1.481	0.475	0.200	1.995	2.394	1.270	0.610
0.225	1.987	2.384	1.265	0.519	0.225	1.679	2.014	1.069	0.657
0.250	1.713	2.055	1.090	0.559	0.250	1.426	1.712	0.908	0.699
0.275	1.487	1.785	0.947	0.596	0.275	1.222	1.466	0.778	0.734
0.300	1.299	1.559	0.827	0.630	0.300	1.053	1.264	0.670	0.766
0.325	1.140	1.368	0.726	0.660	0.325	0.912	1.095	0.581	0.793
0.350	1.004	1.205	0.639	0.688	0.350	0.794	0.953	0.506	0.816
0.400	0.785	0.943	0.500	0.735	0.400	0.609	0.730	0.387	0.854
0.450	0.619	0.743	0.394	0.773	0.450	0.471	0.566	0.300	0.883
0.500	0.490	0.588	0.312	0.805	0.500	0.367	0.441	0.234	0.904
0.550	0.388	0.466	0.247	0.831	0.550	0.287	0.345	0.183	0.921
0.600	0.306	0.368	0.195	0.852	0.600	0.224	0.269	0.143	0.935
0.650	0.240	0.288	0.153	0.870	0.650	0.174	0.209	0.111	0.945

Table 23.6 Network variables for $S(\max) = 1.20$ (reproduced with permission, Levy, R. and Helszajn, J. (1993) Synthesis of series STUB-R complex gyrator circuits using half-wave long impedance transformers, *IEE Proc. Microwave Antennas and Propagation*, 426–432)

23.8 *t*-plane Synthesis of Short-Line Matching Network

It is also possible to employ short-line impedance transformers to match a complex load such as met in connection with the complex gyrator circuit of a typical junction circulator. In order to ensure, for the purpose of synthesis, that all line lengths are commensurate, it is necessary to replace the quarter-wave long stub in the gyrator circuit by two stubs in parallel, one short-circuited and the other open-circuited. The equivalent circuit obtained in this way is indicated in Figure 23.8. The admittance of the load is then described by

$$Y_{in} = G + j(Y_C \tan \theta - Y_L/\tan \theta) \qquad (23.44)$$

Y_C and Y_L are the characteristics impedances of its open- and short-circuited stubs, respectively, and θ_t is the commensurate electrical length

$$\theta_t = \tan^{-1} \sqrt{\frac{Y_L}{Y_C}} \qquad (23.45)$$

It is observed that the commensurate length at resonance is now an arbitrary parameter, not limited to 90° and this is the key to a precise synthesis technique.

An equivalence between this load network and the usual single-stub representation is obtained by equating the susceptance slope parameters of the two arrangements. One definition of the susceptance slope parameter of a load in terms of its susceptance is

$$B' = \frac{\omega_t}{2} \frac{dB}{d\omega} \bigg|_{\theta = \theta_t}$$

Evaluating the proceeding equation in terms of the original variables gives

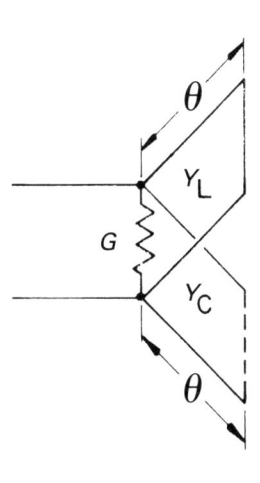

Figure 23.8 Topology of R-stub load using *t*-plane inductance and capacitance

$$B' = \frac{\omega_t}{2}\left(Y_C \sec^2\theta + Y_L \csc^2\theta\right)\frac{d\theta}{d\omega}\bigg|_{\theta=\theta_t} \qquad (23.47)$$

or

$$B' = (Y_C + Y_L)\tan^{-1}\sqrt{\frac{Y_1}{Y_C}} \qquad (23.48)$$

Expressions for Y_C and Y_L in terms of θ_t and B' may be obtained, but are of no value here.

The alternative representation of the load network shown by the equivalence of Figure 23.9 is also instructive. The admittance values of the compound short-circuited stub are

$$Y_1' = Y_C + Y_L \qquad (23.49)$$

$$Y_2' = (Y_C + Y_L)Y_L/Y_C \qquad (23.50)$$

and the resonant length is the same as given in equation (23.45). When $\theta_t = 45°$ then $Y_C = Y_L$, and the compound stub degenerates to a 90° stub of uniform admittance $2Y_L$, as expected.

23.9 Network Parameters of Short-line Impedance Transducers

Sample results for these short-line matching networks of order n = 4 (two unit element matching) are presented in Tables 23.7, 23.8 and 23.9. These are given

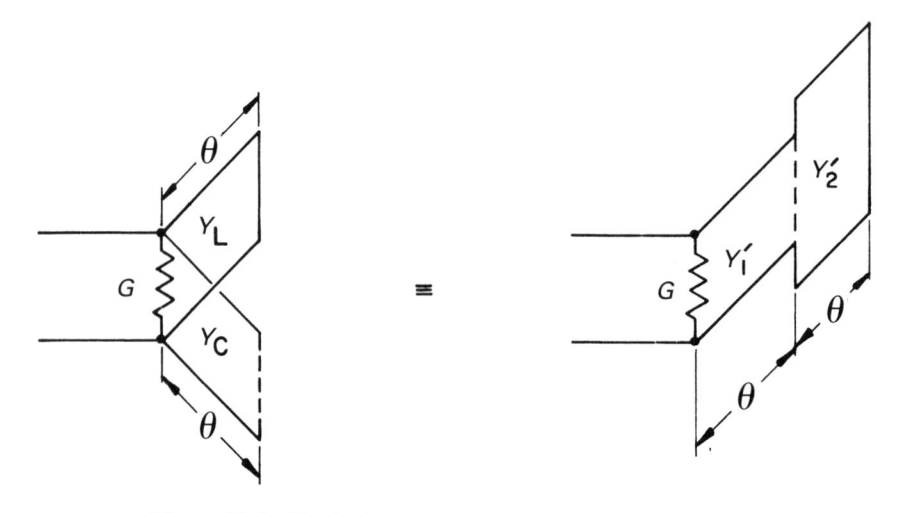

Figure 23.9 Equivalence between degree-2 complex loads

23.9 NETWORK PARAMETERS OF SHORT-LINE IMPEDANCE TRANSDUCERS

$S_{max} = 1.1$, $S_{min} = 1$

W	Y_1	Y_2	G	Y_C	Y_L	B'	Q-factor	θ_l
0.150	6.427	1.746	10.331	51.850	12.810	29.827	2.887	26.43
0.200	4.878	1.368	6.200	23.319	5.558	13.114	2.115	26.02
0.250	3.960	1.155	4.290	12.891	2.942	7.056	1.645	25.53
0.300	3.357	1.023	3.254	8.132	1.767	4.317	1.327	24.99
0.350	2.933	0.937	2.630	5.620	1.157	2.887	1.098	24.41
0.400	2.621	0.879	2.227	4.150	0.807	2.059	0.925	23.80
0.450	2.383	0.840	1.951	3.218	0.590	1.541	0.790	23.19
0.500	2.196	0.813	1.755	2.592	0.448	1.198	0.683	22.57
0.550	2.046	0.796	1.610	2.149	0.350	0.958	0.595	21.97
0.600	1.924	0.785	1.501	1.823	0.279	0.784	0.523	21.37
0.667	1.793	0.778	1.392	1.506	0.213	0.618	0.444	20.60
0.700	1.737	0.776	1.350	1.381	0.188	0.554	0.410	20.23

$S_{max} = 1.1$, $S_{min} = 1.04$

W	Y_1	Y_2	G	Y_C	Y_L	B'	Q-factor	θ_l
0.150	5.755	1.575	8.368	42.848	10.508	24.534	2.932	26.35
0.200	4.375	1.243	5.073	19.464	4.582	10.862	2.141	25.88
0.250	3.558	1.057	3.550	10.881	2.439	5.890	1.659	25.33
0.300	3.023	0.943	2.725	6.946	1.474	3.634	1.334	24.73
0.350	2.648	0.871	2.229	4.856	0.971	2.450	1.099	24.09
0.400	2.373	0.824	1.909	3.626	0.681	1.762	0.923	23.44
0.450	2.163	0.793	1.691	2.841	0.501	1.329	0.786	22.78
0.500	1.999	0.773	1.537	2.309	0.382	1.040	0.677	22.13
0.550	1.868	0.762	1.423	1.931	0.300	0.837	0.588	21.50
0.600	1.761	0.756	1.338	1.650	0.240	0.689	0.515	20.89
0.667	1.647	0.754	1.254	1.374	0.184	0.547	0.436	20.10
0.700	1.599	0.755	1.221	1.265	0.163	0.491	0.403	19.72

$S_{max} = 1.1$, $S_{min} = 1.08$

W	Y_1	Y_2	G	Y_C	Y_L	B'	Q-factor	θ_l
0.150	4.339	1.227	4.969	22.087	5.242	12.389	2.493	25.97
0.200	3.325	0.996	3.141	10.458	2.334	5.646	1.797	25.29
0.250	2.730	0.873	2.300	6.110	1.271	3.158	1.373	24.52
0.300	2.343	0.804	1.846	4.072	0.785	2.010	1.089	23.71
0.350	2.075	0.765	1.575	2.963	0.528	1.395	0.886	22.89
0.400	1.880	0.743	1.402	2.292	0.378	1.030	0.735	22.10
0.450	1.733	0.734	1.286	1.852	0.283	0.795	0.619	21.34
0.500	1.619	0.731	1.204	1.545	0.219	0.635	0.527	20.63
0.550	1.529	0.734	1.145	1.319	0.174	0.520	0.454	19.96
0.600	1.456	0.740	1.101	1.147	0.141	0.434	0.395	19.33
0.667	1.380	0.751	1.058	0.972	0.109	0.350	0.331	18.55
0.700	1.348	0.757	1.042	0.901	0.097	0.317	0.304	18.19

Table 23.7 Short-line matching transformers; (theta midband) $\theta_m = 27°$ (reproduced with permission, Levy, R. and Helszajn, J. (1983) Synthesis of short line matching networks for resonant loads, *IEE Proc. Microwave Antennas and Propagation*, 385–390)

$S_{max} = 1.1$, $S_{min} = 1$

w	Y_1	Y_2	G	Y_C	Y_L	B'	Q-factor	θ_l
0.150	4.856	2.642	11.718	35.925	18.342	33.668	2.873	35.55
0.200	3.706	2.060	6.992	15.917	7.934	14.663	2.097	35.22
0.250	3.028	1.727	4.805	8.640	4.185	7.798	1.623	34.84
0.300	2.587	1.518	3.618	5.338	2.504	4.710	1.302	34.41
0.350	2.280	1.378	2.903	3.605	1.634	3.105	1.069	33.95
0.400	2.055	1.281	2.439	2.597	1.136	2.181	0.894	33.48
0.450	1.886	1.212	2.122	1.963	0.828	1.607	0.757	33.00
0.500	1.755	1.161	1.895	1.539	0.625	1.228	0.648	32.52
0.550	1.651	1.124	1.728	1.241	0.487	0.967	0.560	32.05
0.600	1.567	1.096	1.600	1.024	0.387	0.773	0.486	31.59
0.667	1.478	1.070	1.474	0.815	0.294	0.600	0.407	30.99
0.700	1.442	1.060	1.424	0.734	0.259	0.532	0.374	30.70

$S_{max} = 1.1$, $S_{min} = 1.04$

w	Y_1	Y_2	G	Y_C	Y_L	B'	Q-factor	θ_l
0.150	4.365	2.381	9.479	29.597	16.036	27.639	2.916	35.48
0.200	3.332	1.868	5.709	13.217	6.535	12.105	2.120	35.11
0.250	2.732	1.577	3.966	7.238	3.466	6.479	1.634	34.68
0.300	2.342	1.395	3.020	4.613	2.085	3.939	1.304	34.21
0.350	2.072	1.276	2.451	3.076	1.369	2.614	1.067	33.71
0.400	1.876	1.194	2.082	2.234	0.956	1.849	0.888	33.20
0.450	1.729	1.137	1.830	1.702	0.701	1.370	0.749	32.69
0.500	1.615	1.097	1.651	1.343	0.532	1.054	0.638	32.19
0.550	1.525	1.067	1.519	1.090	0.416	0.833	0.549	31.70
0.600	1.453	1.046	1.419	0.905	0.332	0.674	0.475	31.23
0.667	1.378	1.026	1.320	0.724	0.254	0.522	0.396	30.62
0.700	1.346	1.019	1.281	0.654	0.224	0.464	0.362	30.33

$S_{max} = 1.1$, $S_{min} = 1.08$

w	Y_1	Y_2	G	Y_C	Y_L	B'	Q-factor	θ_l
0.150	3.305	1.847	5.599	15.048	7.481	13.835	2.471	35.19
0.200	2.560	1.486	3.506	6.943	3.315	6.203	1.769	34.65
0.250	2.128	1.288	2.541	3.935	1.796	3.405	1.340	34.04
0.300	1.853	1.172	2.019	2.536	1.104	2.123	1.051	33.41
0.350	1.664	1.099	1.706	1.781	0.739	1.442	0.845	32.79
0.400	1.530	1.053	1.505	1.329	0.526	1.042	0.692	32.19
0.450	1.431	1.023	1.369	1.035	0.392	0.787	0.575	31.62
0.500	1.356	1.003	1.273	0.832	0.302	0.616	0.484	31.08
0.550	1.297	0.991	1.202	0.686	0.240	0.494	0.411	30.58
0.600	1.251	0.983	1.150	0.575	0.194	0.404	0.352	30.12
0.667	1.204	0.978	1.098	0.466	0.150	0.318	0.289	29.55
0.700	1.184	0.976	1.078	0.423	0.133	0.284	0.263	29.28

Table 23.8 Short-line matching transformers; (theta midband) $\theta_m = 36°$ (reproduced with permission, Levy, R. and Helszajn, J. (1983) Synthesis of short line matching networks for resonant loads, *IEE Proc. Microwave Antennas and Propagation*, 426–432)

23.9 NETWORK PARAMETERS OF SHORT-LINE IMPEDANCE TRANSDUCERS

$S_{max} = 1.1$, $S_{min} = 1$

W	Y_1	Y_2	G	Y_C	Y_L	B'	Q-factor	θ_l
0.150	3.928	3.928	14.024	25.578	25.578	40.178	2.865	45.00
0.200	3.020	3.020	8.293	11.014	11.014	17.300	2.086	45.00
0.250	2.491	2.491	5.640	5.779	5.779	9.078	1.609	45.00
0.300	2.149	2.149	4.199	3.438	3.438	5.400	1.286	45.00
0.350	1.914	1.914	3.330	2.230	2.230	3.503	1.052	45.00
0.400	1.744	1.744	2.767	1.541	1.541	2.420	0.875	45.00
0.450	1.618	1.618	2.380	1.116	1.116	1.754	0.737	45.00
0.500	1.521	1.521	2.103	0.839	0.839	1.318	0.627	45.00
0.550	1.445	1.445	1.899	0.649	0.649	1.020	0.537	45.00
0.600	1.385	1.385	1.743	0.515	0.515	0.808	0.464	45.00
0.667	1.322	1.322	1.588	0.388	0.388	0.610	0.384	45.00
0.700	1.296	1.296	1.526	0.341	0.341	0.535	0.351	45.00

$S_{max} = 1.1$, $S_{min} = 1.04$

W	Y_1	Y_2	G	Y_C	Y_L	B'	Q-factor	θ_l
0.150	3.529	3.529	11.323	20.948	20.948	32.905	2.906	45.00
0.200	2.725	2.725	6.750	9.057	9.057	14.226	2.108	45.00
0.250	2.258	2.258	4.634	4.775	4.775	7.500	1.618	45.00
0.300	1.958	1.958	3.486	2.855	2.855	4.485	1.287	45.00
0.350	1.753	1.753	2.793	1.862	1.862	2.924	1.047	45.00
0.400	1.606	1.606	2.345	1.293	1.293	2.031	0.866	45.00
0.450	1.497	1.497	2.037	0.941	0.941	1.479	0.726	45.00
0.500	1.414	1.414	1.818	0.711	0.711	1.117	0.614	45.00
0.550	1.350	1.350	1.656	0.553	0.553	0.868	0.524	45.00
0.600	1.299	1.299	1.533	0.440	0.440	0.691	0.451	45.00
0.667	1.246	1.246	1.411	0.334	0.334	0.524	0.371	45.00
0.700	1.224	1.224	1.363	0.293	0.293	0.461	0.338	45.00

$S_{max} = 1.1$, $S_{min} = 1.08$

W	Y_1	Y_2	G	Y_C	Y_L	B'	Q-factor	θ_l
0.150	2.701	2.701	6.631	10.378	10.378	16.301	2.458	45.00
0.200	2.122	2.122	4.092	4.565	4.565	7.170	1.752	45.00
0.250	1.792	1.792	2.918	2.453	2.453	3.853	1.320	45.00
0.300	1.584	1.584	2.282	1.495	1.495	2.349	1.029	45.00
0.350	1.446	1.446	1.900	0.994	0.994	1.561	0.822	45.00
0.400	1.349	1.349	1.654	0.703	0.703	1.104	0.668	45.00
0.450	1.278	1.278	1.488	0.520	0.520	0.817	0.550	45.00
0.500	1.226	1.226	1.366	0.399	0.399	0.627	0.459	45.00
0.550	1.186	1.186	1.279	0.315	0.315	0.494	0.386	45.00
0.600	1.155	1.155	1.213	0.253	0.253	0.398	0.328	45.00
0.667	1.124	1.124	1.148	0.195	0.195	0.306	0.267	45.00
0.700	1.111	1.111	1.123	0.172	0.172	0.271	0.241	45.00

Table 23.9 Short-line matching transformers; (theta midband) $\theta_m = 27°$ (reproduced with permission, Levy, R. and Helszajn, J. (1983) Synthesis of short line matching networks for resonant loads, *IEE Proc. Microwave Antennas and Propagation*, 426–432)

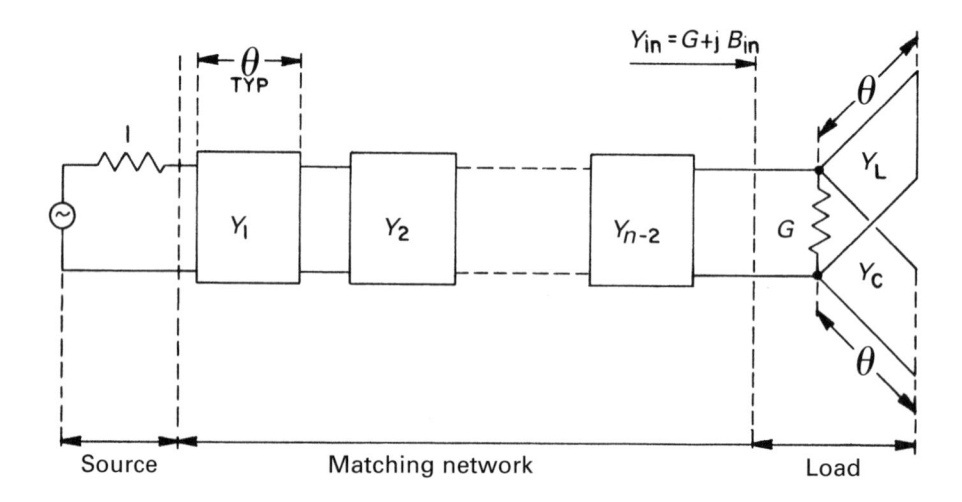

23 SYNTHESIS OF STEPPED IMPEDANCE TRANSDUCERS

Figure 23.10 Topology of $(n - 2)$ U.E.'s in cascade with degree-2 complex load

to illustrate the general features of this type of network. There are even more possibilities than the quarter-wave network presented previously because the electrical length of the unit elements θ_t represents yet another variable in addition to $S(\min)$ (for given bandwidth and $S(\max)$). Both θ_t and $S(\min)$ may be varied to control the load-admittance level G and susceptance slope parameter B' while leaving the Q-factor essentially unchanged. Actually as θ_t is reduced the bandwidth increases somewhat, but a disadvantage of carrying this too far is that the admittance levels may become impractical. The topology of a cascade arrangement of $(n - 2)$ U.E.'s in cascade with degree-2 complex load is indicated in Figure 23.10.

The cases where $Y_2 = 1$ allow the characteristics of loads obtained using measured results in 50 Ω lines to be retained. The matching networks may commence at the load itself, and it is not necessary to use reference planes some distance away from the load to avoid perturbing the junction.

It is interesting to observe that $Y_1 = Y_2$, when $\theta_t = 45°$, and the double unit element degenerates to the classic single quarter-wave transformer. One can also observe that here $Y_L = Y_C$, and the susceptance slope parameter (equation 23.48) degenerates to give $B' = \pi Y_L/2$.

This is related to the susceptance slope parameter of the single-stub-resistor load network by a factor of 2, which is due to the fact that the stubs in our present case are of half the electrical length and hence have twice the effective bandwidth. Incidentally, these identities for the degenerate cases represent a good test of the validity of the synthesis developed in this chapter.

References and Further Reading

Akaiwa, Y. (1974) Bandwidth enlargement of a millimetre-wave Y circulator with half-wavelength line resonators, *IEEE. Trans. Microwave Theory Tech.*, **MTT-22**, 1283–1286.

Allanson, J. G., Cooper R. and Cowling, Z. G. (1949) The theory and experimental behaviour of right-angled junctions in rectangular-section waveguides, *J. Inst. Elec. Eng.*, **93**, Part III, 177–187.

Auld, B. A. (1962) The synthesis of symmetrical waveguide circulators, *IRE Trans. Microwave Theory Tech.*, **MTT-7**, 137–146.

Bramham, B. (1961) A convenient transformer for matching coaxial lines, *Electron. Eng.*, **33**, 42–44.

Carlin, H. J. and Kohler, W. E. (1965) Direct synthesis of band-pass transmission line structures, **MTT-13**, 283–297.

Carlin, H. J. and Friedenson, R. A. (1968) Gain bandwidth properties of a distributed parameter load, *IEEE Trans. Circuit Theory*, **CT-15**, 455–464.

Fano, R. M. (1950) Theoretical limitations on broadband matching of arbitrary impedances, *J. Franklin Inst.*, **249**, 57–98; also 139–154.

French, G. N. and Fooks, E. H. (1069) Double section matching transformers, *IEEE Trans.*, **MTT-17**, 719.

Hauth, W., Lenz, S. and Pivit, E. (1986) Design and realization of very-high-power waveguide junction circulators, *Frequency*, **40**(1), 1–24.

Helszajn, J. (1972) The synthesis of quarter-wave coupled circulators with Chebyshev characteristics, *IEEE Trans. Microwave Theory Tech.*, **MTT-20**, 764–769.

Helszajn, J. and Levy, R. (1993) Synthesis of series STUB-R complex gyrator ciruits using half-wave long impedance transformers, *IEE Proc. Microwave, Antennas and Propagation*, 426–432.

Kiyuzaki, T. and Yada, M. (1963) An investigation on bandwidth enlargement of a Y-circulator. In Preprint 1963, *Nat. Conf. Institute of Electronics and Communications Engineers of Japan*, 55–7.

Levy, R. (1970) General rational function approximation in finite intervals using Zolotarev functions, *IEEE Trans.*, **MTT-18**, 1052–1064.

Levy, R., (1972) Synthesis of mixed lumped and distributed impedance matching networks, *IEEE Trans.*, **MTT-20**, 223–233.

Levy, R. and Helszajn, J. (1982) Specific equations for one and two section quarter-wave matching networks for stub-resistor loads, *IEEE Trans. Microwave Theory Tech.*, **MTT-30**, 55–62.

Levy, R. and Helszajn, J. (1983) Synthesis of short line matching networks for resonant loads, *IEE Proc. Microwaves, Antennas and Propagation* (H).

Matthaei, G. L. (1966) Short-step Chebyshev impedance transformers, *IEEE Trans.*, **MTT-14**, 372–383.

Mokari-Bolhassan, M. E. and Ku, W. H. (1977) Transfer function approximations for a new class of bandpass distributed network structures, *IEEE Trans.*, **MTT-25**, 837–847.

Mokari-Bolhassan, M. E. and Ku, W. H. (1977) Gain-bandwidth limitations and synthesis of single-stub bandpass transmission-line structures, *IEEE Trans.*, **MTT-25**, 848–852.

Reeder, T. M. and Sperry, W. R. (1972) Broad-banked coupling to high-Q resonant loads, *IEEE Trans. Microwave Theory Tech.*, **MTT-20**, 453–458.

Treuhaft, M. A. (1956) Network properties of circulators based on the scattering concept, *Proc. IRE*, **44**, 1394–1402.

Tucker, R. S. (1973) Gain-bandwidth limitations of microwave transistor amplifiers, *IEEE Trans.*, **MTT 21**, 322–327.

Wenzel, R. J. (1971) Synthesis of combline and capacitively loaded inter-digital bandpass filters of arbitrary bandwidth, *IEEE Trans.*, **MTT-19**, 678–686.

Youla, D. C. (1964) A new theory of broad-band matching. *IEE Trans. Circuit Theory*, **CT-11**, 30–50.

24

EXPERIMENTAL EVALUATION OF JUNCTION CIRCULATORS

24.1 Introduction

For commercial purposes the specification of a junction circulator is usually described by its return-loss ($|S_{11}|$), isolation ($|S_{31}|$), and insertion loss ($|S_{21}|$). For design purposes the nature of its three eigen-networks is of more interest in that these circuits are directly displayed by the complex gyrator circuit of the device. The first method, requires 2-port measurements of both phase and amplitude, and the second three different generator settings at each port. In order to avoid some inherent difficulties with these classic techniques, a number of other methods have evolved over the years. The complex gyrator circuit at one port is separately obtained at one frequency by decoupling one of the output ports by adjusting the load at the other output port with a stub-tuner. It probably still represents the most accurate description of this type of device. Swept frequency measurements at one port with the other two ports terminated with different load conditions have also been described. The elements of the complex gyrator of a weakly magnetized junction may be separately related to the difference between the split frequencies of the split counter-rotating eigen-networks or quality factor of the circuit and its susceptance slope parameter. These quantities therefore provide still another means of describing the equivalent circuit of this sort of problem. There are undoubtedly more such relationships. Since the description of the 3-port circulator is closely related to some features of the demagnetized junction its description is also of some interest. It may be deduced from measurements at port 1 with the other two appropriately terminated. One possibility is obtained by making one measurement with ports 2 and 3 terminated in matched terminations and then by making a second one which consists of decoupling port-3 by placing a variable short-circuit at port-2. This latter measurement directly gives the degenerate counter-rotating eigenvalues of the junction. The latter parameters may be separately deduced, at least in planar circuits, by constructing each 1-port eigen-network. Other possibilities have also been described in the literature.

While there is always some dissipation in any practical device the relationship in this chapter are all based on an idealized model with no dissipation.

24.2 Eigenvalue Test Sets

The two classic methods which may be used to completely characterize a symmetrical junction circulator either consist of the direct measurement of its scattering parameters or its eigenvalues. The scattering parameters of the junction are related to its eigenvalues by

$$S_{11} = \frac{s_0 + s_+ + s_-}{3} \tag{24.1a}$$

$$S_{21} = \frac{s_0 + \alpha s_+ + \alpha^2 s_-}{3} \tag{24.1b}$$

$$S_{31} = \frac{s_0 + \alpha^2 s_+ + \alpha s_-}{3} \tag{24.1c}$$

The reflection eigenvalues are separately related to the entries of the scattering matrix by

$$s_0 = S_{11} + S_{21} + S_{31} \tag{24.2a}$$

$$s_+ = S_{11} + \alpha^2 S_{21} + \alpha S_{31} \tag{24.2b}$$

$$s_- = S_{11} + \alpha S_{21} + \alpha^2 S_{31} \tag{24.2c}$$

where

$$\alpha = 1 \cdot \exp(j\,240^\circ) \tag{24.3a}$$

$$\alpha^2 = 1 \cdot \exp(j240^\circ) \tag{24.3b}$$

Scrutiny of these relationships indicates that the eigenvalue description of the device may be deduced from a knowledge of its scattering description. The definition of the scattering matrix is illustrated in Figure 24.1. These may also, of course, be obtained by directly exciting the junction one at a time by its three possible eigenvectors.

$$\overline{U}_0 = \frac{1}{\sqrt{3}}\begin{bmatrix}1\\1\\1\end{bmatrix},\ \overline{U}_+ = \frac{1}{\sqrt{3}}\begin{bmatrix}1\\\alpha\\\alpha^2\end{bmatrix},\ \overline{U}_- = \frac{1}{\sqrt{3}}\begin{bmatrix}1\\\alpha^2\\\alpha\end{bmatrix} \tag{24.4}$$

24.2 EIGENVALUE TEST SETS

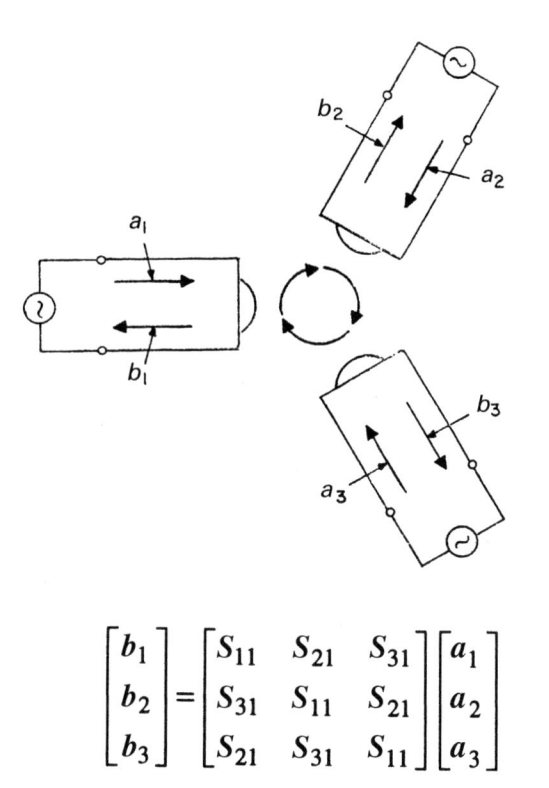

$$\begin{bmatrix} b_1 \\ b_2 \\ b_3 \end{bmatrix} = \begin{bmatrix} S_{11} & S_{21} & S_{31} \\ S_{31} & S_{11} & S_{21} \\ S_{21} & S_{31} & S_{11} \end{bmatrix} \begin{bmatrix} a_1 \\ a_2 \\ a_3 \end{bmatrix}$$

Figure 24.1 Schematic diagram showing definition of scattering variables

The required generator arrangements are given in Chapter 3. Scrutiny of these diagrams indicates that the in-phase eigen-network may be represented by an open-circuited stub and those of the counter-rotating ones by short-circuited stubs. The required eigen-networks are shown in Figure 24.2.

Each of these possible eigen-networks is uniquely characterized once its resonant frequency (or electrical length) and reactance (susceptance) slope parameter are at hand. These quantities define the slopes of the reactance and susceptance slope parameters of the immittances at the midband frequencies by

$$X' = \frac{\omega_0}{2} \frac{\partial X}{\partial \omega} \bigg|_{\omega = \omega_0} \quad (24.5a)$$

$$B' = \frac{\omega_0}{2} \frac{\partial B}{\partial \omega} \bigg|_{\omega = \omega_0} \quad (24.5b)$$

respectively.

474   24 EXPERIMENTAL EVALUATION OF JUNCTION CIRCULATORS

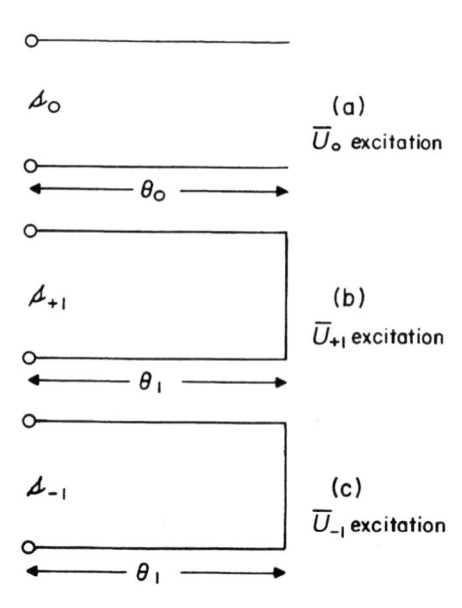

Figure 24.2 In-phase and counter-rotating eigen-networks of junction circulator

The reactance notation is appropriate when dealing with the in-phase eigen-network whereas the susceptance one is more suitable when dealing with the counter-rotating ones.

If the eigen-networks are realized in terms of quarter-wave transmission lines then a knowledge of the absolute values of the characteristic immittances is all that is required to define these quantities

$$X' = \frac{\pi Z_0}{4} \quad (24.6a)$$

$$B' = \frac{\pi Y_0}{4} \quad (24.6b)$$

24.3 Degenerate Counter-Rotating Eigen-Network (s_1)

One possible way in which the degenerate eigenvalue s_1 of a reciprocal 3-port junction can be experimentally determined without the need to fabricate its eigen-network consists of placing a sliding short-circuit at port-2 and a matched load at port-3. The variable short-circuit is varied until there is total reflection at port-1. The reflection coefficient at port 1 is then the eigenvalue s_1. This technique is especially appropriate in the case of a waveguide junction for which the reference terminals are usually ill defined. The experimental arrangement is shown in Figure 24.3.

24.3 DEGENERATE COUNTER-ROTATING EIGEN-NETWORKS (s_1) 475

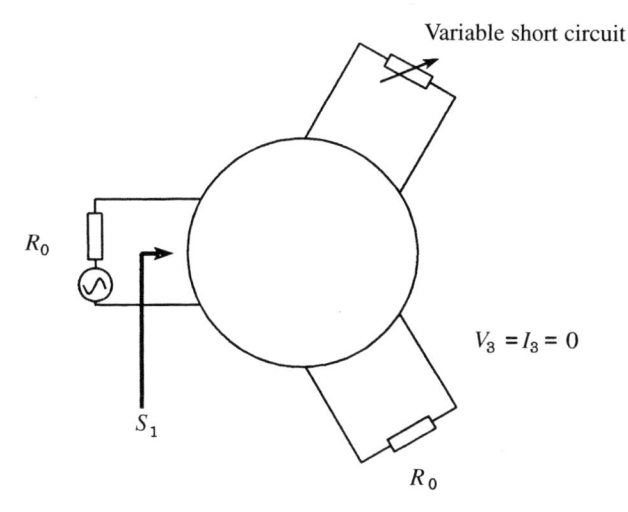

Figure 24.3 Experimental arrangement for measurement of degenerate reflection eigenvalue s_1

The derivation of this technique starts with the relationships between the incoming and outgoing waves at the three ports of the junction.

$$\begin{bmatrix} b_1 \\ b_2 \\ b_3 \end{bmatrix} = \begin{bmatrix} S_{11} & S_{21} & S_{21} \\ S_{21} & S_{11} & S_{21} \\ S_{21} & S_{21} & S_{11} \end{bmatrix} \begin{bmatrix} a_1 \\ a_2 \\ a_3 \end{bmatrix} \tag{24.7}$$

where for a reciprocal 3-port junction

$$S_{11} = \frac{s_0 + 2s_1}{3} \tag{24.8}$$

$$S_{21} = \frac{s_0 - s_1}{3} \tag{24.9}$$

In the arrangement considered here a variable short-circuit at port-2 is located at a suitable plane in order to decouple port-3 from the other two. This gives

$$b_2 = a_2 \exp -j2\left(\phi + \frac{\pi}{2}\right) \tag{24.10}$$

and

$$a_3 = b_3 = 0 \tag{24.11}$$

ϕ is the electrical length between the position of the short circuit and the reference plane of b_2/a_2. Introducing the last two conditions into equation (24.7) gives

$$b_1 = S_{11}a_1 + S_{21}a_2 \tag{24.12}$$

$$b_2 = S_{21}a_1 + S_{11}a_2 \tag{24.13}$$

$$0 = S_{21}a_1 + S_{21}a_2 \tag{24.14}$$

Scrutiny of these three equations indicates that the waves a_1 and a_2 at ports 1 and 2 are out-of-phase there.

$$\frac{a_1}{a_2} = -1 \tag{24.15}$$

Examination of the first and second equations separately indicates that the network between ports 1 and 2 formed in this way is symmetrical. Substituting the above condition into equations (24.12) and (24.13) indicates that

$$\frac{b_1}{a_1} = S_{11} - S_{21} = s_1 \tag{24.16}$$

and

$$\frac{b_1}{a_1} = \frac{b_2}{a_2} \tag{24.17}$$

respectively. So

$$s_1 = \exp -j2\left(\phi + \frac{\pi}{2}\right) \tag{24.18}$$

This measurement, therefore, provides a means of measuring s_1 with reference to the short circuit position. The planes at port 1 at which the standing wave is zero are known as the characteristic planes of the junction. The characteristic plane is defined by the short-circuit positions given by $\phi = 0$, π, 2π, etc.

It is of note that the condition $a_3 = b_3 = 0$ is compatible with that met in the development of the degenerate counter-rotating eigen-network; namely $V_3 = I_3 = 0$.

24.4 In-Phase Eigen-Network

The phase angle of the in-phase eigen-network of a demagnetized junction may be directly measured using the eigenvalue approach by applying equal in-phase signals at the three ports of the junction or by making use of the relationship between the scattering variable S_{11} and the in-phase and degenerate counter-rotating eigenvalues s_0 and s_1

24.4 IN-PHASE EIGEN-NETWORK

$$S_{11} = \frac{s_0 + 2s_1}{3} \qquad (24.19)$$

In a lossless device for which the in-phase eigen-network may be synthesized by an open-circuited network and the degenerate counter-rotating ones by short-circuited ones.

$$s_0 = 1 \cdot \exp(-j2\theta_0) \qquad (24.20)$$

$$s_1 = 1 \cdot \exp -j2(\theta_1 + \pi/2) \qquad (24.21)$$

and

$$S_{11} = |S_{11}| \cdot \exp -j2(\phi_{11}) \qquad (24.22)$$

where

$$|S_{11}| = \frac{(VSWR) - 1}{(VSWR) + 1} \qquad (24.23)$$

The phase angle θ_0 may be evaluated using (24.19) by forming one of four possible relationships between the independent variables $|S_{11}|$, ϕ_{11} and θ_1.

$$2\theta_0 = \cos^{-1}(3|S_{11}|\cos 2\phi_{11} + 2\cos 2\theta_1) \qquad (24.24)$$

$$2\theta_0 = 2\theta_1 + \pi + \cos^{-1}\left(\frac{9|S_{11}|^2 - 5}{4}\right) \qquad (24.25)$$

$$2\theta_0 = 2\phi_{11} + \cos^{-1}\left(\frac{3|S_{11}|^2 - 1}{2|S_{11}|}\right) \qquad (24.26)$$

The first identity is constructed by taking a linear combination of s_0 and s_0^*, and the other two are obtained by forming the products of $S_{11}S_{11}^*$ and $s_1 s_1^*$. The first relationship for θ_0 requires a knowledge of all three independent variables, whereas the second two need only a measurement of $|S_{11}|$ and a knowledge of θ_1 or ϕ_{11}.

The fourth relationship between θ_0, θ_1 and ϕ_{11} may also be formed by eliminating $|S_{11}|$ by equating (24.24) and (24.25). The result at $2\theta_1 = \pi$ involves θ_{11} and ϕ_{11} only

$$\cos^2 2\theta_0 + (4 - 4\cos^2 2\phi_{11})\cos 2\theta_0 + (4 - 5\cos^2 2\phi_{11}) = 0 \qquad (24.27)$$

S_{11} may be evaluated by terminating the two output ports by matched loads in the manner indicated in Figure 24.4, and s_1 may be determined by decoupling port-3 from port-1 by placing a variable short-circuit at port-2 as already described

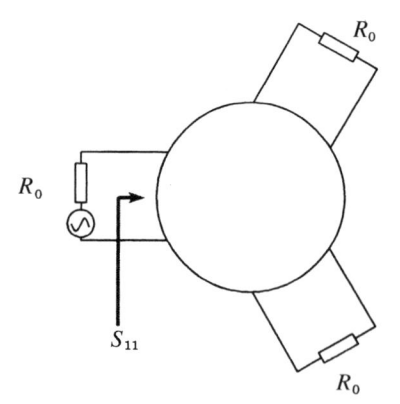

Figure 24.4 Experimental arrangement for measurement of S_{11}

in connection with Figure 24.3. The angles of the scattering variables may be located at the reference plane of the junction ($d_{s/c}$) by replacing the resonator by a metal plug. The result for ϕ_{11} is

$$2\phi_{11} = \frac{4\pi}{\lambda_g}(d_{s/c} - d_{min}) \qquad (24.28)$$

where d_{min} is the position of a minimum in the *VSWR* along the line. A similar relationship applies to $2\theta_1 + \pi$.

$$2\theta_1 = \frac{4\pi}{\lambda_g}(d_{s/c} - d_{min}) \qquad (24.29)$$

24.5 Split Eigen-Networks of Junction Circulator

An especially simple method whereby the split admittance eigen-networks of a circulator for which the in-phase eigen-network may be idealized by a short-circuit boundary condition at the terminals of the junction can be deduced from a knowledge of its complex gyrator circuit. This quantity is defined by

$$y_{in} = g + jb \qquad (24.30)$$

where

$$g = -j\sqrt{3}\left(\frac{y_+ - y_-}{2}\right) \qquad (24.31)$$

$$b = \left(\frac{y_+ + y_-}{2}\right) \qquad (24.32)$$

24.5 SPLIT EIGEN-NETWORKS

The split admittance eigenvalues may be now evaluated by taking linear combinations of the real and imaginary parts of the complex gyrator admittance. The result is

$$y_+ = b + j\frac{g}{\sqrt{3}} \tag{24.33a}$$

$$y_- = b - j\frac{g}{\sqrt{3}} \tag{24.33b}$$

where b represents the susceptance of the degenerate counter-rotating eigenvalues.

$$b = y_1 \tag{24.34}$$

The split frequencies may also be determined from this sort of data by recognizing that these coincide with the frequencies at which the split admittance eigenvalues are zero

$$y_\pm = 0 \tag{24.35}$$

The two required conditions are obtained by having recourse to equations (24.32) and (24.33)

$$g = \mp j\frac{b}{\sqrt{3}} \tag{24.36}$$

One experimental arrangement, in keeping with the definition of the complex gyrator circuit, is obtained by decoupling port-3 from port-1 by adjusting a triple stub tuner at port 2. This arrangement is illustrated in Figure 24.5.

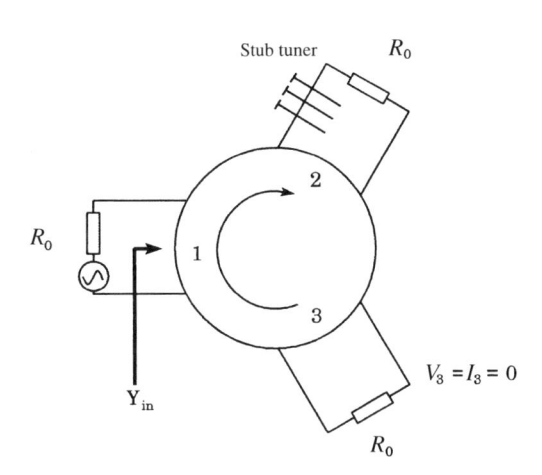

Figure 24.5 Experimental definition of complex gyrator circuit

24.6 The Constituent Resonator

The degenerate counter-rotating eigenvalue (s_0 or s_1) of a reciprocal junction may also be approximately determined by assuming that it consists of the connection of three constituent resonators each one being mutually coupled to the others. This resonator may be defined by either open circuiting two of the three ports of the junction in a stripline resonator or shortcircuiting two of the ports of a waveguide junction. The constituent resonator of a waveguide junction is illustrated in Figure 24.6. If the degenerate counter-rotating eigenvalue is known it may be employed to measure the in-phase one. If the frequency variation of the latter variable may be neglected compared to the former, then it may be employed to evaluate s_1. If the discussion is restricted to the waveguide arrangement then the relationship between the reactance of the constituent resonator and that of the circulator is a constant which may be deduced by comparing the reactance slope parameter of each arrangement.

While the notion of the constituent resonator is more appropriate to stripline circuits it is included here for completness sakes. Its property will now be demonstrated.

The derivation of the required result starts with the current voltage relationship of a reciprocal, symmetrical three-port junction

$$\begin{bmatrix} I_1 \\ I_2 \\ I_3 \end{bmatrix} = \begin{bmatrix} Y_{11} & Y_{12} & Y_{12} \\ Y_{12} & Y_{11} & Y_{12} \\ Y_{12} & Y_{12} & Y_{11} \end{bmatrix} \begin{bmatrix} V_1 \\ V_2 \\ V_3 \end{bmatrix} \qquad (24.37)$$

The constituent resonator is defined by applying electric walls at ports 2 and 3 of the reciprocal 3-port network. Thus

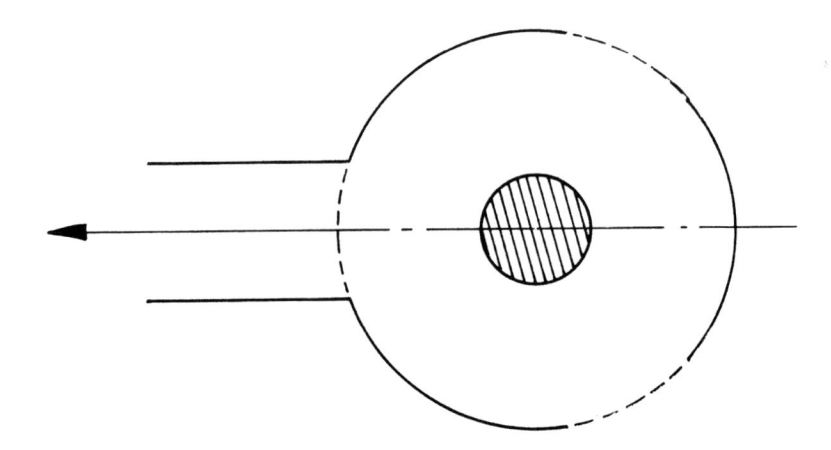

$$V_2 = V_3 = 0 \qquad (24.38)$$

Figure 24.6 Schematic diagram of constituent resonator

The input admittance of the constituent resonator is readily expressed as

$$Y_{in} = Y_{11} = \frac{y_0 + y_+ + y_-}{3} \qquad (24.39)$$

If the splitting between the counter-rotating admittance eigenvalues is symmetric about the degenerate ones and if the in-phase is taken as zero, then

$$Y_{in} = \frac{2y_1}{3} \qquad (24.40)$$

The input impedance (reactance) of the constituent junction is given by inverting (24.40) as

$$Z_{in} = jX_0 = \frac{3z_1}{2} \qquad (24.41)$$

X_0 is the susceptance of the constituent resonator and z_1 the degenerate admittance eigenvalue of the counter-rotating eigen-networks.

A comparison of the reactance of the constituent resonator in (24.40) and that of the junction circulator indicates that the two are related by

$$X_1 = \frac{2}{3} X_0 \qquad (24.42)$$

The preceeding expression also indicates that the reactance slope parameter of the two arrangements are related by

$$X_1' = \frac{2}{3} X_0' \qquad (24.43)$$

This condition is a standard result.

24.7 Split Frequencies

Useful information about junction circulators may also be deduced from a knowledge of the frequency variation of the reflection coefficient at port-1 with ports 2 and 3 terminated in matched load. This method is particularly attractive with swept frequency instrumentation. The two split frequencies of the device coincide with the frequencies for which

$$|S_{11}| = \frac{1}{3} \qquad (24.44)$$

which corresponds with a *VSWR* of 2 : 1 or a return loss of 9.5 dB. The derivation of this conditions begins by making use of the standard relationship between the scattering parameters and its eigenvalues

$$S_{11} = \frac{s_0 + s_+ + s_-}{3} \qquad (24.45)$$

It continues by replacing s_0 by

$$s_0 = -1 \qquad (24.46)$$

and by recognizing that the required result coincides with the conditions for which either s_+ or s_- are in antiphase with s_0.

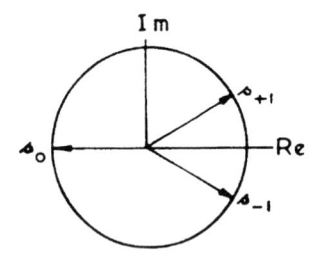

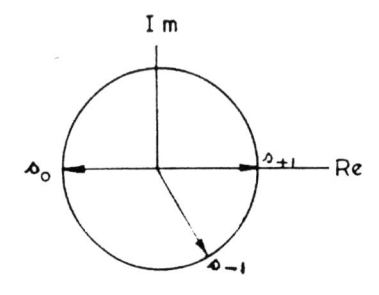

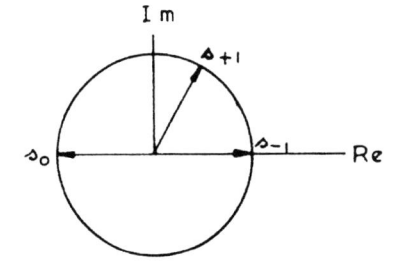

Figure 24.7 Eigenvalue diagrams of magnetized circulator (reprinted with permission, Helszajn, J. (1973) Microwave measurement techniques for junction circulators, *IEEE Trans. Microwave Theory Tech.*, **MTT-21**, 347–351)

The first condition gives

$$|S_{11}| = \left|\frac{s_+}{3}\right| \qquad (24.47)$$

and the second one gives

$$|S_{11}| = \left|\frac{s_-}{3}\right| \qquad (24.48)$$

The required result is obtained by recalling that

$$|s_+| = |s_-| = 1 \qquad (24.49)$$

Figure 24.7 depicts the required eigenvalue diagrams.

24.8 Gyrator Conductance of Circulator

The derivation of the relationship between S_{11} and g with ports 2 and 3 terminated in 50 Ω loads starts with a definition of the reflection eigenvalues.

$$s_0 = 1 \cdot \exp(-j2\theta_0) \qquad (24.51a)$$

$$s_+ = 1 \cdot \exp\left[-j2\left(\theta_1 + \theta'_+ + \frac{\pi}{2}\right)\right] \qquad (24.51b)$$

$$s_- = 1 \cdot \exp\left[-j2\left(\theta_1 - \theta'_- + \frac{\pi}{2}\right)\right] \qquad (24.51c)$$

At midband

$$\theta_0 = \theta_1 = \frac{\pi}{2} \qquad (24.52)$$

and for symmetrical splitting

$$\theta'_- = -\theta'_+ \qquad (24.53)$$

If the bilinear relationships between the reflection and admittance eigenvalues are employed in constructing S_{11} then

$$S_{11} = \frac{1 - 3\tan^2(\theta'_+)}{3 + 3\tan^2(\theta'_+)} \qquad (24.54)$$

Likewise

$$g = \sqrt{3} \tan(\theta'_+) \tag{24.55}$$

It is now possible to eliminate θ'_+ between S_{11} and g. The result in terms of the VSWR is

$$(VSWR) = \frac{|3 + g^2| + |1 - g^2|}{|3 + g^2| - |1 - g^2|} \tag{24.56}$$

For g larger than unity this relationship reduces to

$$g = \sqrt{2(VSWR) - 1} \tag{24.57}$$

For g smaller than unity it is given by

$$g = \sqrt{\frac{2 - (VSWR)}{(VSWR)}} \tag{24.58}$$

The relationship between (VSWR) and g is indicated in Figure 24.8. Whether g is larger or smaller than unity is determined by whether there is a voltage maxima or minima at the load. The condition on this curve for which g equals zero and the VSWR equals 2 coincides with that of a demagnetized junction.

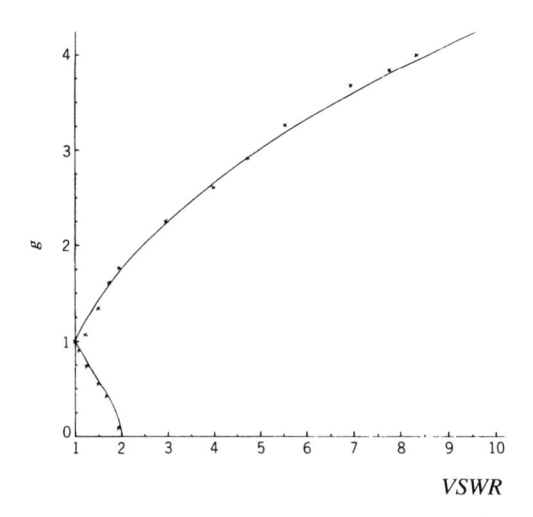

Figure 24.8 Relationship between VSWR and gyrator conductance of junction circulator (reprinted with permission, Helszajn, J. (1973) Microwave measurement techniques for junction circulators, *IEEE Trans. Microwave Theory Tech.*, **MTT-21**, 347–351.) (© 1973 IEEE)

24.9 Susceptance Slope Parameter

One straightforward way to deduce the susceptance slope parameter (b') of a 3-port junction for which the frequency variation of the in-phase mode may be neglected compared to that of the degenerate counter-rotating ones is to measure its *VSWR* in the vicinity of its midband frequency. The derivation of the required result starts by assuming that the input admittance of the device in the vicinity of $g = 1$ coincides with the gyrator one

$$y_{in} \approx 1 - jy_1 \tag{24.59}$$

where

$$y_1 = 2\delta b' \tag{24.60}$$

and

$$\delta = \frac{\omega_0 - \omega}{\omega_0} \tag{24.62}$$

The corresponding reflection coefficient is then given by having recourse to the standard bilinear transformation between reflection and admittance on a uniform line.

$$|\rho|^2 \approx \frac{(2\delta b')^2}{4 + (2\delta b')^2} \tag{24.62}$$

If $|\rho|$ is written in terms of the *VSWR* then

$$|\rho| = \frac{(VSWR) - 1}{(VSWR) + 1} \tag{24.63}$$

The required result is now obtained by equating the two proceeding relationships for $|\rho|$. This gives

$$b' \approx \frac{(VSWR) - 1}{2\delta\sqrt{(VSWR)}} \tag{24.64}$$

This quantity may also be established from a knowledge of the frequency response of the demagnetized junction. The required result is

$$b' \approx \frac{\dfrac{2}{3}\left[\dfrac{(VSWR)^2 - 2.5(VSWR) + 1}{2(VSWR)}\right]^{1/2}}{2\delta} \tag{24.65}$$

24.10 Swept Frequency Description of Eigenvalues

The details of the eigenvalues of a junction circulator may be extracted, as already demonstrated, from a knowledge of the frequency behaviour of various 1-port immittance statements. These types of measurements may also be separately generalized by either forming the input impedance or reflection coefficient at one port of the device with the other two terminated in various load conditions. The required relationship is given by.

$$\rho_{in} = S_{11} + \frac{(\rho_2 + \rho_3)A + \rho_2\rho_3(B - 2S_{11}A)}{1 - (\rho_2 + \rho_3)S_{11} + \rho_2\rho_3(S_{11}^2 - A)} \tag{24.66}$$

where

$$A = S_{12}S_{13} \tag{24.67a}$$

$$B = S_{12}^3 + S_{13}^3 \tag{24.67b}$$

ρ_2 and ρ_3 and the reflection coefficients at ports 2 and 3

The unknown quantities S_{11}, S_{21} and S_{31} may be deduced from measurements of ρ_{in} at port-1 by selecting three different load conditions at ports 2 and 3 and thereafter constructing suitable linear equations from which the unknown quantities may be calculated. If, for instance, $\rho_2 = \rho_3 = 0$ in the description of ρ_{in} then the impedance is given for completeness as $\rho_{in} = S_{11}$, as is readily understood.

The input impedance at port 1 corresponding to the condition $\rho_2 = \rho_3 = 0$ at ports 2 and 3 is separately specified by

$$Z_{in} = Z_{11} + \frac{(Z_{12})^3 - (Z_{12}^*)^3 + 2Z_{12}Z_{12}^*(Z_{11} + 1)}{(Z_{11} + 1)^2 + Z_{12}Z_{12}^*} \tag{24.68}$$

The derivation of this result has been separately dealt with in Chapter 4.

References and Further Reading

Aitken, F. M. and McLean, R. (1963) Some properties of the waveguide Y Circulator, *Proc. IEE*, **110**(2), 256–260.
Anderson, L. K. (1967) An analysis of broadband circulators with external tuning elements, *IEEE Trans. Microwave Theory Tech.*, **MTT-15**, 42–47.
Auld, B. A. (1959) The synthesis of symmetrical waveguide circulators, *IRE Trans. Microwave Theory Tech.*, **MTT-7**, 238–246.
Bosma, H. (1971) Junction circulators, *Advances in Microwaves*, **6**, 125–257.
Fay, C. E. and Comstock, R. L. (1965) Operation of the ferrite junction circulator, *IEEE Trans. Microwave Theory Tech.*, **MTT-13**, 15–27.
Goebel, U. and Schieblich, C. (1983) A unified equivalent circuit representation of H and E-plane junction circulators. In *Eur Microwave Conf. 1983*, 803–808.

Hagelin, S. (1966) A flow graph analysis of 3 and 4-port junction circulators, *IEEE Trans. Microwave Theory Tech.*, **MTT-14**, 243–249.

Helszajn, J. (1970) The adjustment of the *m*-port single junction circulator, *IEEE Trans. Microwave Theory Tech.*, **MTT-18**(10), 705–711.

Helszajn, J. (1970) Wideband circulator adjustment using the $n = \pm 1$ and $n = 0$ electromagnetic field patterns, *Electron. Lett.*, **6**(23).

Helszajn, J. (1972) Synthesis of quarter wave coupled circulators with Chebyshev characteristics, *IEEE Trans. Microwave Theory Tech.*, **MTT-20**, 764–769.

Helszajn, J. (1973) Microwave measurement techniques for junction circulators, *IEEE Trans. Microwave Theory Tech.*, **MTT-21**, 347–351.

Helszajn, J. and Sharp, J. (1983) Resonant frequencies, Q-factor and susceptance slope parameter of waveguide circulators using weakly magnetized open resonators, *IEEE Trans. Microwave Theory Tech.*, **MTT-31**, 434–441.

Helszajn, J. and Sharp, J. (1985) Adjustment of in-phase mode in turnstile circulator, *IEEE Trans. Microwave Theory Tech.*, **MTT-32** 339–343.

Knerr, R. H. (1969) A thin film lumped element circulator, *IEEE Trans. Microwave Theory Tech.*, **MTT-17**(12), 1152–1154.

Knerr, R. H. and Barnes, C. E. (1970) A compact broad-bank thin film lumped element l-bank circulator, *IEEE Trans. Microwave Theory Tech.*, **MTT-18**(12), 1100–1108.

Milano, U., Saunders, J. H. and Davis L. Jr. (1960) A *Y*-junction strip-line circulator, *IRE Trans. Microwave Theory Tech.*, **MTT-8**, 346–351.

Montgomery, C. G., Dicke, R. H. and Purcell, E. M. (1948) Principles of microwave circuits, McGraw-Hill, New York.

Owen, B. (1972) The identification of modal resonances in ferrite loaded waveguide Y-junction and their adjustment for circulation, *Bell Syst. Tech. J.*, **51**(3).

Riblet, G. P. (1977) The measurement of the equivalent admittance of 3-port circulators via an automated measurement system, *IEE Trans. Microwave Theory Tech.*, 401–405.

Schaug-Patterson, T. (1958) Novel design of a 3-port circulator, Norwegian Defence Research Establishment, January.

Schwartz, E. (1968) Broadband matching of resonant circuits and circulators, *IEEE Trans. Microwave Theory Tech.*, **MTT-16**, 158–165.

Simon, J. (1965) Broadband strip-transmission line Y junction circulators, *IEEE Trans. Microwave Theory Tech.*, **MTT-13**, 335–345.

25

CIRCULATOR SPECIFICATIONS

25.1 Introduction

An ideal circulator is usually specified by its return loss (dB) at any port, its insertion loss (dB) between any two ports in the direction of circulation and its isolation (dB) between any two ports in the other direction. The equipment maker is usually, however, more interested in the overall system specification rather than that of the circulator between ideal termination's. These include, in addition to the aforementioned quantities, the peak and average power ratings of the device, its temperature stability, insertion phase shift, phase dispersion and other specific requirements such as cooling provisions and the like. The purpose of this chapter is to briefly outline some of the considerations that enter into the choice of practical circulators with an arbitrary load at port-2 under the assumption that any dissipation loss within the device can be disregarded. Simple approximate closed form expressions for the upper and lower bounds on the VSWR at port-1 of a 3-port circulator with port-2 and port-3 terminated in arbitrary loads are derived. One way to cope with the general problem is to absorb the external load conditions inside the terminals of the circulator. The augmented matrix description of the device at these new reference terminals is now no longer associated with a symmetric junction as it is readily appreciated. However, if dissipation cannot be neglected, then the situation becomes even more complicated. The chapter includes some remarks about the overall insertion loss between ports 1 and 3 of a 3-port single junction circulator with port-2 terminated in a variable short-circuit. Such a situation is met in connection with 1-port negative devices and some duplexer arrangements.

25.2 Specifications

In every day engineering it is usual to express the isolation (I), the return loss (RL) and the insertion loss (L) of a 3-port circulator in decibels instead of as fractional quantities. The necessary definitions are

$$RL \text{ (dB)} = 10 \log_{10}\left(\frac{\text{reflected power at port-1}}{\text{incident power at port-1}}\right) \quad (25.1)$$

$$L \text{ (dB)} = 10 \log_{10}\left(\frac{\text{transmitted power at port-2}}{\text{incident power at port-1}}\right) \quad (25.2)$$

$$I \text{ (dB)} = 10 \log_{10}\left(\frac{\text{transmitted power at port-3}}{\text{incident power at port-1}}\right) \quad (25.3)$$

Each of these quantities is negative in keeping with the usual convention adopted throughout the chapter.

The ratios appearing in these quantities are related to the scattering parameters of the circulator by

$$|S_{11}|^2 = \left(\frac{\text{reflected power at port-1}}{\text{incident power at port-1}}\right) \quad (25.4)$$

$$|S_{21}|^2 = \left(\frac{\text{transmitted power at port-2}}{\text{incident power at port-1}}\right) \quad (25.5)$$

$$|S_{31}|^2 = \left(\frac{\text{transmitted power at port-3}}{\text{incident power at port-1}}\right) \quad (25.6)$$

It is recalled that S_{21} implies transmission between ports 1 and 2 and not between ports 2 and 1. A similar convention applies to S_{31}.

In the absence of dissipation the unitary condition indicates

$$|S_{11}|^2 + |S_{21}|^2 + |S_{31}|^2 = 1 \quad (25.7)$$

In the vicinity of an ideal circulator it also indicates that

$$|S_{11}| \approx |S_{31}| \quad (25.8)$$

These two conditions suggest that the reflection loss at both ports 1 and 2 contribute to the overall insertion loss of the circulator.

$$|S_{21}|^2 \approx 1 - 2|S_{11}|^2 \quad (25.9)$$

25.2 SPECIFICATIONS

Writing the unitary condition in terms of the practical variables gives

$$10^{L(\text{dB})/10} + 10^{RL(\text{dB})/10} + 10^{I(\text{dB})/10} = 1 \tag{25.10}$$

The insertion loss is therefore given in terms of the return loss at a typical port by

$$10^{L(\text{dB})/10} \approx 1 - 2(10^{RL(\text{dB})/10}) \tag{25.11}$$

or

$$L(\text{dB}) \approx 10\log_{10}[1 - 2(10^{RL(\text{dB})/10})] \tag{25.12}$$

It is understood that L, RL and I are negative quantities in keeping with the usual convention adopted here.

The relationship between the VSWR and isolation of an ideal symmetrical circulator is indicated in Figure 25.1.

One means of obtaining circulators with more than 3 ports is to connect to a number in tandem in the manner indicated in Figure 25.2. Some mechanical configurations in the case of the 4-port ones are depicted in Figure 25.3.

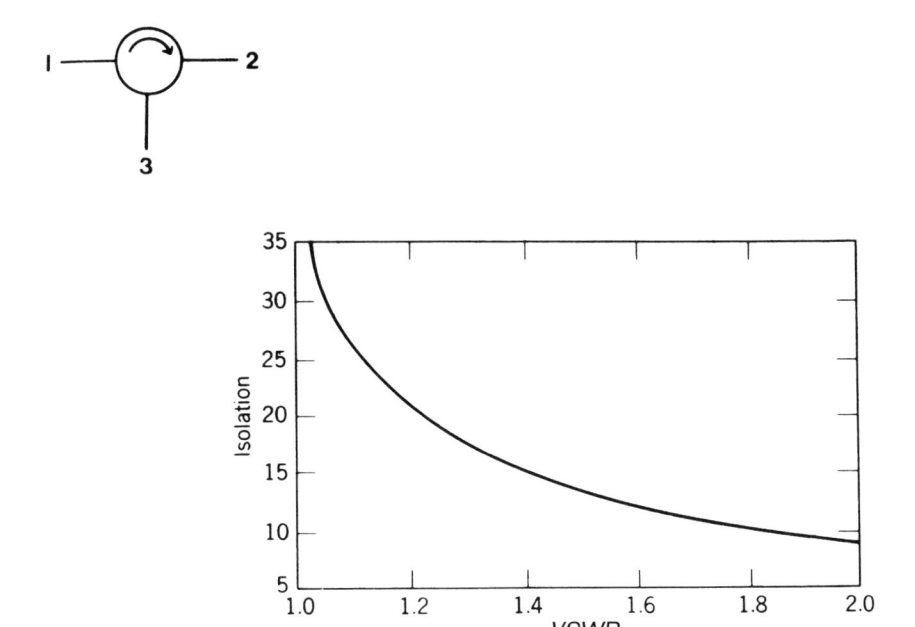

Figure 25.1 Relationship between VSWR and isolation of an ideal 3-port junction circulator

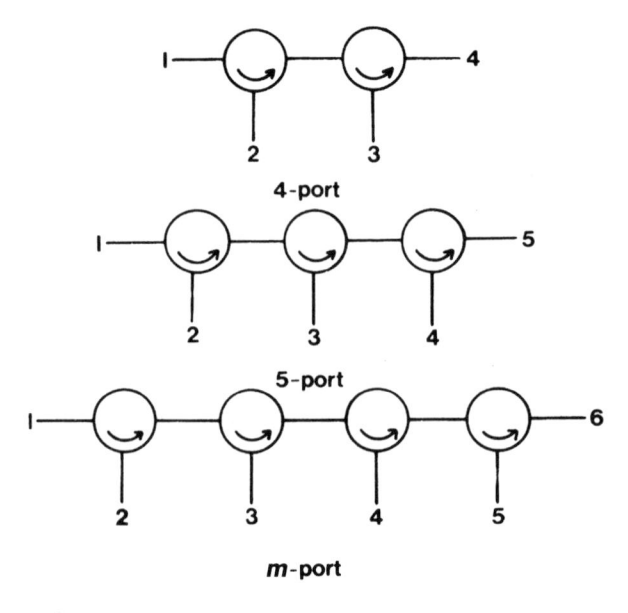

Figure 25.2 Schematic diagrams of 4-, 5- and m-port circulators

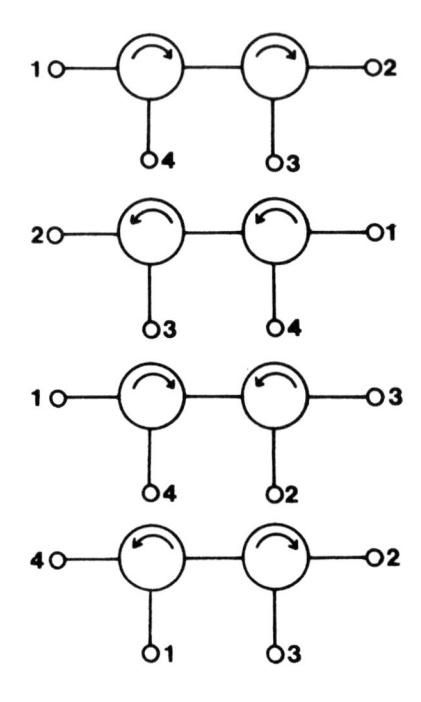

Figure 25.3 Schematic diagrams of different port arrangements of 4-port circulators

25.3 Voltage Standing Wave Ratio

The specification of a microwave device is also sometimes stated in terms of its voltage standing wave ratio (*VSWR*) instead of its return loss or reflection coefficient. The two quantities are related by a bilinear transformation of the form

$$VSWR = \frac{1 + |\rho|}{1 - |\rho|} \qquad (25.13)$$

It is also apparent that $|\rho|$ in this relationship can be inverted to give a bilinear transformation between reflection and *VSWR*

$$|\rho| = \frac{VSWR - 1}{VSWR + 1} \qquad (25.14)$$

Some useful identities between the *VSWR* and the reflection coefficient are

$$\rho = 0 \quad VSWR = 1 \qquad (25.15)$$

$$\rho = \pm 1 \quad VSWR = \infty \qquad (25.16)$$

It is separately recalled for completeness sake that ρ is in general a complex quantity specified by

$$\rho = \frac{Z_L - Z_0}{Z_L + Z_0} \qquad (25.17)$$

The relationship between the *VSWR* and the *RL*.(dB) is

$$RL \text{ (dB)} = 10 \log_{10}\left(\frac{VSWR - 1}{VSWR + 1}\right) \qquad (25.18)$$

The corresponding relationship between the *RL*.(dB) and the reflection coefficient is given by equation (25.1)

25.4 Summation of Reflection Coefficients

In order to proceed with a calculation it is necessary to establish the rules whereby periodic reflection coefficients along a uniform transmission line combine. It is also necessary to establish those met in forming the resultant *VSWR* of two discontinuities in terms of the individual ones. A scrutiny of the first problem indicates that the maximum and minimum values of the overall reflection coefficient due to two separate small discontinuities are merely the sum and differences between the two.

The total reflection coefficient at the input plane of one of two discrete discontinuities spaced by ϕ radians can be shown to be given in terms of the individual ρ_1 and ρ_2 by

$$\rho_t = \frac{\rho_1 + \rho_2 \exp(-j2\phi)}{1 + \rho_1\rho_2 \exp(-j2\phi)} \qquad (25.19)$$

If the product $\rho_1\rho_2$ of the two reflection coefficients is small compared to unity then

$$\rho_t \approx \rho_1 + \rho_2 \exp(-j2\phi) \qquad (25.20)$$

The maximum coefficient is obtained when the spacing between the discontinuities is

$$\theta_1 - \theta_2 - 2\phi = 0, 2\pi, \text{ etc} \qquad (25.21)$$

This gives

$$\rho_{max} \approx |\rho_1| + |\rho_2| \qquad (25.22)$$

The minimum one coincides with

$$\theta_1 - \theta_2 - 2\phi = \pi, 3\pi, \text{ etc} \qquad (25.23)$$

This gives

$$\rho_{min} \approx |\rho_1| - |\rho_2| \qquad (25.24)$$

θ_1 and θ_2 are the phase angles of ρ_1 and ρ_2 respectively.

The maximum and minimum values of the net reflection coefficient are therefore the sum and differences of the individual ones respectively.

25.5 Summation of *VSWRs*

Since the specifications of a typical component is usually stated in terms of its *VSWR* or return loss rather than in terms of its reflection coefficient it is desirable to directly work with these quantities. In order to establish upper and lower bounds on the *VSWR* of two discontinuities in terms of the individual ones it is first necessary to reduce these to equivalent reflection coefficients by having recourse to the appropriate bilinear transformation between the two. The corresponding reflection coefficients are then combined by having recourse to the upper and lower bounds for which rules have already been established as a preamble to calculating the resultant *VSWR* by once more having recourse to the bilinear transformation between the two. This indicates that the maximum value of the *VSWR* is equal to the product of the individual ones and that the minimum value

25.5 SUMMATION OF VSWRS

is equal to the ratio of the two. The derivations of these two conditions begin by recalling to the bilinear transformations between reflection and voltage standing wave ratio in equations (25.13) and (25.14).

It continues by expanding S in the vicinity of unity. This gives

$$S \approx 1 + 2\rho \qquad (25.25)$$

and

$$|\rho| \approx \left(\frac{S-1}{2}\right) \qquad (25.26)$$

The derivation of the required result now starts by forming $|P_1|$ and $|P_2|$ from a specification of S_1 and S_2.

$$|\rho_1| \approx \left(\frac{S_1 - 1}{2}\right) \qquad (25.27a)$$

$$|\rho_2| \approx \left(\frac{S_2 - 1}{2}\right) \qquad (25.27b)$$

The maximum and minimum values of the reflection coefficient of two typical discontinuities are then given by

$$|\rho_{max}| \approx |\rho_1| + |\rho_2| \qquad (25.28a)$$

$$|\rho_{min}| \approx |\rho_1| - |\rho_2| \qquad (25.28b)$$

The corresponding resultant voltage standing wave ratios are now deduced by having recourse to the appropriate bilinear transformation

$$S_{max} \approx 1 + 2|\rho_{max}| \qquad (25.29a)$$

$$S_{min} \approx 1 + 2|\rho_{min}| \qquad (25.29b)$$

If the ratio S_1/S_2 or S_2/S_1 and the product $S_1 S_2$ are formed then it may also be shown, by having recourse to the approximate bilinear transformation in equation (25.25), that

$$S_{max} \approx S_1 S_2 \qquad (25.30a)$$

$$S_{min} \approx S_1/S_2 \text{ or } S_2/S_1, \ S_{min} \geq 1 \qquad (25.30b)$$

The upper and lower bounds of the *VSWR* are therefore the product and the ratio of the individual values respectively.

If this derivation is now repeated by having recourse to the exact upper and lower bounds on the reflection coefficient then it may be readily demonstrated that the above rules are applicable to any values of S_1 and S_2.

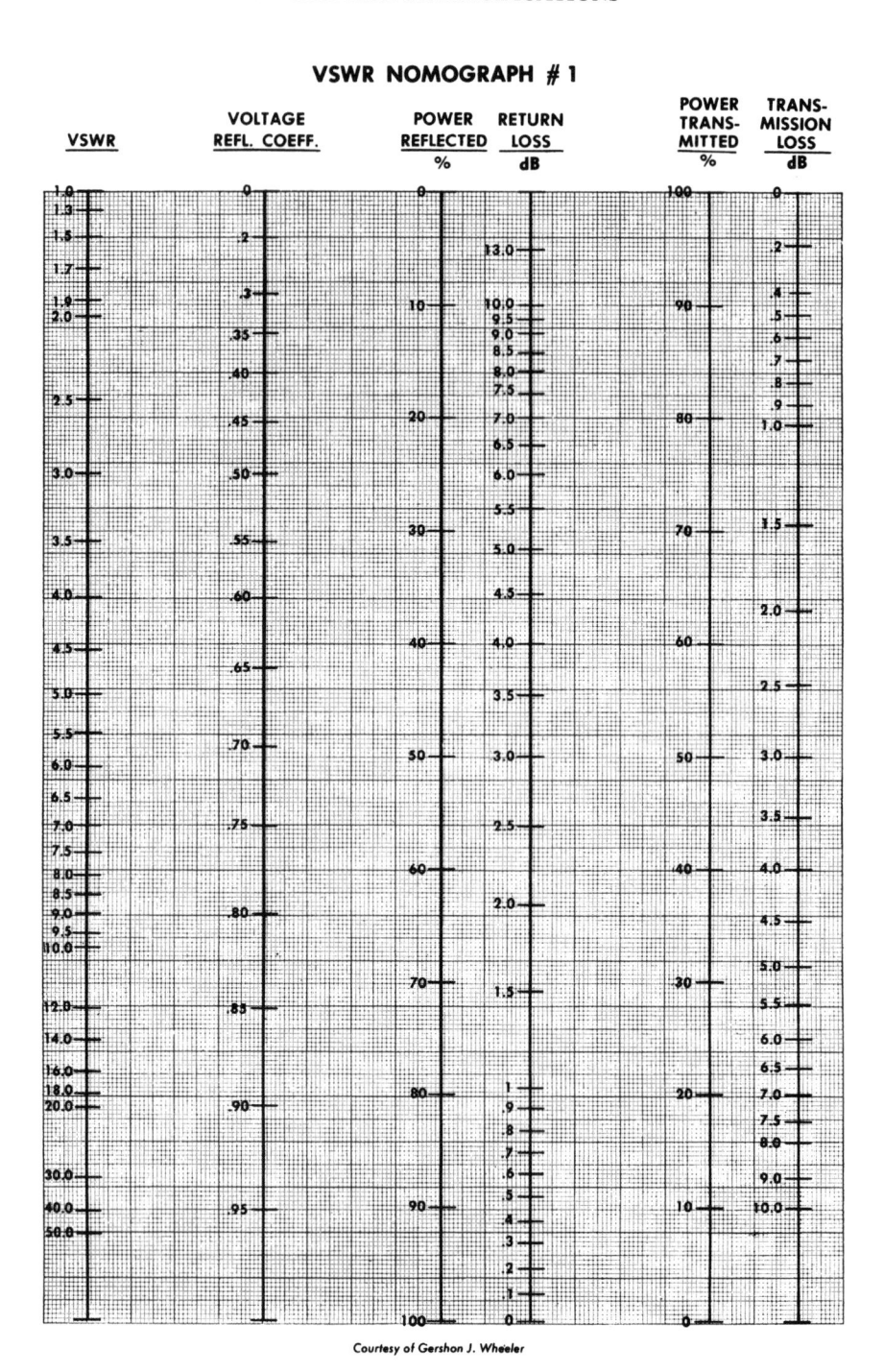

Figure 25.4 Nomogram of transmission variables on 2-port network

25.5 SUMMATION OF VSWRS

VSWR NOMOGRAPH #2

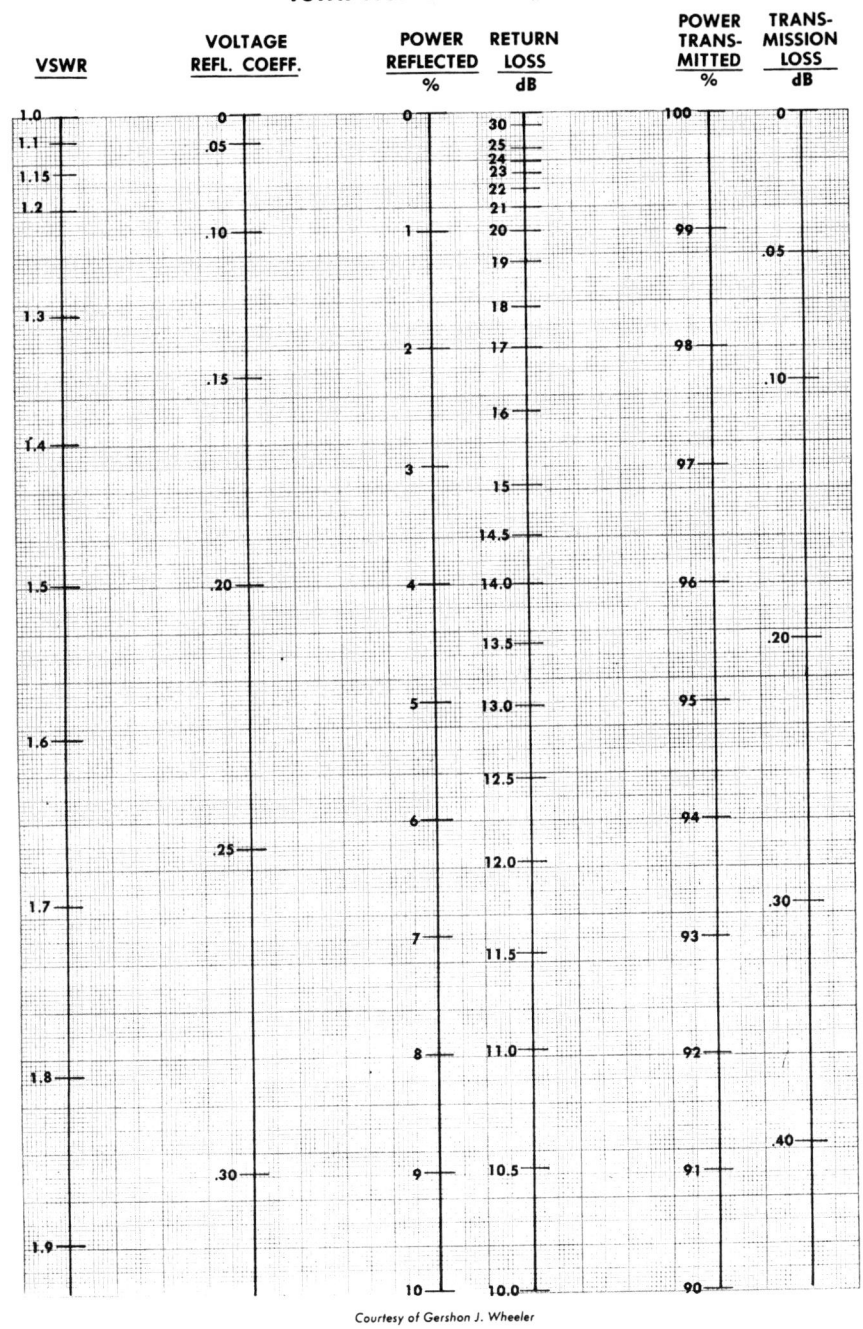

Figure 25.4 *cont.*

Taking $S_1 = 1.07$ and $S_2 = 1.11$ by way of an example gives $\rho_1 = 0.035$ and $\rho_2 = 0.055$, so that $\rho_{max} \approx 0.09$ and $\rho_{min} \approx 0,02$. The corresponding lower and upper bounds on the return loss are

$$RL_{min} \text{ (dB)} = 20.91 \text{ dB}$$

$$RL_{max} \text{ (dB)} = 33.98 \text{ dB}$$

It is also possible to separately evaluate S_{max} and S_{min} at port-2 by making use of the bilinear transformation between the reflection coefficient and the voltage standing wave ratio. This gives, $S_{max} \approx 1.18$ and $S_{min} \approx 1.04$. These two conditions are in agreement with the results obtained by taking the product and ratio of S_1 and S_2 respectively. Figure 25.4 gives a useful monogram between the transmission variables of a 2-port network.

25.6 Isolator Configuration

The most common application of the 3-port junction circulator is in the isolator configuration obtained by terminating port-3 by a matched load. This arrangement is shown schematically in Figure 25.5(a). The isolator is the most commonly used of all microwave ferrite devices. It is a passive nonreciprocal 2-port transmission line which has no or little attenuation in one direction of propagation and finite attenuation in the other. In the terminated circulator configuration the signal in the reverse direction is absorbed in the load placed at the third port of the device. One advantage of this isolator is that the average power at which it can operate is fixed by the rating of the load connected at the third port rather than by that of the junction itself. In such an isolator the input *VSWR* is determined by the isolation of the device and the *VSWR* of the load.

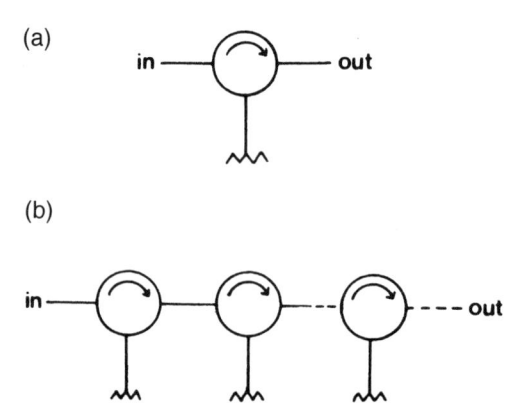

Figure 25.5 (a) Isolator configuration obtained by terminating one port of 3-port circulator. (b) Tandem connection of terminated circulators to form high-ratio isolator

One way to increase the isolation of the single junction is to use a number of these in tandem. A schematic representation of an m-port junction is illustrated in Figure 25.5(b). A typical application of the ferrite isolator is as a buffer stage between a microwave tube, such as a magnetron of klystron, and the rest of the microwave system. Such a buffer stage prevents frequency pulling in these tubes due to long line effects since the tube essentially works into a constant impedance.

The input *VSWR* or *RL* at port-1 or -3 of a circulator when used as an isolator with a finite load at port-2 and a termination at port-3 is therefore of some interest in the layout of practical hardware. This topic is discussed in some detail in the next section.

25.7 Specification of 3-port circulators with non-ideal loads

The *VSWR* at port-1 of a 3-port circulator is in practice not only dependent upon its specification but also upon the load condition at port-2 and the specification of the termination at port-3. While an exact formulation of the problem is in principle available it is too complicated for engineering practice.

Upper and lower bounds to the problem in question may be derived by disregarding any reflection at port-1 in constructing the incident wave at port-2 and by neglecting any secondary reflections at port-1. Since practical components are usually described in terms of a *VSWR* specification this is the notation adopted here. The development introduced here retains the original assumption whereby the incident wave at port-2 is taken as unity instead of $\sqrt{1 - \rho_C^2}$ and whereby any secondary reflections at port-1 are neglected.

The required derivation begins with the rules governing the combinations of two discrete *VSWR*'s S_1 and S_2 associated with two neighbouring discontinuities. The exact upper and lower bounds on the resultant *VSWR*'s are defined in terms of the individual ones by equations (25.29a) and (25.29b).

If the upper bound on the *VSWR* at port-1 is formulated to start with then the development begins by forming the worst case network parameters at port-2

$$S^{(2)}_{\max} = S_C S_A \tag{25.31}$$

$$\rho^{(2)}_{\max} = \frac{S_C S_A - 1}{S_C S_A + 1} \tag{25.32}$$

The corresponding variables at port-3 are separately given by

$$S^{(3)}_{\max} = S_C S_L \tag{25.33}$$

$$\rho^{(3)}_{\max} = \frac{S_C S_L - 1}{S_C S_L + 1} \tag{25.34}$$

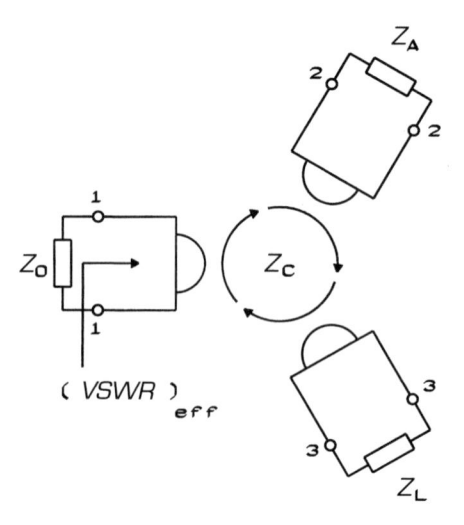

Figure 25.6 Schematic diagram of 3-port junction circulator with non-ideal loads

S_C is the *VSWR* of the circulator at any port, S_A is that of the antenna at port-2 and S_L is that of the load at port-3. A schematic diagram of the arrangement considered here is indicated in Figure 25.6.

The derivation now proceeds by forming the overall reflected wave produced at port-1 due to an incident wave at the same port

$$\rho_{\text{refl}} = 1 \cdot \rho_{\max}^{(2)} \cdot \rho_{\max}^{(3)} \tag{25.35}$$

The *VSWR* at the input port corresponding to this reflection coefficient is given by

$$S_{\text{refl}} = \frac{1 + \rho_{\max}^{(2)} \rho_{\max}^{(3)}}{1 - \rho_{\max}^{(2)} \rho_{\max}^{(3)}} \tag{25.36}$$

Writing this quantity in terms of the original variables gives

$$S_{\text{refl}} = \frac{(S_C S_A)(S_C S_L) + 1}{(S_C S_A) + (S_C S_L)} \tag{25.37}$$

The worst case effective *VSWR* at port-1 is now obtained by forming the product of the *VSWR* due to the reflected wave at port-1 and that of the circulator.

$$S_{\text{eff}} = S_C S_{\text{refl}} \tag{25.38}$$

The required result is

25.7 SPECIFICATION OF 3-PORT CIRCULATORS

$$S_{\text{eff}} = \frac{(S_C S_A)(S_C S_L) + 1}{S_A + S_L} \tag{25.39}$$

This equation may also be readily solved for either S_A S_C or S_L in terms of S_{eff}

$$S_A = \frac{-1 + S_{\text{eff}} S_L}{-S_{\text{eff}} + S_C^2 S_L} \tag{25.40}$$

$$S_L = \frac{-1 + S_A S_{\text{eff}}}{-S_{\text{eff}} + S_C^2 S_A} \tag{25.41}$$

and

$$S_C = \left[\frac{S_{\text{eff}}(S_A + S_L) - 1}{S_A S_L}\right]^{1/2} \tag{25.42}$$

The derivation of the best possible situation at port-1 may be deduced without ado by replacing $S_{\text{max}}^{(2)}$ and $S_{\text{max}}^{(3)}$ by

$$S_{\text{min}}^{(2)} = S_A/S_C, \quad S_A \geq S_C \tag{25.43}$$

$$S_{\text{min}}^{(3)} = S_C/S_L, \quad S_C \geq S_L \tag{25.44}$$

and forming the ratio S_{refl}/S_C instead of the product $S_{\text{refl}} S_C$. Taking the effective VSWR, that of the circulator and those of the load and termination's one at time as the dependent variable gives

$$S_{\text{eff}} = \frac{S_C^2 + S_A S_L}{S_A + S_L}, \quad S_{\text{refl}} \geq S_C \tag{25.45}$$

$$S_A = \frac{S_C^2 - S_L S_{\text{eff}}}{S_{\text{eff}} - S_L}, \quad S_{\text{refl}} \geq S_C \tag{25.46}$$

$$S_L = \frac{S_A S_{\text{eff}} - S_C}{S_A - S_{\text{eff}}}, \quad S_{\text{refl}} \geq S_C \tag{25.47}$$

$$S_C = [S_{\text{eff}}(S_A + S_L) - S_A S_L]^{1/2}, \quad S_{\text{refl}} \geq S_C \tag{25.48}$$

Figure 25.7 indicates the relationship between the best and worst bounds on the VSWR at port-1 and the antenna load at port-2 for two different circulator specifications and one typical load specification. The lower bounds in these curves are left blank between the origin and $S_A < S_C$ in keeping with the condition in (25.43). The robustness of the approximations utilized in this work has been verified by

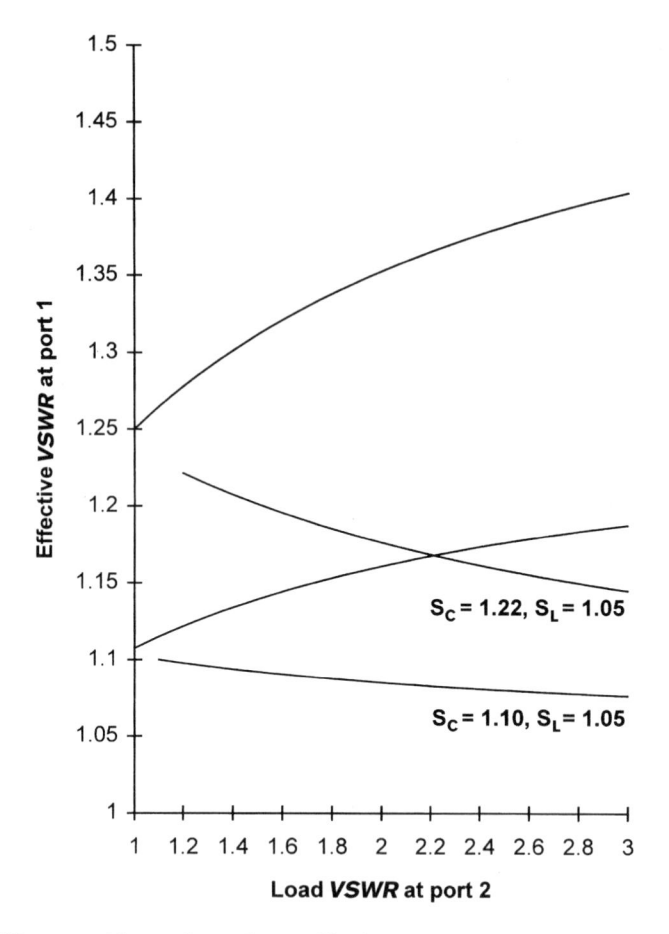

Figure 25.7 Upper and lower bounds on effective *VSWR* of load at port-1 versus VSWR at port-2 for different circulator and termination specifications

introducing either the condition $S_A = S_C$ or $S_L = S_C$ in the lower bound relationship. A scrutiny of each case one at a time indicates that S_{eff} equals S_C in keeping with the exact result.

$$S_{\text{eff}} = S_C. \tag{25.49}$$

The lower bound obtained here is in keeping with the definition of the complex gyrator circuit of this sort of junction when either port-2 or port-3 is terminated in its complex conjugate load.

Taking the particular case for which $S_A = S_L = S_C = 1.22$ indicates that S_{eff} is bracketed between 1.22 and 1.32.

25.8 The General Problem

One way to visualise the actual problem is to replace it by an ideal circulator connected to 2-port reciprocal networks which embody the port conditions. The augmented scattering matrix of this new circuit has no longer the 3-fold symmetry of the original network as is readily understood. The general result for this sort of arrangement has been noted on Chapter 24. It is reproduced here for the sake of completeness

$$\rho_{in} = S_{11} + \frac{(\rho_2 + \rho_3)A + \rho_2\rho_3(B - 2S_{11}A)}{1 - (\rho_2 + \rho_3)S_{11} + \rho_2\rho_3(S_{11}^2 - A)} \tag{25.50}$$

where $A = S_{21}S_{31}$ and $B = S_{21}^3 + S_{13}^3$. S_{11}, S_{21} and S_{31} are the finite scattering parameters of the circulator under ideal load conditions, ρ_2 and ρ_3 are the reflection coefficients of the actual loads at ports 2 and 3.

25.9 Data Sheet

While transmission and reflection are important quantities in the description of any circulator other requirements also enter into its specifications. These include temperature, stability, peak and average power ratings, insertion phase shift, third order intermodulation products and size to mention but a few considerations. Figure 25.8 depicts a typical commercial data sheet of one communication circulator. An insertion of 0.08 dB is not unusual in the design of a high-quality device.

25.10 Insertion Loss of 3-port Circulator with One Port Terminated in a Variable Short Circuit

The insertion loss of a 3-port circulator between ports 1 and 3 with port-2 terminated in a variable short circuit is of some interest in a number of situations. One such application is where a circulator is used to connect a filter and an equaliser. Another is in phasors using circulators with one port terminated in a *p-i-n* diode switch or a section of a uniform short-circuited line as in the situation here. An interesting feature of this arrangement (Figure 25.9) is that the insertion loss between ports 1 and 3 of a 3-port circulator with port-2 terminated in a short circuit varies between one and three times the single-path loss. This result is summarised by the following relationships which may be derived by omitting the dissipation in the in-phase eigen-network and assuming that those of the counter-rotating ones are equal:

$$L_{min} \approx L_0\left[1 + \left(1 - \frac{L_0}{2}\right)^2\right] - L_0\left(1 - \frac{L_0}{2}\right) \tag{25.51}$$

$$L_{max} \approx L_0\left[1 + \left(1 - \frac{L_0}{2}\right)^2\right] + L_0\left(1 - \frac{L_0}{2}\right) \tag{25.52}$$

EUROWAVE Data Sheet

X-Band Waveguide Circulators and Isolators

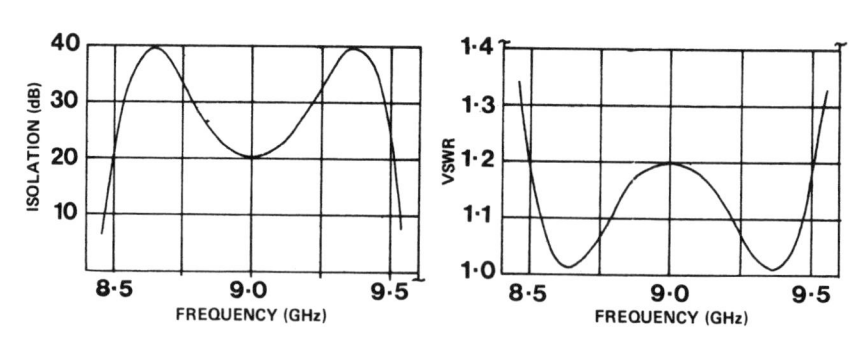

Model C859630

Frequency:	8.5–9.6 GHz
VSWR:	1.07 max
Isolation:	30dB min
Insertion loss:	0.20dB max
Size LxWxH:	55 x 55 x 32.5mm
Waveguide:	WR90 (R100)
Weight:	125g
Temperature	+10°C to +70°C

Model C859625

Frequency:	8.5–9.6 GHz
VSWR:	1.15 max
Isolation:	25dB min
Insertion loss:	0.25dB max
Size LxWxH:	55 x 55 x 32.5mm
Waveguide:	WR90 (R100)
Weight:	125g
Temperature	+10°C to +70°C

Model C859620

Frequency:	8.5–9.6 GHz
VSWR:	1.25 max
Isolation:	20dB min
Insertion loss:	0.35dB max
Size LxWxH:	55 x 55 x 32.5mm
Waveguide	WR90 (R100)
Weight:	125g
Temperature:	+10°C to +70°C

Eurowave, 4 Easter Belmont Rd, Edinburgh, Scotland

March 1973

Figure 25.8 Commercial data sheet

25.10 INSERTION LOSS OF 3-PORT CIRCULATOR

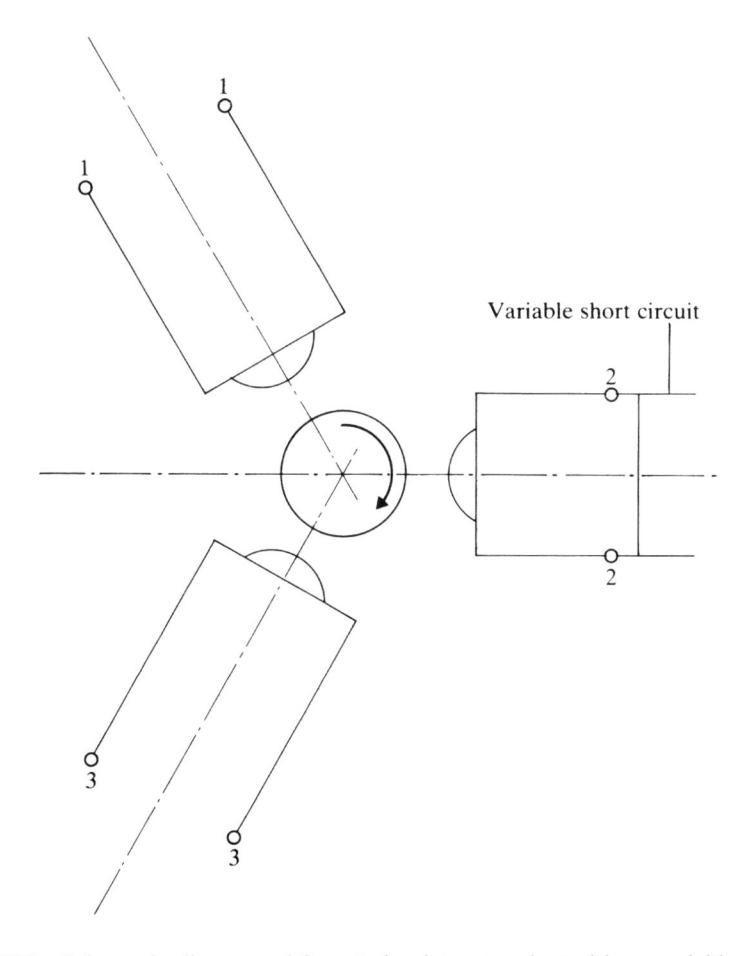

Figure 25.9 Schematic diagram of 3-port circulator terminated in a variable short circuit at port-2

where L_0 is the transmission parameter between ports 1 and 2 and L_{max} and L_{min} are the maximum and minimum values of this parameter between ports 1 and 3 as the short-circuit piston is varied at port-2.

For instance, the result in decibels when $L_0 = 0.10$ ($L_0(dB) = 0.50$ dB) is $L_{min} = 0.475$ dB and $L_{max} = 1.445$ dB. The following experimental results have been obtained for a waveguide circulator consisting of a simple ferrite post at the junction of three rectangular waveguides: $L_0 = 0.36$ dB; $L_{min} = 0.38$ dB; and $L_{max} = 1.14$ dB.

The result indicates that the insertion loss between ports 1 and 3 of a 3-port junction circulator terminated in a variable short circuit at port-2 varies by approximately between one and three times that between adjacent ports. A phenomenological explanation of this behaviour may be understood by taking a combination of standing wave patterns in an ideal circulator. The minimum attenuation condition is obtained by superimposing two such solutions in

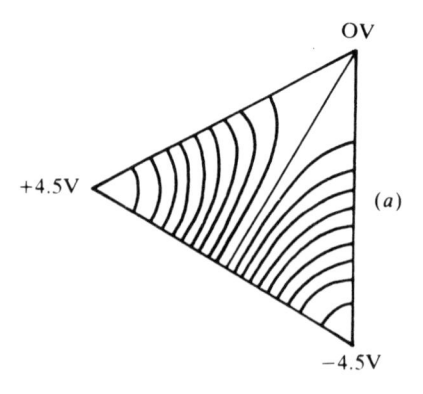

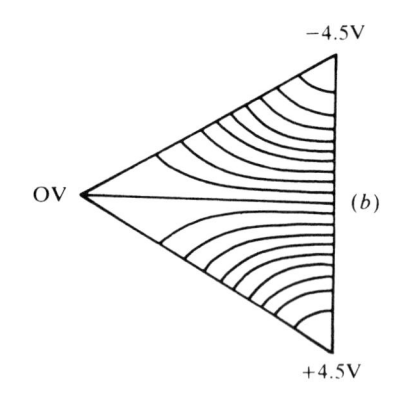

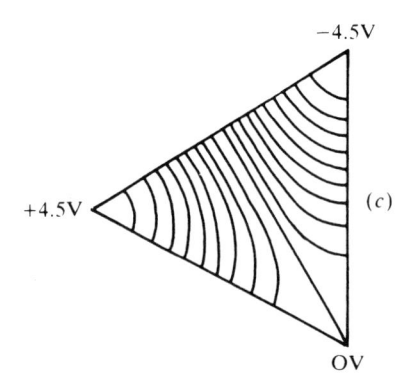

Figure 25.10 Equipotential lines in an ideal circulator using a triangular resonator. (a) Single input at port-2 in phase to that at port-1. (b) Single input at port- 2 in phase to that at port-1. (c) Equal-amplitude in-phase inputs at ports 1 and 2 (reproduced with permission, Helszajn, J. (1979) Standing-wave solution of 3-port junction circulator with 1-port terminated in a variable short circuit, *Proc. IEE, Part H (Microwaves, Optics and Acoustics)*, **126**, March)

25.10 INSERTION LOSS OF 3-PORT CIRCULATOR

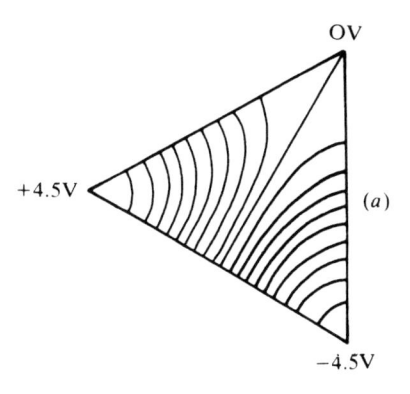

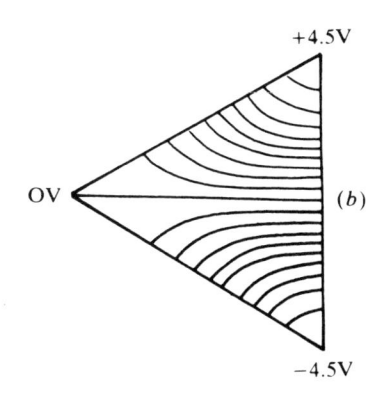

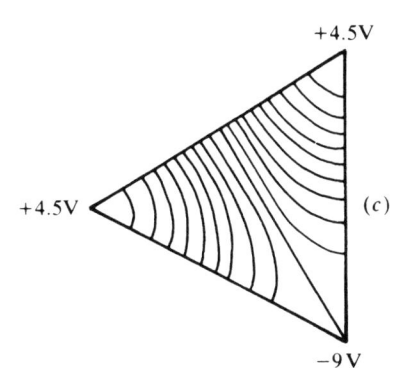

Figure 25.11 Equipotential lines is an ideal circulator using triangular resonators. (a) Single input at port 1. (b) Single input at port 2 in antiphase to that at port 2. (c) Equal-amplitude out-of-phase inputs at ports 1 and 2 (reproduced with permission, Helszajn, J. (1979) Standing-wave solution of 3-port junction circulator with 1-port terminated in a variable short circuit, *Proc. IEE, Part H (Microwaves, Optics and Acoustics)*, **126**, March)

antiphase. Such an arrangement leads to a reversal in the direction of circulation in the junction. The maximum attenuation is obtained by adding the two standing wave solutions in phase.

Figure 25.10(a) depicts the standing wave solution of an ideal circulator using a triangular resonator with a single input at port-1. Figure 25.10(b) illustrates a similar standing wave solution due to an incident travelling wave at port-2 in antiphase to that at port-1, the phase of the latter wave being set by the position of the short-circuit piston at port-2. Figure 25.10(c) indicates the standing wave pattern obtained by taking a linear combination of those in Figure 25.10(a) and (b).

It is observed that the lines of equipotential of the resultant pattern have the same amplitude as those of the original circuit, but are associated with a circulator solution that rotates in the opposite direction to that of the original one. Thus, the insertion loss of this arrangement is consistent with the single-path attenuation rather than the double-path loss obtained by assuming a simple sequence of travelling waves between ports 1 and 2 and 2 and 3.

Figure 25.11 summarizes the case where the incident standing waves are in phase at ports 1 and 2. If the dissipation within the junction is assumed to be proportional to the sum of the squares of the amplitudes at the corners of the resonator, it is evident that the overall dissipation is now three times that obtained with the two travelling waves out of phase 1 and 2, in agreement with equations (25.51) and (25.52).

It is observed that the output waves at port-3 in Figures 25.10 and 25.11 are out of phase as expected.

In the presence of circuit losses the following empirical relations apply:

$$L_{min} \approx 2(L_C + L_f) - L_f \qquad (25.53)$$

$$L_{max} \approx 2(L_C + L_f) - L_f \qquad (25.54)$$

where L_f is the single-path ferrite loss and L_C is the single-path circuit loss.

Such an arrangement may therefore be employed to separate ferrite and circuit losses in three-port junction circulators.

References and Further Reading

Bex, H. and Schwartz, E. (1971) Performance limitation of lossy circulators, *IEEE Trans. Microwave Theory Tech.*, **MTT-19**, 493–494.
Buffler, C. R. (1967) Nomogram for determining effective junction circulator VSWR, *Electronic Communicator*, **2**(6), 119.
Dye, N. E. (1963) Circulator mismatch curves, *Microwaves*.
Hagelin, S. (1969) Analysis of lossy symmetrical three-port networks with circulator properties, *IEEE Trans. Microwave Theory Tech.*, **MTT-17**(6), 328.
Helszajn, J. (1972) Dissipation and scattering matrices of lossy junctions, *IEEE Trans. Microwave Theory Tech.*, **MTT-20**, 779–782.

Helszajn, J., Riblet, G. P. and Mather, J. P. (1975) Insertion loss of 3-port circulator with one port terminated in variable short circuit, *IEEE Trans. Microwave Theory Tech.*, **MTT-23**, 926–927.

Helszajn, J. (1979) Standing-wave solution of 3-port junction circulator with 1-port terminated in a variable short circuit, *Proc. IEE, Part H (Microwaves, Optics and Acoustics)*, **126**.

Humphreys, B. L. and Davies, J. B. (1962) The synthesis of N-port circulators, *IRE Trans. Microwave Theory Tech.*, **MTT-10**, 551–554.

Linkhart, D. K. (1989) Microwave circulator design, Artech House Inc., Norwood, MA.

Solbach, K. (1982) Equivalent circuit representation for the E-plane circulator, *IEEE Trans. Microwave Theory Tech.*, **MTT-30**, 806–809.

26

GYROMAGNETIC EFFECT IN MAGNETIC INSULATOR

26.1 Introduction

The origins of the magnetic effects of magnetization in magnetic insulators are due to the effective current loops of electrons in atomic orbits, and from effects of electron spin and atomic nuclei (Figure 26.1). Each of the features produces a magnetic field which is equivalent to that arising from a magnetic dipole; the total magnetic moment is the vector sum of the individual moments. In ferromagnetic insulators the predominant effect is due to the electron spin. The motion of the total number of magnetic dipoles per unit volume due to electron spins in the presence of a direct magnetic field is described from the linearized equation of motion of the magnetization vector. The characteristic property of such a magnetized insulator at microwave frequencies is that while it has in general a scalar dielectric constant, it has a tensor permeability. Inside a ferrite medium Maxwell's equations must therefore be solved in conjunction with this tensor permeability. The presence of imaginary off-diagonal components having opposite signs in this tensor is the basis for a number of important nonreciprocal effects not usually encountered in a medium whose permeability is a scalar. An important property of a gyromagnetic medium is that whereas it is in general characterized by a tensor permeability it displays a scalar one with one value if the alternating magnetic field rotates in the same sense as the electron spin, another value if it rotates in the opposite sense and a value of unity if it is aligned with the axis of the electron spin. If the frequency of the r.f. magnetic field coincides with the natural procession frequency, the amplitude of the precession becomes particularly large, and the energy absorbed from the alternating magnetic field displays a maximum. Coupling into and out of the YIG system is particularly easy under this condition. The direction of the electronic procession depends on that of the steady magnetic field and its frequency is determined by the product of the gyromagnetic ratio and the amplitude of the direct magnetic field.

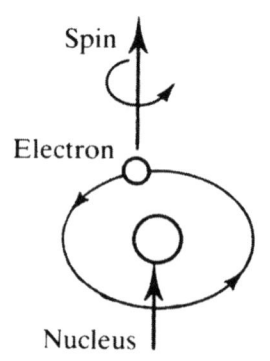

Figure 26.1 Spin motion in magnetic insulators

26.2 Equation of Motion of the Magnetization Vector

Much of the microscopic theory of microwave ferrite devices is based on the equation of motion of the magnetization vector. The derivation of this relationship starts by considering an elementary magnetic dipole having a dipole moment $\bar{\mu}$ placed in a direct magnetic field $\bar{H}_0$; the dipole moment is a measure of the magnetic effects of the electron spin and is orientated perpendicular to the plane

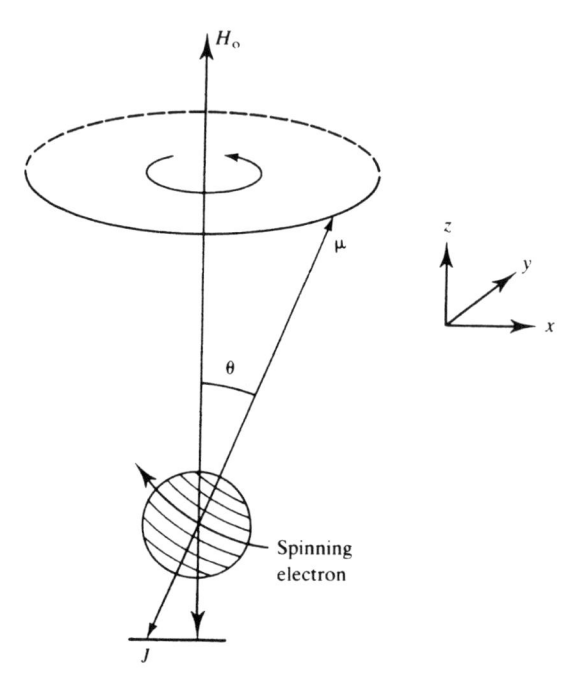

Figure 26.2 Magnetic moment precession about a static magnetic field

of the orbit. The magnetization (M_0) of a magnetic insulator is the total magnetic moment (μ) per unit volume. Associated with the magnetic dipole $\bar{\mu}$ there is an angular momentum $\bar{J}$ defined by

$$\bar{\mu} = \gamma \bar{J} \tag{26.1}$$

γ is the gyromagnetic ratio 2.21×10^5 (rad/sec)/(A/m)

Under equilibrium conditions, the dipole moment vector is oriented along the direction of $\bar{H}_0$ which is usually taken to be in the z-direction. If it is now tilted by a small external force so that it makes an angle θ with H_0 as indicated in Figure 26.2 then if the only field acting on $\bar{\mu}$ is $\bar{H}_0$ the torque (τ) exerted on it is

$$\tau = \mu \times \bar{H}_0 \tag{26.2}$$

Since the torque is the time rate of change of the angular momentum $\bar{J}$, it can also be written in terms of equation (26.1) as

$$\tau = -\frac{1}{\gamma} \frac{d\bar{\mu}}{dt} \tag{26.3}$$

The equation of motion of a single dipole is then defined by combining (26.2) and (26.3).

$$\frac{d\bar{\mu}}{dt} = -\gamma(\bar{\mu} \times \bar{H}_0) \tag{26.4}$$

If N is the number of unbalanced spins per unit volume then the total magnetic moment per unit volume M_0 is defined by

$$M_0 = N\mu \quad \text{(Tesla)} \tag{26.5}$$

and the equation of motion of the magnetization vector is given by

$$\frac{d\bar{M}_0}{dt} = -\gamma(\bar{M}_0 \times \bar{H}_0) \tag{26.6}$$

Much of the classical theory of microwave ferrites is based on the equation of motion of the magnetization vector described by the preceding equation.

26.3 Susceptibility Tensor in Infinite Medium

Because damping is present the amplitude of the precession decreases until the magnetization comes into line with the direct magnetic field. This damping is related to the magnetic losses of ferrite devices. The precession can, however, be maintained by superimposing a small r.f. magnetic field in the plane transverse to the steady magnetic field. The tensor susceptibility or permeability can be obtained from this arrangement.

In the most simple microwave case the total effective magnetic field in equation (26.6) consists of a direct magnetic field $\bar{H}_0$ an alternating magnetic field $\bar{h}$

$$\bar{H} = \bar{H}_0 + \bar{h} \qquad (26.7)$$

The total magnetization consists of the direct magnetization $\bar{M}_0$ and alternating magnetization $\bar{m}$

$$\bar{M} = \bar{M}_0 + \bar{m} \qquad (26.8)$$

If the direct magnetization and magnetic field are taken along the positive z-direction then

$$\bar{H}_0 = \begin{bmatrix} 0 \\ 0 \\ H_0 \end{bmatrix} \qquad (26.9)$$

$$\bar{M}_0 = \begin{bmatrix} 0 \\ 0 \\ M_0 \end{bmatrix} \qquad (26.10)$$

$$\bar{h}_0 = \begin{bmatrix} h_x \\ h_y \\ h_z \end{bmatrix} \qquad (26.11)$$

$$\bar{m} = \begin{bmatrix} m_x \\ m_y \\ m_z \end{bmatrix} \qquad (26.12)$$

Introducing these quantities into the equation of motion gives

$$\frac{dm_x}{dt} = -m_y \gamma (H_0 + h_z) + h_y \gamma (M_0 + m_z) \qquad (26.13)$$

$$\frac{dm_y}{dt} = -m_x \gamma (H_0 + h_z) + h_x \gamma (M_0 + m_z) \qquad (26.14)$$

$$\frac{dm_z}{dt} = -m_x \gamma h_y + m_y \gamma h_x \qquad (26.15)$$

26.3 SUSCEPTIBILITY TENSOR IN INFINITE MEDIUM

In the small signal approximation higher order terms of m and h are neglected. The result is

$$\frac{dm_x}{dt} = -m_y \gamma H_0 + h_y \gamma M_0 \tag{26.16}$$

$$\frac{dm_y}{dt} = -m_x \gamma H_0 + h_x \gamma M_0 \tag{26.17}$$

$$\frac{dm_z}{dt} = 0 \tag{26.18}$$

Re-arranging the preceding equations yields

$$\frac{d^2 m_x}{dt^2} + \omega_0^2 m_x = \mu_0 \omega_m \omega_0 h_x + \mu_0 \omega_m \frac{dh_y}{dt} \tag{26.19}$$

$$\frac{d^2 m_y}{dt^2} + \omega_0^2 m_y = \mu_0 \omega_m \frac{dh_x}{dt} + \mu_0 \omega_m \omega_0 h_y \tag{26.20}$$

$$m_z = 0 \tag{26.21}$$

where

$$\omega_m = \frac{\gamma M_0}{\mu_0} \tag{26.22}$$

$$\omega_0 = \gamma H_0 \tag{26.23}$$

If the time dependence of the alternating quantities is of the form $\exp(j\omega t)$ a susceptibility tensor [X] can be defined which relates the r.f. magnetization to the r.f. magnetic field intensity.

$$\overline{m} = \mu_0 [X] \overline{h} \tag{26.24}$$

where

$$[\chi] = \begin{bmatrix} \chi_{xx} & \chi_{xy} & 0 \\ \chi_{yx} & \chi_{yy} & 0 \\ 0 & 0 & 0 \end{bmatrix} \tag{26.25}$$

and

$$\chi_{xx} = \chi_{yy} = \frac{\omega_m \omega_0}{-\omega^2 + \omega_0^2} \tag{26.26}$$

$$-\chi_{yx} = \chi_{yx} = \frac{j\omega_m \omega}{-\omega^2 + \omega_0^2} \tag{26.27}$$

The components of the susceptibility tensor have a singularity at $\omega = \omega_0$. This denotes a resonance condition and is the principal feature used in the design of YIG filters and other resonance devices.

26.4 Damping

To stabilize the motion of the magnetization vector at the resonance a damping term must be introduced into the equation of motion. A form of phenomenological damping that is often used for engineering purposes is due to Gilbert. This is given in vector form by

$$\frac{d\overline{M}}{dt} \approx -\gamma (\overline{M} \times \overline{H}) - \frac{\alpha}{|M|} \left(\overline{M} \times \frac{d\overline{M}}{dt} \right) \tag{26.28}$$

and is illustrated in Figure 26.3.

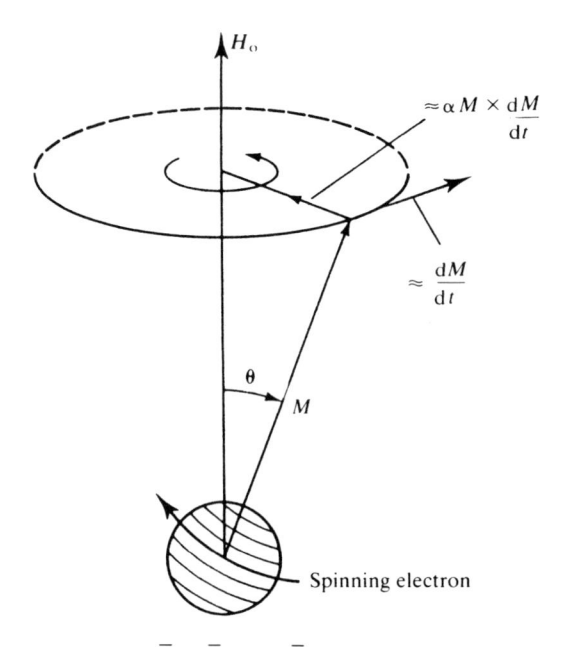

Figure 26.3 Relation between $\overline{M}$, $d\overline{M}$, and $\overline{M} \times dM/dt$. The loss term tends to decrease and damp out the precession

26.4 DAMPING

The components of the tensor susceptibility based on the small signal solution of equation (26.28) are

$$\chi_{xx} = \chi_{yy} = \frac{\omega_m(\omega_0 + j\omega\alpha)}{-\omega^2 + (\omega_0 + j\omega\alpha)^2} \tag{26.29}$$

$$-\chi_{yx} = \chi_{xy} = \frac{j\omega_m\omega}{-\omega^2 + (\omega_0 + j\omega\alpha)^2} \tag{26.30}$$

The only difference between the quantities in equations (26.26) and (26.27) and those in (26.29) and (26.30) is that an imaginary frequency term ($j\omega\alpha$) has been added to the resonant frequency (ω_0). The damping term can therefore always be introduced by replacing ω_0 by $\omega_0 + j\omega\alpha$ in the loss free components. The real and imaginary parts of the entries of the susceptibility tensor are defined by

$$\chi_{xx} = \chi'_{xx} + j\chi''_{xx} \tag{26.31}$$

$$\chi_{xy} = j(\chi'_{xy} + j\chi''_{xy}) \tag{26.32}$$

where

$$\chi'_{xx} = \frac{\omega_m\omega_0(\omega_0^2 - \omega^2) + \omega_m\omega_0\omega^2\alpha^2}{[\omega_0^2 - \omega^2(1 + \alpha^2)]^2 + 4\omega_0^2\omega^2\alpha^2} \tag{26.33a}$$

$$\chi''_{xx} = \frac{-\omega_m\omega\alpha[\omega_0^2 - \omega^2(1 + \alpha^2)]}{[\omega_0^2 - \omega^2(1 + \alpha^2)]^2 + 4\omega_0^2\omega^2\alpha^2} \tag{26.33b}$$

$$\chi'_{xx} = \frac{-\omega_m[\omega_0^2 - \omega^2(1 + \alpha^2)]}{[\omega_0^2 - \omega^2(1 + \alpha^2)]^2 + 4\omega_0^2\omega^2\alpha^2} \tag{26.33c}$$

$$\chi''_{xx} = \frac{-2\omega_m\omega_0\omega^2\alpha^2}{[\omega_0^2 - \omega^2(1 + \alpha^2)]^2 + 4\omega_0^2\omega^2\alpha^2} \tag{26.33d}$$

The relationships between the real and imaginary parts of the susceptibility tensor and ω_m/ω, ω_0/ω and α are plotted in Figures 26.4 to 26.7.

The phenomenological damping factor α can be related to the linewidth of ferrites in the following way. This quantity is usually defined as the difference between the values of the magnetic filed at a constant frequency where χ''_{xx}, the imaginary parts of the diagonal component χ_{xx} of the susceptibility tensor, attains a value that is half of its value at resonance (Figure 26.8). In view of this definition, equation (26.33a) leads to the following equality.

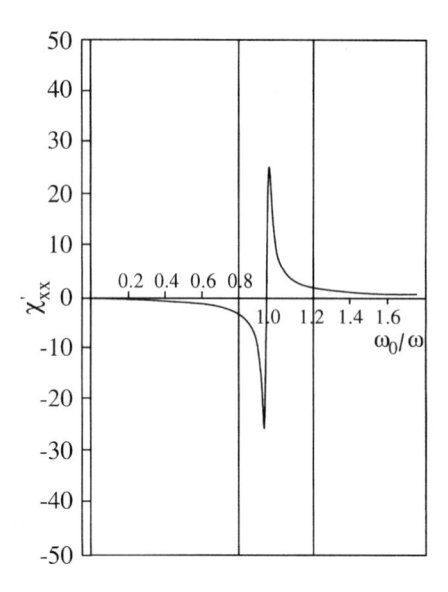

Figure 26.4 Real part of χ_{xx} component for $\omega_m/\omega = 1$ and $\alpha = 0.001$

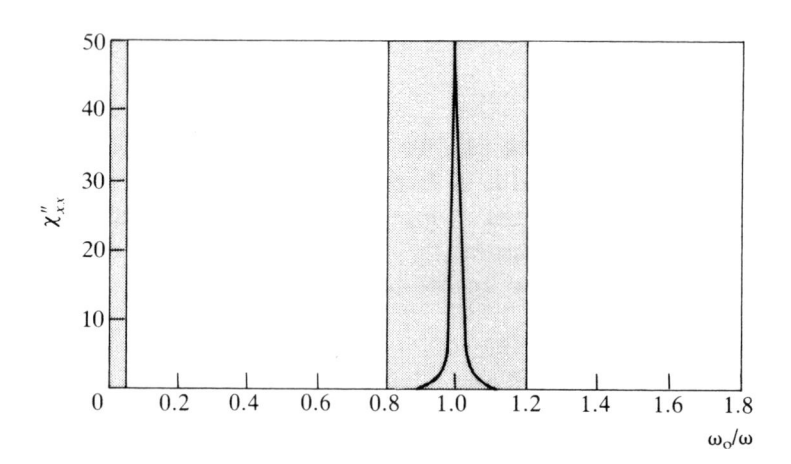

Figure 26.5 Imaginary part of χ_{xx} component for $\omega_m/\omega = 1$ and $\alpha = 0.001$

26.4 DAMPING

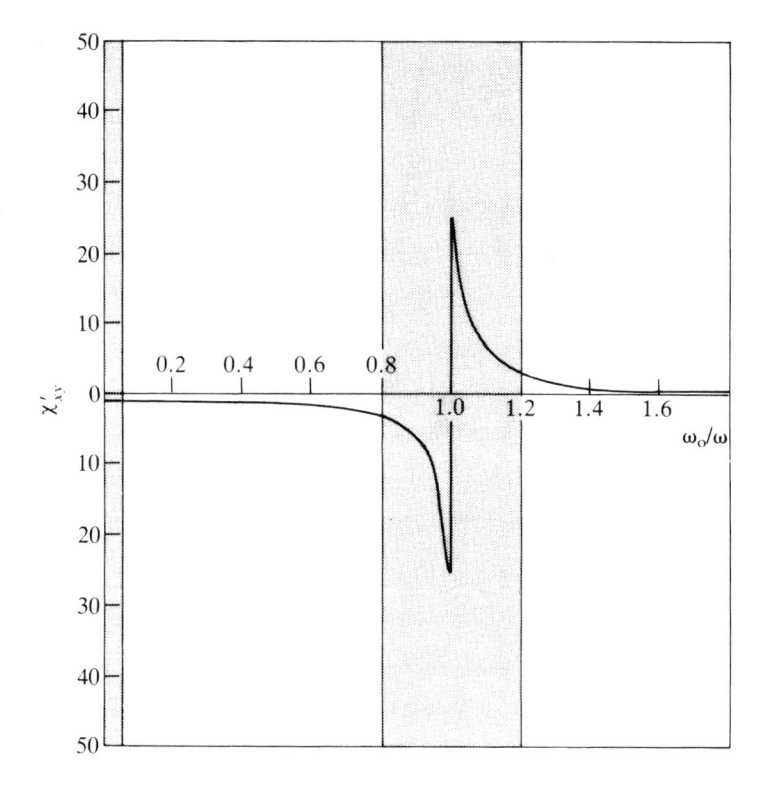

Figure 26.6 Real part of χ_{xx} component for $\omega_m/\omega = 1$ for $\alpha = 0.001$

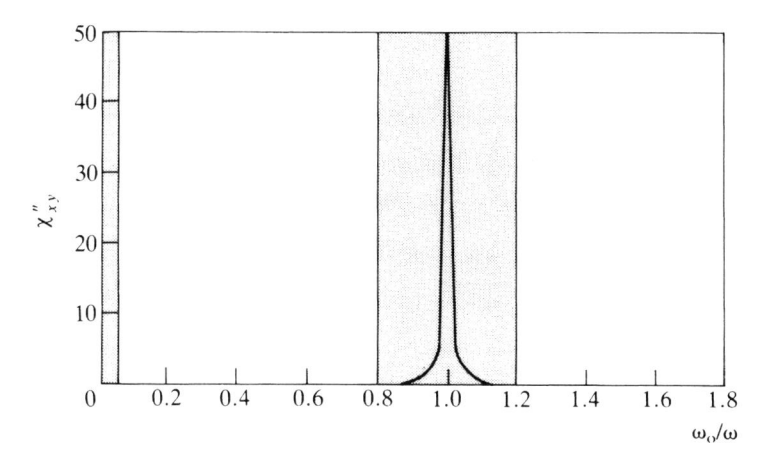

Figure 26.7 Imaginary part of χ_{xx} component for $\omega_m/\omega = 1$ and $\alpha = 0.001$

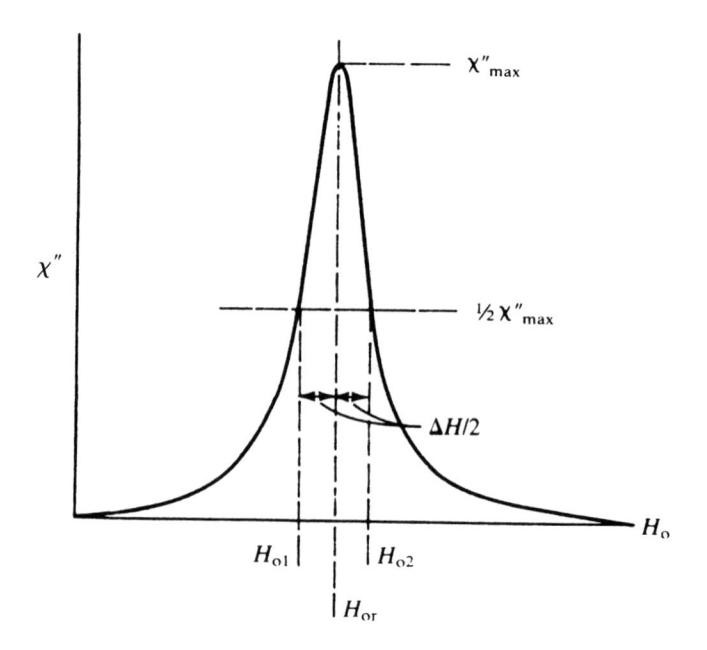

Figure 26.8 Linewidth of ferrites

$$\frac{\omega_m \omega \alpha (\omega_{01,2}^2 + \omega^2)}{(\omega_{01,2}^2 - \omega^2) + 4\omega_{01,2}^2 \omega^2 \alpha^2} = \frac{1}{2} \left| \frac{\omega_m}{2\omega\alpha} \right| \tag{26.34}$$

Rearranging this equation gives a quadratic in $\omega_{0,1}^2$

$$\omega_{01,2}^4 - 2\omega^2 \omega_{01,2}^2 + \omega^4(1 - 4\alpha^2) = 0 \tag{26.35}$$

The roots of this equation are given by

$$\omega_{01,2} - \omega(1 \pm 2\alpha)^{1/2} + \omega \pm \omega\alpha \tag{26.36}$$

Hence

$$\omega\alpha = \frac{\omega_{02} - \omega_{01}}{2} \tag{26.37}$$

If the linewidth of the material is now defined by

$$\Delta H = \frac{\omega_{02} - \omega_{01}}{\gamma} \tag{26.38}$$

then the phenomenological damping factor (α) is related to the linewidth (ΔH) by

$$\alpha \omega = \frac{\gamma \Delta H}{2} \tag{26.39}$$

The linewidth quoted by the material makers therefore provides a useful description of magnetic losses in ferrites or garnets at resonance, although it is noted that this linewidth is not too worthwhile outside the skirts of the main resonance.

26.5 Scalar Susceptibilities

The relationship between $\bar{m}$ and $\bar{h}$ is a scalar quantity provided $\bar{h}$ is adjusted to correspond to one of the normal modes of the system. These modes may be determined by first forming the eigenvalues (scalar susceptibilities) of the tensor susceptibility. The eigenvalue equation to be solved is

$$\chi \bar{H} = [X]\bar{H} \tag{26.40}$$

χ is the eigenvalue, $\bar{H}$ is an eigenvector and [X] is the susceptibility tensor. The components of the magnetic fields are proportional to the corresponding entries of the eigenvectors.

The characteristic equation associated with this equation has a nonvanishing value for $\bar{H}$ provided.

$$\begin{bmatrix} (\chi_{xx} - \chi) & \chi_{xy} & 0 \\ -\chi_{xy} & (\chi_{xx} - \chi) & 0 \\ 0 & 0 & -\chi \end{bmatrix} \tag{26.41}$$

Its three eigenvalues or roots are

$$\chi_1 = \chi_+ = \chi_{xx} - j\chi_{xy} \tag{26.42}$$

$$\chi_2 = \chi_- = \chi_{xx} + j\chi_{xy} \tag{26.43}$$

$$\chi_3 = 0 \tag{26.44}$$

The eigenvectors are now determined from a knowledge of the eigenvalues by substituting each one at a time into (26.40); each root corresponding to one of the eigenvectors. The result is

$$\bar{H}_1 = \bar{h}_+ = \frac{1}{\sqrt{2}} \begin{bmatrix} h_0 \\ -jh_0 \\ 0 \end{bmatrix} \tag{26.45}$$

$$\overline{H}_2 = \overline{h}_- = \frac{1}{\sqrt{2}}\begin{bmatrix} h_0 \\ +jh_0 \\ 0 \end{bmatrix} \tag{26.46}$$

$$\overline{H}_3 = \overline{h}_z = \begin{bmatrix} 0 \\ 0 \\ h_0 \end{bmatrix} \tag{26.47}$$

The magnetic fields described by the last three equations are orthonormal

$$\overline{H}_i \overline{H}_j^* = \delta_{ij} h_0^2 \tag{26.48}$$

δ_{ij} is the Kronecker delta

$$\delta_{ij} = 1 \quad \text{if } i = j$$

$$\delta_{ij} = 0 \quad \text{if } i \neq j$$

The eigenvectors $\overline{h}_+$ and $\overline{h}_-$ define clockwise and anti-clockwise circularly polarised r.f. fields in the transverse plane respectively. The eigenvector h_z describes an r.f. magnetic field parallel to the direct field; no gyromagnetic properties are exhibited by the ferrite in the latter situation.

The scalar susceptibilities $\chi_\pm$ are given in terms of the original variables by

$$\chi_\pm = \frac{\omega_m}{\omega_0 \mp \omega} \tag{26.49}$$

Damping may be introduced into the last relationship by adding an imaginary frequency to the resonant one in the usual way

$$\chi_\pm = \frac{\omega_m}{(\omega_0 + j\alpha\omega) \mp \omega} \tag{26.50}$$

The real and imaginary parts of the two scalar susceptibilities corresponding to the two possible circularly polarized waves are illustrated in Figure 26.9 in terms of ω_m/ω, ω_0/ω and α. The positive susceptibility has a singularity at $\omega = \omega_0$ and coincides with the circularly polarized wave which rotates in the same sense as the precessional motion of the magnetization vector. The negative susceptibility describes the response of the system to a circularly polarized field that rotates in the opposite direction. In this instance there is no singularity. This result is responsible for the non-reciprocal behaviour in microwave ferrite devices.

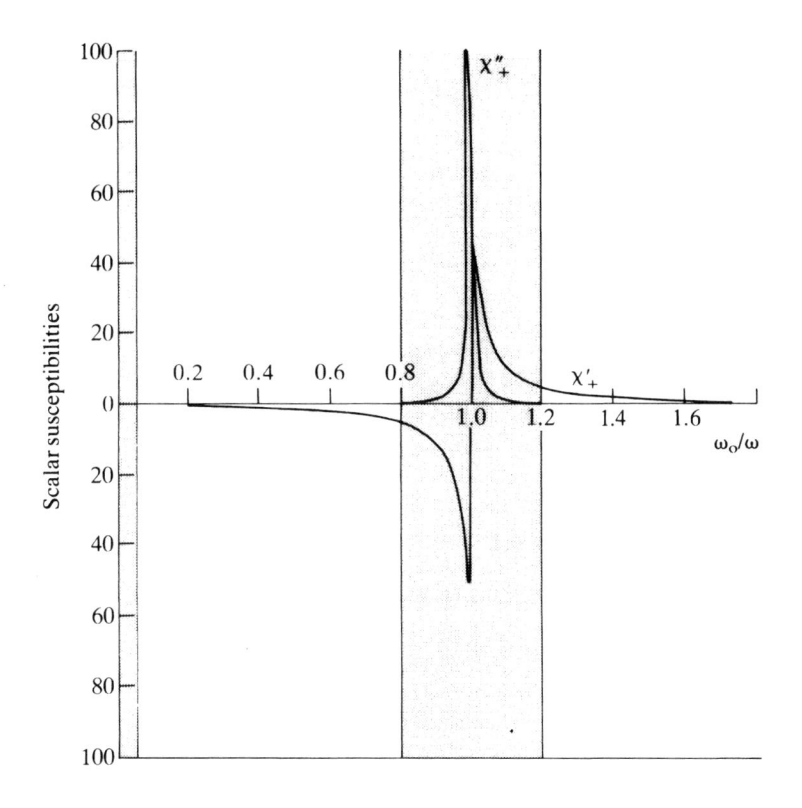

Figure 26.9 Real and imaginary parts of scalar susceptibilities with $\omega_m/\omega = 1$

26.6 Tensor Permeability

A tensor permeability can also be defined by relating the r.f. flux density $\bar{b}$ to the r.f. magnetic field intensity $\bar{h}$. Such a permeability is useful in problems involving Maxwell's equations.

$$\bar{b} = \mu_0 \bar{h} + \bar{m} \tag{26.51}$$

or

$$\bar{b} = \mu_0 [\mu_r] \bar{h} \tag{26.52}$$

where

$$[\mu_r] = [1] + [\chi] \tag{26.53}$$

Making use of (26.25) in the preceding equation leads to the required result

$$[\mu_r] = \begin{bmatrix} \mu & -j\kappa & 0 \\ j\kappa & \mu & 0 \\ 0 & 0 & 1 \end{bmatrix} \quad (26.54)$$

where

$$\mu = 1 + \chi_{xx} \quad (26.55)$$

$$-j\kappa = \chi_{xy} \quad (26.56)$$

The variation of κ with magnetic parameters is the same as that of χ_{xy} in Figure 26.6. The variation of μ with magnetic parameters may be determined from Figure 26.4 by noting that $\chi_{xy} = \mu - 1$.

The permeability is again a scalar quantity for the normal r.f. magnetic fields defined by equations 26.45 to 26.47.

$$\mu_1 = \mu_+ = \mu - \kappa \quad (26.57)$$

$$\mu_2 = \mu_- = \mu + \kappa \quad (26.58)$$

$$\mu_3 = \mu_z = 1 \quad (26.59)$$

The result is derived quite simply be determining the eigenvalues of the permeability tensor in equation (26.54). Damping can again be introduced into (26.57) through (26.59) by adding an imaginary frequency $j\alpha\omega$ to ω_0.

In keeping with the behaviour of the scalar susceptibilities the positive scalar permeability has a singularity at $\omega = \omega_0$, whereas the negative scalar permeability does not exhibit a resonance there.

If the direct field is perpendicular to the direction of propagation it is sometimes more convenient to take the direction of the direct field along the y-coordinate instead of the z-coordinate. Expanding the equation of motion with

$$\overline{H}_0 = \begin{bmatrix} 0 \\ H_0 \\ 0 \end{bmatrix} \quad (26.60)$$

$$\overline{h} = \begin{bmatrix} h_x \\ 0 \\ h_z \end{bmatrix} \quad (26.61)$$

$$\overline{M}_0 = \begin{bmatrix} 0 \\ M_0 \\ 0 \end{bmatrix} \quad (26.62)$$

26.6 TENSOR PERMEABILITY

$$\overline{m} = \begin{bmatrix} m_x \\ 0 \\ m_z \end{bmatrix} \qquad (26.63)$$

readily gives

$$[\chi] = \begin{bmatrix} \chi_{xx} & 0 & \chi_{xz} \\ 0 & 0 & 0 \\ \chi_{zx} & 0 & \chi_{zz} \end{bmatrix} \qquad (26.64)$$

and

$$[\mu] = \mu_0 \begin{bmatrix} \mu & 0 & -j\kappa \\ 0 & 1 & 0 \\ j\kappa & 0 & \mu \end{bmatrix} \qquad (26.65)$$

The eigenvalues of the latter tensor permeability are again described by

$$\mu_1 = \mu_+ = \mu - \kappa \qquad (26.66)$$

$$\mu_2 = \mu_- = \mu + \kappa \qquad (26.67)$$

$$\mu_3 = \mu_z = 1 \qquad (26.68)$$

but the counter-rotating eigenvectors are now in the x–z plane.

If the direct magnetic field is taken along the x-coordinate then

$$[\mu] = \begin{bmatrix} 1 & 0 & 0 \\ 0 & \mu & -j\kappa \\ 0 & j\kappa & \mu \end{bmatrix} \qquad (26.69)$$

and the eigenvectors are in the y–z plane.

If the direction of the direct field is taken along the negative z-direction then the signs of the off diagonal elements of the tensor permeabilities are interchanged and the eigenvalues become

$$\mu_1 = \mu_+ = \mu + \kappa \qquad (26.70)$$

$$\mu_2 = \mu_- = \mu - \kappa \qquad (26.71)$$

$$\mu_3 = \mu_z = 1 \qquad (26.72)$$

but the direction of rotation of the eigenvectors are unchanged. This situation is summarized in Figure 26.10.

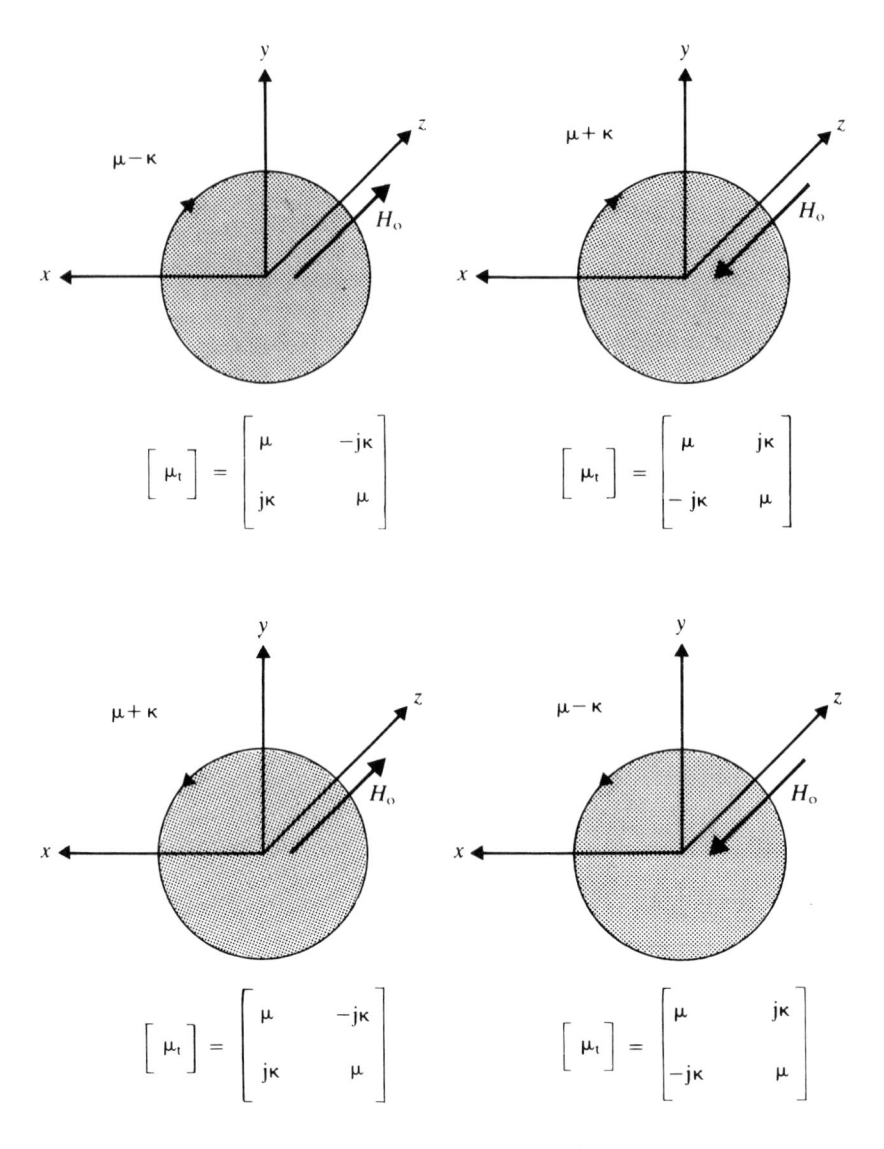

Figure 26.10 Eigensolutions of uniform mode

26.7 Susceptibility Tensor in Ellipsoidal Medium

In order to evaluate the magnetic field inside a finite ferrite medium it is necessary, in principle, to solve Maxwell's equations in conjunction with the Polder tensor permeability subject to the boundary conditions. However, this approach is not always simple or convenient. For an ellipsoidal sample in a uniform direct field the problem can be simplified by introducing demagnetizing fields, as is done in the direct magnetic field case.

26.7 SUSCEPTIBILITY TENSOR IN ELLIPSOIDAL MEDIUM

For the r.f. magnetic field the demagnetizing field is defined by

$$\bar{h}_{dem} = -\begin{bmatrix} N_x(m_x/\mu_0) \\ N_y(m_y/\mu_0) \\ N_z(m_z/\mu_0) \end{bmatrix} \tag{26.80}$$

and for the d.c. magnetic field by

$$\bar{H}_{dem} = -\begin{bmatrix} 0 \\ 0 \\ N_z M_0/\mu_0 \end{bmatrix} \tag{26.81}$$

These two equations can also be written as $\bar{h}_{dem} = -[N]\bar{m}/\mu_0$ and $\bar{H}_{dem} = -[N]\bar{M}_0/\mu_0$ where [N] is a diagonal demagnetizing tensor.

The relationship between the shape demagnetizing factors is

$$N_x + N_y + N_z = 1 \tag{26.82}$$

There are several shapes that are of particular interest in the discussion of microwave ferrite devices. For a sphere, the usual YIG filter resonator,

$$N_x = N_y = N_z = \frac{1}{3} \tag{26.83}$$

while for a long cylinder of radius R and length L,

$$N_z \approx 1 - \left[1 + \left(\frac{2R}{L}\right)^2\right]^{-1/2} \tag{26.84}$$

$$N_x \approx N_y = \frac{1}{2}(1 - N_z) \tag{26.85}$$

and for a thin disc

$$N_z \approx 1 - \left(\frac{L}{2R}\right)\left[1 + \left(\frac{L}{2R}\right)^2\right]^{-1/2} \tag{26.86}$$

$$N_x \approx N_y = \frac{1}{2}(1 - N_z) \tag{26.87}$$

Figure 26.11 gives the demagnetizing factors for some simple ferrite shapes.

26 GYROMAGNETIC EFFECT IN MAGNETIC INSULATOR

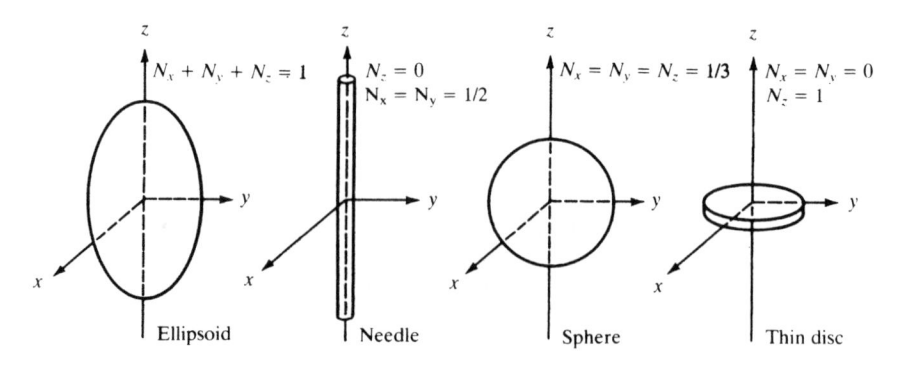

Figure 26.11 Demagnetizing factors

The small-signal approximation of the equation of motion in terms of the external fields is now given with

$$\bar{H} = \bar{H}_0 + \bar{H}_{\text{dem}} + \bar{h}^e + \bar{h}_{\text{dem}} \tag{26.88}$$

by

$$\frac{d^2 m_x}{dt^2} + \omega_r^2 m_x = \mu_0 \omega_m \omega_y h_x^e + \mu_0 \omega_m \frac{dh_y^e}{dt} \tag{26.89}$$

$$\frac{d^2 m_y}{dt^2} + \omega_r^2 m_y = \mu_0 \omega_m \frac{dh_x^e}{dt} + \mu_0 \omega_m \omega_x h_y^e \tag{26.90}$$

$$m_z = 0 \tag{26.91}$$

where

$$\omega_x = (\omega_0 - N_z \omega_m + N_x \omega_m) \tag{26.92}$$

$$\omega_y = (\omega_0 - N_z \omega_m + N_y \omega_m) \tag{26.93}$$

and

$$\omega_r = \sqrt{\omega_x \omega_y} \tag{26.94}$$

The last equation is the well-known Kittel resonance relationship for an ellipsoid. The singularity of the susceptibility components now occurs at $\omega = \omega_r$, and involves the demagnetizing factors. The corresponding external susceptibility tensor is defined by

26.7 SUSCEPTIBILITY TENSOR IN ELLIPSOIDAL MEDIUM

$$\chi^e_{xx} = e_0 \left(\frac{\omega_m \omega_r}{-\omega^2 + \omega_r^2} \right) \qquad (26.95)$$

$$\chi^e_{yy} = \frac{1}{e_0} \left(\frac{\omega_m \omega_r}{-\omega^2 + \omega_r^2} \right) \qquad (26.96)$$

$$-\chi^e_{yx} = \chi^e_{xy} \left(\frac{j\omega_m \omega_r}{-\omega^2 + \omega_r^2} \right) \qquad (26.97)$$

where

$$e_0 = \sqrt{\frac{\omega_y}{\omega_x}} \qquad (26.98)$$

e_0 is defined as the ellipticity of the normal modes of the uniform precession (Figure 26.12). For the infinite medium $e_0 = 1$ and $\omega_r = \omega_0$. The ellipticity is also unity whenever the transverse demagnetizing space factors are equal. The advantage of expressing the entries of the external susceptibility tensor in terms of the uniform mode ellipticity is that the quantities inside the brackets are the same as the corresponding elements of the internal susceptibility tensor, except that the resonant frequency is now specified by ω_r instead of ω_0. The presence of damping is again catered for by introducing an imaginary frequency term $j\omega\alpha$ to ω_x and ω_y. This is equivalent to introducing an imaginary frequency $j\omega\alpha$ to ω_r provided

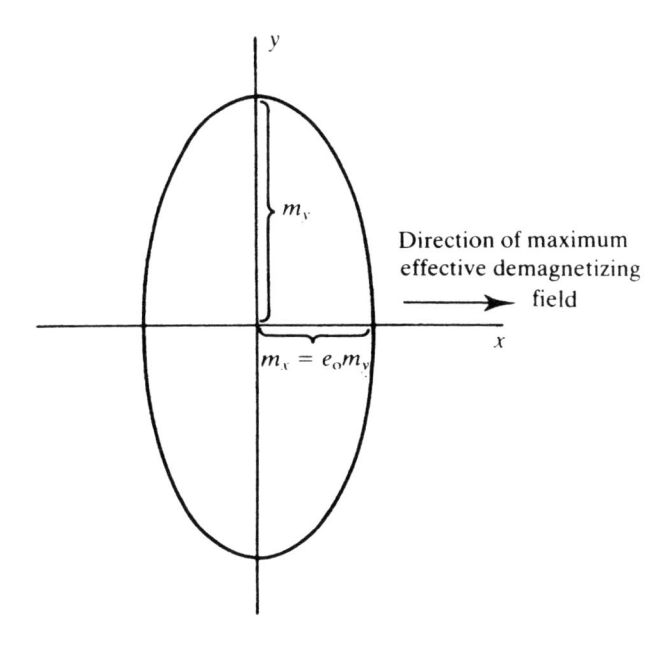

Figure 26.12 The uniform mode ellipse

$$\frac{\omega_x + \omega_y}{2} \approx \sqrt{\omega_x \omega_y} \qquad (26.99)$$

The components of the external susceptibility tensor with damping are therefore approximately described by

$$\chi_{xx}^e = e_0 \left[\frac{-\omega_m(-\omega_r + j\omega\alpha)}{-\omega^2 + (-\omega_r + j\omega\alpha)^2} \right] \qquad (26.100)$$

$$\chi_{yy}^e = \frac{1}{e_0} \left[\frac{-\omega_m(-\omega_r + j\omega\alpha)}{-\omega^2 + (-\omega_r + j\omega\alpha)^2} \right] \qquad (26.101)$$

$$\chi_{yx}^e = \chi_{xy}^e \left[\frac{j\omega_m \omega}{-\omega^2 + (-\omega_r + j\omega\alpha)^2} \right] \qquad (26.102)$$

The real and imaginary parts inside the brackets of equations (26.100) and (26.101) have the same form as equations (26.36) and (26.37) with ω_0 replaced by ω_r. The real and imaginary parts of equation (26.102) are identical with those in equations (26.38) and (26.39) except that ω_0 is replaced by ω_r.

26.8 Partially Magnetized Ferrite Medium

The real and imaginary parts of the entries of μ, μ_z and κ in a partially magnetized ferrite material are of significant importance in the design of practical devices. This section summarizes some empirical relationships for the real parts of these quantities.

Using averaging techniques, Rado has proposed that the real part (κ') of the off-diagonal element of the tensor permeability (κ) may be described by

$$\kappa' = \frac{\gamma M}{\mu_0 \omega}$$

Figure 26.13 illustrates the agreement for a thin rod of YIG; in obtaining this result the parameter κ' has been varied by altering ω.

The dependence of the real part (μ') of the diagonal element (μ) on M and M_0 may also be empirically described on the basis of measurement on a thin rod using a cavity technique:

$$\mu' \approx \mu_d' + (1 - \mu_d') \left(\frac{M}{M_0} \right)^{3/2}$$

26.8 PARTIALLY MAGNETIZED FERRITE MEDIUM

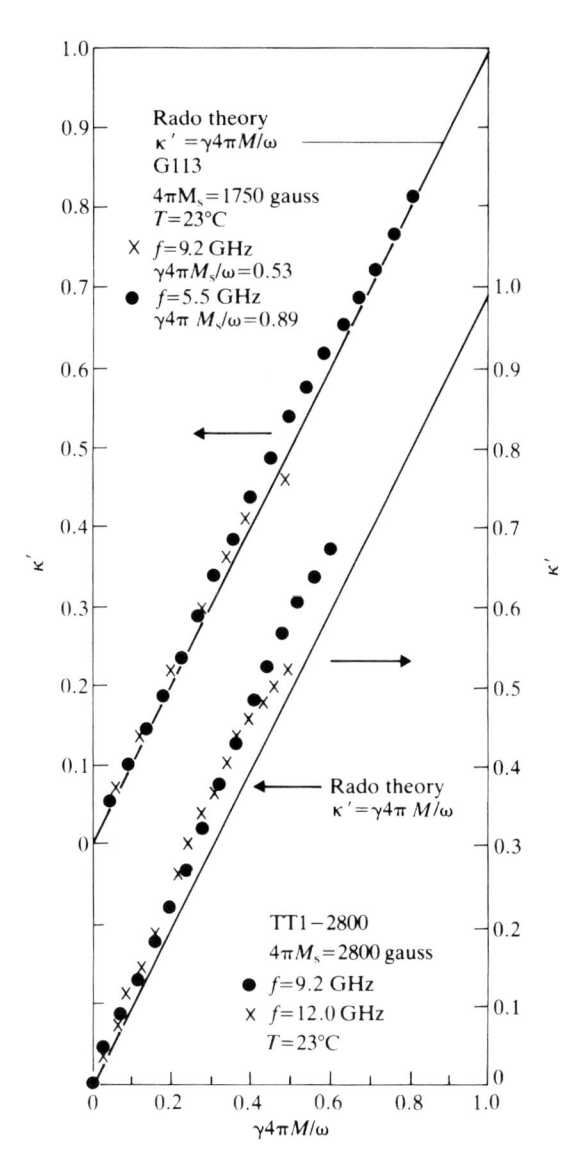

Figure 26.13 κ' versus $\gamma M/\mu_0 \omega$ for G113 at 5.5 and 9.5 GHz and for TT1–2800 at 9.2 and 12.0 GHz. (source: Green, J. J. and Sandy, F. (1974) Microwave characterization of partially magnetized ferrites, *IEEE Trans. Microwave Theory Tech.*, **MTT-22**, 541–645)

Figure 26.14 indicates the agreement between this relationship and experiment for a typical garnet material.

The real part (μ'_{eff}) of the effective permeability (μ_{eff}) is related to those of μ and κ by

$$\mu'_{\text{eff}} \approx \frac{(\mu')^2 - (\kappa')^2}{\mu'}$$

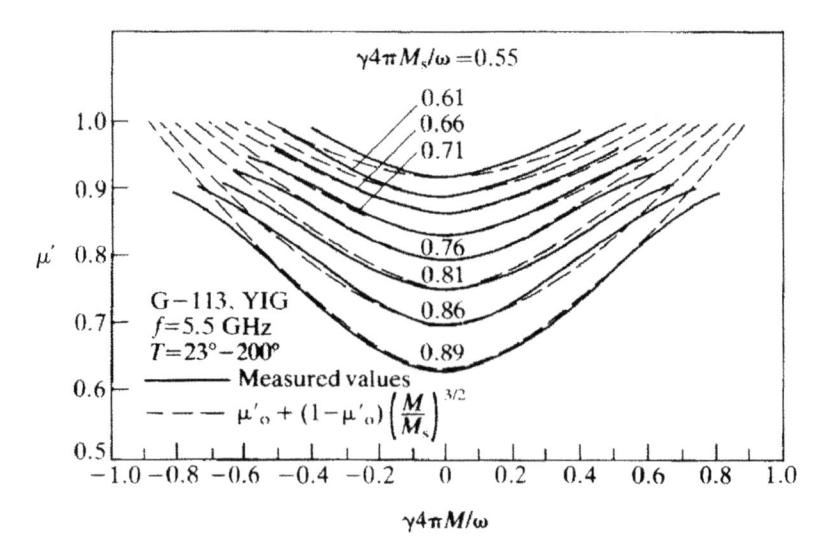

532   26 GYROMAGNETIC EFFECT IN MAGNETIC INSULATOR

Figure 26.14 μ' versus $\gamma M/\mu_0\omega$ for YIG at 5.5 GHz for various temperatures (Source: see Fig. 26.13)

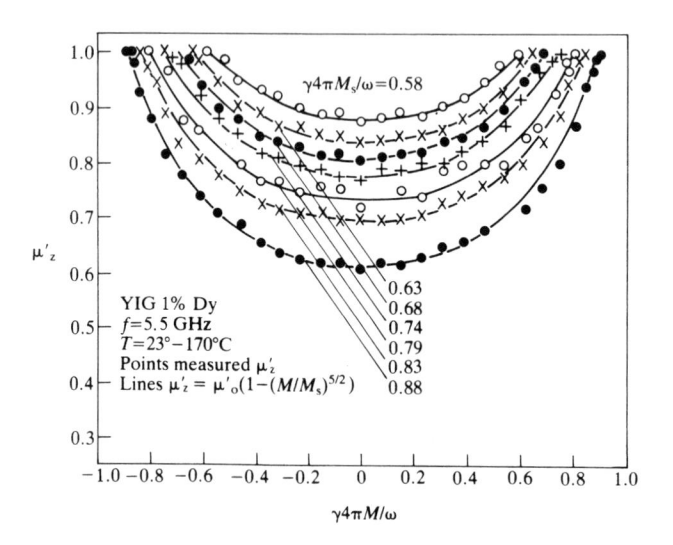

Figure 26.15 μ'_z versus $\gamma M/\mu_0\omega$ for YIG with 1% Dy at 5.5GHz for various temperatures (Source: see Fig. 26.13)

26.8 PARTIALLY MAGNETIZED FERRITE MEDIUM

In a partially magnetized medium the real part of the longitudinal element μ_z is not equal to unity and one empirical relationship is

$$\mu'_z \approx \mu'_d \left(1 - \frac{M}{M_0}\right)^{5/2}$$

Figure 26.15 illustrates the fit between this equation and some experimental results obtained on a spherical sample.

A widely used expression due to Schlomann for the initial or demagnetized permeability of a magnetic insulator consisting of a number of randomly orientated single magnetic domain is

$$\mu'_d = \frac{2}{3}\left[1 - \left(\frac{\omega_m}{\omega}\right)^2\right]^{1/2} + \frac{1}{3}$$

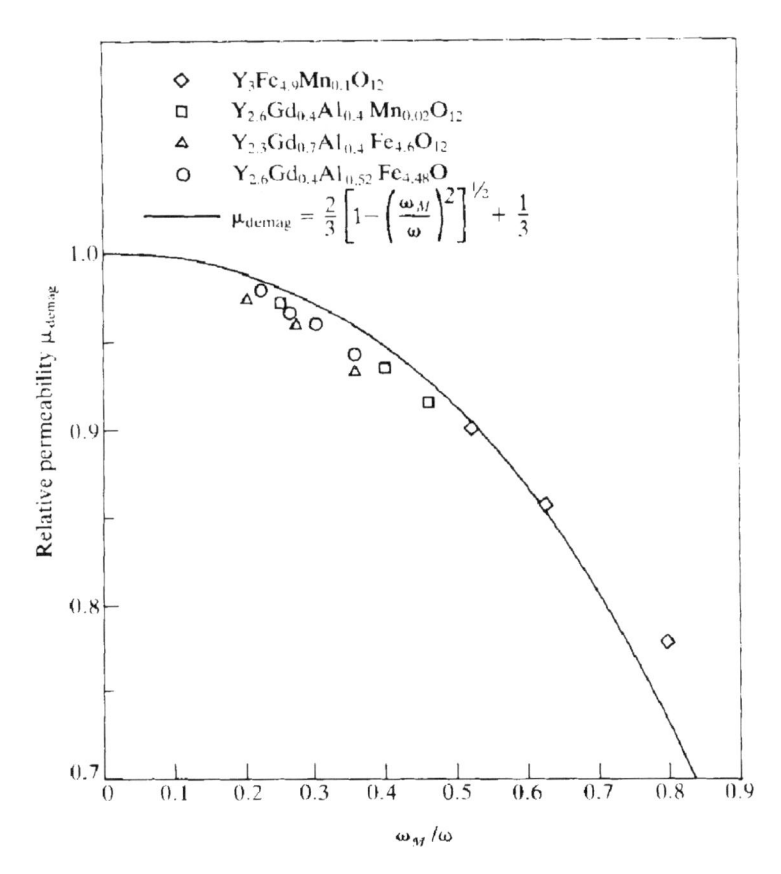

Figure 26.16 Variation of demagnetized permeability as a function of p for various garnet compositions. (Source: Courtney, E. (1970) Analysis and evaluation of a method of measuring the complex permittivity and permeability of micro wave insulators, *IEEE Trans. Microwave Theory Tech.*, **MTT-18**, 476–485)

Figure 26.16 displays the variation of the initial permeability as a function of the normalized saturation magnetization. The coordinate points plotted in this illustration are experimental ones.

The permeability defined by the last equation is real for $\omega > \omega_m$ and imaginary for $\omega > \omega_m$. It thus accurately describes the onset of so called low field loss.

References and Further Reading

Artman, J. O. (1957) Microwave resonance relations in anisotropic single-crystal ferrites, *Phys. Rev.*, **105**, 62.
Belson, H. S. and Kreissman, C. J. (1959) Microwave resonance in hexagonal ferromagnetic single crystals, *J. Appl. Phys.*, Supplement to **30**(4), 175S.
Bloembergen, N. (1956) Magnetic resonance in ferrites, *Proc. IRE*, **44**, 1259.
Bolle, D. M. and Lewin, L. (1973) On the definition of parameters in ferrite electromagnetic wave interactions, *IEEE Trans. Microwave Theory Tech.*, **MTT-21**, 118.
Buffler, C. R. (1962) Resonance properties of single crystal hexagonal ferrites, *J. Appl. Phys.*, Supplement to **33**(3), 1350–1362.
Clogston, A. M., Suhl, H., Walker, L. R. and Anderson, P. W. (1956) Possible source of line width in ferromagnetic resonance, *Phys. Rev.*, **101**, 903.
Clogston, A. M., Suhl, H., Walker, L. R. and Anderson, P. W. (1956) Ferromagnetic resonance linewidth in insulating materials, *J. Phys. Chem. Solids*, **1**, 129–136.
Courtney, W. E. (1980) Analysis and evaluation of a method of measuring the complex permitivity and permeability of microwave insulators, *IEEE Trans. Microwave Theory Tech.*, **MTT-18**, 476–485.
Green, J. J. and Sandy, F. (1974) Microwave characterization of partially magnetized ferrites, *IEEE Trans. Microwave Theory Tech.*, **MTT-22**, 641–645.
Green, J. J. and Sandy, F. (1974) A catalog of low power loss parameters and high power thresholds for partially magnetized ferrites, *IEEE Trans. Microwave Theory Tech.*, **MTT-22**, 645–651.
Helszajn, J. and McStay, J. (1969) External susceptibility tensor of magnetised ferrite ellipsoid in terms of uniform-mode ellipticity, *Proc. IEE*, **116**(12).
Kittel, C. (1948) On the theory of ferromagnetic resonance absorption, *Phys. Rev.*, **73**, 155–161.
Kohane, T. and Schlomann, E. (1968) Linewidth and off-resonance loss in poly-crystalline ferrites at microwave frequencies, *J. Appl. Phys.*, **39**, 720–721.
Landau, L. and Lifshitz, E. (1935) On the theory of the dispersion of magnetic permeability in ferromagnetic bodies, *Physik. Zeit Sowjetunion*, **8**, 153.
Lax, B. (1956) Frequency and loss characteristics of microwave ferrite devices, *Proc. IRE*, **44**, 1368–1386.
LeCraw, R. C. and Spencer, E.G. (1956) Tensor permeability of ferrites below saturation, *IRE Convention Record*, Part 5, 66.
Melchor, J. L. and Vartanian, P. H. (1959) Temperature effects in microwave ferrite devices, *IEEE Trans. Microwave Theory Tech.*, **MTT-7**, 15.
Osborn, J. A. (1945) Demagnetizing factors of the general ellipsoid, *Phys. Rev.*, **67**, 351–357.
Patton, C. E. (1969) Effective linewidth due to porosity and anistropy in poly-crystalline yttrium iron garnet and Ca-V substituted yttrium iron garnet at 10 *Ghz, Phys. Rev.*, **179**, 352–358.
Polder, D. (1949) On the theory of ferromagnetic resonance, *Philosophical Magazine*, **40**, 100–115.

Rado, G. T. (1953) Theory of microwave permeability tensor and Faraday effect in non-saturated ferromagnetic materials, *Phys. Rev.*, 89.

Schlomann, E. (1970) Microwave behaviour of partially magnetized ferrites, *J. Appl. Phys.*, **41**, 204–214.

Schlomann, E. (1971) Behaviour of ferrites in the microwave frequency range, *J. Phys.*, **32**, C1–443.

Schlomann, E. and Jones, R. V. (1959) Ferromagnetic resonance in polycrystalline ferrites with hexagonal crystal structure, *J. Appl. Phys.*, Supplement to **30**, 177S.

Smit, J. and Wijn, H. P. J. (1959) *Ferrites*, John Wiley, New York.

Vrehen, Q. H. F., Beljers, H. G. and de Lau, J. G. M. (1969) Microwave properties of fine-grain Ni and Mg ferrites, *IEEE Trans. Microwave Theory Tech.*, **MAG-5**, 617–621.

INDEX

ABCD parameters
  ideal transformer, 239
  radial line, 219, 235
  uniform line, 307
admittance eigen-values
  definition, 62
  post-circulator, 232
admittance matrix
  post circulator, 231
attenuation, 431
  circulator, 431, 439
  dissipation, 431
  gyromagnetic waveguide, 13
  insertion loss, 431
  non-linear, 431, 442
  return loss, 431
  semi-ideal circulator, 440
  transmission cavity, 437

bandwidth
  see network tables

characteristic equation
  ideal circulator, 43
  reciprocal junction, 41
characteristic plane, 248
  definition, 234
circular polarization
  alternating magnetic field, 1
  circulator, 340
  cylindrical waveguide, 133
  definition, 131
  eigen-vector excitation, 47, 59
  electron spin, 26
circulator
  adjustment, 39

definition, 27, 32
eigen-solutions, 47
first circulation condition, 41
gyrator impedance, 74
properties, 27
schematic diagram, 66
second circulation condition, 43
see 4-port
see degree-1
see degree-2
see eigen-values
see eigen-vectors
see E-plane
see experimental techniques
see finline
see H-plane
see impedance matrix
see Okada
see planar
see post
see scattering matrix
see turnstile
semi-ideal, 33, 440
complex gyrator
  definition, 62, 471
  see degree-1
  see degree-2
  see E-plane junction
  see H-plane junction
  see Okada junction
  see planar junction
composite resonator, 14
  frequency, 192
  geometry, 192
  gyromagnetic, 198

538 INDEX

constituent resonator
 definition, 250
 measurement, 480
counter-rotating eigen-networks
 eigen-network, 69
 measurement, 471, 474
 post-circulator, 240
coupled resonators
 even mode solution, 168
 odd mode solution, 161

degree-1
 adjustment, 392
 admittance, 386
 definition, 379, 380, 384
 eigen-value diagrams, 380
 equivalent circuit, 379
 frequency response, 393
 gyrator circuit, 384
 impedance, 384
degree-2, 379
 adjustment, 390, 394
 admittance, 389
 definition, 379, 380, 383
 eigen-value diagrams, 380, 392
 equivalent circuit, 379
 frequency response, 396
 gyrator circuit, 387
 impedance, 389
demagnetizing factor
 disk, 527
 plate, 527
 rod, 527
 sheet, 527
demagnetizing field, 526
 alternating, 527
 direct, 527
diagonalization
 definition, 50
 matrix, 61
dielectric constant
 effective, 149
dielectric loss tangent
 definition, 433
 effective, 433
 relation to quality factor, 433
dielectric resonator
 closed, 158
 equivalent circuit, 164
 even mode, 157, 168
 frequency, 158, 163, 169, 175
 geometry, 156
 half-wave, 174
 metal enclosure, 173
 mode, 159
 odd mode, 157, 161, 164

 planar, 166
 quarter-wave, 156, 176
dielectric waveguide, 139
 anisotropic, 143
 boundary conditions, 141
 characteristic equation, 142
 dominant mode, 142
 effective dielectric constant, 149
 effective permeability, 152
 modes, 143
 phase constant, 143
 symmetric mode, 142
direct field
 external, 527
 internal, 527
dissipation matrix
 definition, 35
 eigen-values, 54, 431
 matrix, 54
duplexing, 19

eigen-value diagram
 3-port, 41, 44
 4-port, 53
 degree-1, 380
 degree-2, 380, 392
 ideal, 44
eigen-networks
 counter-rotating, 69
 definition, 43
 dissipation, 431
 H-plane circulator, 311, 313, 322, 383
 ideal circulator, 43
 in-phase, 71
 post circulator, 232, 240
 unloaded Q-factors, 434
eigen-solutions
 symmetric junction, 47, 59
eigen-values
 3-port, 39, 318
 4-port, 51
 admittance, 61, 64, 231
 circulator adjustment, 39, 41
 definition, 40
 diagram, 42
 dissipation, 54, 440
 impedance, 58, 66, 233
 poles, 380
 reciprocal junction, 42
 scattering, 41, 472
 test sets, 472
eigen-vectors
 3-port circulator, 60
 3-port, 46
 4-port, 52
 definition, 40

degenerate, 69
in-phase, 71
reciprocal junction, 43
ellipticity
electron spin, 529
E-plane circulator, 339
adjustment, 341, 342
complex gyrator, 347
conductance, 347
counter-rotating eigen-network, 344
degenerate eigen-networks, 341
eigen-values, 346
frequency response, 352
frequency, 341, 344, 373
geometry, 10, 339
in-phase eigen-network, 341, 345, 383
operation, 339, 341
prism resonator, 368
quality factor, 347
Smith chart, 342
solution, 342, 347
split frequencies, 375
susceptance slope, 340, 347, 348, 377
terminals, 343, 345
turnstile version, 340
evanescent junction, 297
conductance, 302
E-plane junction, 339, 341
frequency, 303
gyrator circuit, 301
modes, 299
Okada resonator, 297
quality factor, 301
quarter-wave coupled, 297
split frequencies, 299
evanescent waveguide, 304
equivalent circuit, 304
experimental techniques, 471
complex gyrator circuit, 471
constituent resonator, 250, 480
degenerate eigen-values, 471, 474
demagnetized junction, 471, 485
eigen-networks, 471
gyrator conductance, 483
in-phase eigen-value, 476
quality factor, 268
reactance slope, 473
split frequencies, 471, 478, 481
susceptance slope, 471, 473, 485
swept frequency, 485
test set, 472
external quality factor
cavity, 438
definition, 432

Faraday rotation
definition, 134
vector representation, 134
field patterns
closed waveguide, 115
disk resonator, 81, 83
open waveguide, 143
triangular resonator, 359, 362, 363
finline circulator, 353
adjustment, 354
frequency, 355
frequency response, 355
geometry, 353
4-port circulator, 399
adjustment, 53, 399, 401, 403, 406
counter-rotating, modes, 404, 408, 410
eigen-values, 53, 399
eigen-vectors, 52
geometries, 399, 401
gyrotropy, 405, 408
higher-order solution, 410
scattering matrix, 51
scattering parameters, 403, 408
standing wave, 399, 400, 402, 407, 410, 412
symmetric modes, 404, 406, 408, 410
frequency response
Chebyshev, 446
E-plane circulator, 342
H-plane circulator, 330, 335, 376
Okada circulator, 287
planar circulator, 267
see network tables

gyrator circuit
circuits, 65, 67
definition, 62, 72, 384
imaginary-part, 64
Okada, 282
quality factor, 283
real-part, 64
schematic diagram, 73
see degree-1
see degree-2
gyromagnetic ratio, 513
gyromagnetic resonator
composite, 190, 192
electric wall, 184
evanescent mode, 297
even mode, 190
geometries, 188
magnetic wall, 180
metal enclosure, 195
modes, 185
odd mode, 187, 192

540 INDEX

gyromagnetic resonator (*cont.*)
  partially magnetized, 183
  perturbation theory, 181
  quality factor, 186
  quarter-wave, 180
  split frequencies, 181, 184, 189, 193

half-wave synthesis, 445
  characteristic plane, 454
  degree-2, 456
  frequency response, 454
  impedance, 455
  radian frequency, 455
  reflection, 455
  ripple, 454
  specification, 445
  synthesis, 454, 456
  tables, 454
  topology, 446, 457
half-wave U.E.
  Okada, 279
H-plane circulator, 311
  adjustment, 321
  characteristic plane, 234, 248
  complex gyrator, 311, 313, 319
  conductance, 311
  counter-rotating eigen-networks, 311, 313
  eigen-value diagram, 318
  first circulation condition, 313, 317, 324
  frequency, 317, 321, 335
  frequency response, 330
  gain-bandwidth, 331
  image plane, 312, 313
  in-phase eigen-network, 311, 313, 322, 383
  quality factor, 311, 321, 367
  second circulation condition, 313, 321
  Smith chart, 318
  susceptance slope, 311, 325
  synthesis, 330
  terminals, 332
  topology, 10, 200, 311, 312
hysteris loop
  definition, 423
  effect of gap, 426
  measurement, 424

ideal transformer
  ABCD parameters, 239
image wall, 314
  definition, 10
immittance inverters, 305
  admittance, 305
  impedance, 305
impedance, 199

eigen-values, 61
power–current, 216
power–voltage, 214
see radial waveguide, 219
see rectangular waveguide
see ridge waveguide
voltage–current, 211
impedance matrix
  circulator, 57, 67
  definition, 57
  diagonalization, 58
  eigen-values, 57
impedance transducers, 455
  half-wave long, 454
  quarter-wave long, 446
  short-steps, 463
  synthesis, 445
  tables, 452, 459, 465
  topologies, 200
in-phase eigen-solution
  eigen-network, 71
  measurement, 476
insertion loss
  see attenuation
isolator, 498

Kittel
  resonance, 528

linewidth
  uniform, 520
loaded quality factor
  cavity, 438
  definition, 432

magnetic loss tangent
  definition, 96
  effective, 95
  relation to quality factor, 96
matrices
  see ABCD
  see admittance
  see dissipation
  see impedance
  see scattering
Maxwell, 109
motion of magnetization, 5
  damping, 516, 522
  demagnetizing factors, 526
  ellipsoid, 526
  ellipticity, 529
  equation, 512
  gyromagnetic ratio, 513
  Kittel resonance, 528
  linewidth, 520
  magnetic moment, 512

INDEX

nonreciprocal behaviour, 511
origin, 511
scalar susceptibility, 521
scalar, permeability, 524
subsidiary resonance, 442
tensor permeability, 523, 525
tensor susceptibility, 513, 525, 526

network tables
  degree-2, 452, 459, 465
  degree-3, 453

Okada circulator, 89, 279
  complex gyrator, 256, 283
  degree-1, 287
  degree-2, 265, 279, 292
  evanescent, 297
  geometry, 279, 290, 293
  gyrator conductance, 283
  half-wave coupled, 290
  insertion loss, 280
  power rating, 280, 289
  quality factor, 283, 288
  quarter-wave coupled, 292
  struts, 102, 279, 284
  susceptance slope, 6, 257, 262, 283, 284, 290
  temperature stability, 288
  topology, 15

Okada resonator
  aspect ratio, 282
  composite, 103
  cooling, 102
  dielectric loss, 95
  effective dielectric constant, 90, 280, 285
  effective gyrotropy, 92, 284
  evanescent mode, 297
  filling factor, 283, 284
  frequency, 257, 287
  geometry, 100, 101, 104, 280, 290
  magnetic loss, 95
  power rating, 99
  quality factor, 257, 260, 268, 288, 302
  radial number, 259, 282
  split frequencies, 106
  unloaded Q-factor, 96
  voltage breakdown, 97

permeability
  demagnetized, 530
  effective, 531
  partially magnetized, 530
  scalar, 524
  tensor, 523

perturbation theory
  gyromagnetic resonator, 181
  gyromagnetic waveguide, 128

prism resonator, 366
phase shift, 36
planar circulator
  frequency, 259
  frequency response, 267
  gain-bandwidth, 274
  gyrator conductance, 266
  power rating, 269
  quality factor, 257, 260
  split frequencies, 260, 273
  susceptance slope, 263
  temperature, 268

planar resonator
  15°, 268
  30°, 268
  60°, 268
  definition, 89
  dielectric loss tangent, 95
  effective dielectric constant, 90, 168
  effective gyrotropy, 92
  effective permeability, 93
  frequency, 91, 166
  geometry, 90
  magnetic loss tangent, 96
  quality factor, 94
  temperature stability, 268
  unloaded quality factor, 94
  voltage breakdown, 96

post circulator
  admittance matrix, 231
  boundary conditions, 236
  characteristic plane, 234, 248
  circulator conditions, 243
  degree-2 solution, 247
  eigen-networks, 233, 240
  eigen-values, 232
  frequency response, 246
  gyrator circuit, 242, 265
  ideal transformer, 239
  quality factor, 420
  standing wave, 3, 234
  topology, 230

post resonator
  definition, 78
  gyromagnetic, 84
  modes, 78, 230
  spectrum, 86
  split frequencies, 85, 260, 420
  wave equation, 78

power
  rating, 17, 280, 289

prism circulator
  E-plane, 376
  gyrator conductance, 375
  H-plane, 374
  susceptance slope, 374, 376, 377

prism resonator, 359
  even-mode, 369
  frequency, 366
  odd-mode, 368, 369
  perturbation theory, 366
  quality factor, 367
  symmetry, 359

quality factor
  definition, 283
  effective, 435
  E-plane junction, 368
  G-stub load, 234
  gyromagnetic resonator, 186
  H-plane junction, 367
  measurement, 268
  Okada junction, 283, 288, 301
  planar resonator, 262, 260
  prism resonator, 359
  see external
  see loaded
  see unloaded
  synthesis problem, 451
  temperature stability, 268
  triangular resonator, 359, 364
quarter-wave synthesis, 445
  characteristic plane, 454
  degree-2, 451,
  frequency response, 447, 448
  impedance, 448
  quality factor, 451
  radian frequency, 447
  reflection, 448
  ripple, 446
  specification, 445, 446
  susceptance slope, 451
  synthesis, 446, 448, 449
  tables, 454
  topology, 446, 450
quarter-wave U.E.
  Okada, 279

radial waveguide
  ABCD parameters, 219
  impedance, 221
  terminals, 221
rectangular waveguide
  evanescent, 304
  impedance, 199, 204
  power flow, 202
resonator
  see dielectric
  see gyromagnetic
  see Okada
  see planar
  see post

  see prism
  see re-entrant
  see triangular
  see turnstile
ridge waveguide, 199
  current–power impedance, 216
  cut-off space, 205
  dimensions, 206
  field pattern, 208
  geometry, 200
  half-wave section, 226
  impedance, 211, 214, 216
  power, 214
  voltage–current impedance, 211
  voltage–power impedance, 214

scattering matrix
  3-port, 48
  4-port, 51
  cavity, 437
  circulator, 39
  definition, 25, 32
  diagonalization, 50
  eigen-values, 39, 42
  m-port, 23,
  reciprocal junction, 43
  semi-ideal circulator, 33, 440, 490
  unitary condition, 30
short-line synthesis
  degree-2, 465
  impedance, 463
  specification, 445
  susceptance slope, 464
  radian frequency, 463
  synthesis, 463
  tables, 465
  topology, 446, 463
Smith chart
  E-plane, 342
  H-plane, 318
specification, 489
  arbitrary load, 489
  circuit losses, 508
  data sheets, 503
  insertion loss, 489, 490, 503
  isolation, 489, 490
  non-ideal load, 499
  phase shift, 489
  reflection coefficient, 493
  return loss, 489, 490
  semi-ideal, 490
  temperature, 489
  standing wave ratio, 489, 493, 494, 496
spin motion
  frequency, 515
spinwave instability, 431, 443

split frequencies, 68
  cavity resonator, 181
  composite resonator, 192, 195
  Okada resonator, 104
  post resonator, 420
  planar resonator, 86, 261, 271
  prism resonator, 89
  cylindrical resonator, 320
  switched resonator, 420
standing wave solutions
  3-port circulator, 230
  4-port circulator, 399, 400, 402, 407, 410, 412
struts
  Okada, 284
susceptance slope
  definition, 68
  E-plane circulator, 340, 347, 348
  H-plane circulator, 325, 375
  Okada circulator, 262
  planar circulator, 257, 262
susceptibility
  complex, 436, 517, 530
  definition, 521
  scalar, 436, 521
  tensor, 513, 525
switches, 415
  eddy currents, 415
  externally actuated, 415, 417, 418
  hysterisis loop, 415, 424
  internally actuated, 415, 418
  magnetic circuit, 423
  power, 415
  quality factor, 420
  resonators, 13
  switched resonator, 419
  switching coefficient, 426
  switching time, 415, 427
  using circulators, 417
synthesis, 445
  gain-bandwidth, 465
  quality factor, 445
  see half-wave
  see quarter-wave
  see short-line
  synthesis, 445
  t-plane 445
  t-plane capacitor, 446
  t-plane inductor, 446
  unit-element, 445

temperature
  stability, 268
tensor permeability
  complex, 435

definition, 523
diagonal element, 524
eigen-solutions, 526
eigen-values, 524
off-diagonal element, 524
partially magnetized, 530
t-plane
  half-wave transformer, 454
  quarter-wave transformer, 465
  short-line transformer, 463
transformer
  H-plane circulator, 327
  post circulator, 239
  ridge, 200
triangular post circulator, 359
  quality factor, 359
  quarter-wave coupled, 365
triangular post resonator, 359
  quality factor, 359
  standing wave, 359, 362, 363
  symmetry, 359
turnstile circulator
  E-plane, 340
  geometry, 5
  H-plane, 311
  operation, 5

unit element, 445
unitary condition, 431
  circulator, 32
  definition, 31
unloaded quality factor
  cavity, 438
  definition, 432
  eigen-networks, 434
  Okada resonator, 96

waveguide
  anisotropic, 109, 119, 143
  circular, 109
  cut-off number, 113, 114, 118, 125
  electric wall, 112
  evanescent, 304
  gyromagnetic, 109, 122, 127
  isotropic, 109
  magnetic wall, 112, 117
  modes, 112, 114
  open, 139
  propagation constant, 120, 121, 124
  see radial
  see rectangular
  see ridge
  see dielectric
  triangular, 366
  wave-equation, 111